Residential Construction Academy
Plumbing

Residential Construction Academy

Plumbing

Michael Joyce

THOMSON

DELMAR LEARNING

Australia Canada Mexico Singapore Spain United Kingdom United States

Residential Construction Academy: Plumbing
Michael Joyce

Vice President, Technology and Trades SBU:
Alar Elken

Editorial Director:
Sandy Clark

Acquisitions Editor:
Alison Weintraub

Development Editor:
Monica Ohlinger

Marketing Director:
Dave Garza

Channel Managers:
Erin Coffin
William Lawrenson

Marketing Coordinator:
Mark Pierro

Production Director:
Mary Ellen Black

Production Manager:
Andrew Crouth

Senior Project Editor:
Christopher Chien

Art/Design Specialist:
Mary Beth Vought

Editorial Assistant:
Stacey Wiktorek

Full Production Services:
Carlisle Publishers Services

Cover Photo:
Comstock

COPYRIGHT © 2005 by Thomson Delmar Learning. Thomson, the Star Logo, and Delmar Learning are trademarks used herein under license.

Printed in the United States of America
3 4 5 CK 06

For more information contact:
Thomson Delmar Learning
Executive Woods
5 Maxwell Drive, PO Box 8007,
Clifton Park, NY 12065-8007
Or find us on the World Wide Web at
http://www.delmarlearning.com

ALL RIGHTS RESERVED. No part of this work covered by the copyright hereon may be reproduced or used in any form or by any means—graphic, electronic, or mechanical, including photocopying, recording, taping, Web distribution or information storage and retrieval systems—without the written permission of the publisher.

For permission to use material from this text or product, contact us by
Tel (800) 730-2214
Fax (800) 730-2215
www.thomsonrights.com

Library of Congress Cataloging-in-Publication Data

Joyce, Michael D.
  Residential construction academy : plumbing / Mike Joyce.
    p. cm.
  Includes bibliography references and index.
  ISBN 1-4018-4891-5
1. Plumbing. 2. House construction. I. Title: Plumbing. II. Title.
TH6123.J69 2005
696'.1—dc22

2004017228

NOTICE TO THE READER

Publisher does not warrant or guarantee any of the products described herein or perform any independent analysis in connection with any of the product information contained herein. Publisher does not assume, and expressly disclaims, any obligation to obtain and include information other than that provided to it by the manufacturer.

The reader is expressly warned to consider and adopt all safety precautions that might be indicated by the activities herein and to avoid all potential hazards. By following the instructions contained herein, the reader willingly assumes all risks in connection with such instruction.

The Publisher makes no representation or warranties of any kind, including but not limited to, the warranties of fitness for particular purpose or merchantability, nor are any such representations implied with respect to the material set forth herein, and the publisher takes no responsibility with respect to such material. The publisher shall not be liable for any special, consequential, or exemplary damages resulting, in whole or part, from the readers' use of, or reliance upon, this material.

Table of Contents

Preface xix

SECTION 1 Tools and Materials 1

Chapter 1 **Plumber's Toolbox 3**
  Glossary of Terms 4
  Hand Tools. 5
    Levels 5
    Tape Measures 6
    Squares 7
    Screwdrivers 7
    Pliers 8
    Adjustable Wrenches 8
    Pipe Wrenches 9
    Hammers 10
    Plastic Pipe Saw 10
    Plastic Pipe Cutter 11
    Inside Plastic Pipe Cutter 11
    Metal Cutting Saw 12
    Wallboard Saws 13
    Aviation Snips 13
    Nail Pullers 13
    Knives 14
    Chisels 15
    Basin Wrench 15
    Basket Strainer Tools 16
    Copper Pipe Cutters (Tubing Cutters) 16
    Copper Flaring Tool 17
    Copper Tubing Bender 18
    Torch Regulator Assembly 18
    Flexible Pipe Crimping Tool 19
    Plumb Bob 20
    Chalk Box 20
    Torque Wrench 21

Personal Protection Equipment 21
   Eye Protection. 21
   Face Protection 22
   Hand Protection 23
   Knee Protection. 23
   Foot Protection 24
   Inhalation Protection. 24
   First Aid Kit 24
   Hard Hats 25
Summary. 25
Procedures. 26
   Level Use 26
   Ruler Reading 28
   Folding Ruler Layout 29
   Framing Square Layout. 30
   Pipe Wrench Use 31
   Plastic Pipe Cutting 32
   Basin Wrench Use 33
Review Questions 34

Chapter 2 Power Tools 36

Glossary of Terms 37
General Safety 38
   Electrical Safety 38
   Ladder Safety 38
   Fall Protection. 40
   Ear Protection 40
Drills. 41
   Right Angle Drill 41
   Pistol Drill 42
   Hole Hawg 42
   Hammer Drill 43
Drill Bits 44
   Twist Bits 44
   Auger Bits 44
   Speed Bits 45
   Power Bore Bits 45
   Hole Saw Bits 46
   Masonry Bits. 46
   Step Bits 47
Saws. 48
   Reciprocating Saws 48
   Circular Saws. 48
   Saber Saws 49
   Portable Band Saws 50
Grinders. 50
Powder-Actuated Tool. 51

Air Compressors 52
Jackhammers 52
Summary 52
Procedures 53
   Hole Saw Finished Surface Drilling 53
   Saw Blade Selection 54
   Thrust Cut 55
Review Questions 56

Chapter 3   Types of Pipe 58

Glossary of Terms 59
Pipe Diameters 60
Plastic Pipe 60
   PVC Pipe. 61
   CPVC Pipe 62
   ABS Pipe. 63
   Polyethylene Pipe 63
   PEX Pipe. 64
Other Pipe Materials 65
   Copper Tube 65
   Cast Iron Pipe 66
   Steel Pipe. 67
   Brass Pipe. 68
   Perforated Pipe 68
   Corrugated Stainless Steel Tube. 69
Summary. 69
Review Questions 70

Chapter 4   Fittings 72

Glossary of Terms 73
Degree of Fittings 74
Various Fitting Designs 74
   Offsets 75
   Tees. 75
   Coupling. 76
   Reducers. 76
   Bushing 77
   Male Adapter. 78
   Female Adapter 78
   Unions 79
Water Distribution. 79
   Water Distribution Fittings 80
   PEX 80
   CPVC 81

　　　　Copper 81
　　　　Polyethylene 82
　　　　Galvanized and Brass 82
　　　　PVC 83
　　Drainage, Waste, and Vent 83
　　　　Wye 84
　　　　Combo 85
　　　　Sanitary Tee 85
　　　　Sanitary Cross 86
　　　　Twin Elbow 86
　　　　Test Tee 87
　　　　Cleanout 87
　　　　Closet Bend 87
　　　　Heel Inlet 90° 88
　　　　Closet Flange 88
　　　　P-trap 89
　　　　Trap Adapter 89
　　　　PVC 90
　　　　ABS 90
　　　　Cast Iron 90
　　Summary 91
　　Procedures 92
　　　　Tee Size Identification Procedure 92
　　　　Insert Reducing-Tee Creation Procedure 93
　　Review Questions 94

Chapter 5　Valves and Devices 96

　　Glossary of Terms 97
　　Isolation Valves 98
　　　　Ball Valve 99
　　　　Gate Valve 99
　　　　Stop 100
　　　　Stop and Waste 102
　　　　Gas Cock 102
　　Hose Outlets 103
　　　　Boiler Drain 103
　　　　Hose Bibb 103
　　Reactionary Valves and Devices 104
　　　　Pressure-Reducing Valve 106
　　　　Relief Valves 107
　　　　Check Valve 107
　　　　Vacuum Breaker 108
　　　　Vacuum Relief Valve 109
　　　　Reduced-Pressure Zone 109
　　Summary 110
　　Review Questions 112

SECTION 2 Fixtures and Equipment 115

Chapter 6 Fixtures.................... 117
- *Glossary of Terms* 118
- *Fixture Types* 119
  - Toilets 119
  - Lavatory Sinks.................. 122
  - Bathtubs....................... 123
  - Showers 124
  - Kitchen Sinks 127
  - Laundry Sinks 128
  - Bidets......................... 131
- *Summary*........................ 132
- *Review Questions*................. 133

Chapter 7 Faucets and Drain Assemblies ... 135
- *Glossary of Terms* 136
- *Faucets* 137
  - Lavatory Faucets 138
  - Bathtub and Shower Faucets 139
  - Kitchen Sink Faucets 143
  - Laundry Sink Faucets 146
  - Bidet Faucets................... 149
- *Drain Assemblies*................. 150
  - Lavatory Drain Assemblies 150
  - Bathtub Drain Assemblies........ 151
  - Shower Drains 154
  - Kitchen Sink Basket Strainer..... 155
  - Laundry Sink Basket Strainer 156
  - Bidet Drain Assembly 157
- *Summary*........................ 159
- *Review Questions*................. 160

Chapter 8 Plumbing Equipment and Appliances................ 162
- *Glossary of Terms* 163
- *Appliance Connections* 164
  - Garbage Disposers 164
  - Dishwashers 167
  - Washing Machine Box............ 168
  - Icemaker Box 169
- *Residential Water Heaters*.......... 170
  - Gas Water Heaters 170
  - Electric Water Heaters 178

Solar Water Heater 182
System Protection 183
    Lined Pipe Nipples 183
    Anode Rod 186
    Expansion Tanks 187
Summary 188
Review Questions 189

SECTION 3 Layout and Installation 191

Chapter 9 Blueprint Reading and Drafting . . 193

Glossary of Terms 194
Plumbing Symbols 195
    90-Degree Offsets 195
    45-Degree Offsets 195
    Tees 196
    Perpendicular Tee Configuration 197
    P-trap 198
    Piping 198
    Cap, Reducer, and Plugs 198
    Valves and Devices 199
    Fixtures 200
    Equipment and Drains 202
Abbreviations 202
Architectural Blueprints 206
    Architectural Symbols 207
Drafting 212
Drafting Tools 213
    Scale Ruler 213
    Drafting Triangles 215
    Symbol Templates 217
    Drafting Paper 217
Isometric Drafting 219
    Riser Diagrams 220
Summary 224
Review Questions 225

Chapter 10 Material Organization and Layout 227

Glossary of Terms 228
Communication 229
    Written Communication 229
    Oral Communication 229

Material Organization 230
   Palletizing. 230
   Bagging and Tagging 230
Material Handling 232
   Vehicle Racks 233
Layout. 234
Underground Layout 235
   Wall Layout. 236
   Trench Layout 238
Above-ground Layout. 239
   Fixture Locations. 240
   Floor Joist Conflicts 240
   Wall Layout. 241
   Manufacturer Rough-in Sheet 243
Summary. 245
Procedures 246
   Edge Form Layout Procedure 246
Review Questions. 248

Chapter 11   Water Service Installation 250

Glossary of Terms 251
Water Source. 252
   Public Water System. 253
   Private Water System 254
EPA Standards 255
   Water Quality 255
   Water Filtration. 256
Water Service Installation. 257
   Trench Safety 258
   Burial-Depth Requirements 260
   Water Meter Connection 260
   Well Connection. 261
   House Connection 263
   Same Trench with Sewer 263
   Same Elevation 265
   Different Elevation. 266
   Perpendicular Installation. 266
Summary. 266
Procedures 267
   Installing a Water Service in the Same
     Trench as a Sewer. 267
Review Questions. 269

Chapter 12   Water Distribution Installation . . 271

Glossary of Terms 272

Layout and Sizing 273
    Pipe Sizing 273
    Sizing Theory 274
    Job Site Sizing 274
    Wall Layout. 275
    Toilets 276
    Lavatory 276
    Bathtubs. 278
    Showers 279
    Kitchen Sinks 280
Drilling and Notching. 281
    Walls 282
    Joists 284
Hangers and Supports 287
    Types 287
Compounds and Sealants 295
    Material Safety Data Sheets. 295
    Sealants 295
    Flux and Solder 296
    Soldering 296
    Brazing. 297
    Flaring 298
    Working with Flexible Tubing 299
    Manifold Systems. 299
    Connection Types. 300
    Working with Plastic Pipe 301
    Cutting. 301
    Solvent-Welding. 301
Testing. 302
Summary. 304
Procedures 305
    Determining the Volume of Different
       Pipe Sizes 305
    Determining Percentages of a Hole
       Diameter and Notch 307
    Using Teflon Tape. 309
    Soldering Copper 310
    Brazing. 314
    Flaring Copper Tubing 316
    Solvent Welding. 320
Review Questions. 323

Chapter 13   Drainage, Waste, and Vent Segments and Sizing. 325

Glossary of Terms 326

Introduction 328
Major Segments of a DWV System 328
   Building Sewer 329
   Building Drain 329
   Waste Stack 329
   Stack Vent 330
   Vent Stack 330
   Cleanout 330
Minor Segments of a DWV System 330
   Fixture Drain 330
   Fixture Branch 332
   Horizontal Branch 332
   Individual Vent 332
   Branch Vent 332
   Circuit Vent 333
   Loop Vent 333
   Relief Vent 334
Special Venting Arrangements 335
   Wet Venting 336
   Combination Waste and Vent 336
   Island Venting 337
   DWV Sizing 337
   Fixture Drain Sizing 342
   Fixture Branch and Horizontal Branch
      Sizing 342
   Waste Stack Sizing 344
   Building Drain and Sewer Sizing 346
   Stack Vent and Vent Stack Sizing 346
   Individual Vent Sizing 348
   Branch Vent Sizing 350
   Circuit and Loop-Vent Sizing 350
   Wet-Vent Sizing 352
   Air Admittance Valve Sizing 352
   Island-Vent Sizing 354
Trap Distance 356
Septic Systems 357
Summary 360
Procedures 361
   Positioning Cleanouts at Base of Stacks . . . 361
   Unlisted Fixture Sizing 363
   Continuous Waste 364
   Vent Stack and Building Drain Connection . 365
   Vent-stack and Waste-stack Connection . . . 366
   Island Vent 367
   Crown Venting 369
Review Questions 370

Chapter 14 Drainage, Waste, and Vent Installation 373

Glossary of Terms 374
Layout Considerations 375
   Scope of Work 376
   Guest Bathroom Layout 376
   Kitchen Sink Layout 377
   Master Bathroom 378
   Hall Bathroom 379
   Laundry Room 380
   Building Drain 381
   Venting System 381
   Fixture Rough-In 384
   Toilets 385
   Lavatory 386
   Bathtub 388
   Kitchen Sink 390
   Washing Machine 392
   Shower Pan 393
Hangers and Supports 393
Pipe Protection 394
Connection to Dissimilar Pipe 396
Testing DWV Systems 397
Summary 401
Procedures 402
   Vent Through Roof Layout Procedure 402
   Shower Pan Liner Installation Procedure . . . 403
   Using Wood for Pipe Support 406
Review Questions 407

Chapter 15 Fixture and Equipment Installation 409

Glossary of Terms 410
Escutcheons and Stops 411
Toilets 413
Lavatories 416
Kitchen Sinks 418
Dishwashers 420
Laundry Sink 420
Bidet 421
Water Heaters 422
Electric Water Heaters 423
Gas Water Heaters 424

Summary.......................425
Procedures....................426
   Escutcheon and Stop Installation onto
     a Copper Stub-Out Pipe............426
   Installing a Two-Piece Toilet..........429
   Cultured-Marble Lavatory Faucet
     and Drain Installations............432
   Installing a Stainless Steel Kitchen Sink
     into a Countertop................438
Review Questions..................442

SECTION 4 Troubleshooting............445

Chapter 16 Plumbing Repairs and Troubleshooting..............447

Glossary of Terms.................448
Electric Water Heaters..............449
High-Limit Device.................449
Upper Thermostat.................450
Lower Thermostat.................451
Heating Elements.................452
Electrical Source..................453
Electrical Tests...................454
Gas Water Heaters.................457
Well Pumps......................459
Toilets.........................466
Summary.......................468
Procedures......................469
   Water Heater Draining..............469
   Screw-In Element Replacement........471
   Thermocouple Replacement...........473
   Gas Regulator Replacement...........474
   Pressure Switch Replacement.........476
   Expansion/Storage Tank Replacement....479
   Ballcock Replacement...............481
   Flush Valve Replacement.............483
Review Questions..................486

Chapter 17 Hydronic Heat............488

Glossary of Terms.................489
Theory of Hydronic Heating Systems.....491
   The Heat Source 491

Aquastat.................................492
Reset....................................492
Low-Water Cutoff.........................493
Expansion Tank...........................493
Centrifugal Pumps........................496
Air Vents and Air Separators.............498
Pressure-Reducing Valve (Water-Regulating
 Valve)..................................498
Pressure Relief Valve....................501
Zone Valves..............................501
Flow-Control Valve.......................502
Balancing Valves.........................503
Series Loop System.......................503
One-Pipe System..........................503
Two-Pipe Direct Return...................506
Two-Pipe Reverse Return..................507

Primary-Secondary Pumping...............508
Primary-Secondary Common Piping..........508
Primary-Secondary Circuit Piping.........508
The Circulator Pumps.....................509
Mixing Valves in Primary-Secondary
 Pumping.................................511
Expansion Tanks in Primary-Secondary
 Systems.................................511

Radiant Heating Systems.................511
The Human Body Is a Radiator.............511
Cold 70..................................512
What Is Ideal Comfort?...................512
The Radiant System.......................513
Radiant Heating Piping...................514
Tubing...................................516
Manifold Station.........................516
Water Temperature and Direct Piping......517

*Installing and Starting the Hydronic
System*...................................518
Installing the Boiler....................518
Installing the Piping....................519
Wiring the System........................519
Filling the System.......................520
Firing the System........................520

Summary.................................520
Procedures..............................521
Estimating the Volume of Water in the
 System, Assuming That Type M Copper
 Piping is Used..........................521

Table of Contents xvii

Sample Calculations for Estimating Volume
of Water in the System 522
Calculating the Minimum Volume, V_t,
for the Expansion Tank 523
Filling and Purging the System 524
Review Questions. 525

Glossary 529
Index 535

Preface

Home Builders Institute Residential Construction Academy: Plumbing

About the Residential Construction Academy

One of the most pressing problems confronting the building industry today is the shortage of skilled labor. It is estimated that the construction industry must recruit 200,000 to 250,000 new craft workers each year to meet future needs. This shortage is expected to continue well into the next decade because of projected job growth and a decline in the number of available workers. At the same time, the training of available labor is becoming an increasing concern throughout the country. This lack of training opportunities has resulted in a shortage of 65,000 to 80,000 skilled workers per year. The crisis is affecting all construction trades and is threatening the ability of builders to build quality homes.

These are the reasons for the creation of the innovative *Residential Construction Academy Series*. The *Residential Construction Academy Series* is the perfect way to introduce people of all ages to the building trades while guiding them in the development of essential workplace skills including carpentry, electrical, HVAC, plumbing, and facilities maintenance. The products and services offered through the *Residential Construction Academy* are the result of cooperative planning and rigorous joint efforts between industry and education. The program was originally conceived by the National Association of Home Builders—the premier association of over 200,000 member groups in the residential construction industry—and its workforce development arm, the Home Builders Institute.

For the first time, Construction professionals and educators created National Standards for the Construction trades. In the summer of 2001, the National Association of Home Builders (NAHB), through the Home Builders Institute (HBI), began the process of developing residential craft standards in five trades. They are carpentry, electrical wiring, HVAC, plumbing, and facilities maintenance. Groups of electrical employers from across the country met with an independent research and measurement organization to begin the development of new craft training Standards. The guidelines from the National Skills Standards Board were followed in developing the new standards. In addition, the process met or exceeded the American Psychological Association standards for occupational credentialing.

Then, through a partnership between HBI and Delmar Learning, learning materials—textbooks, videos, and instructor's curriculum and teaching tools—were created to effectively teach these standards. A foundational tenet of this series is that students *learn by doing*. A constant focus of the *Residential Construction Academy* is teaching the skills needed to be successful in the Construction industry and constantly applying the learning to real world applications.

Perhaps most exciting to learners and industry is the creation of a National Registry of students who have successfully completed courses in the *Residential Construction Academy Series*. This registry or transcript service provides an opportunity for easy access for verification of skills and competencies achieved. The Registry links construction industry employers and qualified potential employees together in an online database facilitating student job search and the employment of skilled workers.

About This Book

A home is an essential part of life. It provides protection, security, and privacy to the occupants. It is often viewed as the single most important thing a family can own. This book is written for students who want to learn how to properly install and service the plumbing, piping, and fixtures that provide comfort and convenience for those who reside in the home.

Plumbing covers the basics of plumbing materials, as well as the processes and skills needed to safely install and service residential plumbing equipment. These required skills are discussed and presented in a manner that not only explains what needs to be done but also shows how to accomplish these tasks. General and task-specific safety issues are addressed throughout the book.

This textbook provides a valuable resource for the areas in plumbing that are required of an entry-level plumber, although those actively involved in the industry will also benefit from the material covered. The basic "hands-on" skills as well as the procedures outlined in the book will help individuals gain proficiency in this ever-changing trade.

In addition to topics such as tools, materials, fixtures, and equipment, the book covers a wide range of topics including water service installation, drainage waste and vent installation, and, of course, repairs and troubleshooting. The concept of hydronic heating systems is also covered in great detail. The format of this material is intended to be easy to read and easy to teach.

Organization

This textbook is organized in a manner so that those new to the industry as well as those already working in the field can gain maximum benefit from its content. The four main sections of the book cover the major aspects of the plumbing industry as they affect residential construction:

- **Section 1: Tools and Materials** discusses the safe and effective use of hand and power tools, types of piping, fittings, and valves.
- **Section 2: Fixtures and Equipment** covers the structure and piping of common fixtures, like tubs, showers, and kitchen sinks, as well as common equipment, including garbage disposals and dishwashers.
- **Section 3: Layout and Installation** introduces the planning aspect of plumbing, beginning with blueprint reading and drafting, layout and material organization, and progressing to various types of installation.
- **Section 4: Troubleshooting** addresses various malfunctions commonly encountered, accompanied by methods of repair. This section also takes an in-depth look at hydronic heating systems.

Features

This innovative series was designed with input from educators and industry and informed by the curriculum and training objectives established by the Standards Committee. The following features aid in the learning process:

Learning Features such as the **Introduction**, **Objectives**, and **Glossary** set the stage for the coming body of knowledge and help the learner identify key concepts and information. These learning features serve as a road map for the chapter. The learner also may use them as a reference later.

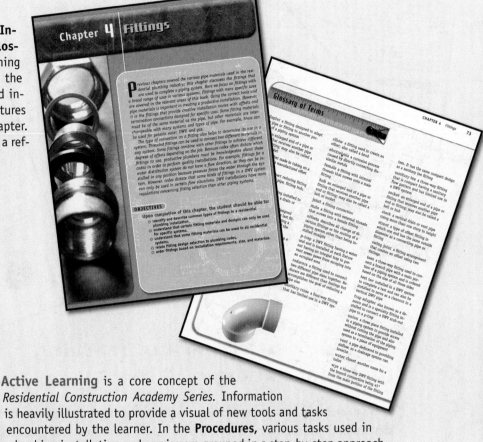

Active Learning is a core concept of the *Residential Construction Academy Series*. Information is heavily illustrated to provide a visual of new tools and tasks encountered by the learner. In the **Procedures**, various tasks used in plumbing installation and service are grouped in a step-by-step approach. The overall effect is a clear view of the task, making learning easier.

Safety is featured throughout the text to instill safety as an "attitude" among learners. Safe job-site practices by all workers is essential; if one person acts in an unsafe manner, then all workers on the job are at risk of being injured, too. Learners will come to appreciate that safety is a blend of ability, skill, and knowledge that should be continuously applied to all they do in the Construction industry.

Caution features highlight safety issues and urgent safety reminders for the trade.

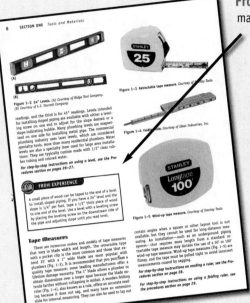

From Experience provides tricks of the trade and mentoring wisdom that make a particular task a little easier for the novice to accomplish.

Review Questions complete each chapter. These are designed to reinforce the information learned in the chapter as well as give the learner the opportunity to think about what has been learned and what they have accomplished.

Turnkey Curriculum and Teaching Material Package

We understand that a text is only one part of a complete, turnkey educational system. We also understand that Instructors want to spend their time on teaching, not preparing to teach. The *Residential Construction Academy Series* is committed to providing thorough curriculum and preparatory materials to aid Instructors and alleviate some of their heavy preparation commitments. An integrated teaching solution is ensured with the text, including the Instructor's e.resource™, print Instructor's Resource Guide, Student Videos, and CD Courseware.

e.resource™

Delmar Learning's **e.resource**™ is a complete guide to classroom management. The CD-ROM contains lecture outlines, notes to instructors with teaching hints, cautions, and answers to review questions, and other aids for the Instructor using this *Series*. Designed as a complete and integrated package, the Instructor is also provided with suggestions for when and how to use the accompanying **PowerPoint, Computerized Test Bank, Video,** and **CD Courseware** package components. There is also a print **Instructor's Resource Guide** available.

PowerPoint®

The series includes a complete set of PowerPoint Presentations providing lecture outlines that can be used to teach the course. Instructors may teach from this outline or can make changes to suit individual classroom needs.

Computerized Testbank

The Computerized Testbank contains hundreds of questions that can be used for in-class assignments, homework, quizzes, or tests. Instructors can edit the questions in the testbank, or create and save new questions.

Videos

The *Plumbing Video Series* is an integrated part of the *Residential Construction Academy Plumbing* package. The series contains a set of four, 20-minute videos that provide step-by-step plumbing instructions. All the essential information is covered in this series, beginning with personal safety and power tools to the installation of major fixtures. Need to know Plumber's Tips and Safety Tips offer practical advice from the experts. The complete set includes the following: Video #1: Personal Safety and Power Tools, Video #2: Working with Copper, Video #3: Working with Plastic Pipe, and Video #4: Installing Toilets, Faucets, and Tubs. An Instructor's Guide is available as well.

CD Courseware

This package also includes computer-based training that uses video, animation, and testing to introduce, teach, or remediate the concepts covered in the videos. Students will be pre-tested on the material and then, if needed, provided with the suitable remediation to ensure understanding of the concepts. Post-tests can be administered to ensure that students have gained mastery of all material.

Online Companion

The Online Companion is an excellent supplement for students. It features many useful resources to support the plumbing book, videos, and CDs. Linked from the Student Materials section of **www.residentialacademy.com,** the Online Companion includes chapter quizzes, an online glossary, product updates, related links, and more.

National Testing

Thomson Delmar Learning is pleased to introduce a new online testing tool to enhance our *Residential Construction Academy Series*. This robust online testing and grading tool will offer testing to students in each of the trade areas of the *RCA Series*. Instructors will be able to obtain a baseline of their student's knowledge by tracking student's grades, completion time, and level of success, while determining whether material has been sufficiently learned at the conclusion of the program. Key codes will provide access to the online test.

Features

- **Flexibility:** Instructors can select content to meet their local building needs
- **Pre- and post-testing capabilities:** Visit www.residentialacademy.com for free pretesting to help plan class time and define student mastery/deficiencies with minimum effort. Then sign up for the National Test.
- **Student Certificates of Achievement:** Student can provide copies of certificates to prospective employers to prove their skill level.
- **Online, self-grading approach:** No wait for grading, results, and remediation.
- **Based on industry standards:** Tests are driven by the National Skill Standards developed by HBI.

About the Author

The author of this book, Michael A. Joyce, is a Master Plumber with over 27 years of experience in the profession. He has participated in major building projects throughout the United States, supervising the large-scale hiring and supervision of plumbers and subcontractors for primary contractors. Mr. Joyce owns a plumbing company and also teaches the plumbing trade via his consulting and training firm, The Plumbing School of Trade, based in Cary, North Carolina. In addition to his trade and teaching experience, he consults for large organizations on their training and construction needs. Mr. Joyce teaches regularly at the Pentagon and has taught at a large hospital organization in New York City. He also teaches Certification and Maintenance Training (CAMT) for the Triangle Apartment Association. His technical credits include authoring *The Plumbing Apprentice Manual, Drilling and Notching Manual, Trade Math and Formulas Manual, Plumbing Blueprint Reading and Drafting Manual, Water Heater Manual, Well Pump Manual,* and the *NC State Plumbing Exam Study Manual.* He is the owner of the *Plumbing Education* Web site, www.plumbingpro.org

Compliance with Apprenticeship, Training, Employer and Label Services (ATELS)

These materials are in full compliance with the Apprenticeship, Training, Employer, and Labor Services (ATELS) requirements for classroom training.

Acknowledgments

Plumbing National Skill Standards

The NAHB and the HBI would like to thank the many individual members and companies that participated in the creation of the Plumbing National Skills Standards. Special thanks are extended to the following individuals and companies:

Michael Dunn, Paul E. Smith Company
John Gallagher, San Diego Job Corps Center
Charlie Gordon, Gordon Plumbing
Larry Howe, Howe Heating and Plumbing, Inc.
Fred Humphreys, Home Builder's Institute
Mark Huth, Thomson Delmar Learning
Michael Joyce, Joyce Company, Inc.
Richard Kerzetski, Universal Plumbing and Heating
Bob Renz, Dynamic Plumbing Systems, Inc.
Ronald Rodgers, Wasdyke Associates
Tom Thornberry, Charlotte Plumbing, Inc.
Ray Wasdyke, Wasdyke Associates

In addition to the standards committee, many other people contributed their time and expertise to the project. They have spent hours attending focus groups, reviewing and contributing to the work. Delmar Learning and the author extend our sincere gratitude to:

Robert Irion, AAA Construction School
Edward Moore, York Technical College
Kevin Standiford, Arkansas State University
Marty Swan, Lincoln Land Community College
Mike Swanson, Cladwell County Career Center
Kevin Ward, McEachern High School

In addition to reviewing and checking the accuracy of the book, Kevin Standiford also developed the PowerPoint slides and the Examview test bank.

Special thanks for the cooperation of Shelhamer Sons of Drums, PA, for their professional assistance in shooting the water tank photos in Chapter 16, as well as members of the Home Builders Institute for their assistance in the photo shoot. A very special thank you goes out to Monica Ohlinger, the developmental editor on this project.

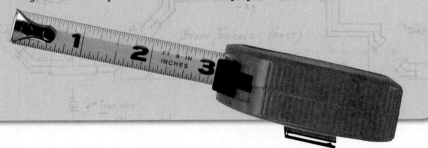

SECTION ONE

Tools and Materials

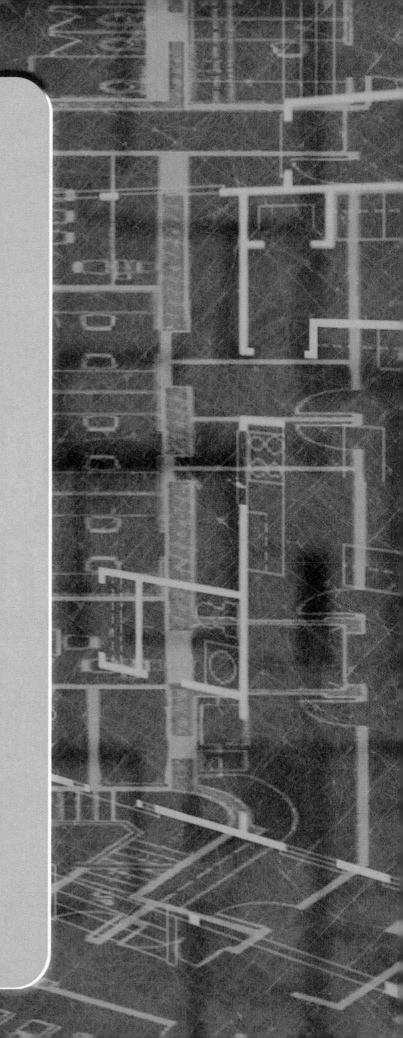

SECTION ONE
TOOLS AND MATERIALS

- Chapter 1: *Plumber's Toolbox*
- Chapter 2: *Power Tools*
- Chapter 3: *Types of Pipe*
- Chapter 4: *Fittings*
- Chapter 5: *Valves and Devices*

Chapter 1 Plumber's Toolbox

Tools fall into two basic categories: hand and power. This chapter introduces you to common hand tools used in the plumbing industry. Many of these are used regardless of the type of plumbing work being done, but specialty tools are used in specific areas, such as making repairs. The use of most specialty tools will be part of job site training, but an employer will expect a student graduating from an apprentice program to have a basic knowledge of common hand tools.

Proper use of hand tools also means their safe use. This area of your training must be a focal point so your respect for safety grows as your career develops. Although a hand tool is manually operated, personal injury can result from improper use, including using it for unintended purposes. Safe use includes remaining attentive to your surroundings and other people in the work area. You will learn additional safety lessons throughout this book and during your career. As you develop your mechanical skills in installing plumbing systems, include the necessary safety steps into your work habits.

OBJECTIVES

Upon completion of this chapter, the student should be able to:
- identify and describe the hand tools the plumber commonly uses.
- use hand tools in a safe and appropriate manner.

Glossary of Terms

band iron thin, perforated metal strapping sold in rolls and used to support pipe

cheater an unsafe method of gaining extra leverage when using a tool

fitting plumbing item used to connect pipes to each other or to fixtures; used to change directions in a piping system

joint term to describe the connection point of a pipe and fittings to each other or to fixtures

nut driver size-specific tool to tighten or loosen nuts, similar to a screwdriver

open-end wrench size-specific tool to tighten or loosen nuts and bolts in confined spaces

slope upward or downward installations used to install drainage or venting piping

socket size-specific tool to tighten or loosen nuts and bolts in confined spaces; uses a ratchet handle for its operation

wallboard material that provides a finished wall surface over the wall structure; gypsum board (drywall) is the most common type used in residential construction

wire cutter tool to cut wire such as electrical wire

Hand Tools

Most hand tools are not as expensive as power tools, but they are equally important in completing a task or project. Most employers expect a plumber to have basic hand tools to even be considered for employment. It may seem expensive to purchase the necessary tools, but it is an investment in your career. "You are only as good as your tools" is an old saying that you will appreciate as your career in plumbing progresses. A plumber's toolbox consists of many common items, and as your skills develop, it will also include specialty tools for specific jobs. The list of tools in Table 1-1 is for plumbers who work in new residential construction. Most of their descriptions are included in this chapter.

The physical size of your toolbox depends on how many tools you have or whether you want more than one toolbox for specific work activities. Because the longest tools you will probably have for a residential job are a 24" level and pipe wrench, you may want a toolbox large enough to house those tools. Some plumbers use tool belts while performing certain tasks, and these can be used as permanent storage for those tools. A toolbox should be lockable and should remain in your possession or in a secure area on a job site.

Levels

A level is one of the most important tools in a plumber's toolbox. It has tubes known as vials that are partially filled with colored liquid leaving a trapped air bubble inside to read desired results. Many levels have a dimensional feature that can be used for measuring distances, similar to a tape measure or ruler. Two types of levels are common: the torpedo level (Fig. 1-1) is a small tool used for quick reading, and the 24" level (Fig. 1-2) is used for more accurate installations. Drainage pipe is installed with **slope** so wastewater flows out of the piping system, but other piping systems and fixtures are installed level. Most quality levels have three different leveling tubes for three different installation positions. One tube is for vertical readings, one is for horizontal

Figure 1-1 Torpedo level. *Courtesy of Ideal Industries, Inc.*

Table 1-1 New Residential Plumber's Tool List

Retractable tape measure (preferably a 25' × 1" type)
Medium Phillips screwdriver (Size #2)
Medium slotted screwdriver
Multi-type screwdriver
Two 10" angled jaw pliers
6" combination pliers
7" locking pliers
8" or 10" adjustable wrench
18" pipe wrench
24" pipe wrench
Smooth jaw pipe wrench
Basin wrench
12" claw hammer
Cat's-paw nail remover
Allen wrench kit
Wood chisel kit
12" concrete chisel
5/16" nut driver or No-hub torque wrench
1/8" to 1-1/8" copper tubing cutters
Copper midget tubing cutters
Copper tubing cutter up to 2" pipe size
Copper flaring tool
Copper tubing bending tool
Plastic pipe saw
Hacksaw
Mini-hacksaw
Flexible piping cutter
Inside PVC pipe cutter
Utility blade knife
Straight-cut type aviation snips or set of three
Pencil
Magic marker
Carpenter's speed square
Basket strainer tool or internal wrench
Torpedo-type level
24" level (may prefer one with slope/grade option)
Torch regulator assembly
Torch striker
1/2" flexible pipe crimping tool
3/4" flexible pipe crimping tool
Plumb bob
Chalk box

SECTION ONE Tools and Materials

Figure 1–2 24" Levels. *(A) Courtesy of Ridge Tool Company. (B) Courtesy of L.S. Starrett Company.*

Figure 1–3 Retractable tape measure. *Courtesy of Stanley Tools.*

Figure 1–4 Folding rule. *Courtesy of Ideal Industries, Inc.*

readings, and the third is for 45° readings. Levels intended for installing sloped piping are available with either a leveling screw on one end to adjust for the slope desired or a slope-indicating bubble. Many plumbing levels are magnetized on one side for installing metal pipe. The commercial plumbing industry uses laser levels, which are considered specialty tools, more than many residential plumbers. Water levels are also a specialty item used for large area installations. They are typically custom made with 1/2" clear rubber tubing and colored water.

For step-by-step instructions on using a level, see the Procedures section on pages 26–27.

 FROM EXPERIENCE

A small piece of wood can be taped to the end of a level to install sloped piping. If you have a 24" level and the slope is 1/4" per foot, tape a 1/2" thick piece of wood to one end of the level. Use a level with a leveling screw by placing the leveling screw on the downstream side of the pipe and adjusting slope until you read level.

Figure 1–5 Wind-up tape measure. *Courtesy of Stanley Tools.*

Tape Measures

There are numerous makes and models of tape measures that vary in blade width and length. The retractable type with a pocket clip is the most common and those that extend 25' with a 1" wide blade are most popular with plumbers (Fig. 1-3). It is recommended that you purchase a quality tape measure; some leading manufacturers offer a lifetime damage warranty. The 1" blade allows a plumber to obtain dimensions over a larger span because the blade extends farther without collapsing in midair. A wooden folding ruler (Fig. 1-4), also known as a rule, offers an accurate reading because it does not sag, and many have an extension slide for internal measuring. They can also be used to lay out certain angles when a square or other layout tool is not available, but they cannot be used for long-distance measuring. An installation—such as an underground piping system—that requires more length from a standard retractable tape measure may dictate the use of a 50' or 100' wind-up tape measure. Wind-up tape measures (Fig. 1-5) are flimsy, and the tape must be pulled tight to avoid incorrect measurements caused by sagging.

For step-by-step instructions on reading a ruler, see the Procedures section on page 28.

For step-by-step instructions on using a folding ruler, see the procedures section on page 29.

FROM EXPERIENCE

Most tape measures highlight numerical increments of 16" in red to indicate the typical dimension from center to center of two wood studs. You can use this feature to locate studs behind finished walls.

Squares

Framing squares are used extensively by carpenters, and plumbers typically use them for layout or when marking large boards to cut (Fig. 1–6). It is not a tool that is used daily. It has two edges that form a 90° angle and is beneficial for laying out angles. A popular small square used to mark boards for cutting is a speed square (Fig. 1–7). It has three edges and one side is at a 45° angle. Both types of squares have dimensions included to use for measuring and layout. As you will learn in the material section of this book, plumbers use three basic offset angles: 90° and 45° angles are the most common, and the less-common 22-1/2° is used for drainage and vent systems.

For step-by-step instructions on using a framing square, see the Procedures section on page 30.

Screwdrivers

Screwdrivers are available in many lengths and shank diameters. Phillips and slotted-head are the two types used by plumbers (Fig. 1–8). The tip of the screwdriver determines the size, and the most common sizes are #1, #2, and #3, with the smallest number indicating the smallest size. The most common type is a multi-screwdriver (Fig. 1–9) that is disassembled and re-assembled for specific uses. Most have combination bits

Figure 1–6 Framing square. *Courtesy of The Stanley Works.*

Figure 1–7 Speed square.

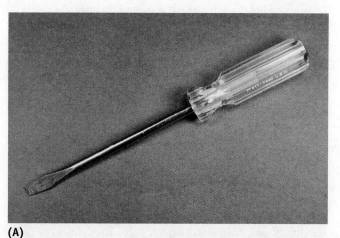

(A)

(B)

Figure 1–8 Slotted and Phillips screwdriver.

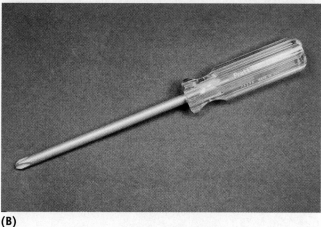

Figure 1–9 Multi-screwdriver. *Courtesy of The Stanley Works.*

that provide four options: a small and a large diameter Phillips and a small and a large diameter slotted screwdriver. The hollow shank of the tool serves as two different-sized **nut drivers.** The small-diameter screwdriver combination bit is housed in the end of the shank that is also a 1/4" nut driver, and the large diameter end of the shank is a 5/16" nut driver. Other multi-screwdrivers are available with replaceable bits ranging up to a 16-in-1 type. Separate Phillips and slotted-type screwdrivers should be included in your toolbox in case your multi-screwdriver is lost or damaged. You should also have a large slotted-type screwdriver for securing or removing large screws.

Figure 1–11 Combination pliers. *Courtesy of Ridge Tool Company.*

 FROM EXPERIENCE

The combination bit from a multi-screwdriver can be used in a drill if a proper drive bit is not available, but do not use the combination bit when the drill is set at high speed to avoid bit damage. A drill with a torque control feature can be used to minimize damage to the screwdriver tip.

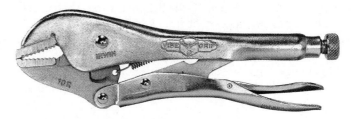

Figure 1–12 Locking pliers. *Courtesy of Irwin Industrial Tool Company.*

Pliers

Pliers are available in various styles and are one of the most practical tools in a plumber's toolbox. The most common type used by plumbers has angled and grooved jaws (teeth) for an adequate grip on pipe. They are sometimes called water pump pliers, or they may simply be referred to as pliers (Fig. 1–10). Most pliers used in the plumbing trade have cushion grip handles. The 10" angled-jaw pliers are widely used and offer an adjustment feature that allows them to open up to 2". Angled pliers are available in both smaller and larger sizes than 10", and you may wish to add various sizes as your toolbox grows. Combination pliers are handy when working with small items in a confined space, but are not used for installing pipe (Fig. 1–11). Locking pliers are useful in removing a nut or bolt that is seized due to corrosion. They typically do not have cushion grips (Fig. 1–12). Needle-nose pliers are not useful in tightening or loosening piping, but are handy for twisting wires or working with small parts. Many also have a **wire-cutting** feature (Fig. 1–13).

Figure 1–13 Needle nose pliers. *Courtesy of Sears Brands, LLC.*

Adjustable Wrenches

Adjustable wrenches are available in various sizes, some with a cushioned grip (Fig. 1–14). The jaw opening is adjustable and smooth so it does not damage metal finishes. An adjustable wrench is a great asset in any toolbox; its adjustment range usually allows it to replace numerous **open-end wrenches** and **sockets**. Its compact design is great in

Figure 1–10 Angled jaw pliers. *Courtesy of Ridge Tool Company.*

Figure 1–14 Adjustable wrench.

Figure 1–15 Pipe wrench. *Courtesy of Ridge Tool Company.*

Figure 1–16 Spud wrench. *Courtesy of Ridge Tool Company.*

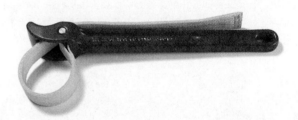

Figure 1–17 Strap wrench. *Courtesy of Ridge Tool Company.*

Figure 1–18 Chain wrench. *Courtesy of Ridge Tool Company.*

a confined space, such as working on water supply connections under a sink. A common size is 9". Using two adjustable wrenches in opposing positions is beneficial when working under sinks.

Pipe Wrenches

As with other popular tools, pipe wrenches come in numerous styles and sizes (Fig. 1-15). Pipe wrenches with grooved jaws (teeth) provide grip for installing or removing metal piping systems or plastic material with designated tightening areas. The least expensive ones have a cast-iron body; aluminum ones are lighter, but more expensive. The two most common sizes are 18" and 24", but smaller and larger sizes are also available. All pipe wrenches have an adjustment range for various pipe or fitting sizes, with larger wrenches accommodating larger pipe sizes. A chrome plated finish is often used in the plumbing industry. Damage to its appearance can be avoided by using a smooth jaw pipe wrench or a more common smooth jaw tool known as a spud wrench (Fig. 1-16). Strap wrenches are also available for working with items that cannot be damaged (Fig. 1-17). Chain wrenches, for specialty work, can be used in place of a pipe wrench for certain applications (Fig. 1-18). Tightening or loosening pipes and fittings requires using two wrenches in opposing positions to eliminate damage to other portions of the system.

For step-by-step instructions on using a pipe wrench, see the Procedures section on page 31.

CAUTION: You may be taught on a job site to place a piece of pipe over a handle of a pipe wrench to create what is known as a "cheater." However, the pipe wrench handle can break under these circumstances. Always use the proper size pipe wrench to avoid injury.

10 SECTION ONE *Tools and Materials*

FROM EXPERIENCE

Most of your working strength comes from using pipe wrenches in the 10-o'clock and two 2-o'clock positions or in the 4-o'clock and 8-o'clock positions.

Hammers

There are numerous hammer types, with different heads, claws, and weights. A plumber uses a hammer mostly for removing nails and for installing wood bracing, pipe supports, and pipe protection shields. It is not as important to focus on the weight or the head type as on the type of claw, because plumbers use hammers less often than carpenters. When removing nails from a wooden structure to allow for holes to be drilled, a straight claw works better than a curved claw hammer (Fig. 1-19). The most common type used by plumbers weighs 16 ounces and has a smooth nailing head (Fig. 1-20). Another common type, a waffle head hammer, is used when framing a building; it is more expensive, but its design offers more security from glancing off when hitting a nail or chisel. Ball peen hammers (Fig. 1-21) were more popular in the plumbing industry in past decades. Some handles are made from wood and others from steel or fiberglass. Some steel handles are solid, and others are hollow. Steel and fiberglass handles have cushioned handle grips, but wooden handles usually do not. It is recommended that you purchase a hammer that has a straight claw, a solid steel handle, and a waffle head.

Figure 1-21 Ball peen hammer. *Courtesy of Sears Brands, LLC.*

Figure 1-19 Curved claw hammer.

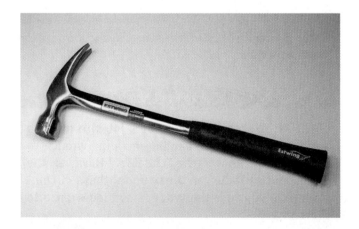

Figure 1-20 Straight claw hammer.

CAUTION: Always wear eye protection while using a hammer, and be aware of others in your work area.

Plastic Pipe Saw

There are saws for almost every use in the construction trade. The two types of plastic piping (ABS and PVC) that you will learn more about in the material section of this book are both cut with the same type of saw (Fig. 1-22). Short and long blade styles are available with cutting teeth that are closer together than on a wood-cutting saw and farther apart than on a metal-cutting saw. The blade is replaceable and is fixed to the handle with a removable screw. The blade can be used to cut small pieces of wood or to cut a hole in drywall, if necessary, but it is best to use tools designed for those purposes to avoid dulling the saw blade. Cutting plastic pipe with a saw leaves a burred edge on the pipe end that

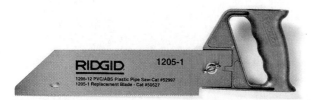

Figure 1–22 Plastic pipe saw. *Courtesy of Ridge Tool Company.*

must be removed before installing the pipe. All pipe ends must be cut square before they are installed into a fitting; correcting uneven cuts wastes pipe and consumes time. Many plumbing apprentices find it difficult to cut larger pipe diameters squarely with a hand saw, but they become more accurate with a little practice. Jobs requiring numerous cuts are typically performed with a reciprocating saw or other power saw.

For step-by-step instructions on plastic pipe cutting, see the Procedures section on page 32.

CAUTION: Use any saw with care and wear leather gloves to avoid cuts and scrapes.

 FROM EXPERIENCE

You can remove burrs with a tool designed especially for that purpose, but it is more common to use a file or pocketknife. Including this step in your normal cutting process allows it to become a natural step in working with plastic pipe.

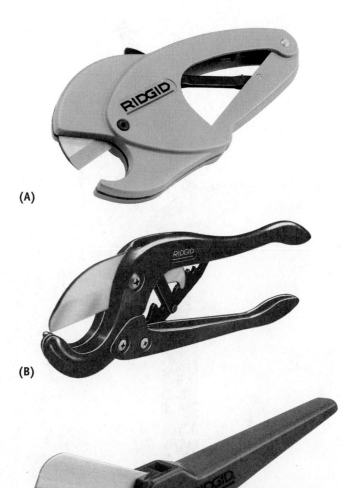

(A)

(B)

(C)

Figure 1–23 Plastic pipe cutters. *Courtesy of Ridge Tool Company.*

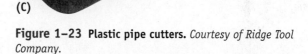

CAUTION: The cutting blade is extremely sharp and can cause serious lacerations

Plastic Pipe Cutter

Smaller-diameter plastic and flexible piping are best cut with a specially designed plastic pipe-cutting tool to ensure a clean, square cut. Designs are available in a variety of prices and sizes (Fig. 1–23). This tool increases productivity, ensures a more accurate cut, and eliminates burrs. The cutting blade is replaceable and fixed to the tool with a screw. Most large tools cut up to 1-1/2" diameter pipe, but the most common are used for smaller pipe sizes. An uneven cut can cause a **joint** to leak, and burrs can cause water flow to be disturbed, which results in pressure loss in a piping system. Burrs can also travel in a water piping system and obstruct filters, screens, and water flow passageways.

Inside Plastic Pipe Cutter

Plastic pipe in a difficult location often must be cut with a drill and an inside pipe cutter (Fig. 1–24). For example, if a drainage pipe is installed below a concrete floor or where the piping cannot be accessed from below a work area, the plastic pipe can be cut from the work area with an inside pipe-cutting tool. Many kinds are available, but the most

12 SECTION ONE *Tools and Materials*

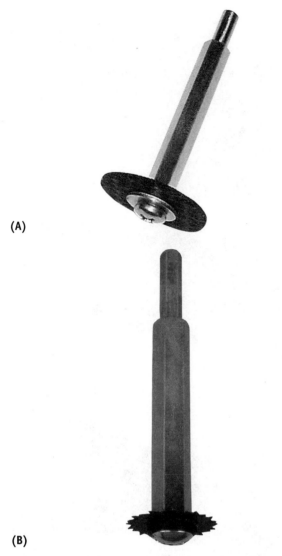

(A)

(B)

Figure 1-24 Inside plastic pipe cutter. *Courtesy of Sioux Chief Mfg.*

popular is an inexpensive, simple rotating type. The cutting wheel has several cutting teeth and is fixed to a drill shaft, which rotates as the drill is operated. The cutting wheel is inserted into the plastic pipe, the drill is activated, and pressure is applied against the pipe wall while the wheel is slowly moved around the inside of the pipe.

CAUTION: The drill should only be activated after the tool is inserted in the pipe to avoid injury from the rotating cutting wheel.

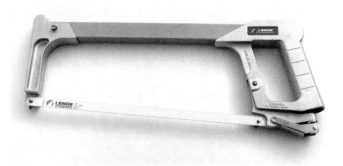

Figure 1-25 Standard hacksaw.

 FROM EXPERIENCE

Temporarily insert a cloth rag into the pipe being cut to act as a safety net in case the screw, cutting wheel, or entire tool falls into the pipe. Be sure the rag does not interfere with the cutting process, and remove it when you complete the work.

Metal Cutting Saw

A hacksaw is used in many areas of construction by almost every trade (Fig. 1-25). This versatile tool, designed to cut through metal, is made with numerous blade types and various numbers of teeth per inch. Any handsaw blade made to cut metal has more teeth per inch than blades designed to cut wood or plastic. A hacksaw blade saws in one direction, and an arrow, located on the side of the blade, indicates correct use. Many different saw frames are available, some of which are made of steel and others of aluminum. Miniature hacksaws are common for cutting bolts that are used to install toilets (Fig. 1-26). Many miniature styles are available, some using a standard hacksaw blade and others requiring specialized blades. If a miniature saw is not available to trim toilet bolts, a standard hacksaw blade removed from its frame can be used.

CAUTION: If you use a hacksaw blade without its frame, wear leather gloves and/or wrap a protective cloth around the blade to protect your hand from the saw blade teeth.

Figure 1-26 Miniature hacksaw. *Courtesy of Superior Tool Company.*

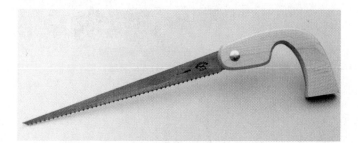

Figure 1-27 Compass saw.

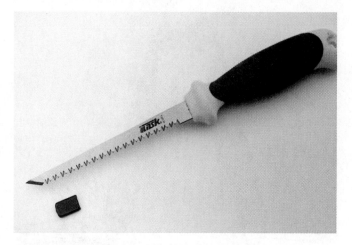

Figure 1-28 Wallboard saw.

Figure 1-29 Types of aviation snips.

Wallboard Saws

Cutting drywall board, after an installation, to repair a leak is common. A **wallboard** saw is used to perform the task. Several styles are available, and some use the same blades as those used with other tools. The most common are compass saws (Fig. 1-27) and wallboard saws (Fig. 1-28). A compass saw cuts circular holes, but can also make square cuts. The wallboard saw is shorter and more rigid than the compass saw, making it preferable for a plumber. If a wallboard saw is not available, the task can be performed with a utility knife (Fig. 1-32) or, in the case of an emergency plumbing repair, with a plastic pipe saw (Fig. 1-22).

Aviation Snips

Aviation shears are also known as tin snips. The three most common styles cut straight, left, or right. Their cushioned grip handles are color-coded indicating their direction of cut. Many manufacturers use a standard color code that indicates yellow is a straight cut, red is a left cut, and green is a right cut (Fig. 1-29). Offset styles place your hand above the material being cut. Plumbers use these tools for cutting a pipe-support material known as **band iron.** Plumbers who also install sheet metal duct or piping may wish to invest in various types of snips, but a straight-cut style is the most popular. The cutting edges of the tool are not replaceable, so the tool must be replaced when it becomes dull or damaged. The user must cut with the tool perpendicular to the item being cut to minimize damage to the aviation snips. Improper use will cause the cutting portion of the tool to become "sprung."

CAUTION: Leather gloves are recommended when cutting sheet metal to avoid cuts. Working with sharp edges of any metal can cause serious lacerations.

Nail Pullers

Removing nails from wood boards before drilling or cutting holes is common to avoid damage to drill bits and saw blades as well as to avoid personal injury. Various methods and tools are used depending on preference and the location of the nail. The two most common tools in the plumbing industry are a flat bar (Fig. 1-30) and a nail paw (Fig. 1-31), both of which have nail-pulling features. The nail paw is also

Figure 1–30 Flat bar.

Figure 1–31 Nail paw.

known as a cat's paw and has two different slotted ends that allow the user options based on the work location. It is often used to pull nails from tight areas or areas that are embedded in a partially drilled hole. A typical flat bar has three options—the two different slotted ends and a nail-pulling slot in the flat portion of the bar.

CAUTION: A hammer is used with a nail-removing tool, so eye protection should be worn when removing a nail. Always discard removed nails in a safe location to eliminate accidentally stepping or kneeling on them.

Knives

A utility knife is another inexpensive tool that is found in most toolboxes on a construction site (Fig. 1–32). It is used to cut wallboard and to cut boxes or tape for material handling purposes. Some models have replaceable blades that retract into the tool body to avoid injury or damage to the blade when it is not in use. Multi-purpose tool sets are very popular and include various styles of blades, small

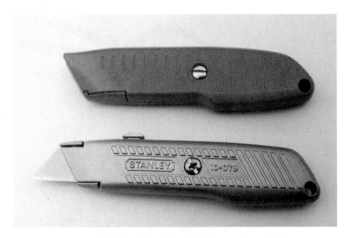

Figure 1–32 Utility knife.

Figure 1–33 Multi-purpose tool. *Courtesy of Leatherman Tool Group, Inc.*

screwdrivers, pliers, and possibly a file. Multi-tool knife sets (Fig. 1–33) increase productivity and are typically sold with a belt carrying case for quick use. A standard pocketknife is also handy and fits into a pocket for quick retrieval.

CAUTION: A utility knife blade is very sharp and can cause serious lacerations. Be sure to retract the blade into the tool housing when it is not in use. Always cut away from your body when using any knife.

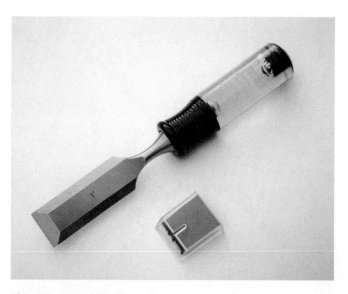

Figure 1–34 Wood chisel.

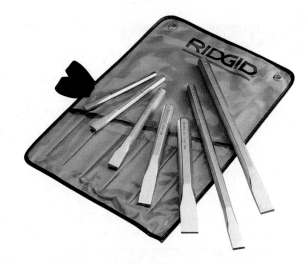

Figure 1–35 Cold chisels. *Courtesy of Ridge Tool Company.*

Chisels

A wood chisel is used in the plumbing industry to notch and split pieces of wood boards. They are available in various widths (Fig. 1–34). Plumbers usually choose 1/2" or 1". A hammer is used to strike the handle of the chisel, which is very sharp. Eye protection must be worn, and hands must be kept away from the sharp cutting edge. Most chisels come with a protective sheath or case and should be stored in a toolbox, not a pocket. Chisels used for chipping concrete, also known as cold chisels, have a cutting edge that is more blunt than that of a wood chisel (Fig. 1–35). A hammer is used to create a forceful blow, so eye protection (and possibly face protection) and leather gloves are required when working. The strike end of a concrete chisel can become flattened, allowing small metal pieces to become airborne and cause injury. The concrete being chipped can also become airborne. Always be aware of other people in your work area.

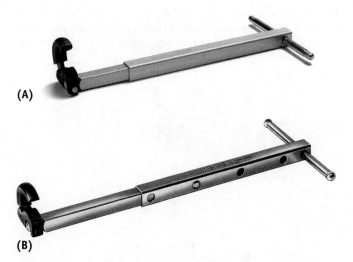

Figure 1–36 Basin wrenches. *Courtesy of Ridge Tool Company.*

CAUTION: The striking end of a chisel can become flared (mushroomed) after repeated use. That flared material must be removed with a grinding power tool to prevent metal fragments from becoming airborne during use.

Basin Wrench

Installing and replacing faucets on an existing sink also means that work space is limited. The tool typically required in this situation is a basin wrench (Fig. 1–36), a specialty tool that should be included in every plumber's toolbox. Two types are available. The least expensive has a non-adjustable shaft, and the most common and more expensive type, known as a telescoping basin wrench, has an adjustable shaft. Most adjustable types come in various lengths. They all have spring-loaded swivel heads to install and remove securing nuts from a faucet and water supply connections in a confined space.

 FROM EXPERIENCE

Many concrete chisels are sold with a protective hand guard to avoid striking your hand with a hammer.

For step-by-step instructions on using a basin wrench, see the Procedures section on page 33.

FROM EXPERIENCE

Do not use another tool on the square-type basin wrench to add extra torque to loosen or tighten nuts. The shaft will become damaged, which eliminates the adjustable length feature.

Basket Strainer Tools

Drainage piping is needed to connect a kitchen sink with a plumbing part known as a basket strainer. Often specialty tools are required to install the drain connection. Numerous tools exist to complete connections to specific fixtures, some of which are multi-purpose. A tool known as a strainer fork (Fig. 1–37) is sometimes used for strainers and tub drains, but its popularity has declined because plumbing material is now manufactured with thinner metals and more plastics than in the past. Many basket strainer tools (Fig. 1–38) are available based on the preference of the installer. A basket strainer can

Figure 1–38 Basket strainer wrench. *Courtesy of Superior Tool Company.*

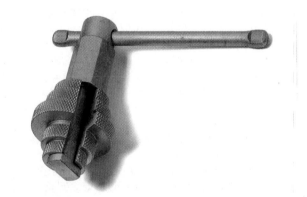

Figure 1–39 Internal wrench. *Courtesy of Ridge Tool Company.*

be accessed after an installation, but a tub drain may not be accessible after the house is complete, which is why proper tightness and correct installation is crucial. An internal wrench (Fig. 1–39) is a multi-purpose tool that inserts into some strainers, tub drains, and other plumbing parts allowing an installer to tighten or remove parts of a plumbing fixture. The internal wrench can be used on interiors with widths ranging from 1″ to 2″.

Copper Pipe Cutters (Tubing Cutters)

Copper pipe is used more often in commercial installations than residential due to the popularity of plastic water piping systems, but it is still very important for a plumber to know how to work with copper. Copper is identified as pipe or tubing, and the cutters are called either tubing or pipe

Figure 1–37 Strainer fork. *Courtesy of Superior Tool Company.*

Most cutters have a device known as a reamer to remove the internal ridge created during the pipe-cutting process. It is important to ream the pipe so the ridge does not cause water flow turbulence within the piping system. A hacksaw or reciprocating saw is often used in confined spaces to cut copper pipe, but, if possible, this practice should be avoided to prevent copper shavings from entering the piping system during the pipe-cutting process. Copper pipe used for medical gas systems or natural gas systems should never be cut with any tool other than a tubing cutter designed for cutting copper, because shavings obstruct critical aspects of the systems. Copper cutters have metal rollers and a cutting wheel that rotate around the pipe as a handle is turned clockwise to advance the cutting wheel through the pipe. The rollers and wheel are replaceable and are secured to the tool frame with specially designed screws or a metal securing clip.

Figure 1–40 Copper mini-cutter. *Courtesy of Ridge Tool Company.*

CAUTION: Sharp edges of copper pipe can cause serious lacerations, so work gloves should be worn.

Figure 1–41 Medium size copper cutter. *Courtesy of Ridge Tool Company.*

 FROM EXPERIENCE

Most mini-cutters have a spare cutting wheel inside the bottom of the handle. A worn tool or loose wheel can cause the wheel to track improperly around the pipe.

Copper Flaring Tool

A flaring tool flares the end of soft copper tubing to create a 45-degree angle that mates with a compatible brass flared **fitting.** A flaring tool is clamped around the copper tubing, and the flaring post is turned clockwise into the tubing to fold the tubing outward, creating the flare. Once the process is complete and the tool is removed from the tubing, the flared end abuts to the flared fitting and is tightened with a flare nut. Many codes dictate that copper tube connections must have a flared joint for certain piping systems. This creates a durable joint that can handle vibration and light jolting without leaking. Most tubing cutters have a built-in reamer, which is a triangular metal accessory that is inserted into the tubing and twisted to remove the internal edge of the tube after cutting. The inside end of the tubing must be reamed with the reamer illustrated in Figure 1–41. The range of tubing that can be flared with a typical flaring tool is 1/8″ to 5/8″. Hammer-type flaring tools are available ranging from 3/8″ to 2″, but are more common for larger pipe sizes used by utility contractors, not by residential plumbers. A more expensive model is the ratchet type shown in Figure 1–43.

Figure 1–42 Quick acting copper cutter. *Courtesy of Ridge Tool Company.*

cutters. Various tool sizes are available for cutting pipe widths ranging from 1/8″ to 4″. The more compact ones are used for smaller pipe sizes in small work areas (Figs. 1–40, 1–41, and 1–42). The smallest type of cutter is called a midget or mini-cutter. Most manufacturers refer to tool sizes based on their model numbers.

18 SECTION ONE *Tools and Materials*

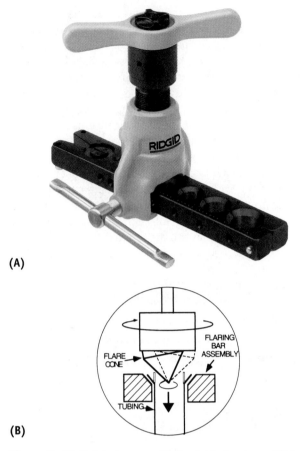

(A)

(B)

Figure 1-43 Ratchet copper flaring tool. *Courtesy of Ridge Tool Company.*

CAUTION: Reaming copper can produce sharp burrs, so leather gloves should be worn.

 FROM EXPERIENCE

Be sure to slide the flare nut over the tubing before clamping the flaring tool around the tubing.

Copper Tubing Bender

A copper tubing bender (Fig. 1-44) allows a plumber to create bends in copper tubes to use in tight spaces and to achieve a professional appearance. Because residential plumbing systems often use flexible tubing, the bending of copper tubing is not widely performed in the residential plumbing industry. However, it is a skill that is required for many exposed connections such as sinks, tank-type toilets,

Figure 1-44 Copper tubing bender. *Courtesy of Ridge Tool Company.*

and other equipment with a small-diameter water supply. Soft copper tubing can kink if it is bent by hand with a too small radius, but the bending tool creates compact offsets. Numerous styles of tools are available, but they all secure the tubing to the tool and have a bending handle to complete the desired offset. Many tools serve pipe sizes ranging from 1/8" to 3/8", but others are for only one pipe size. The tubing is inserted into a grooved portion of the body, and the handle, which is also grooved, is used to manually achieve the desired bend. It will take practice to ensure accuracy of a bend.

 FROM EXPERIENCE

Bending exposed copper with a bending tool is a sign of craftsmanship and is beneficial when connecting water supply tubing to fixtures such as one-piece toilets.

Torch Regulator Assembly

The most common method of connecting copper pipe is with a welding process known as soldering. A typical torch assembly consists of a regulator that controls the amount of flammable gas to be ignited and a torch tip with an orifice designed specifically for the type of gas and torch assembly. Small one-piece torch assemblies that screw directly to a gas cylinder are more popular in the residential industry. They offer more mobility than three-piece assemblies that use a torch connected to a gas cylinder with a hose (Figs. 1-45 and 1-46).

Figure 1–45 One-piece torch assembly with igniter. *Courtesy of TurboTorch.*

Figure 1–46 Three-piece torch assembly with igniter. *Courtesy of TurboTorch.*

Most employers purchase gas cylinders and miscellaneous consumables required for an installation, but they may expect the plumber to own a torch regulator assembly. Use and safety concerns are directly related to the specific type of torch assembly and must be addressed before you even connect to the gas cylinder. A fire extinguisher must be present before igniting a torch, and it is important to pay close attention to the surroundings to ensure no flammables are present. Plumbers use solvent cement and primers to assemble plastic piping and both are dangerous to have in a work area when an open flame is present. This precautionary step also includes soldering adjacent to plastic pipe that was previously assembled and when flame is aimed toward any flammable or combustible materials. An inexpensive tool known as a striker (Fig. 1–47), with a replaceable flint, is the only safe way to ignite the gas. Some torch tips or one-piece assemblies have an igniter built in and do not require a separate striker.

Figure 1–47 Torch striker. *Courtesy, Uniweld Products.*

CAUTION: Serious burns and explosions can result from unsafe use of a torch assembly, so all safety procedures must be known and followed. Always ignite a torch with a striker and never use a cigarette lighter. Never place a cigarette lighter in your shirt, jacket, or pants pockets while soldering.

CAUTION: All gas cylinders must be securely fastened at all times and remain in the vertical position. If a cylinder known as a B-tank tips over, its regulator and stem area can be damaged and an explosion can result.

A residential plumber is best served with a one-piece regulator assembly to increase productivity.

Flexible Pipe Crimping Tool

Flexible water piping systems have replaced copper systems in many residential plumbing applications. The pipe and fittings are joined with a crimping ring that slides over the piping; when the pipe is slid over a barbed fitting (Chapter 4), the ring is crimped in place with a crimping tool (Fig. 1–48). The process is quick and simple and has drastically increased productivity on a job site. The tools are offered in several

Figure 1–48 Crimping tool. *Courtesy of Rostra Tools.*

Figure 1–49 Compact crimping tool. *Courtesy of Mil 3.*

sizes, with the most common being those for 1/2" and 3/4" pipe. A dual-size tool is available in addition to the compact version for 1/2" and 3/4" pipe sizes (Fig. 1-49). The tools require calibration to ensure an adequate crimp is achieved, so a crimp gauge is sold with each tool.

CAUTION: An improper crimp can cause a leak and result in extensive property damage.

 FROM EXPERIENCE

Keep the calibration gauge in your truck or other area away from your toolbox to keep it from being damaged or lost, and check your crimp installations frequently.

Plumb Bob

A plumb bob (Fig. 1–50) is an accurate method of establishing a vertical point of reference to a lower work area from an upper work area. A string is tied to the end of a plumb bob, which is then suspended from an upper location. When the pendulum action stops, it reflects a true vertical position. This is more accurate than transferring a vertical line using a level. A plumber uses a plumb bob to establish the center of a pipe passing from one floor to another for drilling holes. It can also be used to transfer a horizontal dimension from a wall or column to the center of a pipe being installed in a trench.

Figure 1–50 Plumb bob.

 FROM EXPERIENCE

The string can be wrapped around a piece of wood to keep it organized between uses, or a specially designed accessory can be purchased for that use.

Chalk Box

A chalk box (Fig. 1–51) houses string, called a chalk line, and chalk powder that is used to mark a straight line for layout or to cut plywood boards. The chalk line is on a reel and is coated with chalk powder when retracted into the chalk box. Red and blue powders are most common and are added to the box when needed. The external portion of the chalk line is tied to a metal tab that is either held by another person or connected to a nail to create the starting point of the line. The chalk line is pulled taut against the work surface, and the user simply snaps the string to establish a mark. If a mark is required for later use, a clear-coat spray can be applied, so the line location becomes more permanent.

CAUTION: Wear a dust mask when using clear coat spray to avoid inhalation.

Figure 1–51 Chalk box.

FROM EXPERIENCE

A chalk box can be used as a plumb bob. Keep the chalk box dry, or the chalk powder will become paste, which will have to be removed and replaced.

Figure 1–52 Torque wrench. *Courtesy of Ridge Tool Company.*

Torque Wrench

A torque wrench (Figure 1-52) is used to tighten clamps used for installing cast-iron pipe as well as for tightening rubber transition connectors used for dissimilar piping. Both connection types have 5/16" hex head worm-drive screws, and the torque wrench is designed for quick assembly. The average clamp used to connect drainage piping requires a maximum of 60 foot-pounds of pressure, and the torque wrench only tightens up to that specification. The most common type of torque wrench features a T-handle design that ratchets (breaks away) when tightening is complete. They can also be adjusted to loosen an assembled connection. These tools are essential for commercial plumbers, but may not be needed in a residential plumber's toolbox.

CAUTION: Not using a torque wrench to properly tighten a clamp can cause a drainage connection to leak, which may result in flood damage.

FROM EXPERIENCE

Using a drill with a nut-driver bit can overtighten and break a clamp, causing an increase in material costs. If a clamp has multiple worm drives, tighten the nuts alternately.

Personal Protection Equipment

Safety is an attitude. Unfortunately, not all workers respect safety standards; however, the fact is that you are ultimately responsible for your own safety. Personal protection equipment (PPE) is provided by an employer. You should provide any specific items your employer does not provide and any that your particular job site conditions warrant. The safety division of the U.S. Department of Labor is the Occupational Safety and Health Administration (OSHA), which regulates safety standards for the workplace. The construction industry is regulated by OSHA Standards 29 CFR Part 1926 (Safety and Health Regulations for Construction). The information in this chapter focuses on items relating to hand tools; other safety information will be included when relevant throughout this book. Ear protection is an important PPE, especially when power tools are being used. As with all safety items, it will be discussed when most relevant. PPE items relevant to hand tools are listed here and then explained individually following the list.

- Eye protection
- Face protection
- Hand protection
- Knee protection
- Foot protection
- Inhalation protection
- First aid kit
- Hard hat

Eye Protection

Eye protection is one of the items most frequently found on a residential job site. Two common types of eye protection are eyeglasses (Fig. 1-53) and eye goggles (Fig. 1-54). All

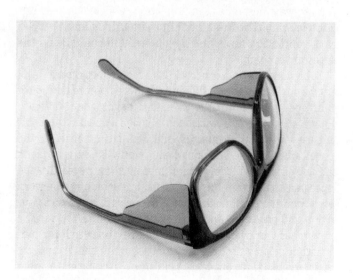

Figure 1–53 Safety glasses.

22 SECTION ONE *Tools and Materials*

Figure 1–54 Safety goggles. *Photo by Bill Johnson.*

MUST BE IDENTIFIED SOMEWHERE ON FRAME TO BE OSHA APPROVED

Figure 1–55 Z87 identification.

PPE must be approved by OSHA and provided by an employer when the potential for danger exists. Safety glasses and goggles approved by OSHA are listed in American National Standards Institute (ANSI) and identified as Z87.1 or simply Z87 (Fig. 1–55). This identification must be readily accessible for review and is located on the frame of the safety glasses. Be sure to locate the identification before assuming you are wearing proper eye protection. Goggles have an adjustable strap that is placed over the head, and many styles have ventilation holes on the top of the frame to keep the lens from fogging.

Drill shavings and concrete dust can collect on the brim of a hat or hard hat and should be removed before taking off your eye protection. Otherwise, this debris could fall into your eyes after eye protection is removed. A neck strap is often used with safety glasses, and debris can settle on the lenses when glasses are not being worn. Be sure to clean the lenses before placing the glasses over your eyes. Shaded lenses are available for use outdoors, but should not be worn while working indoors. Scratches on lenses can obstruct your vision and create an unsafe working condition.

Purchase an eyeglass holder similar to ones sold for prescription glasses to protect the lenses when the glasses are not being used. If you wear prescription glasses, either wear goggles over your glasses or purchase side shields for your prescription glasses. Not all prescription glasses are acceptable as safety glasses; they must be approved before proceeding with any work.

CAUTION

CAUTION: Eye protection should be worn constantly while you are working on a job site and not just when working at a particular work area. Eye damage can also result from other workers in your immediate work area.

FROM EXPERIENCE

Most glasses are sold in a plastic bag. You can cover the exterior of the bag with duct tape to create a slip-in case to store your safety glasses.

Face Protection

Though face protection is more common in the residential industry when removing concrete with power tools, it may be a necessary or particular task on a job site. A full-face shield (Fig. 1–56) is the most common method of protecting your face from flying debris. It is sold with an adjustable headgear to be either placed on your head or fixed directly to a hard hat. Most types allow the user to lift the lens area upward between uses. They have clear lenses that must be kept clean and replaced when scratched. The shield is usually sold separately from the headpiece in a plastic protective bag that can be placed over the lens when it is not in use. Most job sites still require eye protection to be worn under the full-face shield, and, if it is not required, it is recommended.

Figure 1–56 Full-face shield.

> **CAUTION**
>
> **CAUTION:** The shield is not designed to protect you against hot objects or flying objects that strike the shield with extreme force.

> **FROM EXPERIENCE**
>
> A face shield protective cover can be made with a small bath towel folded in half and sewn together, or by covering the plastic bag sold with the shield with duct tape to create an envelope-shaped cover.

Hand Protection

Gloves may seem cumbersome when working with small plumbing items, but they must be worn when necessary to protect you from injury. Metal piping and sheet metal can cause serious lacerations, and working with hot surfaces can cause burns. Gloves are most often made of leather (Fig. 1-57) or cotton. Having a pair of each kind allows you to wear the one most suited for a particular task. Small cuts and scrapes can allow chemicals used in the plumbing industry to enter your bloodstream. Most leather and cotton gloves only protect up to your wrist. Engineer-style gloves provide more protection than cotton gloves and protect the wrist area, but they are difficult to wear when assembling small plumbing items. Wearing gloves that do not provide proper grip during certain tasks can be unsafe. Non-slip gloves have rubberized dots applied to their exterior, which

Figure 1-58 Grip gloves. *Courtesy of Memphis Glove.*

are useful when additional grip is required (Fig. 1-58). Surgical gloves or other rubberized gloves should be worn when working around existing plumbing fixtures, especially if sewage or blood is present.

> **CAUTION**
>
> **CAUTION: Rubber gloves can be penetrated with sharp objects and can provide a false sense of security. Be sure to have a surplus supply. Wear more than one layer or wear a pair under another kind of glove if required.**

> **FROM EXPERIENCE**
>
> Mesh glove liners can be purchased and worn under protective gloves to provide warmth in cold weather.

Knee Protection

Plumbers spend a lot of time in a crouched position and often on their knees. Padded kneepads (Fig. 1-59) should be worn to prevent bone and joint injury. Kneeling on sharp objects can cause them to penetrate your skin and chip your kneecaps. Most kneepads have Velcro straps to adjust to your leg girth and to go around your work pants. The padding inside provides comfort, and many have a hard exterior shell to protect against sharp objects. Several styles are available, and it is recommended that you purchase a quality pair to ensure that they are comfortable.

Figure 1-57 Leather gloves. *Courtesy of Memphis Glove.*

24 SECTION ONE *Tools and Materials*

Figure 1-59 **Kneepad.** *Courtesy of Allegro Industries.*

FROM EXPERIENCE

Knee protection pads might not be part of an OSHA requirement of an employer. It is wise to purchase them yourself to avoid future medical problems.

Foot Protection

Safety shoes or boots must be worn, and a steel-toed version may be required on a specific job site or by an employer (Fig. 1-60). There are numerous acceptable styles of footwear, and an employer usually dictates which can be worn. Many plumbers wear boots that lace up to the lower or middle portions of the shin area. Athletic shoes do not provide adequate protection. Nails and screws are present throughout a job site, and many wood boards that have been removed from a structure still have exposed nails. Even the best footwear will not stop a nail from penetrating its sole and causing foot injury.

CAUTION: Remove nails and screws from wood boards to avoid foot injury.

FROM EXPERIENCE

Because you spend at least one-third of your day wearing work boots or shoes, you should purchase high-quality safety footwear and consider steel-toes even if it is not mandatory.

Inhalation Protection

Job sites are often dusty, and drilling into concrete can create airborne particles and dust that can be inhaled and cause lung problems. Therefore, inhalation protection is needed. A simple paper mask (Fig. 1-61) may be sufficient to prevent inhalation of most construction-related particulates. When working with chemicals or solvents, read all of the safety precautions and data included on product labels and use appropriate inhalation protection. OSHA mandates that a Material Safety Data Sheet (MSDS) for every harmful substance used be available for review by every worker on a job site. You should review the data provided to ensure that you have adequate inhalation protection and that safety procedures are known and followed. Dust masks do not protect an employee from chemical or other harmful substances and are considered nuisance masks.

First Aid Kit

The new residential plumbing industry is fairly mobile because installations do not require a long presence in one workplace. Therefore, main assembly areas are not common. On large projects, there may be a construction trailer or of-

Figure 1-60 Steel-toe work boot.

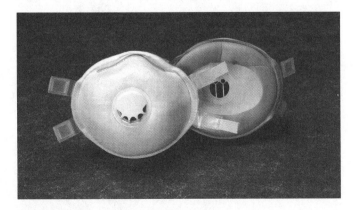

Figure 1-61 **Dust mask.** *Courtesy of Gerson.*

Figure 1-63 Hard hat. *Courtesy of Bullard.*

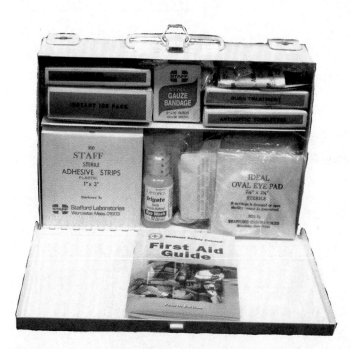

Figure 1-62 Portable first aid kit. *Courtesy of Custom Kits Company, Inc.*

fice with a master first aid station. A small first aid kit (Fig. 1-62) should be readily available in every work vehicle or for every work crew. Single-use eye wash bottles, small bandages, a tourniquet, surgical gloves, wound cleanser, and antibacterial ointment should be included in a portable first aid kit. Basic first aid procedures, location of the nearest medical treatment facilities, and emergency phone numbers should be written inside a first aid kit. If basic first aid is not part of an employee training program, you should learn it on your own. If a co-worker is bleeding, be sure to protect yourself with surgical gloves before providing assistance. A typical portable first aid kit meets ANSI Z308.1-1998 standards.

CAUTION: Not having a first aid kit readily available violates OSHA regulations.

 FROM EXPERIENCE

People working on a construction site should carry spare finger bandages in their wallets for quick use. If adhesive first aid tape is not readily available, duct tape can be wrapped around a sterile first aid gauze pad to maintain pressure on a wound until adequate medical supplies and attention are provided.

Hard Hats

Residential single-family construction plumbers are usually not required to wear hard hats, but that decision is made by each employer or by job site coordinators. It is more likely that employees would be expected to wear hard hats (Fig. 1-63) on multifamily residential construction job sites because of the numerous floors above the lowest work areas. Though most job sites have safety personnel to ensure that hard hats are worn, it is ultimately the responsibility of the employee to wear one when required. Hard hats are available in numerous styles and various colors. Stickers, company logos, and approved accessories can be placed on a hard hat, but the shell cannot be altered, or it will violate OSHA regulations. Hard hats must meet ANSI Z89.1-1997 standards. All hard hats have an internal head rest to raise the hat off the top of your head, and all are adjustable to fit your head size. All hard hats must be worn correctly and according to manufacturer instructions.

Summary

- Hand tools are required for installing various parts of a plumbing system.
- A plumber might be responsible for purchasing hand tools.
- Specialty hand tools are required more often for repair than for new installation work.
- Everyone on a job site is responsible for safety.
- Personal protection equipment (PPE) is usually provided by an employer.
- A Material Safety Data Sheet (MSDS) lists all safety hazards and medical attention requirements for a specific product.
- An MSDS must be available for all hazardous products and kept on file at the job site.

Procedures

Level Use

A Most plumbing levels offer three options for achieving levelness; the air bubble must be contoured within the lines.

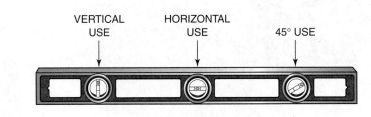

VERTICAL USE — HORIZONTAL USE — 45° USE

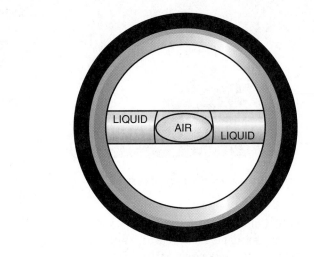

B On sloped piping, the air bubble will touch the lines or go past the lines, depending on the amount of slope on the pipe.

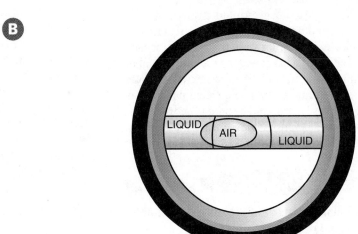

C On a horizontal pipe, place the level on the pipe that is being installed as level, and read the middle tube. Adjust the pipe as required.

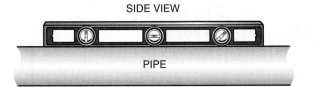

SIDE VIEW

PIPE

D On a vertical pipe that is being installed as straight (plumb), place the level on the pipe and read the appropriate leveling tube (top in this view). Adjust the pipe as required.

E On pipe that is being installed in a 45° position, place the level on the pipe and read the appropriate tube (bottom in this view). Adjust the pipe as required.

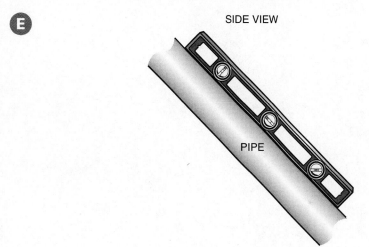

F Set the distance of the slope adjustment screw based on the desired slope per foot of pipe. Place the level on the pipe, and read the middle leveling tube. Adjust the pipe as required. If the length of the level is 24" and the slope is 1/4" per foot, the set distance of the slope adjustment screw is 1/2".

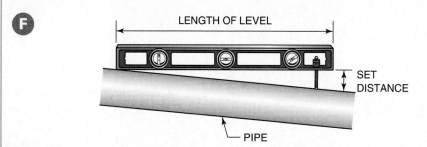

Procedures

Ruler Reading

A Plumbers rarely use dimensions that are reduced to less than 1/8" increments. Use a tape measure in all 1/8" and know that:

- 1/4" is also 2/8"
- 1/2" is also 4/8"
- 3/4" is also 6/8"
- 1" is also 8/8"

A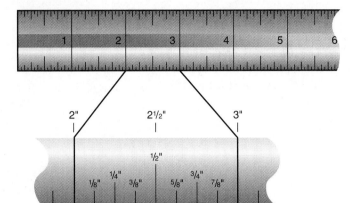

Procedures

Folding Ruler Layout

A To figure angles, place the end of the wooden rule on a given number. For example, if you place the end on 23", the degree offset of the first 6" portion of the rule will be 45° from the second 6" portion of the rule.

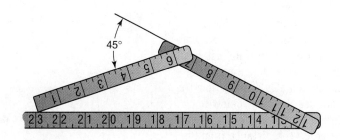

B You can fold the rule back to a compact form or leave it at a desired length, but do not move the first 6" portion, which remains at a 45° angle. You can use this to lay out offsets in a piping system or on a floor, wall, or ceiling.

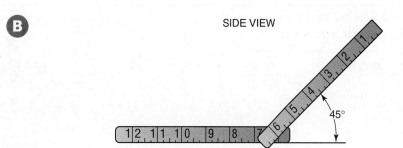

C Other angles can be achieved using the same method but with different starting points.

Starting Point	Angle of First 6" Portion
23-3/4"	22-1/2°
23"	45°
22-1/4"	60°
20-1/4"	90°

Procedures

Framing Square Layout

A To lay out an angle using a framing square, place the square against a wall or other established point. With the framing square positioned at 12" away from the wall, use the chart below to determine the angle of the pipe based on the distance from the framing square to the offset. This distance is measured along the wall.

B Other angles can be achieved using different dimensions.

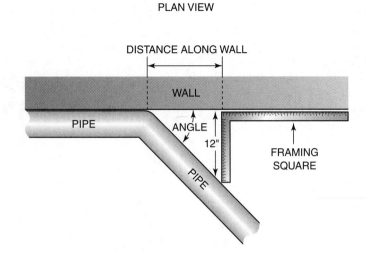

Distance along Wall	Angle
29"	22 1/2°
12"	45°
7"	60°

Procedures

Pipe Wrench Use

A Regardless of the actual position at which you are using the two wrenches, the farther they are spread past a certain point, the more your strength is minimized. In the position illustrated, the maximum spread is from 10 o'clock to 2 o'clock.

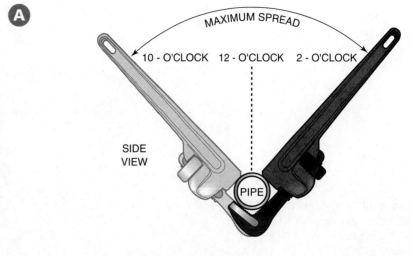

B This illustration shows the maximum spread from 4 o'clock to 8 o'clock.

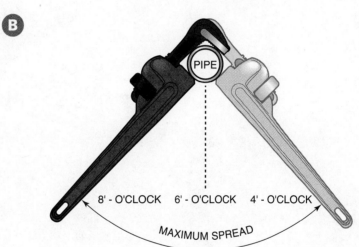

Procedures: Plastic Pipe Cutting

A Small diameter pipe can usually be cut through in one pass without rotating the pipe.

B If preferred, you can rotate the pipe as you partially cut it until the starting point and ending point connect to ensure a straight cut.

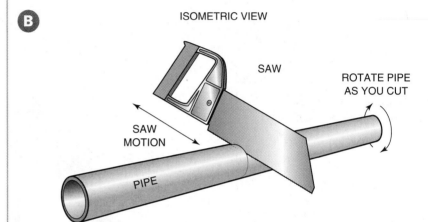

Procedures

Basin Wrench Use

A The head swivels to allow for tightening and loosening. The handle is adjustable to accommodate the various heights required. A T-handle slides through the tool, but is not removable.

A

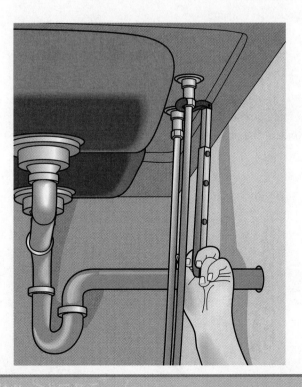

SIDE VIEW
HEAD SWIVELS 90°
PUSH IN BUTTON TO ADJUST
HEIGHT ADJUSTMENT
T-HANDLE
HANDLE SLIDES

B Once the height is adjusted and the head is rotated for the desired use (to loosen in this photo), use one hand to apply rotation.

B

Review Questions

1. **Safety is the responsibility of**
   a. A company safety officer
   b. My co-workers and me
   c. A project supervisor
   d. All of the above are correct

2. **A typical plumber's level is used to install piping**
   a. In the 45° position
   b. In the vertical and horizontal position
   c. With slope (grade)
   d. All of the above are correct

3. **A level for checking accurate pipe slope has a**
   a. Jack screw
   b. Magnetic strip
   c. Protective case
   d. Handle

4. **Two types of squares for layout and for marking boards to cut are a framing square and**
   a. A metal square
   b. An aluminum square
   c. A speed square
   d. A plumber's square

5. **A 25' retractable tape measure is the most common type, but for long distances it is best to use**
   a. A 30' type
   b. A wind-up type
   c. A folding type
   d. None of the above are correct

6. **A multi-purpose screwdriver provides a Phillips and a slotted head bit and a**
   a. 1/4" nut driver
   b. 5/16" nut driver
   c. 1/4" and 5/16" nut driver
   d. 3/8" nut driver

7. **Pliers are an essential tool for plumbers, and the most popular type is the**
   a. Angled jaw
   b. Combination
   c. Locking
   d. Needle nose

8. **Pipe wrenches are used for threading pipe and fittings together, and the most commonly used sizes are**
   a. 8" and 14"
   b. 18" and 24"
   c. 12" and 18"
   d. 14" and 24"

9. **To work with finished surfaces, you can use a strap wrench or a**
   a. Smooth jaw pipe wrench
   b. Chain wrench
   c. Standard pipe wrench
   d. Locking pliers

10. **Plastic piping is cut with a plastic pipe saw, plastic pipe cutting tool, and**
    a. Compass saw
    b. Wallboard saw
    c. Inside pipe cutter
    d. None of the above are correct

11. **The tool to install a faucet under a sink is called a**
    a. Sink wrench
    b. Basin wrench
    c. Faucet wrench
    d. Pipe wrench

12 Tools commonly used when working with copper are the tubing cutter, tubing bender, and

   a. Flaring tool
   b. Torch regulator
   c. Torch striker
   d. All of the above are correct

13 The OSHA approval identification for safety glasses is ANSI

   a. Z87
   b. Z78
   c. 87Z
   d. 78Z

14 A chalk box uses a string and chalk powder to mark a line in either

   a. Green or yellow
   b. Red or blue
   c. Blue or green
   d. Red or yellow

15 To review safety information about a product you should refer to

   a. MSDS
   b. SDSM
   c. DSMS
   d. SMSD

Chapter 2 — Power Tools

Power tools increase productivity for specific tasks and entire projects. Every trade involved in building construction uses some power tools that relate specifically to their work and others that are not trade specific. Residential plumbers use power tools to cut and drill wood more often than commercial plumbers, but they all use power tools to work with concrete. Battery-operated power tools increase productivity, but batteries must be charged frequently, and tools are not available to complete every task.

A power tool is the vehicle to perform a task; sharp cutting or drilling accessories are what makes the tool productive. Employers will expect you to know the basic types and uses of the drill bits, saw blades, and other accessories that are discussed in this chapter. Many of the safety items discussed in Chapter 1 are also relevant for power tool use. Other safety concerns are addressed throughout this chapter.

Each drill type is described separately to clarify its use. Saw blades are discussed within each relevant tool description. Some tools and equipment may not be used frequently, but they are discussed so that you will be able to operate them safely on a job site. These include the portable band saw, grinder, jackhammer, powder-actuated tool, and air compressor. Cutting edges must remain sharp, both as a safety measure and to increase productivity.

OBJECTIVES

Upon completion of this chapter, the student should be able to:

- identify and describe power tools often used by a plumber.
- identify which drill or saw is used relevant to the work location.
- identify which drill bit or saw blade is used relevant to a system installation.
- use power tools in a safe and appropriate manner.
- know the proper personal protection equipment (PPE) required for power tools.

Glossary of Terms

Allen screw a size-specific screw to secure drill bits and saw blades to their operating component; requires a compatible Allen wrench to loosen and tighten

bit an abbreviated term to describe a drill bit

chuck the rotating portion of a hand drill in which a drill bit is inserted

chuck key a tool designed to loosen and tighten the drill chuck

flashing a weather-tight sealing component installed around vent pipes penetrating a roof

floor joist a horizontal wood board used to support plywood or other flooring material

lanyard an approved shock protective line attached to a safety harness and fixed to a secure point or lifeline rope

OSHA Occupational Safety and Health Administration; mandates safety and health regulations

shank the portion of a drill bit that is secured in a drill chuck

wall stud vertical wood board or other material used to build a wall

General Safety

Safety is always a priority while working on a job site. The use of power tools elevates the risk of serious injury for their users and everyone working near a power tool who must be aware of flying debris and unsafe operation. Safety is the responsibility of everyone on a job site; you must remain aware and notify a supervisor if unsafe conditions cannot be corrected immediately. Safe use of power tools also means being prepared for unexpected occurrences on a construction site.

Plumbers often use power tools in confined areas and while working from ladders. Even the most experienced power tool users must be prepared for any reaction of a tool during its operation. For example, a knot in a wood board can cause an abrupt and violent reaction by a power tool. Safety concerns are discussed for each tool described in this chapter, but the actual location where a power tool is used on a job site determines most safety-related issues. A safe approach to using power tools includes positioning your body in relation to how the tool will react if an unsafe condition develops. Because electric power tools use motors, they create airflow to cool the tool during operation, which causes airborne dust and debris. Use of eye and face protection is crucial to avoid personal injury.

Electrical Safety

Electrical power tools are connected directly to a power source, so they should never be used when wet conditions exist. A ground fault circuit interrupter (GFCI), typically called a GFI on a job site, is required by OSHA everywhere a tool is connected to a power source. Many job site activities occur when a building is not protected from weather. If there are puddles, do not route extension power cords in or through them. If it is raining or snowing, do not begin or continue to work in the area, and place the tools where it is dry. If any power tool gets wet, have it inspected by proper authorities or service personnel before placing it back in service. Most contractors use separate GFI cords (Fig. 2–1) rather than long power cords with a GFI end, so cord replacement is not as expensive. A GFI must be tested with a portable GFI tester frequently to ensure proper protection. GFCI have a reset button and a test button to manually check if the mechanical aspect is functioning, but that test does not indicate whether they will work under unsafe conditions.

CAUTION: Using a power tool that is not protected with a GFCI can result in electrocution.

Figure 2–1 Ground fault circuit interrupter. *Courtesy of Coleman Cable.*

 FROM EXPERIENCE

Purchase a GFCI tester at a local electrical wholesale supply outlet or a home center. The tester simulates a tool that is malfunctioning and activates the interruption capabilities of the GFCI receptacle, which can be manually reset. The same test is performed by OSHA when inspecting a job site, and a penalty can result from using faulty GFCI equipment.

Ladder Safety

Working from a ladder exposes you to the possibility of serious injury or death. Working with power tools while on a ladder increases the risk. Drills can react violently when an obstruction is encountered, causing loss of balance. Weather can also play a role in safely working from a ladder. Mud on the soles of work boots, wet surfaces of the steps, and ladders placed on unstable or wet ground are all safety concerns. Stepladders (Fig. 2–2), usually 6' or 8' for residential construction, are usually intended for interior work. The top of a stepladder should not be used for standing and violates OSHA regulations if used in that manner. All ladders have a maximum weight limit, so the weight of materials and tools must be calculated with your body weight as the total weight load. Plumbers use an extension ladder to climb on a roof to install **flashing** around vent pipes that terminate through a roof. An extension ladder (Fig. 2–3) must be set up with the top of the ladder at least 3' above the point where the ladder contacts the roof edge. The angle of the ladder in relation to the ground is important to ensure a safe climb to the top. The bottom should be placed out a distance approximately one-fourth the length of the extended ladder to ensure that a safe angle is established (Fig. 2–4). For example, if the ladder is ex-

CHAPTER 2 Power Tools **39**

Figure 2–2 **Stepladder.** *Courtesy of Louisville Ladders.*

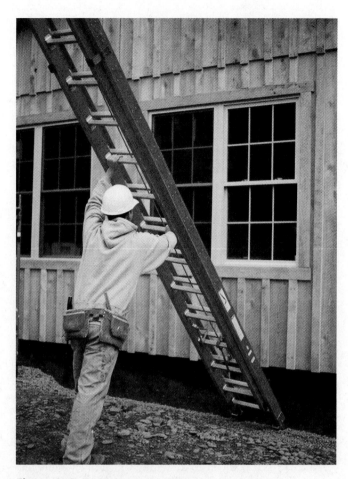

Figure 2–3 **Extension ladder setup.**

tended 20′ from the ground to the top of the ladder, the bottom of the ladder should be around 5′ from the wall or other structural reference point that the top of the ladder touches.

CAUTION: OSHA regulations are sometimes revised, so always stay updated on new regulations and revisions by contacting the Labor Department in your state.

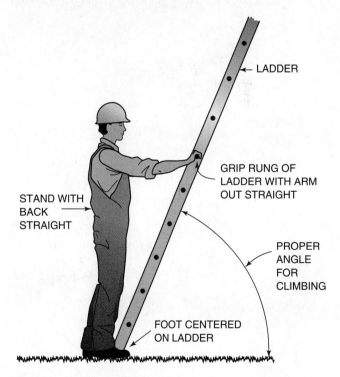

Figure 2–4 **Extension ladder angle.**

CAUTION: Never leave tools or materials on top of a stepladder. The items could fall and cause injury.

FROM EXPERIENCE

Clean the bottom of your work shoes before climbing a ladder, but do not clean them using the steps of a ladder. Place a clean piece of plywood or cardboard a few feet from the base of the ladder to use as a stepping location after your work shoes are cleaned and before stepping on the ladder.

Fall Protection

OSHA dictates that fall-protection measures must be based on the type of work performed and the location of a specific task. Whenever your feet are 6′ or more above the ground while working, OSHA dictates that fall-protection methods must be used. The most common method is a body harness and a securing lanyard (Fig. 2–5). A lanyard is secured to a fixed point or to a lifeline, which is an approved rope or metal cord fixed at two secure points to which the lanyard is connected to provide mobility while working. Not all locations can accommodate a body harness, and the lanyard that connects the harness to a fixed point can actually hamper your ability to perform safely in some situations. If a body harness is not feasible, it is acceptable to use other protective measures such as handrails or safety nets. The type of fall protection used is the responsibility of the employer and the employee. Typically work from a stepladder does not meet the minimum requirements for fall protection set forth by OSHA.

CAUTION: Lack of safety training has serious consequences. OSHA does not accept lack of knowledge as an excuse for being unsafe. You should remain focused on all safety issues for every task you perform. It is your right as a worker to request safety training and the employer's responsibility to provide that training.

FROM EXPERIENCE

Safety training programs, typically known as "tool box lessons," can be purchased, but most information about specific safety items is free where you purchase safety equipment.

Ear Protection

PPE covered in Chapter 1 remains relevant when using power tools. Ear protection is one of the items needed to protect you against long-term injury. Because damage to your inner ear may not be apparent until several years after you actually perform work, you should appreciate why workers are encouraged to wear earplugs or earmuffs. Many styles of earplugs (Fig. 2–6) are available; some are soft foam, others are rubberized. Earmuffs (Fig. 2–7) provide the best protection when working with a jackhammer or other loud power tools.

FROM EXPERIENCE

A ringing sound in your ears during or after power tool use is a sign of inadequate hearing protection. Soft foam earplugs are disposable and can be purchased in large quantities to ensure that you have an adequate supply on a job site.

Figure 2–5 Fall protection components.

Figure 2-6 Earplugs. *Courtesy of North Safety.*

Figure 2-7 Earmuffs. *Courtesy of E.A.R. Inc.*

Drills

As with all power tools, all PPE must be worn when operating a drill, which is essentially an electric motor housed within a tool design. The motor is manually activated to rotate an internal shaft connected to a drill bit **chuck** that rotates a drill **bit.** The drill bit is secured in the chuck with a specialty tool called a **chuck key.** Drills have various amounts of horsepower and have a forward and reverse rotating feature; many have variable speed adjustments or at least a high and low speed. Drills bore holes in a variety of materials including wood, concrete, and steel. Many are designed for specific uses; others have multiple uses. Three drills often used by plumbers to drill wood products are the pistol, the right angle, and the hole hawg. Plumbers often bore large diameter holes to install drainage and vent piping through wood boards. This, along with the location of a hole, helps dictate the drill selection.

Using a drill safely usually requires holding it close to your body, rather than extending your arms too far away from your body. Where your body is located in relation to the drill is important to ensure that you do not become trapped in the workspace if the drill bit encounters an obstruction and your finger cannot be quickly removed from the activating control (trigger). Operating a drill over your head or in front of your face can be dangerous. However, many work locations require you to be in difficult positions, so you must become comfortable with each drill type and its use before working in that manner. Knowing how the drill will react if it encounters an obstruction is as important as knowing the correct body position. Proper training is important and is usually the result of actually working with drills on a job site. You must always notify a supervisor if you are inexperienced with a specific tool or when you are uncomfortable drilling in a certain workplace.

Most drills have a removable side handle or top handle that must be installed in one of various designated locations on the tool prior to using. It is easy to overlook installing the handle before use, and, in many instances, the handle can conflict with the proper drilling of a hole. Selecting a different drill or bit can solve this problem. Productivity should never take priority over safety; becoming injured only reduces productivity. Every time you use a power tool, you should remember that you are holding a dangerous tool that could cause serious injury or even death.

CAUTION

CAUTION: An electric drill can catch loose clothing and long hair. Always keep long shirt sleeves buttoned and long hair out of the drilling area. Always wear proper PPE.

CAUTION

CAUTION: Never drill holes if your back is in a rotated or awkward position. The reaction of a drill when hitting an obstruction can cause back injury if your body position is incorrect.

 FROM EXPERIENCE

Using a drill with fully extended arms is unsafe. Keep arms partially bent and close to your body so your torso strength can help control the drill's power.

Right Angle Drill

Working space between two **floor joists** or **wall studs** is typically a maximum of 14-1/2". Therefore, the best selection to drill holes for plumbing piping is a right angle drill.

Figure 2–8 Right angle drill. *Courtesy of Milwaukee Electric Tool Corporation.*

A right angle drill's (Fig. 2–8) chuck is placed at a 90° or right angle from the body of the tool. Most right angle drills allow the user to rotate the placement of the chuck 360° to adjust to any workspace. Its large body allows a user to grasp it firmly with both hands, and its removable side handle can be located in several places on the tool. Its low-speed operation, compared with other drills, minimizes the violent reaction that can occur when encountering obstructions such as hidden nails or knots in the wood.

CAUTION: Never use a drill with one hand, as it can break your wrist. A drill that reacts violently to encountering an obstruction can trap you in a workspace if you are not properly positioned. Always wear proper PPE.

 FROM EXPERIENCE

A right angle drill is usually the best selection when installing plumbing in a new house because of its flexibility in adjusting the chuck location.

Pistol Drill

A pistol drill (Fig. 2–9) gets its name because of its design, which looks like a handgun. It is typically not used to drill large diameter holes because of its small body size. It would be unsafe and would expose the user to wrist injury if the tool reacted to any obstruction in the work area. Larger versions of this tool are available that can be used for larger diameter holes. It is widely used by plumbers to install wood blocks for supporting pipes or fixtures and to drill small diameter holes through steel or thin materials. This power tool is also offered as battery powered, which increases productivity. The most common style is the 1/2" type, which can handle larger diameter drill bit shanks used with other drill types; it offers more power for drilling holes than the 3/8" type. The smaller

Figure 2–9 Pistol drill. *Courtesy of Milwaukee Electric Tool Corporation.*

3/8" type is limited to use with twist bits (Fig. 2–13) and other small-diameter shank-fastening accessories.

CAUTION: Using a pistol drill with one hand can allow the drill to twist, causing wrist injury. The side handle should be used at all times. Always wear proper PPE.

 FROM EXPERIENCE

A pistol drill can drill a small-diameter hole, known as a pilot hole, to assist in guiding large diameter bits used with other drill types.

Hole Hawg

Another right angled drill, known as a hole hawg (Fig. 2–10), has a fixed drill chuck position tucked further in its body for drilling in small spaces. It is a compact and powerful drill that drills through wood easily, which increases productivity. However, it can be difficult to control if an obstruction is encountered. As with all power tools, users must become comfortable with drilling operations and tool reactions while working on the floor before attempting to work from a ladder. Drilling wall studs and floor joists is often done from a ladder, and users can easily lose their balance. It is vital that you use this drill close to your body with stable body positioning to maintain control of the drill.

CHAPTER 2 *Power Tools* 43

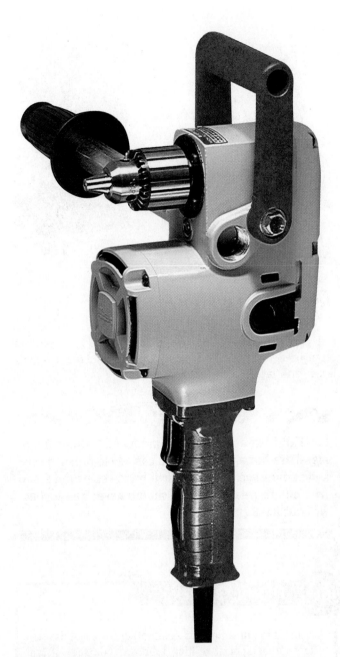

Figure 2–10 Hole hawg drill. *Courtesy of Milwaukee Electric Tool Corporation.*

CAUTION: The thick body design of this tool does not allow the user to hold the body of the drill with both hands; use the top handle provided. Always wear proper PPE.

Hammer Drill

Plumbers often anchor supports to install piping or fixtures into concrete floors, walls, and ceilings. Drilling into or through concrete requires a hammer drill (Fig. 2–11). Many

Figure 2–11 Hammer drill. *Courtesy of Milwaukee Tool Corporation.*

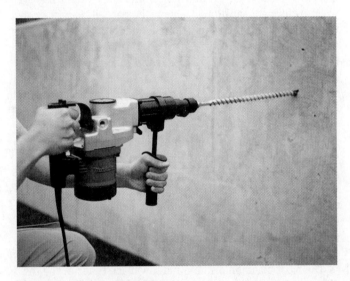

Figure 2–12 Hammer drill use.

small pistol-type drills hammer as they rotate, but the most productive tools for drilling concrete are specially designed for that purpose (Fig. 2–12). A hammer drill rotates as it hammers to act like a spinning chisel. Metal rods known as rebar or metal mesh material are installed to strengthen concrete. If a drill bit encounters this metal while drilling, the tool can react violently and cause injury to your wrists and hands. However, some hammer drills have a safety feature that allows the drill bit to remain lodged in a hole and the drill to remain stable even when the tool is still activated.

CAUTION: Never proceed in drilling a hole when an obstruction is present. It is common for water piping and electrical conduits to be installed below concrete floors and in concrete walls. Always wear proper PPE.

Drill Bits

New residential plumbing installations require drilling, cutting, and notching activities. A variety of drill bits and saw blades are designed for specific purposes, and others have various uses. Certain types are more common than others, based on the product or location of the holes being drilled. Drilling holes requires sharp bits. Many can be manually sharpened, but others are discarded when worn or damaged. Dull bits decrease productivity and can damage the motor of a drill. A plumber may be responsible for providing drill bits on a job site to ensure that they are not abused by drilling through nails and screws. Nails must be removed from drilling areas to prevent damage to bits and drills and to avoid injury when a drill reacts violently. Eye protection must be worn when drilling holes, and, depending on the work performed, ear protection may be needed. Because certain drilling locations may require longer bits than are available, shaft extensions are used for many drill bit types. Concrete bits are sold in various lengths, so the use of shaft extensions is not common.

Saw blades are offered in a variety of styles and lengths, and each power saw has a specific blade type. The blade selection for each tool is based on the type of material being cut and the length or diameter of blade needed to achieve a safe and accurate cut. A plumber uses reciprocating saw blades more often than any other power saw. Additional information is included with the description of each power tool. Many saw blades are multitask or general purpose. Each blade's cutting capabilities vary and must be known before assuming that a specific saw blade can be used on a specific material.

The most common drill bit and saw blade types in the residential plumbing industry are listed and then explained individually following the list:

- Twist bit
- Auger bit
- Speed bit
- Power bore bit
- Hole saw bit
- Masonry bit
- Step bit

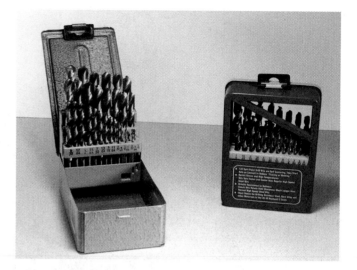

Figure 2–13 Twist bit indexes.

drill chucks because they have a round shaft. Twist bits are often used to drill a hole, known as a pilot hole, to guide larger drill bits in a specific location.

CAUTION: Material shavings produced while drilling metal are hot and sharp, and can cause minor burns and serious eye injury. All drill bits become hot from friction after each use and should never be touched without hand protection.

FROM EXPERIENCE

Drilling through stainless steel or other thick metals usually requires special drilling lubricant (cutting oil) to avoid damage to the bit and to expedite the drilling process.

Twist Bits

Twist bits are often called metal bits because they are designed primarily to drill holes through metal, but they are also used to drill holes in wood, fiberglass, and plastic materials. They are sold separately or in a set known as a drill index (Fig. 2–13) that includes a variety of drill bit sizes. A typical large index consists of bit sizes from 1/16" to 1" with sizes increasing in increments of 1/16". They may require sharpening after several uses and should not be forced through material, because of their brittleness. They are used primarily with a pistol drill but are compatible with most

Auger Bits

Auger bits (Fig. 2–14) are specifically designed to drill through wood products and are very productive. Called self-feeding bits, they shave the wood as they progress into the drilling area. The tip or center point of the bit has a non-replaceable threaded self-feeding worm drive (Fig. 2–15) that pulls the bit through the wood. Two of the threaded-worm-drive center points are classified as coarse and fine thread. The worm drive must remain sharp to be productive. The wood shavings are guided out of the hole through the spiral **shank** known as a flute and are discarded in your work

CHAPTER 2 Power Tools   45

Figure 2–14 Auger bit. *Courtesy of Milwaukee Electric Tool Corporation.*

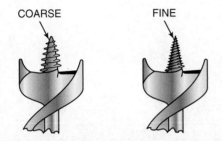

Figure 2–15 Auger bit worm drive types.

area. They are available in a variety of diameters and lengths, and shaft extensions are used if an available bit is too short to perform a task. A plumber typically uses 3/4" to 1-1/4" sizes in 6" to 12" lengths. They are used for installing water piping systems and are very useful when drilling through exterior boards to install hose faucets and for drilling other thick wood structures.

 FROM EXPERIENCE

Drill holes in thick wooden structures in several steps by removing the bit at short intervals. Remove shavings from the hole, so the bit does not become lodged in the wood.

Speed Bits

Speed bits (Fig. 2–16), often called spade bits, are not used as often as other bits for drilling wood in the plumbing trade. Plumbers sometimes use them to drill holes in wood products such as cabinets to install piping to a dishwasher, but a hole saw bit (Fig. 2–20) may be more desirable for that purpose. It does not have a threaded worm drive as a center point, which means the user is the driving force behind its progress. Due to the looseness of the particle wood construction of most cabinets, drill bits that use a worm drive to self-feed cannot be used. The speed bit is a more productive selection than an auger bit, because it shaves the wood product quickly. Because this bit is often used for finished areas, a two-step process for completing a hole is recommended to avoid damage to finished areas. The hole should be drilled partially through one side of the work area until

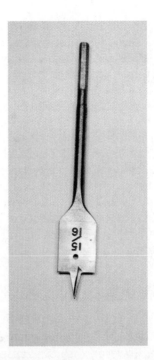

Figure 2–16 Speed bit.

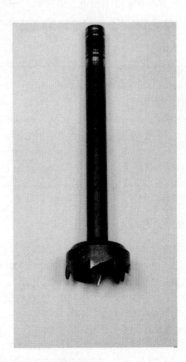

Figure 2–17 Power bore bit.

its center point penetrates the opposite side; then the hole should be completed from the opposite side.

Power Bore Bits

Power bore bits (Fig. 2–17), the most common drill bits in new residential plumbing installations, are specifically designed for drilling wood products. This self-feeding bit has a worm-drive center point similar to an auger bit, but the center point is replaceable. The most popular sizes used by a plumber are 1-1/8", 2", 2-9/16", and 3-5/8". Each drill bit size is used for a certain pipe size in the plumbing trade. The length of power bore bits can be extended with a bit extension accessory, because they have a fairly short shaft and can typically only penetrate 4". The shavings created by a power bore bit are discharged from the drilling area, but, when drilling deep holes through thick boards, the shavings can remain in the drilling area. Drilling may have to be performed in increments when penetrating thick boards. Removing the bit from the hole and then removing the shavings prevents the bit from being lodged in the wood. It is important to maintain sharp drilling edges on the bit with a file so it remains productive and so the drill is not damaged.

> **CAUTION:** Power bore bits are self-feeding and sharp, and well-maintained bits do not require much force to bore a hole. However, they can cause a more violent reaction than other drill bit types if a nail is encountered during the drilling process.

 FROM EXPERIENCE

> Purchase a sharpening file designed for working with metal, and store it with your bits or drill. Sharpen your bits before they become dull to help maintain a safe and productive workplace.

Hole Saw Bits

Plumbers often use hole saw bits (Fig. 2–18) to drill through a variety of thin materials and finished surfaces, such as fiberglass shower walls, countertops, or cabinets. As their name indicates, they saw through the work area as they bore a hole, which creates sawdust instead of wood shavings. The material being drilled with a hole saw is retained in the bit housing. The user must remove the resulting core of the drilled area after the hole is completed or before penetrating a board that is thicker than the remaining depth of the bit housing. The hole saw uses a smaller, specialized twist bit as a pilot bit during the initial drilling process to keep it centered in the desired location. Each bit is removable from a shaft known as an arbor. Two common arbor types are used, based on the hole saw size. The twist bit is replaceable and has a flat-sided shaft that is secured within the arbor with an **Allen screw.** The most common sizes used by a plumber range from 1/2" to 2" and can be purchased separately or as a kit with a storage case. Larger sizes, used to drill through a

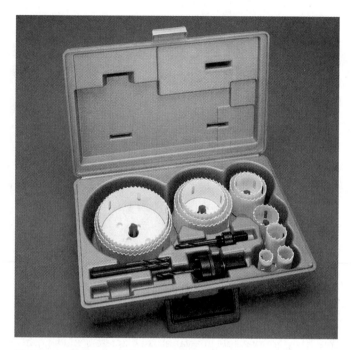

Figure 2–18 Hole saw kit. *Courtesy of Lenox.*

fiberglass shower wall when installing a single-handle faucet, are typically purchased separately.

For step-by-step instructions on drilling a finished surface with a hole saw, see the Procedures section on page 53.

> **CAUTION:** Damage to the work area can result if the hole saw contacts the work area at a high rate of speed. It is recommended that a pilot hole be drilled separately in finished surfaces to eliminate this problem.

 FROM EXPERIENCE

> Duct tape or masking tape applied to finished work areas can minimize the possibility of chipping the work area while drilling with a hole saw.

Masonry Bits

Some masonry bits are used with pistol drills and have a carbide tip, but they are typically used for anchoring small fasteners in hollow concrete blocks and ceramic tile. Masonry bits used with a hammer drill for penetrating concrete floors and walls are specifically designed for that purpose and are known

CHAPTER 2 Power Tools 47

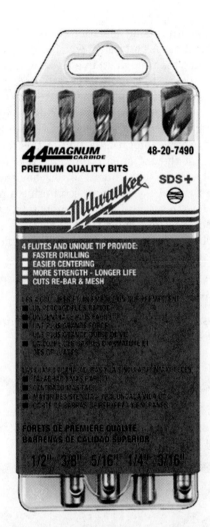

Figure 2-19 **Hammer drill bits.** *Courtesy of Milwaukee Electric Tool Corporation.*

Figure 2-20 **Step bit kit.** *Courtesy of Lenox.*

CAUTION: Metal reinforcement bars encountered while drilling into concrete can cause the drill to twist violently, which results in wrist and arm injury.

A torpedo level can be fastened to the drill with duct tape to align the drill and ensure that the hole is drilled correctly.

as hammer drill bits (Fig. 2-19). Hammer drill bits are designed for installing heavy-duty anchors to support pipe or pieces of equipment, and for penetrating solid concrete. They are expensive, and drilling solid concrete can often lead to hitting obstructions such as metal reinforcement bars (rebar) embedded in the concrete. A new drilling location must be established when obstructions occur to avoid damage to the drill bit and to avoid injury from the violent reaction of the drill when encountering obstructions. Not all drill bit shanks are compatible with every drill type, so they are purchased based on the drill type used. However, shank adapters can be purchased for some bits to allow compatibility with various drill types. Some hammer drills have a leveling feature on the top of the tool to ensure that holes are drilled level into concrete. Many hammer drills can also be used as chipping hammers, and a variety of chisel bits are used for that purpose. In addition to wearing eye protection, face protection, and work gloves, a dust mask or other breathing apparatus must be worn to prevent inhalation of concrete dust while using masonry bits. Ear protection is also necessary to avoid internal ear damage from loud noise produced with drilling and chipping concrete.

Step Bits

Step bits (Fig. 2-20) are also called vari-bits because they are designed to replace various individual twist bits with one tool. They are used to drill holes in metal like twist bits but increase productivity by drilling holes in increments that would normally require using several individual twist bits. The two most common sizes used by plumbers range from 1/4" to 1/2" and from 1/4" to 1". One disadvantage of this tool is that the entire bit must be replaced if the tool is damaged or can no longer be sharpened. Eye protection is crucial when using any drill bit, but face protection and a long-sleeved shirt may be required when drilling through steel so hot, sharp metal shavings do not contact bare skin.

CAUTION: The drill can react violently when the drill bit steps to each different bit size.

FROM EXPERIENCE

Using a drilling lubricant (cutting oil) fluid lengthens the life of the drill bit.

Saws

The power saws most often used by plumbers are discussed in this chapter. Saws are used extensively in the plumbing industry for multiple purposes. The popularity of power tools has minimized the use of handsaws for many tasks, but small diameter plastic pipe is still usually cut with a specially designed handsaw. The popularity of battery-powered tools has allowed plumbers to use power saws in more confined spaces, such as in a crawlspace, and eliminates having to use an extension cord as a source of electrical power. This is one of the main reasons handsaws are used less frequently now than in the past to cut plastic piping. Power saws were always used extensively to cut metals because using a hacksaw is not productive. Safe use of any power tool is vital and must be a focal point every time you prepare to activate one. Always wear PPE, inspect your work area for water or other dangerous conditions, and remain aware of those working around you.

Reciprocating Saws

The word *reciprocate* means to return or to counter, so a reciprocating saw (Fig. 2–21) operates in a piston-like motion by thrusting a saw blade forward and backward in a sawing motion. Because it is a power tool, its sawing motion is rapid, and it increases productivity. A reciprocating saw is often called a sawzall, and it is usually used by residential plumbers to notch wood boards, to cut plywood floors, and to cut plastic pipe. It can cut most construction material, and numerous blades are available for specific materials. Saw blades are categorized by the number of teeth per inch and the total length desired. Blades used for cutting wood have fewer teeth per inch than blades designed to cut metal.

This tool can also be battery powered, which increases mobility and productivity. The reciprocating saw has a thrust shoe that must remain in contact with the product being cut to ensure a smooth cutting operation. To start a cut in a flat surface, a thrust cut is performed, which is common for cutting the opening for a tub drain in a plywood floor. This can cause the tool to react violently. It is better to drill a hole in the work area and proceed with the thrust shoe against the work surface. This tool requires two hands to operate and can be dangerous because the blade is not protected with a safety guard. The reciprocating action of this tool requires enough space at the end of each forward stroke for the saw blade to fully thrust away from the tool before it returns back toward the tool. Overconfidence in using a reciprocating saw can lead to serious injury, not only from the cutting operation, but also from being stabbed with the end of the blade. The blade also becomes extremely hot from the rapid cutting action and can cause burns.

For step-by-step instructions on saw blade selection, see the Procedures section on page 54.

For step-by-step instructions on performing a thrust cut, see the Procedures section on page 55.

CAUTION: Never use the power cord to lower the saw to the floor when you are working from a ladder. The tool could slip and stab your leg or foot.

CAUTION: Always know what is located on the opposite side of a cutting area. Cutting wires or piping can result when cutting through walls, floors, or ceilings.

FROM EXPERIENCE

Most saws allow the blade to be used facing two different directions. If a difficult cutting location is encountered with the blade in its normal direction, turn the blade to face the opposite direction to complete the cut.

Circular Saws

Carpenters use circular saws more than plumbers do, but, when used, this tool increases productivity and ensures square cuts. Circular saws can also be battery powered (Fig. 2–22). A plumber normally uses this tool to cut wood

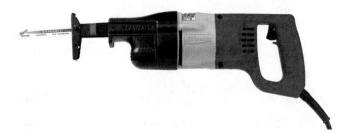
Figure 2–21 Reciprocating saw. *Courtesy of Milwaukee Electric Tool Corporation.*

CHAPTER 2   Power Tools   49

Figure 2–22 **Battery powered circular saw.** *Courtesy of Milwaukee Electric Tool Corporation.*

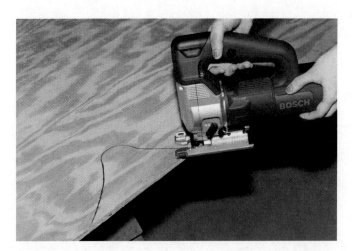

Figure 2–23 Saber saw.

boards for pipe supports, or plywood for strengthening floor joists that have been drilled when required by code. Though a plumber may cut small boards with a sawzall, the circular saw has a removable guide bar to create a straighter cut. As with all power tools, there are specific safety concerns. The saw blade can be pinched by the wood being cut, causing the saw to thrust backward violently toward the user. The blade is protected by a safety guard that should never be removed or fixed in a manner that disables it. Blade selection varies with the material being cut, and the diameter of the blade must match the saw used. A plumber typically uses a general-purpose wood blade.

CAUTION: A circular saw blade can cause serious lacerations, including amputation of fingers. Stand away and to one side of the back of the saw, and keep your fingers away from the saw blade. Always let the saw rotation stop before placing the saw on the ground after use. Always wear the proper PPE.

FROM EXPERIENCE

Because most plumbers use circular saws for small depth cuts, a small-diameter battery-powered circular saw increases productivity.

Saber Saws

A saber saw (Fig. 2–23) is also called a jigsaw. It is used by plumbers primarily to cut holes in countertops to install sinks, and it is capable of completing circular cuts. It has a reciprocating cutting action with very short strokes. This compact handheld tool offers smooth operation, and various blade types are available depending on the material being cut. Blades are ordered based on the number of teeth per inch, with metal-cutting blades having more teeth per inch than wood-cutting blades. The saber saw has a base plate that must remain firmly on the work surface while cutting. The blade must be allowed to complete its full stroke cycle, which means that the material being cut must be thinner than the blade length, or the tool will react violently. The blades are thin and break easily when used incorrectly. As with the sawzall, the blade is unprotected and can cause injury. It becomes hot when used due to friction and should only be removed from the tool after it cools or with hand protection.

CAUTION: Though a saber saw is compact, two hands must be used during operation to protect against a violent reaction if an obstruction is encountered.

FROM EXPERIENCE

Finished surfaces such as countertops can be damaged while cutting. To avoid this, tape the outline of the cut and lay out the cut line on the tape. Tape can also be applied to the bottom of the base plate to avoid scratching the surface being cut.

Figure 2–24 Portable band saw. *Courtesy of Milwaukee Electric Tool Corporation.*

Figure 2–25 Small grinder. *Courtesy of Milwaukee Electric Tool Corporation.*

Figure 2–26 Large grinder. *Courtesy of Milwaukee Electric Tool Corporation.*

Portable Band Saws

A portable band saw (Fig. 2–24) is often called a porta-band. It is designed for cutting metal and must be used with extreme caution because it can easily amputate a finger. A plumber uses this tool for cutting threaded rod to install hangers or other metal support materials. The circular blade inserts around two circular rotating wheels with rubber belts that grip the blade. The saw has a metal thrust bar that must be in contact with the material being cut to ensure safe operation. Because this tool is designed to cut metal, blade selection is not as varied as it is for tools designed to cut both wood and metal. The number of teeth per inch is still the criteria used to select the proper blade; the more teeth per inch, the smoother the cutting process. The portable band saw is designed for two-hand operation, but a plumber must often hold the material being cut with one hand while using the tool with the other hand. This is unsafe and should be avoided by finding an alternate approach to completing the task. The blade is very thin and can break when pinched by the material being cut. The cutting portion of the blade is unprotected, so this tool should be housed in a case when it is not in use, or the blade should be removed after each use.

CAUTION: A band saw is used in machine shops to cut thick steel and in butcher shops to segment meat, which shows why this tool is capable of serious injury. Never place your hands near the cutting blade, and never remove the thrust bar. Do not leave a discarded band saw blade on the ground as it can cause a tripping hazard because of its circular design.

FROM EXPERIENCE

The portable band saw increases productivity when cutting threaded rod to install hangers. The 24 teeth-per-inch blade can cut cast-iron pipe when the pipe location does not allow for a cast iron cutter.

Grinders

Portable grinders are more common in the pipefitting trade, but a residential plumber should know how to use them safely. The diameter of the abrasive wheel on the two common types is 4-1/2" and 7". The maximum operating speed of a grinding wheel must be compatible with the grinder being operated. The 4-1/2" grinder (Fig. 2–25) is used for working with small-diameter piping or in confined work areas. The 7" grinder (Fig. 2–26) is used for working with large-diameter piping or for removing large quantities of material. The wheels can be for either grinding or cutting. The grinding wheels are used to grind metal burrs from cut pipe or to bevel a pipe to prepare it for welding. Cutting wheels are thin and designed to cut through metal piping. Abrasive wheels are selected based on the material to be ground or cut. Wheels are also available for grinding concrete. Read all manufacturers' information before using any wheel to ensure that you have the correct one for the specific material and use.

A grinder has a safety guard to protect you against a violent reaction during use, but the grinding face and cutting edge are not protected and can cause serious lacerations. A removable side handle must be installed before each use, and two hands must be used at all times to operate a grinder to maintain control of the tool. Sparks are produced when used, so eye, ear, and face protection is required. A long-sleeved shirt is recommended, and loose clothing should be secured. Flammable or explosive materials or fuels must not be present in the work area, and protective measures must be taken to eliminate sparks from injuring other workers. Small metal pieces can also become airborne and fly in the same direction as the sparks. Your work area should be located away from other workers, and a designated person may need to stand watch and notify others not to enter.

Figure 2-27 Powder-actuated tool use.

CAUTION: A grinder rotates at high speed and can grab loose clothing. Always button long-sleeved shirts and keep shirttails tucked into pants.

 FROM EXPERIENCE

Purchase grinding wheels designed to work with specific materials. Using a carbon steel wheel on stainless steel material will cause rust to form on the stainless steel. Using a concrete wheel on steel can cause the wheel to break into pieces.

 FROM EXPERIENCE

Diamond-tipped cutting wheels are available for specialty cutting, such as ceramic tile or concrete. Always use the correct blade for the specific material being cut and follow manufacturer instructions.

Powder-Actuated Tool

Using a powder-actuated tool (Fig. 2-27) requires that a user become certified by the manufacturer because an explosive charge is used to drive an anchor into concrete and steel. Such a tool operates similar to a handgun by forcing the anchor out of the tool like a bullet when a trigger is activated. A safety feature of this tool requires the barrel to be pressed firmly against the work surface before it can fire to avoid accidental operation. Explosive caps are color coded to indicate their firepower, and a selection chart shows the correct charge for a specific application. Plumbers typically use this tool to install hanger rods, which have nail point ends and threaded ends. The nail point end is driven into the concrete, leaving the threaded portion exposed for a plumber to connect a hanger rod. The most common anchor size for supporting plumbing piping is 3/8" diameter, but 1/4" can be used for small copper water piping if required. Ear and eye protection must be used at all times, and goggles are recommended over safety glasses.

CAUTION: Misuse of this tool can cause serious injury or death. Always follow all manufacturer's instructions when using this tool. Know the proper charge to use, and make sure the opposing side of a wall, ceiling, or floor is clear of other workers in case the anchor accidentally penetrates the work area.

 FROM EXPERIENCE

Powder-actuated tools are more effective than hammer drills for installing many kinds of anchors. Clean and lubricate this tool after each day's use or when stated by the manufacturer to maintain its safe operation.

Figure 2-28 Portable air compressor. *Courtesy of Porter Cable.*

Air Compressors

A residential plumber uses an air compressor (Fig. 2-28) to pressure-test water piping systems. Water is used with a hydrostatic test, but air testing is more common, especially in cold climates where pipes could freeze during construction. There are various kinds of air compressors. Portable ones are most common, because they can remain on a work truck between uses. An air compressor is connected to an electrical system and then turned on so air flows into a self-contained storage tank until it achieves a set pressure that has been adjusted by the user. Most codes dictate that the pressure for the test must be at least 100 pounds per square inch (psi), which can create a dangerous situation for everyone on the job site. If a pipe fitting leaks, the air pressure must be relieved from the piping and a repair made. If piping is damaged while under pressure, the release of air can cause injury. After a piping system is tested and inspected, the air pressure should be relieved or lowered to ensure that the job site is safe. A warning sign should be placed on the job site notifying other workers that a high-pressure test is occurring and to proceed with caution. Safety glasses and ear protection should be worn in case of an accidental release of air from the piping or air compressor. Hoses that supply air from the compressor to the piping system are under the same test pressure and should be inspected to ensure that they are in proper condition.

Jackhammers

Concrete is removed with either a chipping hammer or a jackhammer (Fig. 2-29). Small amounts of concrete can be removed around existing piping with a chipping hammer, or, in many instances, a rotary hammer with dual purpose capabilities. If large areas of concrete must be removed, a jack-

Figure 2-29 Electric jackhammer. *Courtesy of Bosch.*

hammer is needed. Either electric or air-activated jackhammers can be used. Air types are more powerful, but electric ones are sufficient for most residential applications. A chisel bit and a point bit remove concrete, but each has specific uses. A chisel point removes large pieces of concrete and a chisel point bit penetrates a small area at a time. Gloves, eye protection, lung protection, and ear protection must be worn. Care should be taken if you do not know if the electrical conduit or water piping is located within or below the concrete floor.

Summary

- Power tools must be used by trained personnel only.
- Serious injury can result from misuse of power tools.
- Personal protection equipment must be worn while using power tools.
- Using power tools from a ladder can create unsafe working conditions.
- A body harness may be required when working from a ladder and is almost always required from a scaffold.
- Ground fault circuit interrupters (GFCI) must be used with all power tools.
- Most drill bits and saw blades have specific uses.
- Various drill types are used for certain work areas.
- Many drill bits must be sharpened periodically.

Procedures

Hole Saw Finished Surface Drilling

A
- Apply masking tape or duct tape to countertop to protect the finished surface from chipping.
- Lay out holes over the tape.

B
- Drill a pilot hole in the center of each using a separate bit or the one on the hole saw.
- Place the hole saw center bit into the drilled pilot hole before activating the drill.
- With the hole saw placed square on the taped area, activate the drill to rotate the hole saw.
- The type of surface being drilled and how the hole saw reacts determines the actual drilling operation.
- Use short drilling duration when establishing a hole; when the hole saw begins drilling into the countertop, bring the drill up to full speed.
- Never begin drilling a hole with the drill at full speed, because the hole saw might dig into the drilling surface. This can cause a violent reaction and possibly damage the countertop.

A
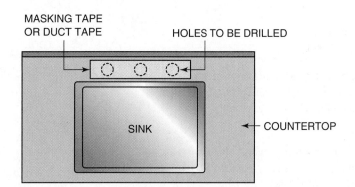

B
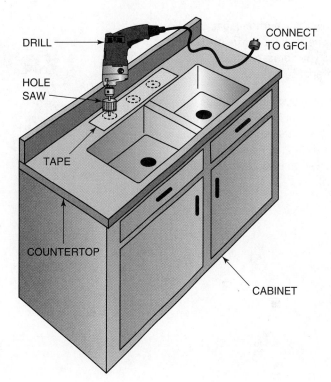

Procedures

Saw Blade Selection

A Identify the type of material being cut.

- Know the basic offerings of saw blades from tool or material suppliers. Many stores only sell a limited selection.

- Understand that wood blades have fewer teeth per inch (TPI) than metal blades.

B Know that the length of a blade refers to its actual cutting length.

- Know the length required to ensure that the entire blade can freely reciprocate through the material to be cut.

- It is better to have a blade that is too long than too short in most cutting situations. A blade that is too short can break if it is not allowed to complete a full cutting stroke.

A
CONNECTS TO TOOL
1"

B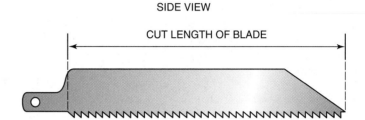
SIDE VIEW
CUT LENGTH OF BLADE

Procedures

Thrust Cut

A
- Place reciprocating saw as close to the floor as possible while activating the saw.
- Place the saw blade against the plywood to begin the cutting procedure.
- The tool may react violently if too much pressure is applied or if the tip of the saw blade is placed at too great an angle.
- The thrust shoe swivels allowing the tool to be used at an angle and still be in contact with the surface.

B
- Once the blade penetrates the plywood floor with a thrust cut, position the saw so the thrust shoe is firmly against the floor.
- Complete the desired cut.
- The saw does not have to remain 90° from the work surface, but it must be kept at an angle that allows the thrust shoe to be firmly against the cutting surface.

A

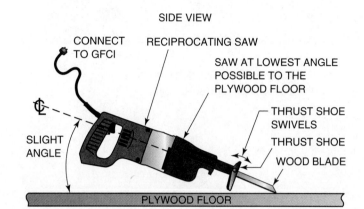

B

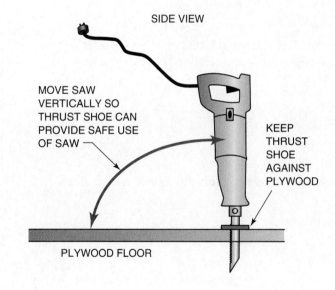

Review Questions

1. An electrical power tool must be connected to an electrical circuit that is protected by a
   a. GFCI
   b. FCIG
   c. CFIG
   d. GCFI

2. Fall protection is required for a work area that places your feet
   a. More than 3' above the ground
   b. 6' or more above the ground
   c. On top of a 3' stepladder
   d. On the ground

3. The top of an extension ladder must extend past the point where the ladder rests on the roof edge at least
   a. 1'
   b. 2'
   c. 3'
   d. 4'

4. Two approved hearing protection items are ear plugs and
   a. Cotton balls
   b. Pieces of clean rags
   c. Ear muffs
   d. None of the above are correct

5. Three common drills used for wood are the pistol drill, hole hawg drill, and
   a. Right angle drill
   b. Left angle drill
   c. Upside down drill
   d. Hammer drill

6. A drill used to install anchors in concrete is a
   a. Concrete drill
   b. Hammer drill
   c. Jackhammer
   d. Right angle drill

7. Though many materials can be drilled with twist bits, they are designed to drill through
   a. Metal
   b. Plastic
   c. Wood
   d. Concrete

8. The two different types of worm drive feed screws of an auger bit are a coarse type and a
   a. Rough type
   b. Smooth type
   c. Fine type
   d. Bald type

9. A power bore bit is the most common type used by plumbers and is
   a. Self-feeding
   b. Also known as an auger bit
   c. Also known as a speed bit
   d. Also known as a hole saw bit

10. A hole saw requires the user to remove
    a. The side handle of the drill
    b. The pilot bit before drilling
    c. The core of material after drilling
    d. A set screw

11. The trade name of sawzall indicates that it is a
    a. Circular saw
    b. Reciprocating saw
    c. Portable band saw
    d. Hole saw

12. The trade name jigsaw indicates that it is a
   a. Saber saw
   b. Reciprocating saw
   c. Circular saw
   d. Hand saw

13. A powder-actuated tool requires that the user be
   a. Certified
   b. A licensed plumber
   c. A supervisor
   d. None of the above are correct

14. A plumber tests a water piping system with either a hydrostatic test or
   a. A performance test
   b. An air test
   c. A smoke test
   d. A gas test

15. The two common types of jackhammer bits are the point type and the
   a. Auger type
   b. Speed type
   c. Chisel type
   d. Twist type

Chapter 3 | Types of Pipe

You have learned about the necessary hand tools and power tools to install a plumbing system. This chapter focuses on the various types of piping used in the new residential plumbing industry. Pipe is used to bring potable water and gas into a building and to allow sewage and wastewater to drain from a building. Piping used in the residential plumbing industry falls into three basic categories: drainage waste and vent (DWV); water distribution; and gas. Some materials are used in all three piping systems; others are only installed in specific systems. Copper and brass can be used in all three systems. DWV piping is designed to operate under less pressure than water distribution piping. Gas piping systems are made from some materials that cannot be used for water distribution or DWV installations, such as black steel and aluminum.

State and local code authorities must approve all piping materials. Many codes dictate that piping used for interior installations be different from piping used on the exterior of a building. Knowing the location of an installation and identifying the type of pipe to be used is crucial to installing a safe, high-quality plumbing system. Several variations of pipe are manufactured from the same material, and each is approved for specific installations based on their unique characteristics

The American Society for Testing and Materials (ASTM) and the National Sanitation Foundation (NSF) are organizations that test and rate piping materials and standards. The accumulated data and recommendations from ASTM and NSF determine what type of pipe is allowed by code. Some considerations in the selection process include codes and products specified by an architect or owner. The product actually chosen and installed is often based on cost control measures and availability.

Pipe and tubing are two different classifications of materials. Plumbers typically refer to hard or rigid materials as pipe and to flexible or soft materials as tubing. Most pipe is classified based on wall thickness and categorized in schedules. Copper tube is also classified by wall thickness, but it is categorized in Types as opposed to Schedules. Common pipe and tubing materials are discussed separately in this chapter.

OBJECTIVES

Upon completion of this chapter, the student should be able to:

- identify and describe common types of pipe and tubing used in a residential plumbing installation.
- understand that certain pipe and tubing materials can only be used for specific systems.
- understand that some pipe and tubing materials can be used in all residential systems.
- relate pipe and tubing selection to plumbing codes.
- order pipe and tubing materials.

Glossary of Terms

drain portion of a drainage system that is installed on the interior of a building

fitting item in a plumbing system that connects to piping or another fitting to achieve a desired offset or specific connection

foam-core type of non-pressure DWV plastic pipe that has a solid outer layer, a cellular foam middle layer, and a solid inner layer

nominal diameter size of pipe and fittings used to order materials; does not indicate exact diameter

offset angle that changes a piping route expressed in degrees of a circle such as 90° and 45°

pipe often used to describe all pipe, tube, and tubing, but is defined as any rigid or hard materials that would break if flexed more than 2% of its diameter

potable water free from impurities that could cause disease; safe for human consumption

radius half the diameter of a circle; relates to the bend of a fitting or tubing

schedule classification of pipe indicating the wall thickness of the pipe

sewer portion of a drainage system that is installed on the exterior of a building

tube less rigid than pipe, but more rigid than tubing; often referred to when ordering copper, i.e., copper tube

tubing flexible or non-rigid materials that can deflect more than 2% of its diameter without breaking; often referred to as pipe

type used to specifically describe different copper tube specifications and to differentiate between various materials, i.e., Type M copper tube or types of plastic pipe

vent pipe that allows airflow in a drainage system

water distribution entire potable water piping system, relates to hot and cold water systems

water service piping on the exterior of a building connecting the potable cold water source, such as a water meter, to the interior water piping system in a building

Pipe Diameters

Pipe in a plumbing system is round and ordered based on its diameter. **Pipe** and **tubing** have an inside diameter (ID) and an outside diameter (OD), but are ordered by a sizing system known as **nominal diameter.** The actual ID and OD dimension of pipe, if measured with a ruler, would vary by material types, so the dimension is simply rounded off to become a nominal diameter. The thickness of a pipe wall varies depending on the material, and this varying thickness determines the difference in the OD and ID. A 2" pipe made from one type of material may have a different diameter than a 2" pipe made from a different material. The ID is important for determining the flow requirements for piping, so in some instances, a plumber orders pipe based on the nominal ID.

Copper is a thin-walled type of **tube** used in the air conditioning and refrigeration (ACR) trades as well as the plumbing trade, and the ordering method varies between the two trades. Plumbers order copper tube using both the OD and ID, but the AC trade orders by the OD only. Steel piping materials are manufactured using the same outside diameter for each pipe size. That sizing system is known as iron pipe size (IPS). Plastic pipe used for DWV installations is manufactured using IPS, with the inside diameter varying based on the wall thickness. Three common wall thickness classifications are **Schedule** (S) 10, 40, and 80, with S-40 being the most common at about 1/4" thickness. A higher schedule rating indicates a greater wall thickness. Figure 3-1 illustrates identifying inside and outside diameters.

FROM EXPERIENCE

Nominal is defined as so-called, which clarifies that the OD and ID are not actual dimensions, but vary with different types of pipe. IPS indicates that different materials have the same OD but does not indicate what the ID is.

Table 3-1 Pipe Types and System Uses

Material Type	DWV	Potable Water
PVC	Yes	Yes
ABS	Yes	No
CPVC	No	Yes
PEX	No	Yes
Copper	Yes	Yes
Galvanized steel	Yes	Yes
Black steel	No	No
Brass	Yes	Yes
Cast iron	Yes	No
Polyethylene	No	Yes
Perforated PVC	No	No

Because codes and preferences allow for a variety of piping materials for a specific system, it is important to know the allowable options for each type. Using the wrong material can result in code violations, property damage, and health risks to a consumer. Table 3-1 is a list of common materials and their acceptable uses in a residential plumbing system as DWV piping and distributing **potable water.** They are described separately in this chapter. Other less common types may be included in job site training based on company preference and code acceptance.

Many piping materials are offered in several forms, known as lengths and rolls (coils). The available length of pipe depends on the type of material, and the length of a roll of tubing is dependent on the type and size. Manufacturers offer certain types and sizes based on code allowances and practical manufacturing processes. Most pipe is sold in 20' lengths, but cast iron DWV pipe is sold in a variety of lengths, with the maximum being 10'. Flexible rolls of tubing for installing water distribution systems are typically sold in 100' lengths, but rolls of soft copper tube range from 40' to 100' in length.

Table 3-2 lists common piping materials used in the residential industry and the lengths usually offered by a manufacturer. Material in roll form varies in length. Wholesale stores offer materials in customized lengths that differ from those in Table 3-2.

Plastic Pipe

The residential housing industry is always striving to build homes at competitive prices to remain affordable. The introduction of plastic piping for drainage, **vent,** and **water distribution** systems has helped achieve this. In the mid-twentieth century, drainage and vent systems used metal piping, such as cast iron and galvanized iron, and water distribution systems were mostly installed with copper tubing. Higher housing costs are a direct result of labor costs

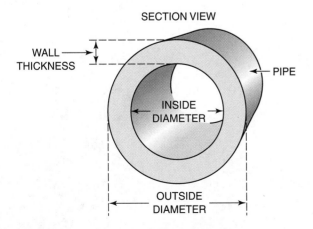

Figure 3-1 Outside and inside pipe diameters.

Table 3-2 Available Lengths of Common Pipe Materials

Material Type	30"	5'	10'	20'	21'	Roll
PVC			√	√		
ABS			√	√		
CPVC			√			
PEX			√			√
Copper			√	√		√
Galvanized steel					√	
Black steel					√	
Service-weight cast iron	√	√	√			
No-hub cast iron			√			
Polyethylene						√
Perforated PVC			√			
Brass					√	

that continue to increase. Plastic pipe products have helped to offset the cost of labor because they are less expensive than metal products and installation time is reduced. One disadvantage of plastic products pertains to their lack of fire-rating capabilities. Most commercial applications still use copper tube for water distribution and cast iron pipe for drainage and vent installations. Plastic piping is the product of choice in the residential industry, however, and reverting back to metal products is not likely. Many custom homes do blend plastic and metal products for some installations that are discussed later.

PVC Pipe

Plastic piping known as polyvinyl chloride (PVC) is offered in a wide variety of types and schedules and is the most widely used product for DWV installations in new residential applications. This very durable product increases productivity because of its lightweight characteristics and method of joining. The residential industry uses a solvent welding cement, called glue, to join pipe and fittings. The outside diameters are manufactured to IPS standards. Some kinds of PVC are rated for pressure systems, and others are used specifically for gravity draining systems. PVC pressure piping is commonly used to provide water service to residential buildings, but most codes do not allow PVC to be used in water distribution systems on the interior of a building. All PVC in residential applications has a maximum temperature rating of 140° Fahrenheit (F). Pipe and tubing for water service installations must be capable of handling 160 psi pressure at 73°F. Pipe or tubing for water distribution systems must be rated to distribute 180°F at 100 psi pressure to enable the product to safely distribute hot water.

DWV PVC piping for residential applications typically uses a Schedule 40 (S-40) type in either solid or cellular-core. Cellular-core PVC is known as foam-core and can only be used for gravity type systems; it is often used for cost savings instead of the more expensive solid-core PVC pipe. **Foam-core** has a zero pressure rating and voids the warranty if subjected to any pressure other than filling the system for a water test inspection. Thin-wall PVC is acceptable for installing the building **sewer,** which is the drainage pipe that connects **drains** from inside the building to their termination point, such as the city sewer system. The rating of the thin-walled PVC is different from the solid-core or foam-core, and using it inside a building for a DWV system is a code violation.

As with all pipe materials, the approved identification markings must be visible on the exterior of the pipe. PVC pipe usually used in residential applications is white. A thin-walled PVC for sewer installations is green and uses a push-on rubber o-ring connection as opposed to a solvent weld connection. A plumber must always review manufacturer's identification markings to ensure proper use of any product. PVC can be subjected to more household chemicals than most other kinds of pipe. Long-term exposure to ultraviolet (UV) rays will damage PVC pipe; therefore, it should be covered with pipe insulation or painted according to manufacturer's instructions, which typically dictate that water-based latex paint should cover the exposed areas. Figures 3-2 through 3-5 show several different kinds of PVC pipe used in the residential plumbing industry.

FROM EXPERIENCE

Foam-core PVC pipe is accepted by most codes for underground and above-ground installations serving a residential building. It is common for foam-core to be used for the entire DWV system in a building and to connect to a public sewer or septic tank.

SIDE VIEW

1120 SCH40 260 PSI NSF-PW ASTM D 1785 ASTMD 2665 NSF-DWV

PLAIN END 260 PSI

1120 SCH40 330 PSI NSF-PW ATMD 1785

BELL END 330 PSI

BELL END

Figure 3-2 Schedule 40 solid core PVC pipe used for pressure and gravity systems.

Figure 3–3 Schedule 40 foam core PVC pipe specifically used for gravity systems.

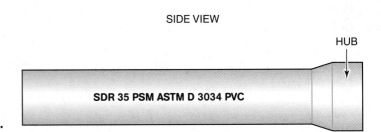

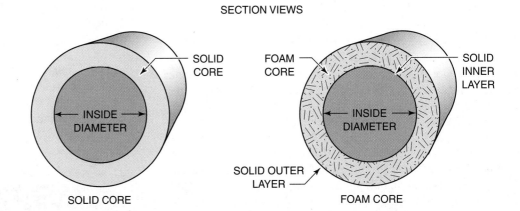

Figure 3–4 SDR PVC pipe used for sewer installations.

Figure 3–5 Section view of solid core and foam core PVC pipe.

CAUTION: Many codes limit the burial depth of many plastic materials due to the weight of soil on the pipe. Too much pressure on a pipe can crush it.

CPVC Pipe

Chlorinated polyvinyl chloride pipe (CPVC) is flexible, but is typically called pipe and not tubing. A variety of material types are allowed by code for use in water distribution systems in residential buildings, CPVC is one that is often used. It is a yellowish-white material and is joined with a solvent weld process (glue). Glue used for CPVC is different from glue used for PVC, because the two products are incompatible. The OD of CPVC is not the same as PVC, and their nominal sizes do not reflect their actual sizes. The CPVC manufactured for residential applications is identified as SDR 11 and ranges in size from 1/2" to 2" diameter. The pressure capabilities of CPVC vary with the temperature of the water being distributed.

The maximum pressure rating of 400 psi for SDR 11 CPVC pipe is based on its being used to distribute water at 73°F. The higher the temperature of the water, the lower the pressure rating of the pipe. CPVC pipe supplying 180°F water has an adjusted rating capability of 100 psi. The specific use of the piping must be known for safe operation. Capability data is available where pipe is purchased. As with all plastic piping, CPVC expands and contracts based on the water temperature. A CPVC system must be tested with water, never air. Long-term exposure to UV rays will damage CPVC, so it must be covered or painted according to the manufacturer's instructions to avoid damage. Figure 3-6 identifies CPVC pipe.

 FROM EXPERIENCE

Some CPVC glue manufacturers may not require that a primer be applied to pipe and fitting surfaces before applying glue, but most codes dictate that primer must be used. Read manufacturer instructions and your local code book for proper techniques.

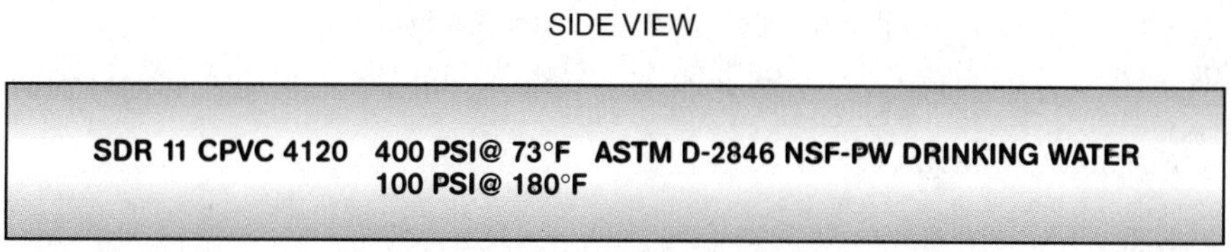

CHAPTER 3 Types of Pipe 63

SIDE VIEW

SDR 11 CPVC 4120   400 PSI@ 73°F   ASTM D-2846 NSF-PW DRINKING WATER
100 PSI@ 180°F

Figure 3-6 CPVC pipe identification.

> **CAUTION**
>
> **CAUTION:** Solvent cement and primer are dangerous chemicals. Read all Material Safety Data Sheets (MSDS) before use and know all safety aspects, which include flammability dangers and first aid procedures.

ABS Pipe

Acrylonitrile butadiene styrene (ABS) pipe is black in color and is used to install DWV systems. ABS pipes must be joined with a solvent weld process designed specifically for that product. It is not compatible with PVC. ABS is offered in a foam-core option, and available sizes range from 1-1/2" to 6". Its maximum water-temperature capability is 140°F, the same as PVC DWV. The OD of ABS pipe is manufactured to IPS standards, and its rating is identified on the side of the pipe. Like CPVC it cannot be tested with air pressure or be exposed long-term to UV rays. Figures 3-7 and 3-8 show identification of ABS pipes.

FROM EXPERIENCE

Distinguishing ABS from PVC is simple; ABS is black and PVC is white. The color does not dictate the rating of the pipe. If a piece of pipe does not have the manufacturer's label, always assume that it is rated not for pressure.

> **CAUTION**
>
> **CAUTION:** All plastic piping is flammable, and, when ignited, causes toxic fumes. Never use an open flame near plastic piping or install plastic piping in areas designed for metal pipe applications.

FROM EXPERIENCE

Plastic pipe can warp if it is left in the sun too long, which can make it difficult to install in a drainage system.

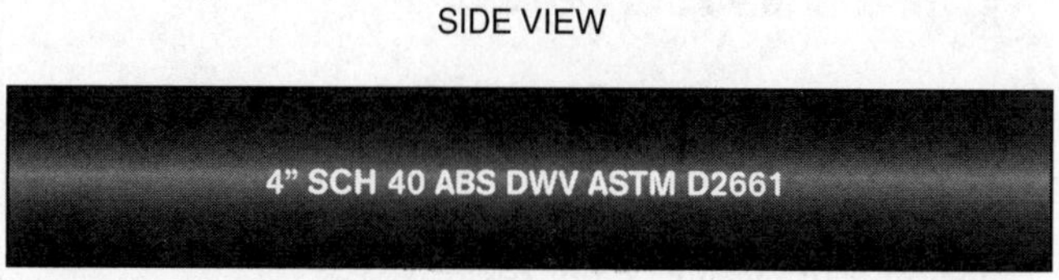

SIDE VIEW

4" SCH 40 ABS DWV ASTM D2661

Figure 3-7 Solid core ABS piping used for DWV installations.

SIDE VIEW

ASTM F-628 ABS DWV NOT FOR PRESSURE

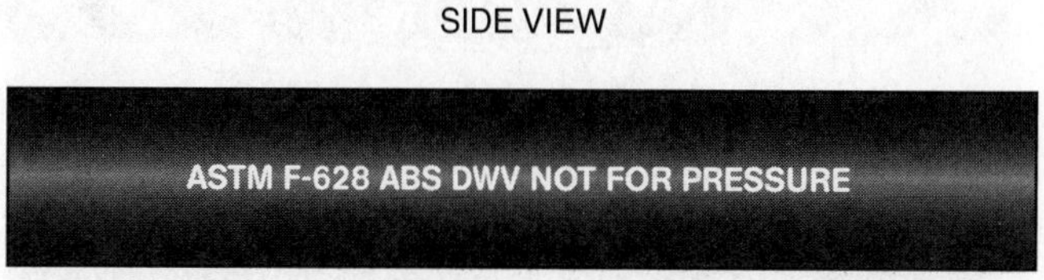

Figure 3-8 Foam core ABS used for DWV installations.

Polyethylene Pipe

Polyethylene (PE) tubing is used for **water service** installations but is not allowed by code to be installed on the interior of a residential building. It is black in color and is typically sold in 100' rolls (Fig. 3-9). It is often used in well-pump installations because of its long lengths and durability in cold conditions. It joins together with specially designed barbed connections and stainless steel hose clamps, which means that it can be pressurized immediately after a connection is complete. It is not used as much today due to the popularity of other flexible piping products. PE is a thin-wall product and, like all plastic piping products, cannot be subjected to long-term UV rays.

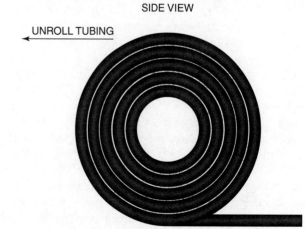

Figure 3-9 Roll of polyethylene tubing used for water service.

The **radius** of a bend must be great enough so the pipe does not kink. Plumbers often heat the pipe with a torch to soften the pipe while joining, but this voids the warranty of the product.

FROM EXPERIENCE

PE pipe is easily repaired, with brass-type, barbed **fittings** being the best selection for joining two pipe ends. Brass does not corrode under normal conditions and provides a rigid connection.

CAUTION: Heating PE piping with a torch creates dangerous fumes and, although this practice may be common, it should be avoided.

PEX Pipe

Cross-linked polyethylene (PEX) tubing sounds like it should be compatible with PE, but the two piping systems use different connection methods. PEX is used for water distribution systems and has become one of the most popular selections because of its cost and labor efficiency. PEX can be joined with a specially designed crimp ring and can be pressurized immediately after the connection is complete. PEX can also be connected by expanding the inside diameter of the tubing, which requires a different tool than a crimping tool. The expanding tool connection method uses different tubing than that used for the crimp method. The two different connection methods also dictate the different compatible tubing used. This typically whitish colored tubing is also available from some manufacturers in red and blue so the installer can differentiate between piping used for hot and cold water systems (Fig. 3–10). The popularity of this product for residential water distribution systems resulted from the discontinued use of another flexible piping known as polybutylene.

PEX is sold in 20′ lengths and 100′ rolls, but longer rolls can be custom ordered. As with all flexible tubing, PEX will kink if the radius created is too sharp. The flexibility of this product also means that it must be supported at closer intervals than is required for pipe. PEX is shipped from the manufacturer with a UV protective cover. Long-term exposure to UV rays will damage the integrity of the product. Plastic tubing should be handled with more care than rigid pipe or copper tube. Scratching the side of the tubing when installing it through drilled holes can shorten the life of the product. Nails that protrude in a drilled hole that the tubing is inserted in can gouge the PEX tubing.

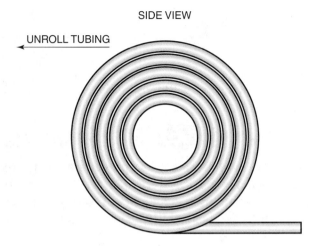

Figure 3–10 Roll of PEX tubing used for water distribution systems.

Another cross-linked polyethylene product that is used widely in the heating industry and is growing in popularity for water distribution systems is Polyethylene—Aluminum—Polyethylene or PEX–AL–PEX tubing. As its name indicates, it is a multi-layered tubing with a PEX interior, aluminum middle layer, and PEX outer layer. It allows a more rigid installation than PEX while maintaining the ease of installation. Many **offsets** created with PEX require support to maintain the desired installation intent, but the aluminum layer maintains the shape of a bend without holding it in place. The tubing is sold in roll form, with the length of the roll depending on the size. PEX–AL–PEX is sold in sizes ranging from 3/8″ ID to 1″ ID. It is ordered based on the ID of the inner PEX tubing. Many codes have not accepted the use of PEX–AL–PEX for use in the plumbing industry. It is more expensive than PEX. As with many new products, changing plumbing codes to accept their use may be slow but many eventually become popular for labor savings or quality reasons.

FROM EXPERIENCE

When ordering PEX material, recognize that the maximum spacing between horizontal supports is 32″, which reflects the distance between every other floor joist. This shows that some codes are based on practical applications and not solely on safety or theory.

CAUTION: To ensure a quality installation and avoid product failure, discard PEX tubing that has been exposed long-term to UV rays or that has been damaged. Property damage claims will result if a contractor installs defective materials.

Other Pipe Materials

Copper tubing and cast iron pipe are still being used in the residential plumbing industry even though most residential projects use plastic products. Copper can be used for connecting faucets or the pieces of equipment that are supplied with plastic water distribution systems. Cast iron can be installed in residential applications for the entire piping system or for sound detention within a wall when the rest of the system is plastic. Brass and galvanized iron piping are only connected to certain items. Perforated plastic piping is discussed here because of its unique application. It is extremely important for a plumber to know about all products and their applications even though they are not widely used. A plumber may encounter the following products during a renovation project or while performing a repair.

Copper Tube

Four basic types of copper tube are used in the plumbing industry and two more are used for medical and industrial piping systems. The four types in residential applications, based on the thinnest wall to the thickest wall, are Type DWV, Type M, Type L, and Type K.

Each type of copper must be permanently marked by the manufacturer by color-coding or scoring on the side of the pipe (Fig. 3-11). The most common diameters of copper tube used in the residential plumbing industry are 1/4" to 1". Copper tube is manufactured to a standard known as copper tube size (CTS). The outside diameter of all four basic types is the same and the inside diameter varies with the wall thickness. The size selection is based on codes and the number of fixtures served. Table 3-3 lists the four types and their basic use for the residential plumbing industry. Copper tube is available in hard and soft forms, with soft copper sold in rolls. The table indicates that not all types of copper tube are available in roll form. Type L is listed as acceptable for underground installations, but your local code may only allow Type K underground.

The labor and material savings of PEX and CPVC has minimized the use of copper in the residential plumbing industry. Using copper for connections to shower faucets, equipment connections, and termination of the water supply serving a plumbing fixture provides rigidity. It is used in many new in-

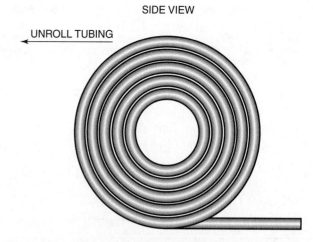

Figure 3–11 Roll of copper tube used for water distribution systems.

stallations. Many expensive residential construction projects still use copper for the water distribution systems. Copper is joined with a welding process known as soldering.

Type L and Type K copper are available in hard and soft forms; 20' is the standard length for hard copper tube, and soft rolls are available in 40', 60', and 100' lengths, depending on their size. Type M is the most common copper tube in the residential plumbing industry and is less expensive than Types L and K. The OD of a specific size remains the same regardless of the type and form. Rolls of soft copper are more common for supplying water to a dishwasher and icemaker. For icemaker connections, 1/4" OD is used and 3/8" OD is used for supplying a dishwasher. Soft copper is more commonly used for underground installations, but is allowed throughout the water distribution system. Plumbers order copper tube using OD or ID, which can cause some confusion when ordering from a plumbing wholesale store that refers to it differently than the plumber does.

Table 3-4 lists the nominal, OD, and ID sizes of copper tube and how a plumber orders the product. Notice that the OD sizes from 5/8" to 2-1/8" are ordered using their nominal sizes, which correlate with their ID sizes. There are two selections for 1/2" nominal sizes—1/2" OD copper is ordered stating the abbreviation OD, and 1/2" nominal size is actually 5/8" OD. If you state 1/2" copper while placing an order, you will actually be ordering 5/8" OD. Use the OD sizes

Table 3–3 Types and Basic Uses of Copper Pipe in a Residential Installation

Copper Type	Potable Water	DWV	Underground	Available in Roll
DWV		√		
Type M	√	√		
Type L	√	√	√	√
Type K	√	√	√	√

Table 3-4 Copper Tube Sizes and How a Plumber Orders Each Common Size

Nominal Size	OD	ID	Order As	Order Using
1/8"	1/4"	1/8"	1/4"	OD
1/4"	3/8"	1/4"	3/8"	OD
5/16"	5/16"	3/16"	5/16"	OD
1/2" OD	1/2"	3/8"	1/2" OD	OD
1/2"	5/8"	1/2"	1/2"	ID
3/4"	7/8"	3/4"	3/4"	ID
1"	1-1/8"	1"	1"	ID
1-1/4"	1-3/8"	1-1/4"	1-1/4"	ID
1-1/2"	1-5/8"	1-1/2"	1-1/2"	ID
2"	2-1/8"	2"	2"	ID

when ordering 1/4" OD, 3/8" OD, and 1/2" OD tubing. A plumber does not usually install 5/16" OD copper tube, but it is important to recognize that it exists to ensure that you do not purchase it by mistake.

 FROM EXPERIENCE

When ordering 3/8" OD and 1/4" OD at a wholesale outlet, you may have to specify their intended use, such as for an icemaker or a dishwasher, to ensure that you are ordering the correct size.

CAUTION: When working with copper, wear gloves to avoid cuts from the sharp tube ends. Clean and bandage cuts immediately to avoid infection.

Cast Iron Pipe

Cast iron pipe is often used in residential DWV systems for vertical installations to allow a quieter draining process in walls. The two most common types of cast iron (CI) used in a plumbing system are no-hub (NH) and service weight (SV). They have different minimum sizes and methods of connection. Because codes do not allow pipe smaller than 2" to be buried below ground and because SV is most often used there, 2" SV is the smallest size available. The largest size typically used in the residential plumbing industry is 6". NH is more common for above-ground installations—1-1/2" is the smallest and 4" is usually the largest. Cast iron pipe is also referred to as soil pipe.

SVCI pipe comes in a variety of lengths with either a single hub (SH) or a double hub (DH). When ordering, you must indicate SH or DH. The abbreviated version for ordering a DH piece of SVCI pipe is DHSVCI. Various lengths of SV pipe are sold to accommodate an installation and to minimize the number of plain end pieces that would otherwise be wasted. The three most common SV lengths are 30", 5', and 10'. SV pipe that is ready for installation has one plain end and a hub on the other end. SV is joined with a black neoprene gasket that is often called an SV gasket. Lubricant is used to slide the plain end into the gasket lined hub.

NH gets its name from having two plain ends, or no hubs. NHCI pipe is only available in 10' lengths. NH is joined with a stainless steel band encasing a neoprene sleeve that is often referred to as an NH clamp or NH coupling. It slides over the pipe ends and is tightened with a specially designed 5/16" torque wrench. Both types of cast iron pipes are cut with chain cutters. Figure 3-12 includes NH and SV cast iron pipe. Figure 3-13 includes an NH coupling and an SV gasket. Figure 3-14 is an SV hub, also referred to as a bell.

Figure 3-12 No hub and service weight cast iron pipe.

Figure 3-13 No hub couplings.

CHAPTER 3 Types of Pipe   67

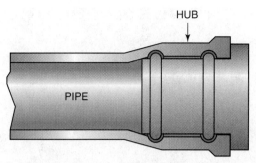

Figure 3–14 Service weight hub that is also known as a bell.

Figure 3–15 Galvanized and black steel pipe.

 FROM EXPERIENCE

Service weight and no-hub cast iron pipe can also be joined with a transition coupling, which uses the same 5/16" torque wrench that tightens NH clamps. It does not have a stainless steel sheathing and may require additional support near the connection of the two horizontal pipes.

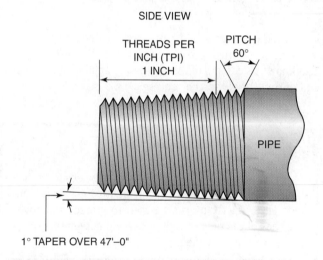

Figure 3–16 National Pipe Thread standard for plumbing systems.

CAUTION: Cast iron pipe is heavy, so proper lifting procedures must be used to eliminate back injury. Always use two people to carry long lengths, and be sure both people are on the same side of the pipe when carrying.

Steel Pipe

Galvanized and black steel pipe are the two kinds usually used in the residential plumbing industry (Fig. 3-15). Galvanizing is a process that coats steel pipe to minimize corrosion, so it can be used to distribute potable water. The pipe may still rust under extreme conditions. Black steel pipe cannot be used for potable water distribution. It is used for residential gas supply piping. Galvanized pipe can also be used for DWV systems and was the material of choice in the early and middle twentieth century. Many gas codes still allow galvanized to be installed for gas systems. Though steel pipe is sold with plain ends, most smaller pipe sizes are manufactured with threads on each end. The most common type of steel pipe is Schedule 40, which indicates its wall thickness; Schedule 10 and 80 are available, but are not relevant in the residential plumbing industry.

The standard length of steel pipe is 21', and the most common sizes in the residential industry range from 1/2" to 1". The IPS standard referred to in the PVC and ABS sections of this chapter originated from this type of material. Threads comply with a National Pipe Thread (NPT) standard (Fig. 3-16). An NPT is tapered with the smallest diameter of the thread at the pipe end to allow proper tightening as the pipe is threaded (screwed) into a fitting. The number of TPI is part of the NPT standard and varies based on the diameter of the pipe. A 3/4" NPT has 14 TPI, and a 1" has 11-1/2 TPI. Short manufactured threaded pieces of pipes known as nipples typically come in lengths up to 6", but some wholesale stores offer custom lengths up to 4' (Fig. 3-17). Pipe nipples are typically available in increments of 1/2". Some threaded steel pipe is shipped from the manufacturer with a plastic protective cap on one end and a shipping coupling on the opposing end. The shipping coupling is discarded before installing the pipe as it does not have tapered internal threads and cannot be used in a plumbing system.

Before a steel pipe is connected, a joint compound known as pipe dope, which is specifically designed for safe use in a plumbing system, is applied to the threads. The pipe is then threaded into a fitting and tightened with a pipe wrench. Steel pipe is cut with a specially designed tool known as a cutter and then the burr created during the cutting process

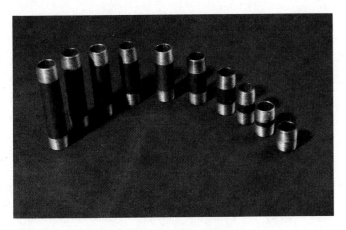

Figure 3-17 Various lengths of pipe nipples.

Figure 3-18 Brass close nipple.

is removed with a reamer. Threads are created on the exterior of the pipe end with threading dies. Specially formulated oil, also known as cutting oil, is constantly applied over the pipe end during the threading process.

 FROM EXPERIENCE

Always protect threads to avoid damage by ensuring that the protective cap and shipping coupling remain on the pipe ends. Duct tape can be used to protect and cover the ends of the pipe so that debris will not enter it while it is stored and handled.

CAUTION: Some codes may restrict the use of Teflon tape when installing gas piping systems. Always know your local codes.

CAUTION: Always read the MSDS before using joint compound and cutting oil. Know all safety precautions and medical treatments in case of accidental consumption of oil or compound.

Brass Pipe

Brass pipe was used extensively in the early to middle twentieth century for water distribution systems, but it is no longer the material of choice. It is more corrosive resistant than galvanized steel and is used to manufacture faucets, valves, and other products used for potable water, such as pipe nipples. The threads comply with NPT standards and are manufactured as IPS. Common brass nipples range from a short length, or close nipple, up to 6" long. A close nipple (Fig. 3-18) has threads throughout its length with no exposed pipe between the opposing threads. The total length of a close nipple depends on the pipe size because of the varying TPI of a specific pipe diameter. A smaller diameter close nipple is shorter than a close nipple of larger diameter.

Brass nipples are used for connections to plumbing equipment, such as a water heater. Brass pipe is chrome plated to use in exposed areas, such as the showerhead connector. A manufacturer performs the chrome plating process, and care must be taken when working with chrome to protect the finished surface. Strap wrenches are used to tighten chrome nipples to keep from scarring the finished surface. Brass pipe is cut, threaded, and assembled with the same methods as galvanized and black steel pipe.

 FROM EXPERIENCE

A wide variety of chrome nipples are usually not readily available in most plumbing wholesale stores, and specific sizes may have to be specially ordered. Identify all quantities and sizes of chrome nipples required at an early stage of construction to minimize material organization problems.

Perforated Pipe

Perforated pipe is most often used in the residential plumbing industry for septic tank systems. Perforated pipe is usually manufactured from PVC. The most common size is 4" in diameter and 10' in length. A thin-wall type of perforated piping is most often used in a septic system design to drain wastewater below ground (subsurface). The pipe has a

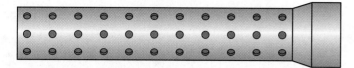

Figure 3-19 Perforated pipe commonly used in a septic tank system.

series of 5/8" diameter holes along its sides that allows the water to drain from the pipe into surrounding stone, gravel, and soil.

Most perforated pipe has a bell end and a plain end, which is joined with a solvent weld process (glued). Perforated pipe can also be useful as a liner in a vertical water well to extract potable water from below ground. The lining process creates a stable vertical shaft and protects the suction piping and pump if a well collapses. Figure 3-19 shows a perforated pipe.

 FROM EXPERIENCE

Well-casing codes may require Schedule 40 pipe to be used, but if perforated Schedule 40 is not available, a series of holes can be drilled along the sides of the pipe.

Corrugated Stainless Steel Tube

Corrugated stainless steel tube (CSST) is a new product for gas distribution that may not be approved by many local plumbing codes. It is a flexible stainless steel tube sheathed with a yellow plastic jacket (Fig. 3-20). Yellow is

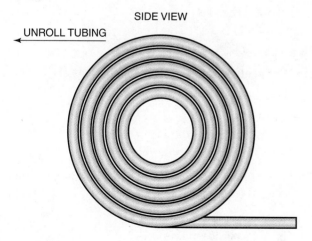

Figure 3-20 Corrugated stainless steel tubing used for gas distribution.

an approved industry-standard color to identify gas piping. As with all flexible tubing, offsets in a CSST system can be created without installing pipe fittings. It is sold on wooden spools in 150' to 250' lengths in sizes ranging from 1/2" to 2" diameter. Connections of CSST are made with a compression-type fitting. With building codes constantly being scrutinized and strengthened throughout the country, flexible tubing is growing in popularity. Rigid gas piping such as black steel can leak after a small earthquake, but flexible tubing can absorb the shock and move with a structure.

 **CAUTION**

CAUTION: When handling and storing flexible gas tubing, always protect the outer sheathing from being nicked by sharp objects. Keep ends of tubing covered so debris cannot enter the tubing.

Summary

- Pipes are manufactured to specific standards.
- American Society for Testing and Materials (ASTM) and National Sanitation Foundation (NSF) are organizations that rate materials.
- Copper tube is rated as Types.
- The most common pipe ratings are known as Schedules.
- Plastic pipe and flexible tubing have maximum temperature ratings.
- Polyvinyl chloride (PVC) can be used for Drainage, Waste, and Vent (DWV) and limited water distribution installations.
- PVC cannot be used for water distribution inside a building.
- Chlorinated polyvinyl chloride (CPVC) can be used for water distribution inside a building.
- Cross linked polyethylene (PEX) is a flexible tubing used for water distribution.
- Two connection methods can be used for installing PEX fittings to the tubing.
- Solvent welding process is known as gluing plastic pipe connections.
- Copper connections are soldered using flux and solder.
- Brazing copper connections is known as silver soldering.
- Brazing copper connections does not require flux.
- Service weight (SV) and no hub (NH) are the two types of cast iron pipes used for DWV systems.
- Black steel pipe is used for gas piping systems.
- Galvanized steel pipe is used for water distribution systems.
- Brass pipe and fittings can be used for all piping systems.

Review Questions

1. **The abbreviation OD stands for**
   a. Odd diameter
   b. Outside diameter
   c. Opposing distance
   d. Over dimension

2. **The most common type of PVC and steel pipe based on their wall thickness is rated as**
   a. Schedule 40
   b. Schedule 10
   c. Schedule 80
   d. Schedule 50

3. **PE is the abbreviation for**
   a. Polybutylene
   b. Cross-linked polyethylene
   c. Polyethylene
   d. Plastic exterior

4. **CPVC is used in a residential installation for**
   a. Non-potable water systems
   b. DWV systems
   c. Gas systems
   d. None of the above are correct

5. **PEX is joined with either an expanding process or**
   a. Crimp ring
   b. Hose clamp
   c. Solvent weld
   d. Soldering process

6. **The color of ABS pipe is**
   a. White
   b. Green
   c. Blue
   d. Yellowish-white

7. **The two common types of cast iron pipe are no hub and**
   a. Service weight
   b. Galvanized
   c. Perforated
   d. Threaded

8. **1/2" ID copper tube is also ordered as**
   a. 1/2" OD copper tube
   b. 5/8" OD copper tube
   c. 3/8" OD copper tube
   d. 7/8" OD copper tube

9. **Polyethylene tubing is joined with hose clamps and**
   a. Crimp fittings
   b. Threaded fittings
   c. Barbed fittings
   d. Soldered fittings

10. **A service weight cast iron pipe with two hubs is ordered as**
    a. DHSVCI
    b. SVCIDH
    c. SVDHCI
    d. SVSHCI

11. **Corrugated stainless steel tube is used for**
    a. Cold water distribution
    b. Gas distribution
    c. Hot and cold water distribution
    d. Drainage waste and vent

12. **Actual pipe sizes vary depending on the material type and are ordered using their**
    a. Nominal size
    b. Actual inside dimension
    c. Actual outside dimension
    d. None of the above are correct

13 Plastic pipe and tubing must be protected from

a. Water
b. Air
c. Ultraviolet rays
d. Wind

14 Three common types of copper tube are

a. M, L, and K
b. A, B, and C
c. X, Y, and Z
d. Schedules 10, 40, and 80

15 A tapered pipe thread used in the plumbing industry complies with

a. National Pipe Thread standards
b. Straight Pipe Thread standards
c. Standard Pipe Thread standards
d. Beveled Pipe Thread standards

16 IPS is the abbreviation for

a. Internal pressure schedule
b. Iron pipe size
c. Inside pipe size
d. Interior plastic standards

17 The smallest length of pipe nipple manufactured is a

a. Close nipple
b. Small nipple
c. Short space nipple
d. Miniature nipple

18 The American Society for Testing and Materials is abbreviated as

a. ASFTAM
b. ASFTM
c. ASTM
d. None of the above are correct

19 The abbreviation psi stands for

a. Per second interval
b. Per square inch
c. Perfectly smooth interior
d. Plastic standards institute

20 Pipe and tubing rated for water distribution must be capable of handling water that is

a. 180°F at 100 psi pressure
b. 140°F at 100 psi pressure
c. 100°F at 100 psi pressure
d. 120°F at 120 psi pressure

Chapter 4 | Fittings

Previous chapters covered the various pipe materials used in the residential plumbing industry; this chapter discusses the fittings that are used to complete a piping system. Here we focus on fittings with a broad range of uses in various systems. Fittings with more specific uses are covered in the relevant areas of this book. Using the correct tools and pipe materials is important in creating a productive installation. However, it is the fittings that provide creative installation routes with offsets and termination connections designed for specific uses. Some fitting materials must be of the same material as the pipe, but other materials are interchangeable with many systems and types of pipe. For example, brass can be used for potable water, DWV, and gas.

The type of connection on a fitting also helps to determine its use in a system. Threaded fittings can be used to connect two different materials in any system. Some fittings combine with other fittings to achieve different degrees of offsets depending on the job. Because codes often dictate which fittings to use, productive plumbers must be knowledgeable about these codes in order to perform quality installations. For example, fittings for a water distribution system do not have a flow direction, so they can be installed in any position because pressure forces the water through the system. However, codes dictate that some kinds of fittings in a DWV system can only be used in certain flow situations. DWV installations have more regulations concerning fitting selection than other piping systems.

OBJECTIVES

Upon completion of this chapter, the student should be able to:

- identify and describe common types of fittings in a residential plumbing installation.
- understand that certain fitting materials and designs can only be used for specific systems.
- understand that some fitting materials can be used in all residential systems.
- relate fitting design selection to plumbing codes.
- order fittings based on installation requirements, size, and materials.

Glossary of Terms

adapter a fitting designed to adapt one pipe or fitting to another portion of a piping system

bell an enlarged end of a pipe or fitting that receives another pipe end or fitting; may also be called a hub or socket

bend an offset made in tubing on a job site; also, a manufactured offset fitting

bushing a compact reducing fitting that inserts into a pipe, fitting hub, or socket

cleanout a DWV fitting installed to clear obstructions from a drain or sewer

closet bend a specially designed 90-degree fitting used as the last fitting of a drainage system serving a water closet; one side is 4" and the other is 3"

combo abbreviated term meaning a combination of a DWV wye fitting and eighth bend (45°) fitting

coupling a sleeve that connects two equal sized pipe ends to form a continuous pipe

drain a pipe that receives discharge from a fixture(s)

DWV abbreviation for drainage waste and vent system

elbow a fitting used to create an offset; also called a bend

electrolysis a corrosion process caused by directly connecting dissimilar metals

female a fitting with internal threads that screws onto a male fitting

hub an enlarged end of a pipe or fitting that receives another pipe end or fitting; may also be called a bell or socket

joint a fitting connection

male a fitting with external threads that screws into a female fitting

offsets describes all change of direction fittings or the routing of a piping system other than being installed straight

p-trap a DWV fitting having a water seal and is installed at each fixture not having an integral trap to prevent sewer gases from escaping into an occupied area

reducers a fitting used to connect two different pipe sizes together. Reducers are different than busings but accomplishes the goal of reducing a pipe size

sanitary cross a four-way fitting that has limited use in a DWV system. It has the same compact design as a sanitary tee fitting

sanitary tee a three-way fitting that is compact having a sanitary flow pattern and has limited use in a DWV system

socket an enlarged end of a pipe or fitting that receives another pipe end or fitting; may also be called a bell or hub

stack a vertical drain or vent pipe rising more than one story in height

street a type of offset fitting in which one end has the same outside diameter as a connecting pipe or fitting

swing joint a fitting arrangement that creates an offset using two fittings

tees a three-way fitting used to connect a branch pipe with a main portion of a piping system and is ordered based on the size of all three sides

test tee installed in a DWV system to complete a test and can also be installed to serve as a cleanout in a vertical DWV pipe

trap adapter also known as a desanco and is a specialty fitting installed to connect a DWV stub-out pipe to a p-trap

union a three piece fitting installed in a piping system to provide access without cutting the pipe and also used as a termination of the piping system to a piece of equipment

vent a pipe dedicated to providing airflow, so a drainage system can breathe

water closet another name for a toilet

wye a three-way DWV fitting with the branch connection being 45° from the main portion of the fitting

Degree of Fittings

The 360-degree circle and the geometric right triangle provide the basis for describing the degree of **offset** of a fitting. The most common offsets used in the plumbing industry are 90°, one-fourth of a circle; 45°, one-eight of a circle; and 22-1/2°, one-sixteenth of a circle. Manufacturers design fittings to achieve numerous other offsets, but plumbers must at least be familiar with these three. A 22-1/2° fitting is used for **DWV** systems, and 90° and 45° fittings are manufactured for most piping systems. The 45° fittings are not used in flexible tubing systems, because flexible tubing allows offsets to be created without fittings.

Each offset fitting is also called an **elbow**, abbreviated as ell, or a **bend.** The degree of a fitting is usually referred to when ordering fittings; however, DWV cast-iron fittings are ordered by the fraction of a circle the fitting represents. A cast-iron 90° fitting is ordered as a quarter bend, a 45° fitting as an eighth bend, and a 22-1/2° fitting as a sixteenth bend. Figure 4–1 shows why an offset fitting is called an elbow or a bend. Figure 4–2 illustrates how fittings relate to the degrees of a circle.

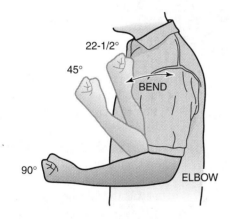

Figure 4–1 An elbow or bend is used to describe an offset fitting.

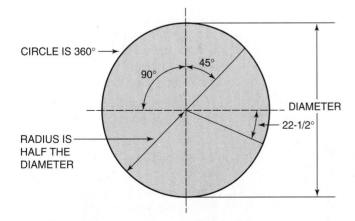

Figure 4–2 Degrees of an offset relate to segments of a circle.

Various Fitting Designs

Before installing any piping system, a plumber must have a basic knowledge of fittings, codes, and job-site conditions, as well as the correct materials to use for the specific system. Water distribution fittings are more compact than DWV fittings, because water flows through the piping with pressure. A DWV system drains by gravity and has unique fitting characteristics that limit the flow restrictions within a piping system. Fittings are classified by their design and are named based on their material type and unique characteristics. Figures 4–3 through 4–12 show various fitting designs. Material types and fittings for specific systems are described throughout this chapter.

Fittings are available with one of two basic connections. Some can either receive pipe (i.e., have a hub) or be inserted into a **hub** or **socket** of a pipe end. Others, such as no-hub, cast-iron (NHCI) fittings, are connected with a specially designed clamp. A **street** fitting has one end that receives pipe and one end with the same kind of connection as the pipe (Fig. 4–3). It is used mostly with offset fittings and is not available in all materials and designs. For example, NHCI is not available with a hub and, therefore, not in a street design.

Even though service-weight, cast-iron (SVCI) fittings are designed like street fittings, they are not called street fittings. Figure 4–3 illustrates socket and street-fitting designs. Flexible tubing does not use fittings with sockets, so it does not use street fittings. Socket depths vary depending on the material used and whether the fitting is serving a pressure or DWV system. Pressure-type plastic piping products that use a solvent weld (glue) **joint** have deeper sockets than do plastic DWV fittings. The smallest plastic DWV pipe and fittings are 1-1/4"; the smallest cast iron are 1-1/2". Plumbers must be familiar with common fitting designs and the sizes available. Abbreviations are common in the plumbing industry, certainly when referring to fitting connections. When ordering fittings with different connecting ends, you must indicate the variations with abbreviations. Many of these abbreviations are included in the relevant sections of this chapter.

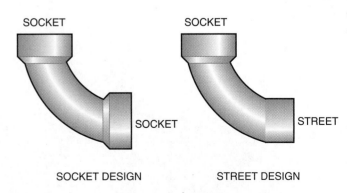

Figure 4–3 Socket fitting and street fitting designs.

FROM EXPERIENCE

The abbreviation St. signifies that you are ordering a street fitting. A brass street 90° is ordered as Brass × St. 90°.

Offsets

The purpose of all offset fittings is to change the direction of piping. Several variations of 90° offsets are designed for specific uses. Figure 4-4 shows a 90° fitting that creates a perpendicular change in direction in a piping system. Pressure systems, such as water and gas, use a short design pattern, and DWVs use a longer-radius design pattern. Figure 4-5 shows a 45° fitting, which is used extensively in most piping systems. Two 45° fittings can be joined to create a 90° offset. Figure 4-6 shows a 22-1/2° fitting, which is common in a DWV system. Two 22-1/2° fittings can be joined to create a 45° offset.

A **swing joint** is a combination of two fittings that create an offset in a piping system. Some difficult piping installations often require a degree of offset that cannot be achieved with the use of certain fittings. A common swing joint configuration uses 90-degree and 45-degree fittings to offset around an obstruction or to create a compact rolling offset.

FROM EXPERIENCE

Many 90° fittings are available as reducing 90°s. One end of the fitting is larger than the other connecting end to reduce pipe size.

Tees

A fitting with three connections used for pressure systems is known as a **tee.** Three basic sizes are available for specific installation requirements and code adherence. DWV systems use a tee design with a different flow pattern than a pressure tee; it is discussed in the DWV section of this chapter. The openings of a tee are identified as either run or branch for sizing and ordering purposes. Run openings are those in the direction of flow through a tee, and a branch is perpendicular to the run (Fig. 4-7).

When ordering a tee, the largest opening is stated first, then the size of the run, and finally the size of the branch. One exception is the bullhead tee, in which the branch is larger than the run. In this case, the branch size is given before the run size and the words *bullhead tee* are included. If all three sides are the same size, only that size is used to order the tee. When both sides of the run are the same size, the single run size and the branch size are used. The same

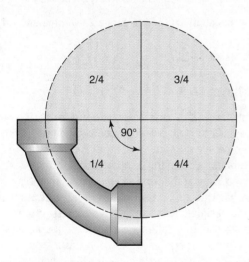

Figure 4-4 A 90° fitting creates a perpendicular offset that is 1/4th of a circle.

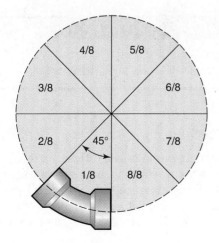

Figure 4-5 A 45° fitting is 1/8th of a circle.

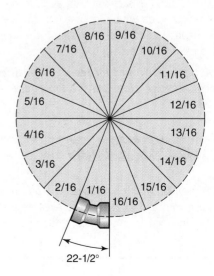

Figure 4-6 A 22-1/2° fitting is 1/16th of a circle.

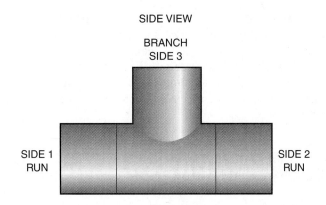

Figure 4-7 Three sides of a tee used for pressure piping systems.

Table 4-1 Common Tee Sizes and How a Plumber Orders Each Tee

Side 1	Side 2	Side 3	Order as
1/2"	1/2"	1/2"	1/2" tee
1/2"	1/2"	3/4"	3/4" × 1/2" bullhead tee
3/4"	3/4"	3/4"	3/4" tee
3/4"	3/4"	1/2"	3/4" × 1/2" tee
3/4"	1/2"	1/2"	3/4" × 1/2" × 1/2" tee
3/4"	1/2"	3/4"	3/4" × 1/2" × 3/4"
3/4"	3/4"	1"	1" × 3/4" bullhead tee
1"	1"	1"	1" tee
1"	1"	3/4"	1" × 3/4" tee
1"	1"	1/2"	1" × 1/2" tee
1"	3/4"	3/4"	1" × 3/4" × 3/4" tee
1"	3/4"	1/2"	1" × 3/4" × 1/2" tee
1"	1/2"	1/2"	1" × 1/2" × 1/2" tee

set of dimensions is used to order a tee, regardless of the material type. Table 4-1 lists common sizes of tees and how a plumber orders each one. Because all three sides of a copper tee have soldered connections, the abbreviation C × C × C is used when ordering.

For step-by-step instructions for tee size identification and insert reducing-tee creation, see the Procedures section on pages 92–93.

 FROM EXPERIENCE

A manufacturer or plumbing wholesale store identifies a tee based on run and branch sizes. Always refer to the largest opening first to ensure that you are ordering the correct tee.

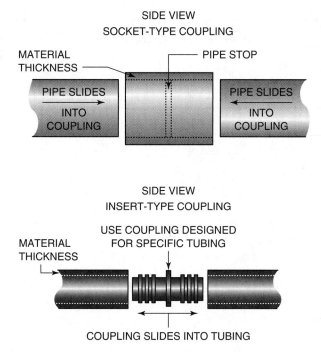

Figure 4-8 Couplings used to connect two pipe ends of equal size.

Coupling

A **coupling**, also known as a sleeve, connects two equal-size pipe ends to form one continuous pipe (Fig. 4-8). Except for SVCI pipe, every piping system uses a coupling to connect two pipe ends. A coupling is not available in a street design. The connecting method for a piping system dictates whether a coupling has threads or a socket, or is inserted into the pipe. A coupling designed for flexible tubing inserts into the tubing. NHCI uses a coupling called a clamp. SVCI double-hub fittings are rarely used because each pipe end has a **bell** to receive a plain pipe end. Some plastic pipe also has a bell on the end of each length of pipe to eliminate the need for a coupling.

 **FROM EXPERIENCE**

The abbreviation for coupling, when ordering, is Coup.

Reducers

A fitting that connects two different-size pipes is a **reducer**. In a reducing coupling, both sides are designed to receive two different-size pipe ends to achieve the same result as a coupling. A fitting reducer is a street fitting. The largest side is the street side, which inserts into a socket of another

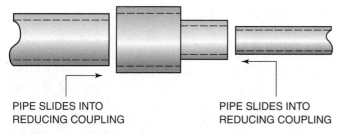

Figure 4–9 Reducing coupling is used to connect two different pipe sizes.

Figure 4–10 Threaded and PVC bushing used to reduce pipe size.

fitting to reduce a pipe size. A fitting reducer is often used when other reducing fittings are not available. Inserting a fitting reducer into a coupling can create a reducing coupling. A reducing tee is created by inserting a fitting reducer into side 2 or side 3 of a tee (see Fig. 4–7). Flexible tubing has reducers that insert into tubing but, because all fittings insert into the tubing, it is not considered a street design.

NHCI uses reducing fittings, but, because it does not use a socket connection, it is not available as a fitting reducer. NHCI reducers are made in long or short versions. Reducing NHCI couplings are also available for pipe reduction. SVCI reducers are designed like fitting reducers, but, because that is the basic connection design of SVCI, they are referred to as reducers and not fitting reducers. Reducers can be used to create fittings that may not be available for purchase. Reducing tees are often created on job sites. Creating a reducing tee increases labor and material costs and often results in a tee that is larger than a manufactured reducing tee. However, if you are sent to a store to purchase a reducing tee and the specific size is not available, a plumber typically expects you to create the desired tee. Fittings with female threads on both ends are referred to as reducers and not reducing couplings. Threaded fittings are not available as fitting reducers. Figure 4–9 shows a reducing coupling and a fitting reducer design.

 FROM EXPERIENCE

When purchasing reducing couplings and fitting reducers in a wholesale store, make sure that you are purchasing the correct fitting. Store employees and customers often misplace items accidentally. Making the wrong selection decreases productivity on a job site. The abbreviation for a reducer is Red., and a fitting reducer is abbreviated as Ftg. Red.

Bushing

A **bushing** connects two different-size pipes. It is similar to a fitting reducer but is more compact. A bushing inserts into a fitting hub and receives a pipe end or street fitting. They are common in threaded and plastic systems, but are not used for NHCI, SVCI, and flexible tubing systems. Plastic piping systems use bushings that solvent weld (glue) into a fitting and some are threaded internally to receive threaded pipe or fittings. Most codes do not allow metal materials to be screwed into plastic materials because expansion and contraction of the piping system can cause the connection to leak or the plastic to break. Many gas codes do not allow bushings, because the bushing can cause the fitting receiving it to crack. Instead, most codes require that the pipe size be reduced with a reducing coupling. If your local code does allow bushings to be installed in a gas piping system, it might dictate that the system remain exposed and not be concealed in a wall or ceiling. Figure 4–10 illustrates a threaded and a DWV polyvinyl chloride (PVC) bushing.

CAUTION: Installing a bushing in a gas piping system in a concealed location can create a potentially explosive situation.

 FROM EXPERIENCE

A PVC bushing for receiving threads from a steel pipe end or fitting should be schedule 80, not schedule 40. Schedule 80 is thicker than schedule 40 and is gray instead of white. The abbreviation for bushing is Bush.

Male Adapter

A fitting with external threads on one end and a socket connection on the other end is known as a **male** adapter. The socket portion is designed to connect to a specific type of material but the threaded portion can connect to any material that is compatible according to National Pipe Thread (NPT) standards. A male **adapter** is manufactured to Iron Pipe Size (IPS) standards. It has the same outside diameter as that of equally sized steel pipe threads, regardless of the material used to produce it. Most fitting manufacturers make a male adapter, and many offer a street male adapter. NHCI, SVCI, and threaded steel systems do not use male adapters. Small copper male adapters are available with a street end. Flexible tubing systems use male adapters that insert the unthreaded portion into the tubing, leaving the threaded portion exposed to screw into the desired location. Plastic piping systems are connected to metal systems with plastic male adapters threaded into metal fittings. A sealing compound called pipe dope or a sealing Teflon tape is applied to the male threads before screwing them into the receiving threads to eliminate leaks. Specialty fittings with male threads are made from many different materials. For example, a brass 90° fitting, with a socket end and a male threaded end can be installed in a copper system. Steel piping systems do not use male adapters. Figure 4-11 shows a male adapter that connects a pipe to a female thread.

FROM EXPERIENCE

The letter *M* signifies a male adapter. The abbreviation used to order a copper male adapter is C × M.

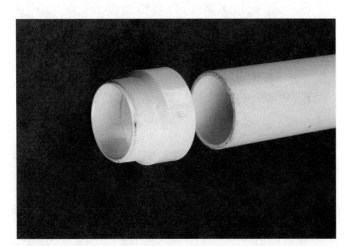

Figure 4-11 A male adapter is used to connect unthreaded piping to female threaded connections.

Female Adapter

A fitting with internal threads on one end and a socket on the other end is a **female** adapter. The threaded portion of a female adapter is the opposite of a male adapter. The socket connects to a specific material, but the threaded portion can receive any type of material that has external tapered threads compatible with NPT standards. Female adapters are manufactured to IPS standards. Most fitting manufacturers offer female adapters, and some make small-diameter adapters as street designs. Threaded steel piping systems do not use female adapters. Female adapters used in NHCI and SVCI systems are called tapped adapters. In flexible tubing systems, the unthreaded portion of a female adapter is inserted into the tubing, leaving the female threads exposed to receive male threads. Plastic female adapters should not be connected to metal male threads because the female adapter can split when the metal piping expands. Pipe dope, which is applied to male threads to seal a threaded connection, should never be applied to the internal threads of a female adapter. Excess joint compound can travel through a piping system, creating an obstruction and blocking the passageways of the devices installed in a system. Specialty fittings with female threads are made from many different materials. For example, a brass 90° fitting with a socket end and a female threaded end can be installed in a copper system. Figure 4-12 shows a female adapter used to connect a pipe to a male thread.

FROM EXPERIENCE

The letter *F* signifies a female adapter. The abbreviation used to order a copper female adapter is C × F.

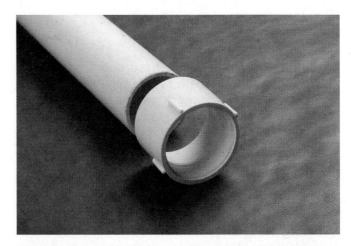

Figure 4-12 A female adapter is used to connect unthreaded piping to male threaded connections.

Figure 4–13 Dielectric union used to connect dissimilar metals.

Unions

A three-piece fitting designed to provide direct access within a piping system is called a **union.** Most codes specify that unions must remain accessible, and most gas codes dictate that a union cannot be installed in a ceiling. Similar and dissimilar materials can be connected in a number of ways. Unions connect inline systems to piping systems. They are used primarily in pressure systems, but can also be used in portions of DWV systems that require sewage pumps. Unions on equipment such as water heaters make replacement easier. On many water heaters, copper tube connects to a steel storage tank, necessitating a connection between two different metals. This connection can result in a process known as **electrolysis,** which causes the metals to corrode. To avoid electrolysis, a five-piece dielectric union is installed to the copper tube and to the steel tank to isolate the connection of the dissimilar metals. A rubber gasket seals the connection to eliminate a water leak, and a plastic composite material isolates the copper tube connection to the union tightening nut. Figure 4–13 shows a dielectric union designed to prevent electrolysis.

The typical three-piece union is not designed to connect dissimilar metals, but to connect pipes of the same material and separate the piping without cutting the pipe. The method of connecting to the pipe varies with the type of pipe. Unions connecting compatible plastic piping materials are solvent welded, but threaded unions can connect to plastic pipe with a male adapter. Steel piping uses threaded unions that connect to the male threads of a pipe end. Figure 4–14 illustrates a PVC union and a threaded union.

 FROM EXPERIENCE

A dielectric union connected to a water heater with a brass nipple is the best installation option. Galvanized nipples are also approved for this use, but they can eventually cause rust to form inside the piping system. Many water heaters are manufactured with a pipe nipple connected directly to the dielectric union.

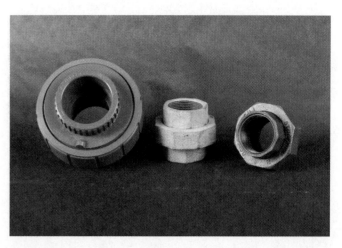

Figure 4–14 PVC and threaded unions are used to separate a piping system.

Water Distribution

Water distribution systems provide safe drinking water to our homes. Water that is safe for human consumption is free from impurities and is known as potable water. No pipe, fitting, or other product installed in a potable water system can consist of more than 8% lead. So far, this chapter has focused on basic fittings and how they are used in various systems. It will now discuss specific systems and the unique fittings used within the systems. Residential water distribution systems can use six types of material. Piping installed outside a residential building can be different from the piping installed on the inside of the building. Some of the materials used outside for distributing water to a house cannot be used for distributing it inside a house. Most codes dictate that any piping material approved only for exterior use must adapt to approved piping no closer than 5′ from the exterior of the house. Table 4–2 lists the six materials that are discussed separately in this chapter as well as the fitting materials used to complete a system.

Table 4–2 Water Distribution Materials

Pipe Type	Fitting Type(s)	Connection Type
PEX	Brass	Crimped/Expanded
CPVC	CPVC	Solvent Welded
Copper	Copper and brass	Soldered
Polyethylene	Nylon, brass, and galvanized	Clamped
Galvanized	Galvanized	Threaded
Brass	Brass	Threaded

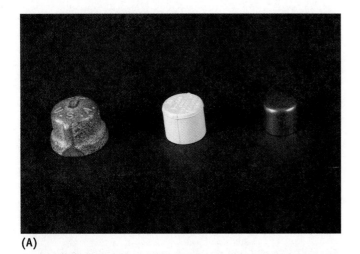

Figure 4–15 Water distribution fittings (A) Galvanized Cap, PVC Pressure Cap, Copper Cap (B) Copper × Male 90°, Copper × Female Drop Ear 90°, Copper × Copper Drop Ear 90° (C) Threaded Galvanized Plug, Brass Threaded Street 90°, Brass Threaded Street 45°.

Water Distribution Fittings

Figure 4–15 shows various fitting designs used specifically in water distribution systems. Many other fittings are also used, but the ones shown are the most common. Specialty fittings are manufactured for a variety of different connections. The use of a brass fitting with a soldered end is determined by the connection rather than the material used. Any connection that is to be soldered is ordered as copper and abbreviated as C. A drop-ear 90° is available with both connections to be soldered and is ordered as a C × C DE 90°. A drop-ear 90° with one connection to be soldered and the other with female threads is ordered as a C × F DE 90°.

 FROM EXPERIENCE

A company's installation practices and preferences might dictate the materials and the number of fittings used to create various offsets. The most important considerations in choosing fittings are whether the materials are approved for use in a potable water system and whether they are compatible with the type of piping installed.

PEX

Cross-linked polyethylene (PEX) has become one of the most popular materials for water distribution systems in residential construction. Fittings and accessories used with PEX are typically brass, but a leading manufacturer does offer them in a new plastic material known as Polysulfone (PLS). The fittings have a unique ribbed pattern on the exterior for holding once they are installed. Depending on the type of tubing being installed, it can be connected either by crimping or by expanding.

In the crimping method, a fitting is inserted into the tubing, secured with a crimp ring, and crimped with a crimping tool. With the expanding method, a PEX ring slides over the tubing, an expander tool is inserted into the tubing, and the tube is expanded. The fitting is selected based on installation requirements, such as an offset in a pipe route or a connection to another type of pipe. Figure 4–16 illustrates the unique features of a PEX fitting. PEX adapts to other approved materials with adapters or connectors designed for that purpose—usually male or female adapter fittings (see Figure 4–17). PEX is easily adapted to copper with a fitting that has compatible design features for both materials. Several variations of connectors are available. Some slide over the copper; others slide into a copper fitting, which allows a more compact connection and eliminates the copper pipe be-

Figure 4–16 PEX fittings are inserted into PEX tubing and crimped in place or expanded.

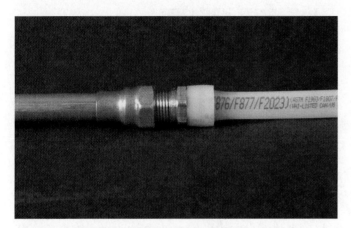

Figure 4–17 PEX can connect to copper using a variety of adapters.

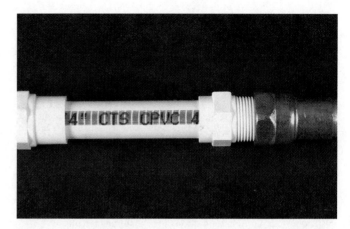

Figure 4–18 CPVC connects to copper with male and female adapters.

tween a copper fitting and a PEX-to-copper adapter. A PEX × copper adapter that slides over the copper tube is ordered as PEX × C.

 FROM EXPERIENCE

One advantage of a PEX installation is that the system can be pressurized immediately; it does not require a curing time like a solvent-welded (glued) system does.

CPVC

Chlorinated polyvinyl chloride (CPVC) is another popular choice for residential water distribution systems. Because both CPVC and PEX are approved by most plumbing codes for water distribution systems, the choice of one over the other is typically a matter of preference. CPVC fittings are manufactured from the same materials as the piping, and connections are solvent-welded (glued) with special glue. CPVC glue cannot be used with other types of plastics. After an installation is complete, a solvent-welded joint must cure for about 24 hours before the system can be pressurized. This curing time is usually not a problem unless it is too hot or too cold. In any event, always follow manufacturer instructions. Repairing CPVC connections does pose a problem for homeowners because they are without water while the joint cures. CPVC is a flexible pipe, but it is not as flexible as PEX, and is usually called pipe and not tubing. Offset fittings are required to create 45° transitions in CPVC pipe routes; offsets cannot be created by bending as with PEX tubing. CPVC is usually connected to other materials with male and female adapters. CPVC male and female adapters are manufactured with brass threads, and some have CPVC male and female threads. Figure 4–18 shows a CPVC male adapter that connects to other piping systems such as copper.

 FROM EXPERIENCE

Installing male and female adapters with brass threads ensures a higher-quality installation.

Copper

Three different grades of copper tube are used for water distribution systems, but the fittings for all three are the same. Copper fittings for water distribution are different from DWV copper fittings. Copper DWV fittings have a greater radius and a shallow socket because a DWV system

Figure 4-19 Copper tube slides into a copper fitting.

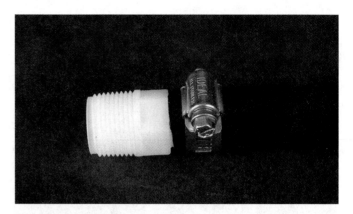

Figure 4-20 Polyethylene fitting is secured to tubing with a stainless steel hose clamp.

drains by gravity rather than pressure like water distribution systems. Copper fittings are joined to copper tube by soldering or brazing. To solder, the tube ends are sanded clean, and the fittings are cleaned with a wire brush. The tube end and fitting sockets are thinly coated with a specially formulated paste, called flux, and assembled. Heat is then applied to the fitting, which melts the flux, and a lead-free filler material, known as solder, is used to weld the fitting to the tube. The soldering process increases labor costs on a job site, which is one of the main reasons that PEX and CPVC have become popular for residential construction. Many fittings used in copper systems are made of brass but are soldered like copper fittings. Figure 4-19 illustrates how a copper fitting connects to a copper tube.

 FROM EXPERIENCE

Copper is still used in many regions of the country for residential installations, and, although PEX or CPVC are the materials of choice, copper is still popular for the final connections to water heaters.

Polyethylene

Polyethylene (PE) tubing uses nylon and galvanized or brass fittings. PE fittings have a unique ribbed design, which is only compatible with PE tubing. The hose clamp is placed over the PE tubing, the fitting is inserted into the tubing, and the clamp is placed near the pipe end and tightened with a screwdriver or nut driver. PE is only approved for use outside, and most codes dictate that it must not be installed less than 5' from a building. The connection from the PE to a material approved for interior use is usually made with a male adapter. Several types of PE tubing are manufactured for water service and irrigation installations, but they all use the same fittings. Figure 4-20 shows a PE fitting connecting with PE tubing.

Galvanized and Brass

Galvanized pipe is no longer used for entire water distribution systems, but fittings and nipples are still installed. Rust is produced when galvanized material is exposed to certain water qualities, which makes it a poor choice for water distribution systems. Therefore, brass materials are more widely used. However, galvanized is often used to cap or plug pipe ends and to test piping systems because it is less expensive than brass. Both materials are approved for use throughout water distribution systems, but brass fittings are used where poor water quality is present and galvanized materials will corrode. Threaded galvanized and brass fittings are manufactured to IPS and NPT standards and are compatible with threaded adapters made from other materials. Galvanized materials cannot legally connect to copper unless a dielectric union is used to avoid electrolysis. Brass connections of many kinds are common in the repair industry. Brass fittings are chrome plated and installed in finished locations. They are also used with copper, PEX, and PE and are designed to be compatible with each pipe or tubing. The residential plumbing industry is competitive, and, although brass may be too expensive for new construction projects, it does provide a higher-quality installation. Figure 4-21 illus-

Figure 4-21 Galvanized fitting is commonly used to test water distribution systems.

fittings are the same for all pressure piping systems. Figure 4-22 illustrates the proper PVC-to-metal connection using male and female adapters.

Drainage, Waste, and Vent

Drainage, waste, and vent (DWV) represents the combined drainage and vent systems. Codes vary concerning the installation of drainage systems and vent systems, but the same materials can be used for both. Fittings in DWV systems are designed to allow wastewater and sewage to flow out of a drainage system with little resistance using only gravity. Horizontal drainage must be sloped downward and away from a fixture, and a vent must slope back to a drain to allow moisture to drain into the drainage system. DWV systems have more fittings with specific uses than any other piping system. Knowledge of DWV codes that dictate the size and positioning of specific fittings is extremely important when selecting and installing DWV fittings. Table 4-3 lists the materials in a DWV system. Each common fitting and material type is explained separately in the following portion of this chapter, and the common offset fittings are explained at the beginning of this chapter.

Figure 4-23 illustrates the three flow directions that determine the proper use of drainage fittings. Table 4-4 lists common fittings installed in DWV systems and their approved flow positions based on plumbing codes. The radius of a fitting dictates its flow position. A steady horizontal-to-horizontal flow of wastewater and sewage in a DWV piping system is interrupted as it changes course; therefore, it requires a long-radius-pattern fitting. A horizontal-to-vertical flow is not restricted so a standard-radius DWV fitting can be used. A vertical-to-horizontal flow can be restricted so it requires a long-radius fitting. The 90° fittings are available in various styles based on their radius.

Plastic DWV fittings come in three basic 90° designs—vent, standard, and long radius. A long-radius 90° is also known as a long-sweep 90°, and a vent 90° might be called a short-sweep 90°. Cast-iron fittings are available in standard, short-sweep, and long-sweep 90° designs. A short-sweep, cast-iron 90° does not have the same radius as a

Figure 4-22 PVC male adapter is connected to a metal female adapter.

trates one type of galvanized fitting used in testing a water distribution system.

PVC

PVC can only be used for the exterior portion of a residential water distribution system. Pressure rating and fitting designs for PVC are different from those for DWV. The flow radius of fittings for PVC are more compact, and the fitting socket is deeper so the solvent-welded joint can be exposed to higher pressure. CPVC male and female adapters are available with both brass and plastic threads, but PVC adapters are only made with plastic threads. A PVC female adapter should not be screwed onto a metal male adapter. The expansion and contraction of piping systems could cause the plastic female adapter to crack and leak. Metal-to-plastic connections should only be performed with plastic male adapters and metal female adapters. Pressure PVC in residential applications is usually 3/4". It often provides water service from a meter to 5' from the exterior of a house, and codes dictate that 3/4" is the minimum size that can be used to serve a single-family home. Pressure fittings are also used in some piping installations from the discharge of a sewage pump or groundwater (sump) pump. The largest pressure fitting typically used by a residential plumber is 2" in diameter. Irrigation systems use a thin-wall PVC pipe, but the

Table 4-3 DWV Fitting Materials

Material	Abbreviation	Residential Use
DWV Polyvinyl Chloride	DWV PVC	Extensively throughout USA
DWV Acrylonitrile Butadiene Styrene	DWV ABS	Some states in USA
No-hub Cast Iron	NHCI	Vertical stacks for noise control
Service-Weight Cast Iron	SVCI	Rarely
Galvanized	Galv.	Rarely
DWV Copper	Type DWV	Rarely

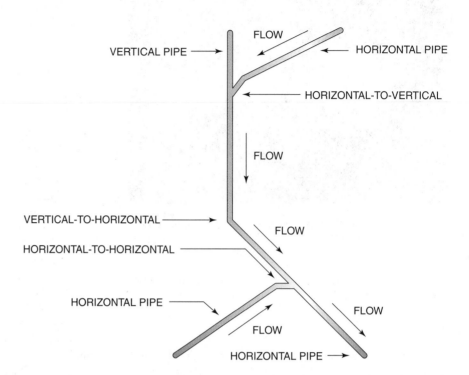

Figure 4–23 Flow directions determine the correct fitting positions required by code.

Table 4–4 DWV Fittings' Approved Flow Position[6]

Fitting	H-H[1]	H-V[2]	V-H[3]
Standard-radius 90°	No[4]	Yes	No[5]
Long-radius 90°	Yes	Yes	Yes
Wye	Yes	Yes	Yes
45°	Yes	Yes	Yes
Sanitary Tee	No	Yes	No
Combo Wye & 1/8	Yes	Yes	Yes
22-1/2°	Yes	Yes	Yes

[1]Horizontal-to-horizontal flow transition
[2]Horizontal-to-vertical flow transition
[3]Vertical-to-horizontal flow transition
[4]Can be used to turn out of a wall as the last fitting serving a fixture
[5]Can be used as the last fitting serving a toilet
[6]Check your local code for exceptions to this list

plastic DWV short-sweep (vent) 90°, indicating that different materials have different radiuses and different identifying names. The cast-iron, short-sweep 90° has a longer radius than a standard-radius, cast-iron 90° and a shorter radius than a cast-iron, long-sweep 90°. A cast-iron, short-sweep can be installed in any flow position according to most codes, but some codes dictate that a long-sweep 90° fitting be used for all positions except horizontal to vertical.

Fittings with 90° transitions by design, but having more than two connections with various radiuses are explained later in this chapter. There are also exceptions to some flow-pattern codes. Because walls and ceilings in residential buildings have limited space, many codes allow a standard-radius 90° fitting to be used as the last fitting to connect to a fixture even when the fitting is in the horizontal-to-horizontal position.

 FROM EXPERIENCE

A DWV system is the most difficult to understand but one that must be thoroughly understood to become a licensed plumber. DWV fitting installation requirements are strictly regulated by codes, and knowledge of all codes is required before selecting and installing any DWV fitting.

Wye

A **wye** has three connections and is named for its similarity to the letter Y. The side outlet connection, known as a branch, is at a 45° angle to the two other inline connections, which are known as the run. The 45° branch creates a direction of flow that eliminates the disturbance of wastewater within a DWV piping system. A wye can be installed with the branch and run in the horizontal and vertical positions for drainage applications. It can also be inverted, placing the run in the vertical position and the branch facing downward, to receive vertical vent piping.

A 45° fitting can be joined with the 45° branch to make a 90° branch, thereby creating the same flow pattern as a

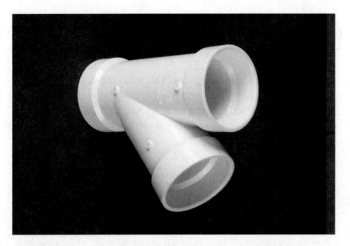

Figure 4–24 A wye fitting is used for drainage and vent systems.

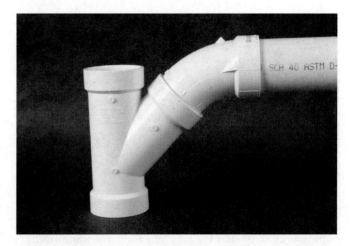

Figure 4–25 A combo is used in all flow positions of a drainage system.

combo fitting (Fig. 4–24). A double-wye fitting has two branches, both on a 45° angle from the run, and on opposite sides of the fitting. Some codes may not allow a double wye to be installed with all four connections in the horizontal position to ensure that one side (branch) is not sloped away from the run. A double wye is often installed horizontally under a double-bowl kitchen sink to connect the two bowls or in the inverted (upside-down) position in a vent system. Figure 4–24 shows a wye fitting that is available in all DWV materials.

 FROM EXPERIENCE

A wye is one of the most versatile fittings in a DWV installation. The 45° branch is recognized by code as the imaginary line between horizontal and vertical flows. Vertical is any position from true vertical to 45° from true vertical. Horizontal is considered to be anywhere from true horizontal to 45° from true horizontal.

Combo

Combo is the trade name for a three-sided fitting that creates a long-radius 90° branch (side inlet) that is perpendicular to the run. This three-connection fitting eliminates the need to create a single fitting by combining a wye and 45°. A cast-iron 45° fitting is also known as a 1/8th bend, and a combo is commonly known as a combo wye and 1/8th bend. A combo is used extensively in the drainage system with the branch and run in the vertical or horizontal positions. The long-radius flow pattern of the branch directs the wastewater and sewage though a drainage system with little resistance. It can be used in the venting system, but a more compact fitting known as a **sanitary tee** is also acceptable and is less expensive than a combo. Figure 4–25 shows a combo fitting, which is available in all DWV materials. A double-combo fitting is also available with two branches, each at a 90° angle to the run, on opposite sides of the fitting. Some codes may not allow a double combo to be installed with all four connections in the horizontal position to ensure that one side is not sloped away from the run. If allowed by code, a double combo can connect two back-to-back fixtures.

 FROM EXPERIENCE

A combo might also be called a long TY (tee wye).

Sanitary Tee

A sanitary tee is a compact fitting with three connections that is used in drainage and vent systems. The side connection, known as the branch, creates a standard radius that is at a 90° angle from the run. The word *sanitary* indicates that it is used for the DWV system. It is similar to a tee in a pressure system, but directs the flow of wastewater and sewage through a drainage system. Because the flow pattern is compact, the location of a sanitary tee in a DWV system is limited by code. Many codes state that a sanitary tee cannot be installed on its back, which means that it can only be used in a flow position of horizontal-to-vertical. A venting system does not handle wastewater, so most codes allow sanitary tees to be used in any position. However, many code officials do not allow sanitary tees for venting applications in any position that is not allowable by drainage codes. A cast-iron sanitary tee, called a tapped tee, has a female threaded branch. Figure 4–26 shows a sanitary tee used in a DWV system.

Figure 4–26 A sanitary tee is installed with the wastewater flow horizontal to vertical.

 FROM EXPERIENCE

Many plumbers refer to a sanitary tee as a short TY (tee wye). A plastic DWV sanitary tee has the same radius as a standard-radius plastic DWV 90°.

Sanitary Cross

A **sanitary cross** has four connections, two of which are branches, and the same flow pattern as a sanitary tee. It is used to connect fixtures that are located side-by-side and back-to-back. Most codes limit the kinds of fixtures that can connect into the two branches. In addition, a sanitary cross is regulated like a sanitary tee. Sanitary crosses are available in all the same materials as those used for DWV installations. A cast-iron threaded version of a sanitary cross, known as a tapped cross, has female threads on both branches. Various sizes are available. In some, all four sides are sized equally; in others, the branches are smaller than the run. Figure 4–27 shows a sanitary cross in a DWV system.

 FROM EXPERIENCE

Some codes dictate that a sanitary cross cannot be used back to back or side by side to serve toilets or kitchen sinks with garbage disposal units. Codes that do allow a toilet to be connected to a sanitary cross dictate that the branch can only be connected to another toilet.

Twin Elbow

A twin elbow has the same flow pattern as a sanitary tee, sanitary cross, and standard 90° elbow. A cast-iron twin elbow is ordered as a double 1/4 bend. It is called a twin elbow because two 90° fittings are made to connect back-to-back fixtures in a confined space. This three-connection fitting is limited to connecting low-volume fixtures, such as sinks, that share a common drain pipe. Some codes do not allow this fitting to be used in the horizontal-to-horizontal flow position. Even though the flow pattern of a twin elbow is the same as a sanitary tee, it can be installed with all three sides in the horizontal position, if code allows. A twin elbow can only be used as the last fitting in a drainage system serving approved fixtures. Figure 4–28 is a photo of a twin elbow used for a DWV installation.

 FROM EXPERIENCE

A twin elbow is used to connect two fixtures to the same horizontal drain. It cannot connect two fixtures to a vertical drain.

Figure 4–27 A sanitary cross is used to connect back-to-back and side-by-side fixtures.

Figure 4–28 A twin elbow is used to connect back-to-back approved low volume fixtures.

CHAPTER 4 Fittings 87

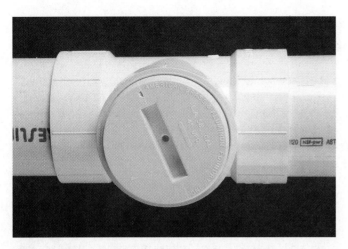

Figure 4–29 A test tee is used for testing a DWV system and as a cleanout at the base of a stack.

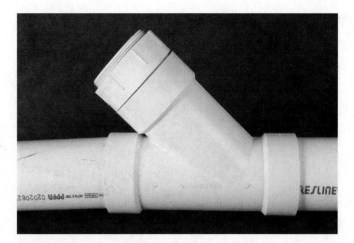

Figure 4–30 A cleanout is installed in a DWV system to provide access to clean a drain.

Test Tee

A **test tee** cannot receive discharge from a **drain**. It is used in a drainage system to provide access for cleaning and is often referred to as a cleanout tee. It also serves as a place to install testing accessories for the required testing and inspection of a DWV system. A test tee has three connections, one with female threads, and with the side connection being 90° from the flow. The sole purpose of the threaded connection is to receive a cleanout plug that is tightened in place when a DWV system is complete. Codes dictate that a cleanout be installed at the base of every vertical stack of a DWV system; a test tee is often used to satisfy that requirement. Test tees are available in all the materials that are used for DWV installations. Figure 4–29 shows a test tee used in a DWV system.

 FROM EXPERIENCE

The threaded portion of a test tee cannot be used to receive flow from a drain, because it does not have the same direction of flow as a sanitary tee.

Cleanout

Cleanouts are installed throughout DWV systems to provide access into the piping to clear obstructions. A female adapter and a threaded plug, often called a cleanout cover, are used together to create a single fitting. Codes dictate the location and size of cleanouts. Plastic DWV cleanouts install over piping, and a street-style version is inserted into a fitting hub. NHCI cleanouts have the same outside diameter as the connecting pipe. Cast-iron cleanouts are called tapped ferrules by manufacturers, but plumbers order them as cleanouts. The threaded cover, available in two distinct styles, screws into the female threads of a cleanout. The most common cover has a raised center portion, often called a lug, to allow removal with a wrench. The other has a countersunk lug so it can be installed flush with a finished surface. Covers for plastic materials are made of the same material as the cleanout, and cast-iron cleanouts have brass covers. Figure 4–30 shows a cleanout in a DWV system.

 FROM EXPERIENCE

Proper sizing of a cleanout is important. Codes dictate the minimum size, based on the size of the drain served by a cleanout.

Closet Bend

The **closet bend,** a reducing 90° fitting, is the last fitting of a drainage system serving a toilet (also known as a **water closet**). Codes dictate that downstream pipe and fittings must be the same size or larger than upstream pipe and fittings. A code exception allows the installation of a closet-bend fitting, which is a standard-radius fitting, with the flow from vertical-to-horizontal. The 4" side of the closet bend can only be installed vertically, with the 3" piping installed horizontally. A plastic street version is available with a 4" street side and a 3" side with a hub. Figure 4–31 shows a closet bend installed in a DWV system.

 FROM EXPERIENCE

This fitting is only offered in a 4" and 3" size, because the minimum size of a pipe serving a toilet is 3", and the maximum size of a flange connecting a toilet to a drainage system is 4".

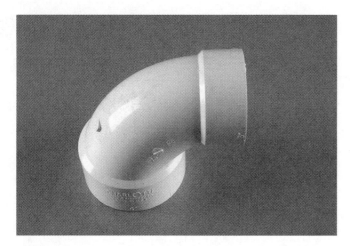

Figure 4-31 A closet bend fitting is used as the last fitting serving a toilet.

Heel Inlet 90°

A heel inlet 90° is a specialty DWV fitting that is often referred to as a heel outlet fitting by a plumber. It is a DWV 90° with a branch located at the heel of the bend that can only be used in specific locations. The standard-radius fitting is only made in 4" × 2", 3" × 2", and 3" × 1-1/2" sizes in plastic, and in 3" × 2" in NHCI. A long-sweep version is available in plastic, but only in 3" × 2" and 3" × 1-1/2" sizes.

The standard-sweep version is the most common. Most codes allow it to be installed in two positions, but some only allow the heel portion to be installed vertically. If the heel inlet can be in the horizontal position, it must serve as a fixture drain and not a dry vent to ensure that the heel inlet connection is cleaned by the flow of water. Figure 4-32 illustrates a heel outlet 90° but does not indicate a flow position. All other codes pertaining to flow positions of 90° fittings apply. Some codes allow this unique fitting to be installed in any position in a vent system. However, many code officials only allow the positions dictated by drainage codes to apply.

FROM EXPERIENCE

Less commonly used 90° fittings have a side inlet and a high heel inlet. They have the same limited installation uses as standard 90°s, so codes must be carefully reviewed before using them.

Closet Flange

The flange that connects the toilet (water closet) to the drainage system is called a closet flange. Closet flanges are often made from PVC, ABS, or cast iron. Plastic drainage materials have compatible flanges that solvent-weld (glue) to the piping system. Cast-iron flanges, which are available in a no-hub version and a type that is placed over the connecting cast-iron pipe, are not typically used in residential construction. Plastic flanges can be installed over the connecting pipe or inserted into a fitting hub. Some can also be inserted into the pipe, but that violates some codes because it reduces the inside diameter of the piping system.

Another closet flange that is illegal according to some codes is an offset flange. If a drain pipe is installed too close to a wall or another fixture, an offset flange can be used to avoid correcting the pipe location; however, it only allows for a 2" offset. All flanges have anchoring holes that secure them to the floor to eliminate movement of the toilet. Closet flanges also have slotted openings with a set of securing bolts (closet bolts) that secure the toilet to the flange. Most flanges have slots that allow adjustment of the closet bolts and also fixed slot locations that do not allow adjustment of the closet bolts. A wax seal prevents harmful sewer gases from entering occupied areas and seals the connection between the toilet and the closet flange. Figure 4-33 includes a plan view and a side view of a closet flange. The side view shows that the flange has depth. Plastic materials are man-

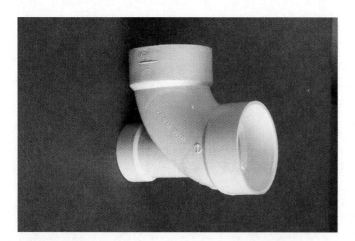

Figure 4-32 A heel inlet 90° has specific and limited use in a DWV system.

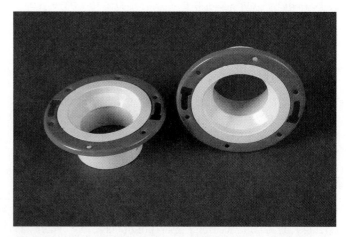

Figure 4-33 A closet flange is used to install a toilet to the drainage system.

ufactured to a standard depth, but cast-iron flanges are made in varying depths.

FROM EXPERIENCE

A closet flange is ordered by the size of the connecting pipe and a standard 4" flange opening. A closet flange for a 3" pipe connection is ordered as a 4" × 3" closet flange, and a 4" pipe connection is ordered as a 4" × 4" closet flange.

P-trap

Every fixture must be served with a protective water seal to prevent harmful sewer gas from entering an occupied space. A **p-trap** gets its name from its similarity to the letter *P*. It receives the outlet flow of water from all residential fixtures except a toilet, which has an integral trap. P-traps are available in a variety of styles, including one-piece and two-piece designs. Cast-iron p-traps are one piece and must be installed directly below a fixture. Plastic p-traps are available in a two-piece design known as a slip-joint, which allows a swivel adjustment to accommodate slight installation variances. In slip-joint p-traps, slip-joint nuts, and washers are tightened to prevent leaks but can be removed for replacement and cleaning. Slip-joints are located under sinks or other accessible fixtures. Other two-piece plastic p-traps can be adjusted but are solvent-welded. They can be concealed in walls, floors, and ceilings to serve such fixtures as bathtubs, showers, washing machines, and floor drains. Another two-piece design often used under a kitchen sink has a slip-joint nut and two solvent-welded hubs. They often have cleanout plugs in the bottom of the trap to allow cleaning. Figure 4-34 shows a two-piece p-trap that would not require access.

FROM EXPERIENCE

An approved p-trap has a 2" minimum and a 4" maximum water seal, but a deep-seal p-trap is available when dictated by code or for unique installation locations.

Trap Adapter

A slip-joint p-trap is tubular and has a smaller outside diameter at the connecting DWV pipe. The fitting that connects tubular sizes to DWV pipe sizes is a **trap adapter**, or desanco. A trap adapter resembles a male adapter, but with a slip-joint nut and washer. It is typically installed at the fixture installation phase of a project. It connects to plastic DWV piping systems with a solvent-welding process (glue) and must be made of the same material as the plastic DWV pipe. Plastic trap adapters have connections that install over the DWV pipe; a street version inserts into a plastic DWV fitting hub. Cast-iron, copper, and galvanized piping systems use brass trap adapters. Copper systems use a brass trap adapter that either is soldered over the copper tube or threaded into a female adapter or over a male adapter. Galvanized piping systems use the same threaded, brass trap adapters as copper. All trap adapters connect with the tubular P-trap, which is a compression-type connection. A slip-joint nut and washer are placed over the tubular p-trap, and the tubular portion of the trap is inserted into the adapter. The nut and washer are tightened onto the male threads to make a watertight connection. The most common tubular sizes are 1-1/4" and 1-1/2". Figure 4-35 shows a typical trap adapter used in the residential plumbing industry.

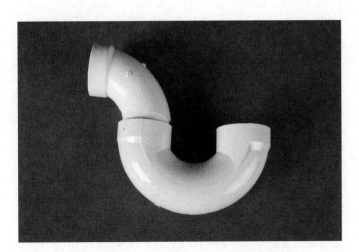

Figure 4-34 A P-trap is installed so sewer gas cannot enter an occupied space.

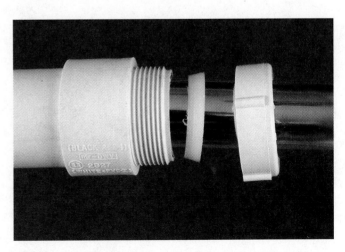

Figure 4-35 A trap adapter connects a tubular P-trap to a DWV piping system.

FROM EXPERIENCE

A plastic trap adapter can be used as a DWV male adapter, but a DWV male adapter cannot be used as a trap adapter. The tubular portion of a trap does not insert into most plastic DWV male adapters, but the threads of a trap adapter are manufactured to NPT and IPS standards.

PVC

Polyvinyl chloride (PVC) is one of the most widely used products for DWV installation in the residential plumbing industry. PVC fittings are manufactured from the same material as the pipe and are sealed with a solvent-welding process (glue). A primer is applied to the pipe and fitting surfaces before the glue is applied. PVC is available in pressure and DWV styles, and each is specifically designed for a different application. All DWV fittings are identified on the side as DWV PVC designs. For potable water installations, the products are identified with a National Sanitation Foundation (NSF) mark of approval. A pressure fitting does not have flow direction characteristics, or the radius required for DWV applications for the adequate flow of wastewater and sewage. The hub of a DWV is not as deep as a pressure fitting because it is designed to flow by gravity and not pressure. Figure 4–36 includes a PVC DWV fitting and a PVC pressure fitting.

FROM EXPERIENCE

Glue used to weld PVC pipe and fittings is specifically designed for PVC and cannot be used with other plastic pipe and fitting materials.

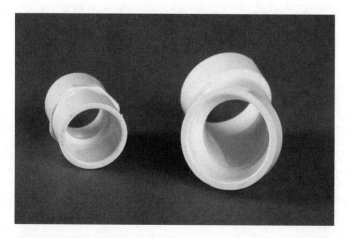

Figure 4–36 PVC DWV fittings and PVC pressure fittings have unique characteristics.

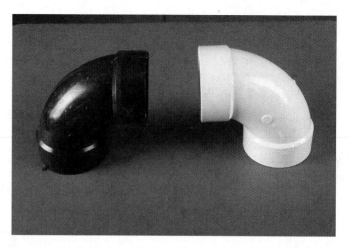

Figure 4–37 ABS fitting has the same flow pattern as PVC, but is black instead of white.

ABS

Acrylonitrile butadiene styrene pipe (ABS) is more common in the western regions of the country. ABS is black and easily distinguished from PVC, which is white. ABS fitting designs and code regulations are the same as those for PVC. The solvent-welding process is also the same as for PVC but uses glue specifically designed for ABS. ABS fittings are designed to be installed only with ABS pipe, but an ABS trap adapter can connect to a PVC slip-joint p-trap. Connections between ABS and other materials must be made with approved adapters, such as male and female adapters. All-purpose glue is available to join PVC and ABS, but it is illegal according to most codes. Figure 4–37 is a photograph of an ABS fitting installed in a DWV system.

FROM EXPERIENCE

ABS is approved by all codes for DWV installations, but its popularity varies by region or state. Most mobile home manufacturers use it extensively for DWV systems.

Cast Iron

Two common types of cast iron (CI) are no-hub (NH) and service weight (SV). Cast-iron pipe and fittings are coated with an asphalt solution to help seal their porous surfaces, which result from the casting process. NHCI, the most common of the two types, is used in residential installations to provide quieter wastewater flow through vertical **stacks** in walls. NHCI products are so named because they do not have hubs to receive pipe ends. Instead, they are connected to pipe with a specially designed clamp. An SVCI fitting has a hub, or bell, to receive a pipe end. It is primarily used for underground installations. Two approved ways to seal an SVCI

Figure 4-38 No-hub, cast-iron fittings are installed in a DWV system.

joint are with an elastomeric (rubber) gasket and, less commonly with lead and oakum.

CI fittings are more expensive than plastic products and often take longer to install. Cast iron is used more frequently in the commercial plumbing industry, but a residential plumber must have knowledge of and basic installation skills concerning cast iron. Cast-iron DWV fittings must comply with all the same code regulations as other DWV material types. Because cast-iron fittings are used more often for commercial installations than for residential applications, many unique fitting designs are made in cast iron that are not made in plastic. Figure 4-38 shows an NHCI fitting used for DWV installations.

 FROM EXPERIENCE

A heavy-weight, cast-iron product is available with all fitting designs and connections the same as SVCI pipe and fittings.

Summary

- Most fittings are manufactured from the same material as the connecting pipe.
- Brass fittings are interchangeable with numerous piping materials.
- Threaded fittings can be used to connect different material.
- Plumbing codes dictate that solvent-welded fittings can only be used with compatible pipe.
- PEX fittings are specifically designed for the type of PEX connections used.
- Cast-iron fittings are specifically designed for either NH or SV pipe.
- Dissimilar metal connections can cause corrosion and must be protected against electrolysis.
- DWV fittings have a flow pattern, and plumbing codes dictate their use.
- DWV fittings are not designed for use with pressure piping systems.
- Water piping fittings have more compact designs than DWV fittings.
- Black steel fittings cannot be installed in a potable water system.
- A male adapter has external threads and a female adapter has internal threads.
- A street fitting has one end that is the same outside diameter as the connecting pipe.
- A fitting socket is also known as the hub.

Procedures

Tee Size Identification Procedure

A Always use the largest side of the tee when ordering. If all sides are the same, use only one side to order.

A

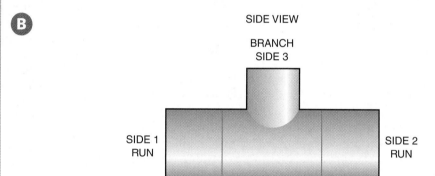

B If side 1 and side 2 are the same, order using only the two different sides.

B

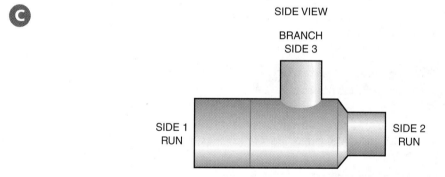

C If side 1 is larger than side 2 and side 3, and even if side 2 and side 3 are equal, you must state all three sides.

C

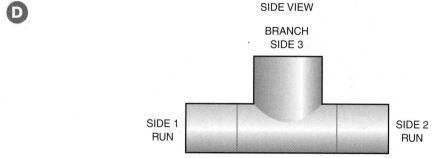

D If side 3 is larger than side 1 and side 2, this is called a bullhead tee. On bullhead tees, side 1 and side 2 are always equal. You must state the largest side, which is side 3, and then the word *bullhead*.

D

Procedures

Insert Reducing-Tee Creation Procedure

- If a reducing tee is not available, one can be created by inserting fitting reducers into a tee.
- Select the tee that is closest in size to the tee you desire.
- Select the appropriate fitting reducer(s).
- **A** Insert the fitting reducer(s) into the socket(s) of the tee to create the desired reducing tee.

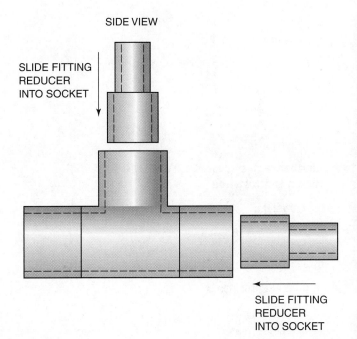

Review Questions

1. Solder and flux used for potable water systems cannot contain more than
   a. 2.0% lead
   b. 0.2% lead
   c. .002% lead
   d. 20% lead

2. A 2″ × 1-1/2″ tee describes a fitting having
   a. Two connections
   b. Three connections
   c. Four connections
   d. One connection

3. Another term to describe a DWV cast-iron 45° fitting is a cast-iron
   a. 1/16 bend
   b. 1/8 bend
   c. 1/4 bend
   d. 1/6 bend

4. A male adapter has
   a. Internal threads
   b. External threads
   c. A slip-joint nut and washer
   d. A threaded cover

5. Without considering any code exceptions, horizontal-to-horizontal flow of a DWV system requires a
   a. Standard-radius-pattern fitting
   b. Short-radius-pattern street fitting
   c. Long-radius-pattern fitting
   d. Street-pattern fitting

6. A desanco is another name for a
   a. Trap adapter
   b. Male adapter
   c. Female adapter
   d. Bushing

7. Every fixture not having an integral trap must be protected by a(n)
   a. S-trap
   b. D-trap
   c. P-trap
   d. Two-way cleanout

8. To avoid the electrolysis process between copper and a dissimilar metal, a
   a. Dielectric union is installed
   b. Grounding rod is installed
   c. Galvanized union is installed
   d. Reducing coupling is installed

9. A closet flange is used to secure a DWV pipe and a
   a. Toilet
   b. Sink
   c. Tub
   d. Bidet

10. To complete a PEX × PEX connection, the fitting is inserted into the PEX tubing and
    a. Soldered
    b. Threaded
    c. Crimped
    d. Solvent-welded

11. Type M, L, and K copper tube use
    a. Different fittings
    b. The same fittings
    c. Some of the same fittings
    d. Plastic fittings

12. A test tee is used to test a DWV system and as a
    a. Sink drain
    b. Tapped sanitary tee
    c. Cleanout
    d. Vent

13. **A sanitary tee can only be used in a drainage system when the flow is from**
    a. Vertical-to-horizontal
    b. Horizontal-to-horizontal
    c. A sink
    d. None of the above is correct

14. **The two names that identify parts of a tee are run and**
    a. Outlet
    b. Branch
    c. 45°
    d. Vent

15. **A 2" × 2" × 1/2" tee is ordered as a**
    a. 2" × 1/2" × 2" tee
    b. 1/2" × 2" × 2" tee
    c. 2" × 1/2" tee
    d. 2" tee

16. **The trade name *combo* refers to the two fittings,**
    a. Wye and 1/4 bend
    b. Wye and 1/16th bend
    c. Wye and 1/8th bend
    d. Wye and 1/6th bend

17. **A threaded fitting that receives a pipe end, but also inserts into another fitting to reduce a pipe size is called a**
    a. Reducing coupling
    b. Fitting reducer
    c. Bushing
    d. Coupling

18. **A fitting that has a hub or socket on one end and the other end the same size as the pipe is called**
    a. A street fitting
    b. A male fitting
    c. An insert fitting
    d. A hub fitting

19. **A PVC DWV cleanout has the same threads as**
    a. A male adapter
    b. A female adapter
    c. A trap adapter
    d. None of the above are correct

20. **A solvent welding process typically relates to**
    a. Priming and gluing plastic pipe and fittings
    b. Only DWV plastic pipe and fitting
    c. A process known as soldering
    d. PEX products

Chapter 5 | Valves and Devices

Isolation and regulating devices are part of our daily lives, but we may not recognize their importance until they fail to operate. Our vehicles, televisions, computers, and plumbing systems are only a few of the operating systems that function without much input on our behalf. You have learned about the tools and materials needed to install residential plumbing systems; this chapter introduces you to the valves and devices in residential installations. A valve can be a manually operated means of isolating a water or gas system. A valve can also be a device that does not require manual operation and does not isolate a system.

Isolation valves in residential piping systems are manually controlled to isolate the flow of water or gas on the downstream side of valves. Some isolation valves are compatible for both water and gas; others can only be used for a specific system. Some devices are mandated by code to provide safe operation of a piping system; others are required in case another device fails. All valves and devices are rated according to their operating capabilities, and the model numbers on many indicate their unique features. Some devices react to pressure, temperature, or unsafe conditions within a piping system. Codes dictate that many safety devices must be inspected periodically and certified annually to ensure they are performing as expected. Some valves and devices that have specific and limited uses are discussed in the relevant chapters of this book.

OBJECTIVES

Upon completion of this chapter, the student should be able to:
- identify and describe valves and devices used in a residential plumbing installation.
- understand that certain valve designs can only be used for specific systems.
- know the safety devices used in residential piping systems and their unique characteristics.
- relate installation of valves and devices to plumbing codes.

Glossary of Terms

air gap unobstructed vertical space from a device outlet to a point where water could backflow into a piping system

anti-siphon a device that prevents siphoning of contaminants into a piping system

backflow the dangerous reversal of flow in a piping system that can contaminate the system

back siphon an occurrence caused by a vacuum in a piping system

ball valve type of isolation valve used in most piping systems

boiler drain drain outlet on a water heater used to drain the storage tank; has a garden hose connection

check valve a one-way directional device installed to protect water from reversing flow in a piping system

flush clean a piping system with air or water pressure

gas cock type of isolation valve used in a gas distribution system

gate valve type of isolation valve used mostly in water distribution systems

hose bibb often called a sillcock, it is available in numerous designs to serve as a water distribution outlet; has a garden hose connection

hose outlet the point where a hose can connect to a faucet or boiler drain

isolate separate a portion of a piping system by turning a valve or device

isolation valve general term describing that a valve isolates a system or portions of a system

lug a designated raised portion of a valve used in place of a handle and operated with a wrench or tool

pressure-reducing valve device used to reduce the incoming pressure to a system or portion of a system

reactionary valves devices that react to certain conditions within a system to provide protection, such as backflow, pressure, and temperature

reduced-pressure zone valve a reactionary device to protect a potable water system against the possible backflow of undesired water from a connected portion of the system

relief valve relieves a storage tank or piping system of dangerous conditions such as pressure and temperature or both.

stop a type of isolation valve

stop-and-waste valve an isolation valve that also has the capability to manually drain the isolated portion of the system.

T&P valve a combination temperature and pressure relief valve

vacuum breaker reactionary device that breaks a vacuum in a piping system when unsafe backflow conditions are present

vacuum relief valve a type of vacuum breaker that is commonly used on a water heater that is piped with a side inlet connection

Isolation Valves

Every residential dwelling is required by code to have at least one **isolation valve**. It must be in a readily accessible location, so the homeowner can shut off the water supply in case of an emergency or to make a repair. Residential construction codes throughout the country include the required location of an isolation valve. It is common for houses with crawl spaces to route the main water distribution piping to a closet or other accessible area within the living space and to install a valve that is readily accessible. Most codes dictate that an isolation valve installed in a crawl space must be within 3′ of the access into the crawl space. In houses with basements, plumbers can install isolation valves where the water service piping enters the basement. The source of a water supply system is also a factor in the location of isolation valves. Municipal water systems require water meters. In many colder regions of the country, the meter is located within the residence; in warmer climates, it is near the street in a water meter box. Municipal water authorities install an isolation valve before a water meter to allow for repair or replacement. A plumber must install a separate isolation valve on the downstream side of the meter in or near the house before the first connection to a fixture in the house.

Isolation valves are made in a variety of types and sizes. Most codes dictate that the minimum size for a residential water service is 3/4″. The maximum size is based on the number of plumbing fixtures. Table 5-1 is a list of common isolation valves that will be explained individually in this chapter. Many codes dictate that the main isolation valve for a house and the isolation valve for a water heater be a full-port design. A full-port valve has the same inside diameter (ID) as the connecting pipe and does not drastically restrict the volume of water that flows through a valve.

The ID of the piping system is a factor in determining the volume of water that might be supplied by a particular pipe size. If you do not state on an order that you want a full-port valve, you will be ordering a restricted-port valve. Many kinds of isolation valves are not available in full port, but they might still be approved by most codes for use within water distribution systems. Some regulating devices and safety devices in piping systems have a restrictive nature and interrupt the flow of water. This causes a pressure loss and lowers the volume capabilities of the water distribution system. Some valves are operated with a wheel-type handle, but others only require that a handle be rotated one-quarter of a turn or 90° from the flow direction. Bronze is an acceptable material for valves and devices for residential potable water systems. Internal parts of valves and devices must also comply with safe drinking water standards. They can consist of various materials, including Teflon, rubber, stainless steel, and various plastics. A plumber must know the correct materials and locations for specific valve types based on codes and valve designs.

As are all materials in a piping system, valves and devices are categorized by their pressure capabilities. The most common classifications of residential valves and devices are 125 and 150 pounds per square inch (psi), but some are classified up to 600 psi. The pressure rating of each valve and device is shown on the body of the item, indicating its safe operating pressure. Often valves can be used for different types of piping systems; many are rated for water, oil, and gas (WOG). If a valve is not approved for potable water, its installation in a water distribution system violates plumbing codes and can contaminate the potable water system.

Some isolation valves and most devices in a piping system have a direction of flow, which requires the installer to know the flow direction of the water or gas before connecting the piping system. All directional valves and devices are marked to indicate the flow either with an arrow or by stating the designated in or out connections of the valve or device. Figure 5-1 is a comparison of the features of a full-port and a restricted-port isolation valve. Some valves and devices must be installed in certain positions, such as vertical or horizontal, but others can be installed in any position. Manufacturer's installation data indicate the correct installation positions and should be reviewed when installing valves and devices. Some valves are sold without instructions; knowing the correct operation of each item avoids incorrectly installing a valve.

Table 5-1 Common Isolation Valves

Type	Residential Uses
Ball valve	Water and gas
Gate valve	Water
Stop valve	Water
Stop and water valve	Water
Gas cock	Gas

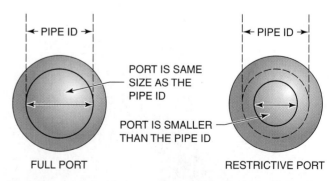

Figure 5-1 A full-port valve has a larger inside diameter than a restrictive port valve.

FROM EXPERIENCE

A full-port valve is larger than a similar type of valve with a restricted port. A full-port valve might have to be special ordered even though most codes require its use.

CAUTION

CAUTION: Always turn the isolation valve to the open position slowly to avoid a water hammer, which can be caused by the sudden flow of water into the piping system.

Ball Valve

A **ball valve** has an internal ball with a hole in its center that creates a flow passageway through the valve and isolates flow when the ball is rotated 90° from the flow direction. Some ball valves have a T-handle, but the most common ones have a lever handle. A vertical stem protrudes from the valve body of the internal ball. The lever handle is secured to the stem with a tightening nut. The lever handle design, which only requires a 90° rotation to close the valve, classifies a ball valve as a quarter-turn isolation valve. A ball valve has become one of the most popular valve designs for isolation in a piping system. Unlike most valves, the most common ball valve design uses a stainless steel ball sandwiched between two Teflon seats, rather than a rubber washer, to isolate the flow. If the lever handle is in the same direction (parallel) as the flow within the piping system, the valve is in the open position. Most ball valve designs are not direction valves and can be installed based on the operating location of the handle instead of the direction of flow of the piping system. A ball valve can be installed in any position in a piping system.

Threaded or soldered connections are the two most common types of brass connections used in the residential industry. Threaded valves have female threads that connect to a piping system with male threads, such as a male adapter or a pipe nipple. Ball valves are also manufactured from plastic with a non-metallic ball to isolate the water flow. All plastic materials must be approved for the fluid or gas being distributed. Their ratings will appear on the body of the valve, similar to those on a bronze valve.

Insulation must often be installed in water distribution piping systems to prevent freezing or condensation, or for energy conservation. If thick pipe insulation is used, a handle extension is installed to raise the handle from the connecting pipe and pipe insulation. Figure 5–2 shows a typical

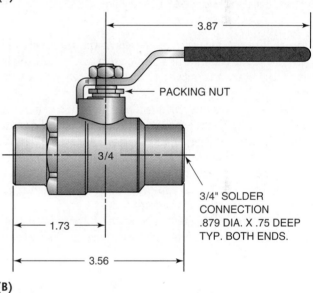

Figure 5–2 A ball valve is a quarter turn isolation valve. *Courtesy of Conbraco.*

ball valve for a residential water distribution system. Figure 5–3 illustrates the manual operation of a ball valve.

Gate Valve

A **gate valve** has a metal gate (disk) that slides vertically to open and close the valve. A wheel handle fixed to a stem raises and lowers the internal gate when it is turned. The handle is turned counterclockwise to open the valve and clockwise to close it. The two basic types of gate valves are the rising-stem type and the non–rising-stem type. The rising-stem design requires additional space in front of the handle, because the stem and wheel handle move outward from the valve when opening. A non–rising-stem gate valve raises the internal disk (gate); the wheel handle and stem do not move outward when opening. A non-rising disk has

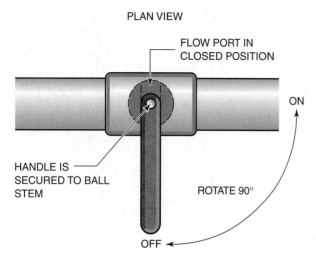

Figure 5-3 A ball valve isolates flow in a piping system when the lever handle is turned 90° from piping.

threaded or soldered end connection is used in a residential application. Figure 5-4 shows a rising stem and non–rising-stem gate valve. It illustrates the disk movement of each stem design. Figure 5-5 illustrates the unique body style of a gate valve.

 FROM EXPERIENCE

You can tell if a rising-stem gate valve is in the open or closed position by the distance from the stem to the piping system. The only way to tell if a non–rising-stem gate valve is open or closed is to turn the wheel handle manually until it stops rotating.

internal threads that ride up and down the threaded stem when operated. The disk of a rising stem is attached to the end of the stem and moves up and down when the stem moves in and out of the valve body.

Most gate valves are not direction valves and can be installed based on desired handle location instead of the direction of flow of the piping system. A gate valve will operate when installed in any position, but if it is installed with the handle directed toward the ground, debris can settle around the stem. It is approved for use in a potable water system but not for gas systems.

A gate valve is designed to be in the fully open or fully closed position and is not used as a regulating valve. A

Stop

A **stop** is a directional flow isolation valve that uses a rubber washer to stop the flow of water. Various designs exist to serve as isolation valves either for an entire water distribution system or for an individual fixture. The popular ball valve has replaced the stop as the main isolation valve for an entire house. The stop has a wheel handle secured to one end of the stem with a screw, and a rubber washer at the opposite end of the stem. It is a restrictive port valve, which is one reason it is no longer widely used in piping systems. A stop is more popular as an individual fixture-isolation valve and is often installed to connect the water distribution system to the fixture tubing connection.

Angled and straight stops for individual fixture isolation are manufactured with chrome and rough brass finishes.

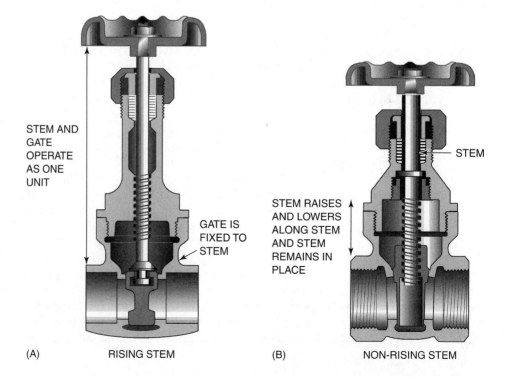

Figure 5-4 Rising stem and non-rising stems are two gate valve designs used in residential applications. *Courtesy of Crane Co., all rights reserved.*

CHAPTER 5 *Valves and Devices* **101**

Figure 5-5 A gate valve has a unique body design to house the internal gate. *Courtesy of A. Y. McDonald Mfg. Co.*

stop. Compression connections are explained in depth in relevant sections of this book. The rubber washer on a stop is replaceable and is either secured to the stem with a screw or pressed over a securing post that is an extension of the stem. Figure 5-6 shows a stop that isolates an entire system or a piece of equipment. Figure 5-7 shows an angled stop that isolates an individual fixture.

 FROM EXPERIENCE

Because a stop restricts the flow of water, it violates most equipment connection codes that dictate that the isolation valve be a full-port type.

Chrome stops are installed in exposed areas, such as under a toilet or a free-standing sink, and rough brass finishes are installed where they are not exposed, such as under kitchen sinks. The handle for a fixture-isolation stop is typically oval shaped instead of round. The connection method varies depending on the type and purpose of the stop. When a stop is installed as an isolation valve for a system or piece of equipment, both connections are the same size and are typically threaded or soldered. Connections to PEX are also available. The two common sizes for stops used to isolate a residential water supply system or piece of equipment are 1/2" and 3/4". Each end of an angled or straight fixture-isolation stop is a different size and has different connection types. Most fixtures are served by 1/2" pipe, with 3/8" tubing connecting the stop to the fixture. This is an industry standard connection. Because different types of pipe or tubing are used in a water supply system, the connection for the 1/2" side of a stop can connect to various materials. The 3/8" side of a fixture-isolation stop is a compression connection that uses a nut and ferrule to seal the tubing to the

Figure 5-6 A stop having the same size connections is used to isolate a water distribution system or piece of equipment. *Courtesy of A. Y. McDonald Mfg. Co.*

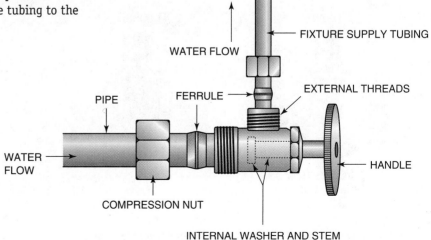

Figure 5-7 A stop with two different connection sizes is used to isolate an individual fixture.

Stop and Waste

A **stop-and-waste valve** is designed like a stop to isolate an entire water distribution system, but it also has a draining feature. When freezing is a concern, or when a small portion of a piping system must be drained, a stop and waste valve is installed. It is a direction valve with a flow direction arrow on the body of the valve. It is not available as an individual-fixture stop with a compression connection. Because of the various materials used for water distribution systems, the stop-and-waste valve is made with various connection types; the most common have soldered and threaded ends.

A stop-and-waste valve cannot be installed where a **backflow** of nonpotable water could enter the water distribution system through the drain port while the system was not under pressure. Many kinds of valves can be installed below ground in specially designed valve boxes, but it is illegal by all codes to install a stop-and-waste valve below ground. Figure 5–8 is a stop-and-waste valve that isolates a portion of a water distribution system and allows the isolated portion to be drained. Figure 5–9 is a sectional view exposing the unique features of a stop-and-waste valve.

Figure 5–8 A stop and waste is used only aboveground.
Courtesy of A. Y. McDonald Mfg. Co.

 FROM EXPERIENCE

Stop-and-waste valves are frequently used to isolate outside hose faucets that are not freeze proof. The stop is turned to the off position, the drain cap is removed and the isolated piping is drained, and then the cap is placed back on the stop. Most outside hose faucets are freeze proof, so stop-and-waste valves are typically not required for freeze protection.

Gas Cock

A valve known as a **gas cock** is used for gas distribution systems. Because ball valves are approved by most codes for isolating individual gas equipment, gas cocks are used more as a matter of preference. Many ball valves for gas isolation are manufactured with a unique T-handle, which differs from a typical WOG ball valve lever handle. A ball valve specifically designed for a gas system is not rated for use with other systems. Gas cocks are usually used to **isolate** entire systems, and utility providers commonly use them to isolate gas meters. Many gas cocks do not have manual handles such as levers or wheel handles. Instead, they must be opened and closed with a wrench. The wrench is placed on a raised lug and turned clockwise to close and counterclockwise to open.

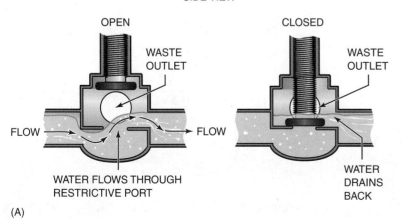

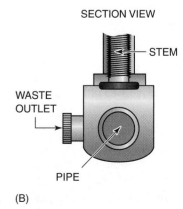

Figure 5–9 A stop-and-waste valve allows a downstream pipe to be drained.

Figure 5–10 A lug-type gas cock requires a wrench to operate. *Courtesy of Conbraco.*

Figure 5–11 A lever-type gas cock operates by hand. *Courtesy of Conbraco.*

Most codes do not allow a ball valve to isolate a system, because it can be operated without a wrench. System isolation valves are usually located on the exterior of a building near a gas meter. Most gas cocks used in conjunction with a gas meter have an alignment hole in which to place a padlock to secure the gas distribution system when not in use. The stem of the gas cock is also the flow channel for the gas. It is a round steel rod with a portion removed to create a flow slot (passageway). The 1/2" and 3/4" sizes, which have female threaded connections, are the most frequently used for residential applications. Figure 5–10 shows a lug-type gas cock, and Figure 5–11 is a lever-type gas cock used for isolation of a residential gas system.

Hose Outlets

Various types of hose outlet connections are used in piping systems to drain equipment and systems and to supply water. The most common hose outlets are categorized as hose bibbs, wall hydrants, and boiler drains. Hose threads are different from pipe threads. Outlets have 3/4" male hose threads. A hose connection to a piping system can provide a primary point of entry for contaminants to pollute a water distribution system. There are strict regulations concerning the design and installation of all hose connections to a water distribution system. Back siphoning can occur if a hose connected to a water supply pipe is placed in a contaminated source and the water system becomes depressurized. The hose can act like a drinking straw and allow backflow into the water distribution system. Direct connection of spraying devices such as for lawn fertilizing can also be a serious threat to a potable water system. To combat the threat of backflow into a drinking water system, all codes dictate that approved methods be installed to prevent backflow. Most manufacturers design safety devices into their products. Figure 5–12 illustrates possible backflow into a potable water system.

Boiler Drain

A hose outlet connection to drain storage tanks is known as a boiler drain. Because a boiler and a water heater are protected with other approved backflow devices, most codes do not require backflow devices in boiler drains. The boiler drain has numerous uses within a system, but its most common residential application is as a drain. It normally remains closed after the system is pressurized and only functions when a piece of equipment is drained. If a boiler drain is used as a water source, all codes require the installation of a separate backflow device similar to the one in Figure 5–21. A boiler drain is available with male and female threads, usually 1/2" and 3/4". Some boiler drains have the hose outlet connection 90° from its pipe connection—others are 45°. Some use a wheel handle; others have a T-handle. A boiler drain has a restrictive internal design feature similar to a stop that uses a rubber washer to stop the water flow. Most washing machine boxes have boiler drains. Figure 5–13 shows a typical boiler drain.

 FROM EXPERIENCE

A hose cap can be installed onto the hose threads of a boiler drain to prevent leakage from the rubber washer.

Hose Bibb

A hose bibb, also known as a hose faucet or a sillcock, allows water flow from a pressurized piping system. Some hose bibbs are similar to boiler drains, and others are freeze-proof. Most hose bibbs are protected with a backflow device. Those that do not provide backflow protection are illegal to

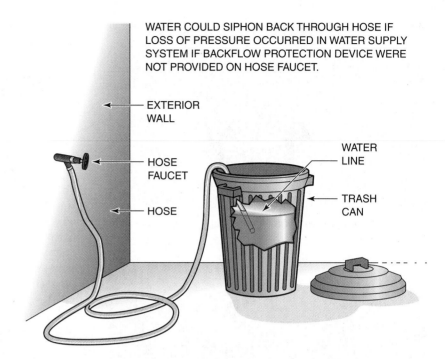

Figure 5-12 Backflow into a drinking water system can occur from a misplaced garden hose.

Figure 5-13 A boiler drain is used to drain a non-pressurized storage tank or boiler. *Courtesy of A. Y. McDonald Mfg. Co.*

install. The freeze-proof types, available in numerous designs and lengths, are meant to be installed in the exterior wall of a residential building. The washer and the handle of a freeze-proof hose faucet are secured to opposite ends of an extended-length stem. Freeze-proof designs extend the rubber washer into the building to prevent the water in the piping system from freezing. Turning the wheel handle clockwise stops the flow of water from the hose faucet. The water remaining in the freeze-proof hose faucet after closing drains to the exterior. To ensure that the water drains from the hose faucet, it must be installed so that it is sloping slightly downward toward the exterior of the building. Garden hoses must be removed after use to ensure the freeze-proof design operates correctly. Many hose faucets can be used for both hot and cold water, but some have a maximum temperature rating of 120°. The pipe serving a hose faucet is 1/2", and the connections include PEX, soldered, and male threads. A freeze-proof hose faucet is available in various lengths, most commonly 4" to 12", with the selection depending on the installation location. Figure 5-14 shows a common freeze-proof hose bibb installed in a residence and equipped with an **anti-siphon** device. The freeze-proof hose bibb in Figure 5-15 is illegal because it does not have an anti-siphon or backflow-prevention device.

 FROM EXPERIENCE

Installing a hose outlet with no backflow prevention device that connects to a potable water supply is a serious code violation. Each state regulates the fine and penalty for violating this code.

Reactionary Valves and Devices

Many valves and devices react automatically to temperature and pressure differences or to the reversal of flow within a piping system. Safety valves and devices can regulate pressure, or discharge high pressure or temperature, and many others protect potable water systems. Numerous devices are installed in piping systems, and some of their operating features are similar regardless of the manufacturer. Most safety and regulating valves and devices have a direction of flow and must be installed according to the manufacturer's requirements. Safety devices are a crucial part of potable water installations, and gas piping systems are strictly regulated by

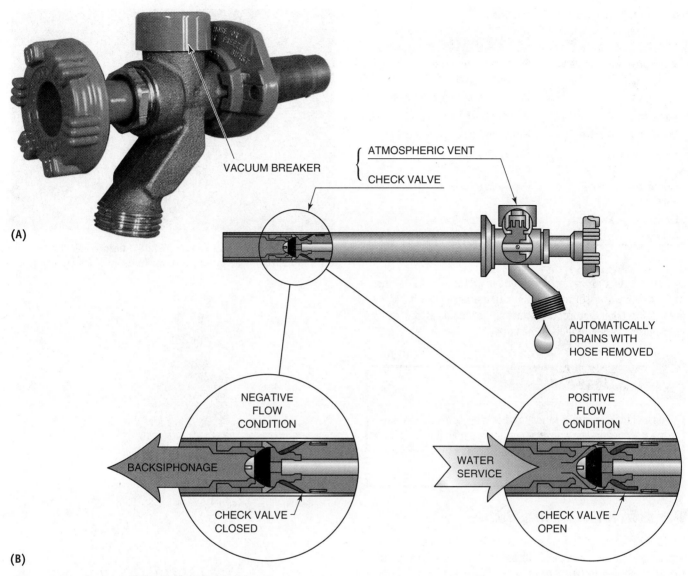

Figure 5–14 A hose bibb must have an anti-siphon or backflow prevention device to be legally installed. *Courtesy of Woodford Manufacturing.*

Figure 5–15 This type of hose bibb is illegal because it does not have an anti-siphon or backflow prevention device. *Courtesy of Woodford Manufacturing.*

all building codes. Plumbers must respect all regulations and adhere to the installation instructions for **reactionary valves** and devices to ensure safe and complete installations.

To make sure that the municipal water supply is protected against failed devices, multiple backflow prevention methods are used within a single residential building. All codes dictate that the unobstructed vertical space below a faucet spout be above the overflow height of the fixture and that it remain unobstructed. The vertical space is known as an **air gap;** it is the only definite means of backflow protection. Most codes dictate that the air gap of a potable water outlet must be at least twice the diameter of the outlet. An air gap cannot serve as a means of backflow prevention for a hose outlet. Because many connections to a plumbing system cannot be installed with an air gap, valves and devices minimize the risk of contamination. Table 5–2 is a list of reactionary valves and devices and their purpose in a plumbing system. A cross

Table 5-2 Reactionary Valves and Devices

Type	Residential Uses
Pressure-reducing valve	Reduce incoming water pressure
Check valve	Prevent reverse flow of water
Vacuum breaker	Prevent back siphoning
Vacuum relief valve	Prevents back siphoning
Relief valve	Relieve excessive pressure/temperature
Reduced-pressure zone valve	Prevents backflow
Double-check valve assembly	Prevent backflow

Figure 5-16 A strainer can be installed before a device to capture debris in the piping system. *Courtesy of Conbraco.*

connection is a piping arrangement in which a potable and a nonpotable water source are illegally connected. A backflow-prevention device protects against siphoning occurrences and eliminates cross-connection hazards.

 FROM EXPERIENCE

There are numerous kinds of backflow-prevention devices. Knowing the correct type for a particular installation is crucial to avoid contamination of a potable water supply.

Pressure-Reducing Valve

Codes typically allow the pressure in water service piping and hose faucets to exceed 80 psi, but the pressure in piping that serves fixtures must be reduced to a maximum of 80 psi. A regulating device known as a **pressure-reducing valve** (PRV) reduces the incoming water pressure to a safe operating range. Many models are available, and residential models can be manually adjusted to accommodate a particular installation. Most residential PRVs have an adjustment range from 25 to 75 psi and are factory set to regulate water pressure to 50 psi. A PRV requires maintenance and possibly future adjustment, so it must be installed in an accessible location. An isolation valve should be installed upstream of the device. If a union is not included in the PRV's connection design, a separate union should be installed, so the device can be easily removed.

PRVs are available in a variety of sizes, but 3/4" is the minimum for reducing the main piping to a house. A water service piping system should be **flushed** clean before installing a PRV to ensure that dirt or debris does not damage its internal regulating components. A strainer is an accessory that prevents particles from traveling through a piping system. Many PRV designs have removable built-in strainers that can be cleaned. A PRV is a restrictive device, so the volume of water is diminished as the flow passes through it. Figure 5-16 shows a strainer that is installed if the pressure-reducing valve does not have a strainer. Figure 5-17 shows a typical residential pressure-reducing valve.

Figure 5-17 A pressure-reducing valve regulates the incoming water pressure to a house. *Courtesy of Wilkins, a Zurn Company.*

 FROM EXPERIENCE

Piping systems and faucets can be damaged if a PRV is not installed where needed. Pressures in many municipal water systems fluctuate and exceed the maximum 80 psi allowed by code. Therefore, a PRV may be required by your local code.

Relief Valves

A **relief valve** protects a piping system and equipment from extreme temperatures and pressures. All relief valves are self-operating and open and close as they react to operating conditions of a system. Many dual-use relief valves provide protection against both temperature and pressure. These are called T&P relief valves. Most codes require that a water heater be equipped with the proper T&P relief valves before being shipped from a factory so the contractor does not install the wrong one. A residential-style T&P relief valve typically has male pipe threads that thread into the water heater storage tank and female pipe threads that connect to the discharge drain. A T&P relief valve is threaded into the factory-designated opening of a water heater, which must be within the top 6" of a water heater to sense the hottest water in the tank.

A T&P relief valve has a rating tag that identifies what kind it is. The temperature-sensing probe of a **T&P valve** is immersed in the water and triggers the opening of the relief valve if the water temperature reaches 210°F in a residential water heater. The pressure-sensing capabilities of a T&P relief valve range from 75 to 150 psi. The rating for a residential water heater is usually factory set at 150 psi. When the pressure and temperature reach the factory-set limits, the relief valve opens to relieve the pressure and temperature in the water heater. The size, pressure rating, and temperature rating of a T&P relief valve are important elements in choosing the correct device. The British thermal unit per hour (BTU/HR) rating of a device is also important when selecting a T&P valve. BTUs are discussed in the water heater section of this book. The discharge side of a T&P serving a water heater is routed to a safe location. This is explained in the water heater installation section of this book.

A well pump system has a pressure-relief valve but not a temperature relief valve, because it only distributes cold water. The relief valve ensures that the piping system and storage tank are protected in case the automatic controls fail to disconnect electrical power to the pump. The relief valve serving a water pump system is typically installed at the same location as the storage tank. A well pump pressure-relief valve for a residential application is usually 1/2". Figure 5–18 is a temperature and pressure-relief valve for a residential water heater.

CAUTION: Selecting the wrong relief valve can be extremely dangerous in a piping system and storage tank. The maximum operating pressure of a storage tank must be known to select the proper relief valve.

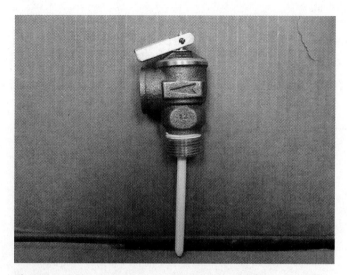

Figure 5–18 A temperature and pressure-relief valve is a required safety device installed with every water heater. *Courtesy of Conbraco.*

 FROM EXPERIENCE

Know the manufacturer's requirements when selecting a relief valve for any system. Do not rely on what is currently installed or on any rule of thumb methods.

Check Valve

A backflow-prevention device known as a **check valve** is installed to ensure that water flows in only one direction. A single check valve is not an approved method for protecting a potable water system from contamination. However, a dual check valve may be allowed by some codes as a form of backflow prevention. Check valves are used in residential applications for hot water circulating pumps and well pumps.

Swing and spring check valves are the most common for residential applications. The swing style is used for most applications, but the spring is employed more often for well pump systems. A swing check valve has an internal disk that swings open when water flows through a system and closes when the flow has stopped. A spring check valve has a soft sealing disk that is held closed with a spring. When water flows, the spring is depressed to allow flow through the check valve. Swing check valves are connected by soldering and by female threads. Spring check valves most often have female threads. It is recommended that a union be installed near a check valve to replace the device. A swing check valve typically has a removable cap to allow entry into the valve for inspection, cleaning, and repair.

Because of its one-directional flow, an arrow is located on the side of the check valve to ensure correct installation. A spring check valve can be installed in any position, but a swing check can only be installed with the disk in the closed position. This limits installation to a horizontal position and an upward-flow vertical position. A swing check is also installed on the discharge piping from a sewage ejector or sump pump to keep the discharge water from flowing back into the pump basin.

The disk in a swing check valve is typically brass or bronze. This can cause a slamming noise when the flow stops quickly and the disk closes. To combat that problem, a non-slam swing check can be installed. A spring check's soft disk operates silently, but it can become obstructed more easily because of its restrictive design. Check valves are sold in all pipe sizes; the most common for residential applications are 1/2" to 1". Figure 5-19 is a swing check valve, and Figure 5-20 is a spring check valve.

FROM EXPERIENCE

Check valves for sewage or sump pumps can be made from different materials than water-distribution check valves, and many types are available for drainage applications. Spring check valves cannot be used for a sewage or sump pump installation.

Vacuum Breaker

When a water distribution system is isolated or loses pressure, a vacuum is created that can allow contaminated water to enter a potable water system. As its name implies, a **vacuum breaker** breaks a vacuum created in a piping sys-

Figure 5-20 A spring check valve provides a spring-loaded closing feature. *Courtesy of Watts Regulator.*

tem. To prevent the possibility of contamination, vacuum breakers in many forms are installed throughout a piping system. If a vacuum occurs in a piping system, air is drawn through the vacuum breaker to equalize the pressure in the piping system, so the potable water outlet does not become a point of entry for contaminants.

Many faucets are designed with built-in vacuum breakers. Handheld showers typically require a vacuum breaker because the hose can reach the bathwater. Code-approved hose faucets have vacuum breakers as part of their design to prevent backflow. A hose-thread vacuum breaker can be installed on the outlet of an unprotected hose bibb or boiler drain to satisfy plumbing code requirements. Hose-thread vacuum breakers typically have a breakaway screw that is tightened after installation so the vacuum breaker cannot be easily removed. Vacuum breakers are also installed on flush valves on commercial toilets and urinals. Figure 5-21 shows vacuum breakers on hose outlets to prevent contamination of potable water piping systems.

FROM EXPERIENCE

Not all vacuum breakers for hose faucets are designed for cold climates. Research manufacturer's data before installing a vacuum breaker to ensure that it is designed for the specific operating conditions.

Figure 5-19 A swing check valve opens when there is flow in a piping system. *Courtesy of Watts Regulator.*

Figure 5-21 Numerous vacuum breakers are used to prevent backflow of water into a water supply system. *Courtesy of Woodford Manufacturing.*

Vacuum Relief Valve

A **vacuum relief valve** or atmospheric relief valve is installed in the cold water piping that serves a water heater. When it senses a loss of water pressure, it opens to equalize the piping system with atmospheric pressure (zero gauge pressure). During normal operating conditions, the vacuum relief valve can withstand pressure without leaking. It only opens when it detects a loss of system pressure. A variety of vacuum relief valves are available for use throughout a piping system.

Most residential water heaters connect at the top of the storage tank and are equipped with a factory-installed funneling tube known as a dip tube. The dip tube is inserted into the cold-water inlet connection and routes the incoming cold water to the bottom of a storage tank. It also has an anti-siphon feature. The manufacturer drills a small diameter hole in the dip tube a maximum of 1" from the top. This breaks a vacuum if a negative situation occurs within the tank. This feature allows only the top 1" of hot water to flow out the storage tank and into the cold-water piping system. That minimal amount of backflow does not warrant the installation of a vacuum relief valve by most codes. A storage tank that does not have a dip tube must have a vacuum relief valve in the cold water supply to the heater. Many bottom-fed storage tanks require a vacuum relief valve because they do not have a dip tube. Figure 5-22 shows a common vacuum relief valve like those installed in some water heaters.

CAUTION: Never install a vacuum relief valve near an electrical source or any other area that can be damaged by water. The operation might spill water onto the surrounding areas and cause an electric hazard. Spill-proof styles are available for installation in sensitive areas.

Figure 5-22 A vacuum relief valve prevents backflow of water from a storage tank. *Courtesy of Watts Regulator.*

FROM EXPERIENCE

Most vacuum relief valves that serve residential water heaters have 3/4" male pipe threads. Because they are brass, they can thread directly into copper female adapters with no corrosion concerns.

Reduced-Pressure Zone

A **reduced-pressure zone valve** (RPZ) is the most reliable device for preventing backflow of contaminated water into a potable water supply system. Several design features are available for various applications. The designs may vary, but the basic concept is the same for most RPZ valves. Essentially, two spring-loaded check valves are manufactured as a single device. Some versions are known as double check-valve assemblies. Most codes do not allow them in many piping situations because they do not discharge water attempting to backflow. A true RPZ completely isolates the water of one zone from the zone supplying the potable water. An RPZ that discharges undesirable water from the piping system is known as a reduced-pressure principle backflow preventer. If a dangerous backflow condition is present, an RPZ discharges the water from the downstream zone so it does not flow back into the potable water system. As its name indicates, it basically senses a reduced pressure in the potable water system and protects that zone. Figure 5-23 illustrates how an RPZ creates two zones.

Some codes require RPZs to be installed in residential buildings, and all codes require them on irrigation systems. Fertilizers used on lawns with recessed lawn sprinklers pose a serious threat to a potable water system. An illegal connection of a contaminated water source to a potable water system can pollute an entire piping system.

An RPZ discharges the water that is attempting to backflow from one zone to another out of a relief valve port. Most codes dictate that the opening for the relief valve discharge port must be at least 12" above the highest surrounding area or ground. The height of the relief valve above the ground is known as the critical level. An installer might overlook the possibility that a low-lying area could puddle during heavy rain or a submerged room could become flooded. This could cause contaminated water to enter a potable water system through the discharge port.

Most codes dictate that an RPZ must be tested and certified annually by a certified technician for correct operation. The test ports required for annual certification are installed at the factory on most devices. Because the annual test can require replacement of internal parts, many codes dictate that the maximum height above a floor is 36" to allow for servicing. However, any dictated maximum height cannot override the minimum critical level requirements. An RPZ has

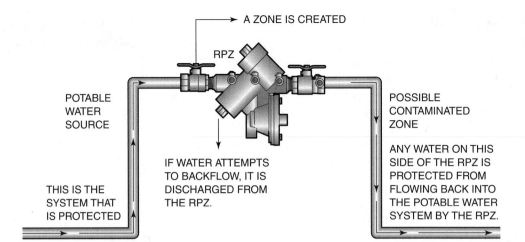

Figure 5-23 An isolated zone is created on the downstream side of an RPZ.

Figure 5-24 An RPZ is the best device to prevent backflow in a piping system. *Courtesy of Wilkins, a Zurn Company.*

Figure 5-25 A double-check valve assembly is a backflow prevention device. *Courtesy of Wilkins, a Zurn Company.*

two factory-installed isolation valves with female threaded connections. A strainer is recommended on the upstream side of an RPZ. This can be an option when ordering the device. Figure 5-24 shows a typical reduced-pressure zone valve for residential applications. Figure 5-25 shows a double-check valve assembly often used in place of an RPZ if local codes permit.

 FROM EXPERIENCE

An RPZ will freeze when exposed to cold climates. Removing it or providing adequate protection will avoid permanent damage to the device.

CAUTION: Removing a required RPZ is illegal and could cause contamination of a municipal water distribution system or a private well. This could endanger a large portion of a community and could cause serious illness or death.

Summary

- Numerous valve designs exist, many with multiple uses.
- Some valves have a full port and others have a restrictive port.
- A valve used under a fixture to isolate a single fixture is known as a stop.
- A valve used specifically to isolate the gas supply to a fixture is known as a gas cock.

- An approved ball valve can be used to isolate water and gas.
- A stop-and-waste valve can drain the isolated portion of a pipe.
- Common gate valve designs are rising and non-rising stem types.
- Many devices and faucets are considered to be valves.
- Valves and devices installed for potable water must be approved by plumbing codes.
- Threaded valves and devices typically have female threads.
- Soldered connections are used for many valves and devices connecting to copper tube.
- Plastic valves and devices are available with solvent welded connections.
- Backflow devices are installed to protect a potable water system.
- An air gap is the only sure way to prevent backflow.
- A pressure-reducing valve reduces the pressure in a piping system.
- A temperature and pressure relief valve is a safety device installed in a water heater storage tank.
- Spring and swing check are the two most common types of check valves.

Review Questions

1. **A valve whose flow passageway is the same diameter as the inside diameter of the pipe is**
   a. A reduced-port design
   b. A full-port design
   c. An inside-diameter port design
   d. An outside-diameter port

2. **A stop valve with a drain port to allow water to drain from the isolated piping is called a(n)**
   a. Stop drain valve
   b. Draining valve
   c. Stop-and-waste valve
   d. Angle stop

3. **The two basic kinds of gate valves used for residential applications are**
   a. Rising stem and non-rising gate
   b. Rising stem and non-rising stem
   c. Non-rising gate and non-rising stem
   d. Outside stem and yoke

4. **A regulating device used to lower the pressure of a potable water system is called a**
   a. Reduced-pressure zone valve
   b. Double check valve
   c. Pressure-relief valve
   d. None of the above are correct

5. **A valve designed to isolate the water supply to an individual fixture such as a sink is**
   a. An angle or straight stop
   b. A sink stop
   c. A gate valve
   d. A gas cock

6. **A valve rated for WOG means it can be used for**
   a. Water, oil, and gas
   b. Water or gas only
   c. Water only
   d. Gas only

7. **The two types of check valves commonly used for residential applications are**
   a. Swing and atmospheric
   b. Spring and vacuum
   c. Swing and spring
   d. Vacuum and atmospheric

8. **The mandatory safety relief valve used on a water heater to protect against extreme conditions is a(n)**
   a. Temperature and pressure relief
   b. Vacuum breaker
   c. RPZ valve
   d. Double check valve

9. **The only legal hose outlet that is not required to have a backflow prevention device is a**
   a. Wall hydrant
   b. Hose bibb
   c. Boiler drain
   d. Freeze-proof type

10. **A reduced-pressure zone valve**
    a. Is adjustable and reduces water pressure
    b. Isolates a zone in a piping system
    c. Typically does not require annual certification
    d. Is not available in a 1" size

⑪ **An operating gate valve**
   a. Uses a rubber washer
   b. Uses a lever handle
   c. Makes it acceptable for gas
   d. None of the above are correct

⑫ **The backflow device used on the water piping supply to a water heater is called**
   a. A double check valve assembly
   b. A reduced-pressure zone valve
   c. An atmospheric relief valve
   d. A reduced-pressure zone valve

⑬ **The only sure backflow prevention method is**
   a. A check valve
   b. An air gap
   c. A vacuum breaker
   d. A reduced-pressure zone valve

⑭ **A water heater with a top piping connection has an anti-siphon hole drilled in the**
   a. Dip tube
   b. Check valve
   c. Isolation valve
   d. Cold water piping

⑮ **A threaded valve or device is manufactured with**
   a. Male threads
   b. Female threads
   c. Soldered ends
   d. PEX connections

⑯ **A device that prevents debris from entering a piping system is a**
   a. Stop and waste
   b. Spring check
   c. Strainer
   d. Debris device

⑰ **The maximum pressure that most codes allow to a fixture is**
   a. 50 psi
   b. 60 psi
   c. 80 psi
   d. 100 psi

⑱ **A valve known as a stop is a**
   a. Full-port valve
   b. Restrictive-port design
   c. Non-directional valve
   d. Rising-stem design

⑲ **A swing check valve**
   a. Is a directional valve
   b. Is a non-directional valve
   c. Can be installed in all positions
   d. Is controlled by a spring

⑳ **A spring check valve cannot be installed in**
   a. All positions
   b. Sewage-pump discharge piping
   c. Well-pump piping
   d. None of the above are correct

SECTION TWO

Fixtures and Equipment

SECTION TWO
FIXTURES AND EQUIPMENT

- **Chapter 6:** *Fixtures*
- **Chapter 7:** *Faucets and Drain Assemblies*
- **Chapter 8:** *Plumbing Equipment and Appliances*

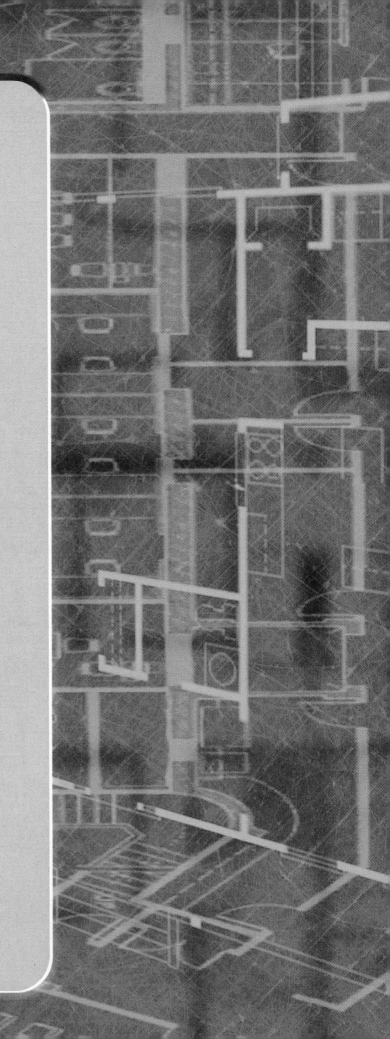

Chapter 6 | Fixtures

Now that you have learned about tools, equipment, and materials, this chapter identifies the common residential plumbing fixtures. Once you possess basic knowledge of the types of fixtures and their variations, you will be on your way to learning how to install the plumbing system that serves the fixtures. The actual installation of fixtures is discussed later in the book.

Plumbing fixtures have evolved along with all other areas of the plumbing industry. Fixtures have changed drastically in the last several decades regarding style, color, and material. Recent trends in the industry have been toward manufacturing replicas of some fixtures from the first half of the twentieth century. Today's fixtures range from basic to extraordinary. Electronics have been introduced into the designs of many faucets along with water saving features. Materials used depend on the type of fixture. Fiberglass, cast iron, and vitreous china are frequently used, as well as stainless steel, which is one of the most often used materials for kitchen sinks. Finishes for faucets and accessories have also evolved.

OBJECTIVES

Upon completion of this chapter, the student should be able to:

- identify the basic types of residential fixtures.
- order each fixture based on type and variations.
- understand that different fixtures are installed during various phases of construction.
- recognize the importance of manufacturer installation information.

Glossary of Terms

custom defines homes with fixture upgrades

pop-up a drain-operating assembly for a lavatory sink, abbreviated as PO

rough abbreviation of the term *drain rough-in;* describes the distance to the center of a toilet from the wall behind the toilet

rough-in the phase of construction before wall finishes are installed

rough-in sheet information pertaining to specific installation requirements or other unique characteristics of a fixture

stub out pipe that serves a fixture installed during the rough-in phase of construction

submittal data indicating the type of fixture to be used and its unique characteristics; sometimes includes installation information

trim out the phase of construction when plumbing fixtures are installed

Fixture Types

Plumbing codes dictate that every house must have at least one toilet, a lavatory sink, a bathtub or shower, and a kitchen sink. Codes dictate that the materials used to manufacture plumbing fixtures must have smooth, impervious surfaces and be defect free. Porcelain enameled surfaces must be able to withstand acid without damage to the fixture. Toilets must be self-cleaning during their flushing cycle and have a toilet seat. Clearances from walls and other fixtures are strictly regulated by code, which will be explained in the pipe installation section of the book. Because projects may use different fixtures, a plumber might have to submit manufacturer data to builders, owners, or architects for approval of a fixture for each project. The data is submitted on a form called a **submittal**, and, once the fixtures are approved, the plumber can install the piping systems according to the specific requirements of the fixtures. Piping systems are installed behind walls, and in floors and ceilings during the phase of construction known as the **rough-in** phase. The piping terminations of rough-ins are connected directly to the fixtures they will serve if the fixtures have been installed. If not, they are piped outward from a wall or floor, in which case they are known as **stub outs**. The stub-out pipes penetrate either the floor or the wall as dictated by the specific fixture requirements and then connect to the fixture during the fixture installation phase. This is known as the **trim-out** phase of construction. Understanding the fixtures is the first step in being able to install the necessary piping systems. Table 6-1 lists common residential fixtures, which will be explained separately in this chapter.

Toilets

Many homes have 2-1/2 bathrooms, which means that three toilets are installed. The toilet selection varies based on the cost of the home and the consumer's preference. The color of a toilet is based on the personal preference of a builder or consumer, with white being the most common. Two-piece combination toilets consisting of a bowl and a tank are installed most frequently; one-piece toilets are used in more expensive homes and often in only one of the bathrooms. A two-piece toilet must be assembled by a plumber on the job site, but a one-piece toilet does not require assembly. A master bathroom is typically the only one in the house that is in the same confines as a bedroom. It often has more expensive fixtures than the other bathrooms in the home, which might only have basic designs and colors. Figure 6-1 shows a standard two-piece toilet, and Figure 6-2 shows two styles of a typical one-piece toilet.

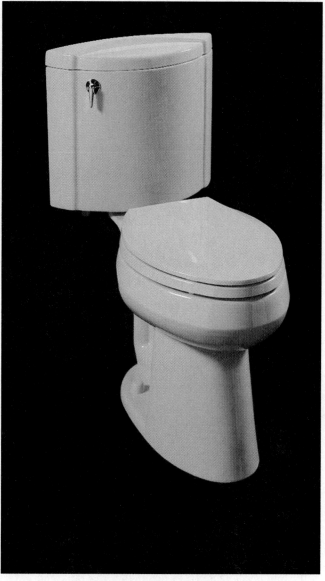

LEFT HAND TANK HANDLE

Figure 6-1 A two-piece toilet is the most common for residential installations. *Courtesy of Kohler.*

Table 6-1 Common Residential Plumbing Fixtures

Fixture Type	Notes
Toilet	Also known as a water closet
Lavatory sink	Also known as a lavatory, lav, or basin
Bathtub	Also known as a tub
Shower	Not part of a tub and shower combination
Kitchen sink	Can also serve a garbage disposal and dishwasher
Laundry sink	Also known as a laundry tray
Bidet	Personal hygiene fixture

(A) STANDARD PROFILE (B) LOW PROFILE

Figure 6–2 A one-piece toilet is common for many custom homes or as a fixture upgrade. *Courtesy of Kohler.*

FROM EXPERIENCE

A toilet tank and bowl must be compatible with one another. The toilet tank lid is usually unique to a particular tank and is not interchangeable with other tank designs.

All new toilets must use a maximum of 1.6 gallons per flush (gpf) to adhere with water conservation regulations. The most common residential toilet bowl design uses a siphon-jet flushing action. When the tank handle activates a flushing cycle, the water flows from the tank and enters the rim of the bowl. Small holes in the rim are angled to allow the water to create a vortex. Another stream of water, known as a jet stream, exits the rim into the passageway of the toilet, providing the initial thrust in the flushing process. The vortex (swirl) begins a siphoning action to evacuate waste from the bowl. Once the siphoning action begins, the wastewater is pulled from the bowl and discharged into the drainage system. Figure 6–3 illustrates the siphon-jet flushing action.

FROM EXPERIENCE

The other type of toilet flushing action is a blow-out design, more common on commercial and industrial sites. A pressure-assisted toilet uses air compressed within a storage tank located in the toilet tank to create a blow-out flushing action.

CHAPTER 6 *Fixtures* **121**

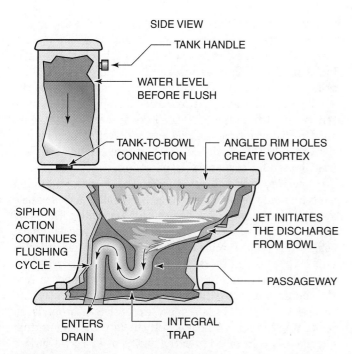

Figure 6–3 Siphon-jet flushing are used for most residential toilets.

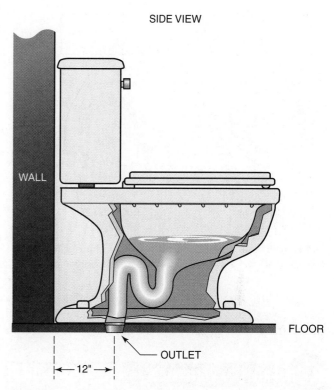

Figure 6–4 A 12″ rough toilet is standard; and other distances from a back wall must be specified.

The tank handle is typically located on the left side of a toilet tank, but it can be ordered on the top or right side. Handicap fixtures must comply with the Americans with Disabilities Act (ADA). The tank handle of an ADA compliant toilet must be on the side of the tank that is farthest from a sidewall. A handle located on the top of a tank also satisfies ADA handle location regulations. The height of the toilet bowl, including the seat, from the floor is regulated by code. Toilets are manufactured to comply with plumbing and ADA floor-height codes. An average non-handicap toilet bowl height is 15″ from the floor. ADA codes dictate that the minimum height from the floor to a toilet seat is 16-1/2″, and the maximum is 19-1/2″.

The location of pipes serving a toilet and specific installation codes are discussed in the piping installation sections of this book. The outlet pipe location depends on the type of fixture and its unique or standard drainage connection location. The outlet distance of a standard toilet is 12″ from the wall behind the toilet (back wall). If you do not specify any other dimension when you order, you will purchase a 12″ **rough** toilet. Ten- and fourteen-inch rough toilets are available in some designs but they are not common. The location of the water pipe serving a toilet is important. Most one-piece toilets have different location requirements than two-piece toilets. You must know whether you are installing a one-piece toilet before you begin the water pipe rough-in. Figure 6–4 illustrates the standard distance a toilet is placed from a back wall.

 FROM EXPERIENCE

When ordering a one-piece toilet, always ask for the manufacturer's data sheet—or **rough-in sheet**—to confirm the installation location of the water and drain pipes.

Residential toilets have round bowls, but code dictates that elongated (oval) bowls must be installed in commercial applications. When an elongated bowl is installed in a home, it is usually considered a fixture upgrade. One benefit of an elongated bowl is that it allows more frontal room for the user than a round bowl does. The code that requires a minimum 21″ clearance from the front of a bowl to a front wall or fixture can be violated if an elongated bowl is installed in a room designed for a round bowl. Check your local codes pertaining to the minimum front clearance before substituting an elongated bowl for a round bowl. Toilet seats must conform to the bowl design; for example, elongated seats are used for elongated bowls. Residential toilet seats have a closed front, but code dictates that commercial seats have an open front. Residential seats have a top lid (cover), but most codes do not allow commercial seats

122 SECTION TWO *Fixtures and Equipment*

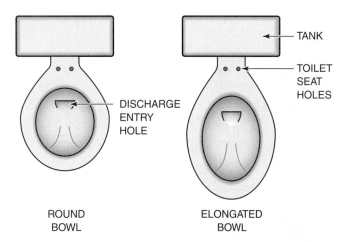

Figure 6–5 Round toilet bowls are standard for residential installations, and elongated bowls are a fixture upgrade.

to have lids. Figure 6–5 compares round and elongated toilet bowl designs. Figure 6–6 compares open-front and closed-front toilet seats.

 FROM EXPERIENCE

Toilet seats are usually plastic, but wooden seats are also available. Plastic seats are less likely to harbor bacteria when they become worn or scratched.

Lavatory Sinks

A sink installed in a bathroom is also called a lavatory. It is abbreviated as lav and is often referred to as a basin. Many types, shapes, and colors are available, and, as with all fixtures, some are more common than others. Many homebuilders install solid surface countertops with pre-molded sink basins, so the plumber does not install a separate sink. Installation and accessories are explained in the fixture installation section of this book, but here we will discuss the variables in ordering a lavatory. ADA codes dictate the countertop height from the floor, the knee space under the sink, and the distance from the side and back walls. Many lavatories are sold specifically for ADA adherence, but they are usually used for commercial applications.

A lavatory sink is ordered by its shape, size, color, and mounting requirements as well as the number and distance apart of the faucet holes. The color is typically selected by the consumer; white is the least expensive and the most common. A typical residential home has lavatories installed into a countertop. These are drop-in sinks and are either round or oval. A drop-in sink requires that a hole be cut into the countertop that is the right size for the particular sink that is to be installed. A more expensive lavatory is an under-counter sink. A hole is still cut into the countertop, but the sink is installed from the underside of a solid surface countertop. Figure 6–7 compares a drop-in and an under-counter lavatory sink.

 FROM EXPERIENCE

A plumber does not usually cut the hole for a sink in the countertop. The manufacturer provides a cut-out template with most sinks to ensure that the hole is cut to exact specifications. The plumber usually provides the template to the contractor, who is responsible for cutting the countertop.

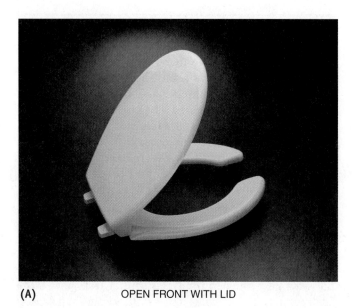

(A) OPEN FRONT WITH LID

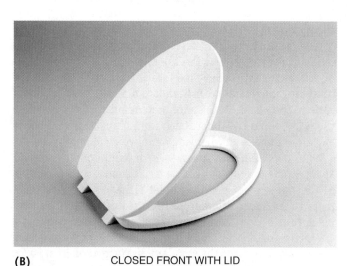

(B) CLOSED FRONT WITH LID

Figure 6–6 A toilet seat must match the bowl design. Residential toilets have a top lid with a closed front. *Courtesy of Kohler.*

CHAPTER 6 *Fixtures* **123**

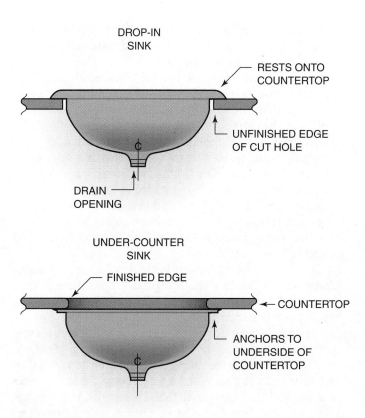

Figure 6–7 A drop-in lavatory sink is more common than an under-counter lavatory sink.

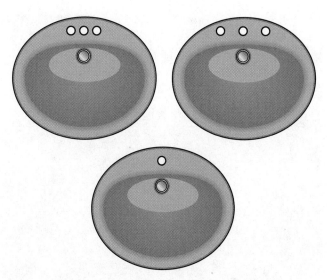

Figure 6–8 When ordering a lavatory sink, the faucet hole quantity and spread must be specified.

 FROM EXPERIENCE

Caulking seals a sink to the countertop. Most sink manufacturers include a small tube of adhesive caulk with each sink so the plumber has the correct type and color.

The faucet hole design (spread) must be compatible with the style of faucet being installed. The spread is the distance between the hot and cold faucet inlets, which is 4" on a standard lavatory. The middle hole of a three-hole lavatory is centered between the inlets to receive the drainage operating assembly (**pop-up**) and/or a faucet spout connection. Faucets that require a greater spread between the hot and cold inlets are often used on a three-hole lavatory with an 8" spread. Less common is a one-hole design that is centered in the sink. Figure 6–8 compares the various faucet hole designs for lavatory sinks.

 FROM EXPERIENCE

The type of faucet dictates the hole design to order for a lavatory sink. Not all faucets are compatible with all sinks.

Most homes have a half-bath in the area where guests are entertained that consists of one lavatory and a toilet. Many homeowners upgrade the fixtures in the guest bathroom by installing a pedestal lavatory sink. A pedestal sink is wall-hung with a decorative vertical leg called a pedestal. It is not designed to be the sole support of the basin (bowl), but instead conceals the piping below the sink while providing a decorative appearance. The bowl is supported with brackets that are anchored to a wood support called backing that the plumber installs during the rough-in phase of construction. The manufacturer's information must be reviewed to correctly install the wood backing. Ordering a pedestal sink for a faucet hole spread is like ordering any other lavatory sink. A sink that is more common in the commercial industry is a wall-hung sink. It also uses wood backing, but it does not have a pedestal. Figure 6–9 is a pedestal sink, and Figure 6–10 is a wall-hung sink.

 FROM EXPERIENCE

Because the plumber usually does not have the actual fixture on a job site during the rough-in phase, the manufacturer's wall-backing information must be requested.

Bathtubs

A bathtub, often called a tub, is used for soaking, but most tubs are used for showering as well. A standard residential tub is 5' in length and averages 30" wide. The depth of water a tub can hold varies by tub design. Some tubs are sold separately, others with wall kits, and others as one-piece tub and shower units, and still others with whirlpool

Figure 6–9 A pedestal sink, common in guest bathrooms, mounts onto a wall and has a vertical leg. *Courtesy of Kohler.*

Figure 6–10 A wall-hung lavatory sink is more common in the commercial industry. *Courtesy of Kohler.*

features. Most tubs are installed during the rough-in phase of a project, but some drop-in tubs are installed on top of tile or other solid surface after the finished surface is complete. A one-piece tub and shower unit is fiberglass, with tub and walls molded as a single unit; it is installed during the rough-in phase. Other wall finishes, including tile, are completed after the tub is installed.

The drain and faucet are typically located at the end of the tub known as the head wall. Drains can be located in areas other than the end of a tub, and the faucet can be installed in the most suitable location. The bottom surface of a tub slopes toward the drain, and every tub has an overflow hole (port) where an assembly known as a bath waste and overflow is installed. Tubs are ordered as either left hand or right hand. Tubs installed during the rough-in phase of construction are secured to walls by an installer. If the open area is to the left of a user, it is a left-hand tub, and vice versa. Figure 6–11 is a comparison of a left-hand and a right-hand tub.

Whirlpool tubs have various installation requirements. Some are similar to a standard tub; others hold a large amount of bathwater. Some large-capacity tubs, known as garden or Roman tubs, do not have whirlpool accessories and are installed either during the rough-in phase or the trim-out phase of construction. Most large-capacity garden tubs and whirlpool tubs are drop-ins installed on top of a tiled platform built by another contractor. Many other tub styles are available, including some that are installed in the corner of a bathroom. Claw-foot tubs were common in the early part of the twentieth century, and the design has again become popular. Many manufacturers offer this old-style replica, which is typically used for bathing, not showering. It can be converted, however, with a wrap-around shower curtain and various tub and shower faucets, to be used as a tub and shower. The claw-foot tub does not anchor to a wall; it is installed over the finished floor during the trim-out phase. Figure 6–12 shows a few non-typical tub designs.

 FROM EXPERIENCE

The physical size of the tub often requires that it be placed in the general work area before walls or doors are installed. One-piece tub and shower units are often delivered to a construction site after the roof is installed to eliminate unproductive handling.

Showers

Though many tubs have shower faucets and are used as a tub and shower unit, a shower is considered to be a fixture that is not used for soaking. Individual shower units are often referred to as shower stalls. The floor is the shower base and slopes toward the drain. The drain location can vary but is usually in the center of the base. Codes dictate that a non-handicap shower must be at least 30" × 30" (900 square inches) and have a threshold so water does not spill onto the area outside the shower. The drain location and the size of

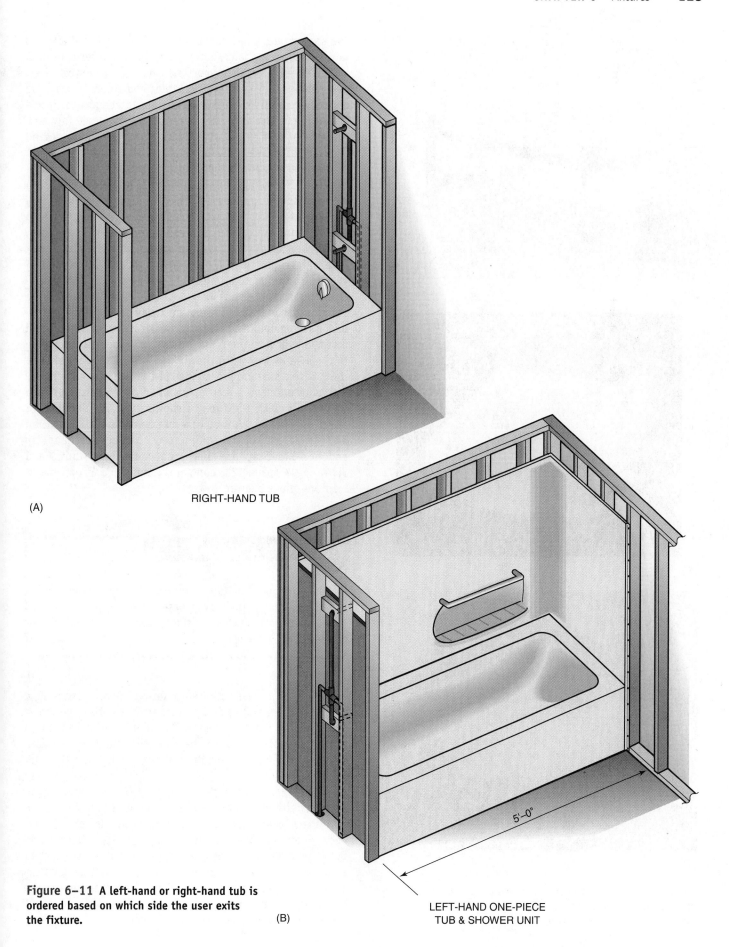

Figure 6–11 A left-hand or right-hand tub is ordered based on which side the user exits the fixture.

the shower must be known before installing the drain and faucet. ADA-compliant shower bases have a lower threshold and a larger area than most non-handicap shower bases. Showers are sold in numerous styles and colors; some have shower doors and others use a curtain to keep water inside the shower. Multi-piece shower units are available in which the shower base is installed during the rough-in phase and the walls after the wallboard is installed.

One-piece shower units are installed during the rough-in phase. Many homes have tile shower bases and walls, in which case the plumber installs the drain, water piping, and faucet. A tiled shower base installed on a wood floor must have a safety pan installed before the tile is placed. A plumber usually installs the safety pan. The most common safety pan is PVC liner material sold in 4' and 6' widths and in custom lengths. In some areas of the country copper safety pans are installed. Many showers have seats, and handicap showers must have a seat or be capable of accommodating a wheelchair. If a seat is constructed in a tiled shower, the plumber must provide waterproofing to the seat as well as to the shower base. One-piece shower units with a seat are pre-molded at the factory. Many handicap showers have a factory-installed folding seat. Shower drains vary with the type of shower base. A fiberglass or other pre-molded shower base uses a different drain than many tiled shower bases. Figure 6–13 compares handicap and non-handicap one-piece tub and shower units, and Figure 6–14 (page 128) compares handicap and non-handicap one-piece shower units.

FROM EXPERIENCE

A fixture manufacturer typically provides compatible shower drains with shower bases or one-piece units. A plumber must purchase a three-piece shower drain assembly for a tile shower base.

A corner shower is often referred to as a neo-angle shower. It is installed in the corner of a bathroom and is available in a variety of widths. The drain location is dictated by the size of the shower base, so the manufacturer data must be reviewed during the rough-in phase. Many corner showers are sold in multi-piece sets, but one-piece sets are also available. In a one-piece unit, the two corner walls are typically either tiled or have fiberglass wall panels, and the two sidewalls and door are glass. This design is very popular because of its appearance and its unique use of floor space. Some benefits of a corner shower are that it does not need additional wall framing to support the glass walls, and its angled doorway makes use of small spaces that otherwise would not adhere to code clearances. Figure 6–15 (page 129) shows a typical one-piece corner shower.

(A)

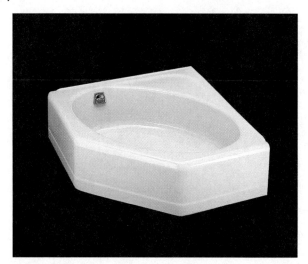
(B)

(C)

Figure 6–12 Many tub designs are available, and their specific installation requirements must be reviewed before pipe installation begins. *Courtesy of Kohler.*

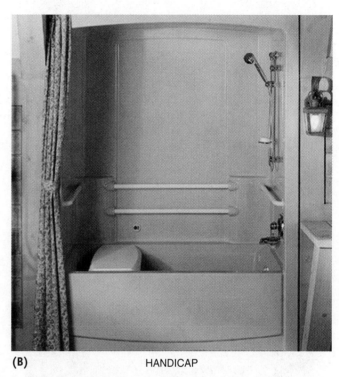

(A) NON-HANDICAP (B) HANDICAP

Figure 6–13 A one-piece tub and shower unit is the most common for residential bathrooms. *Courtesy of Kohler.*

Kitchen Sinks

Most residential kitchen sinks have either a single or a double bowl and are available in several different materials. Stainless steel is the most competitively priced material, and cast iron is durable and often considered to be a fixture upgrade. Cast iron is available in a variety of colors, and both materials come in a variety of styles that vary in size, depth, and shape. Another popular lightweight material is soapstone. Most kitchen sinks are surface mounted and installed into a countertop, but solid-surface countertops can have sinks mounted from underneath. Surface-mounted sinks, also known as self-rimming sinks, are manufactured with holes for installing the faucet directly onto the sink. Holes must be drilled into the countertop surface to install faucets for most under-counter sinks. Plumbers typically do not cut or drill the required holes for a sink or faucet, but they do provide the information needed to do so. Figure 6–16 compares sink styles that install onto and under a countertop. Figure 6–17 and Figure 6–18 (page 130) compare a stainless steel and a cast iron sink installed in a countertop.

 FROM EXPERIENCE

A single-bowl residential kitchen sink is usually 25" × 22", and a double-bowl sink is typically 33" × 22". A self-rimming stainless steel sink is secured to the underside of a countertop, and a typical cast-iron sink rests on the countertop with no attachments.

A kitchen sink is usually installed in a standard-width cabinet known as a sink base. The piping serving the sink is fairly typical for most sinks; the piping during the rough-in phase of construction can often be done before the sink is selected. Most kitchen sinks have a garbage disposal unit. The size of the drain opening is standard for all kitchen sinks. Most kitchens are equipped with an automatic dishwasher installed adjacent to the kitchen sink by a plumber. The hot water supply to the kitchen sink also serves the dishwasher. The drain for the dishwasher connects to the drain serving the kitchen sink, or it can drain into a designated port of a garbage

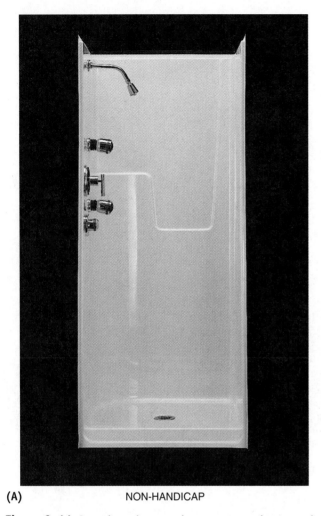

(A) NON-HANDICAP (B) HANDICAP

Figure 6–14 One-piece shower units are common in master bathrooms. *Courtesy of Kohler.*

disposal. Many kitchens have a specialty sink located adjacent to the larger sink or installed in an isolated island within the kitchen. Some specialty sinks, such as a bar sink, are installed for entertaining; others are for food preparation. A basket strainer connecting the piping to the sink keeps objects that could obstruct the drain from entering the drainage system. Many smaller sinks, such as a bar sink, have smaller drain openings and require a smaller basket strainer. Installation of a basket strainer, garbage disposal, dishwasher, and other accessories for a kitchen sink are explained in the fixture installation section of this book. Figure 6–19 compares some specialty sinks often located in a kitchen.

FROM EXPERIENCE

Specialty sinks are often considered an upgrade and are not usually installed in a standard residential home. Many sinks have a three-bowl design that includes a specialty bowl and two standard bowls.

The faucet hole spread for a kitchen sink is typically 8″ from the hot to the cold faucet connections. Most kitchen sinks are offered as a three-hole design, but, if a separate handheld sprayer is to be used, the sink must be ordered as a four-hole design. Many kitchen faucets have pull-out spouts that are also handheld sprayers. These are typically installed in a three-hole design sink. The center hole is aligned with the center of the sink, which allows the spout to flow into the desired bowl of a double-bowl sink. A common accessory on kitchen sinks is a soap dispenser, which may necessitate a four-hole design. Some faucets only require a centered single hole, which means the sink must be special ordered. Figure 6–20 (page 131) compares two common layouts for kitchen sinks with faucet hole options. The hole locations serve particular designs, including those with different dimensions between holes.

Laundry Sinks

A laundry sink is also called a laundry tray or utility sink. They are typically installed in the room with the washing machine, but they might also be installed as a

CHAPTER 6 Fixtures 129

Figure 6–15 A corner shower provides a unique appearance and accommodates design challenges in smaller bathrooms. *Courtesy of Kohler.*

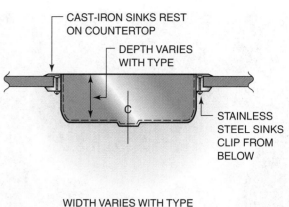

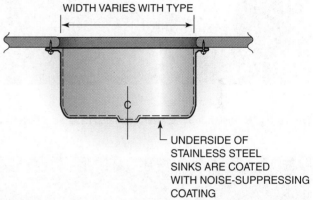

Figure 6–16 Some sinks are installed into a countertop, and others install under a countertop.

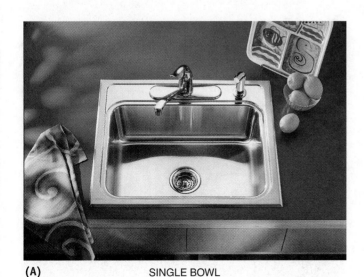

(A) SINGLE BOWL

(B) DOUBLE BOWL

Figure 6–17 Stainless steel is one of the most common materials for sinks installed in residential kitchens. *Courtesy of Kohler.*

130 SECTION TWO *Fixtures and Equipment*

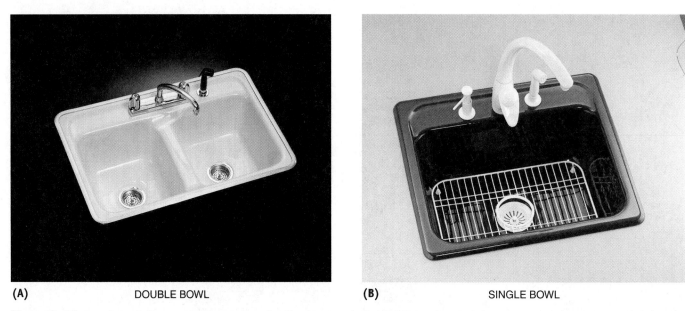

(A) DOUBLE BOWL (B) SINGLE BOWL

Figure 6–18 Cast-iron sinks are usually considered a fixture upgrade from stainless steel sinks. *Courtesy of Kohler.*

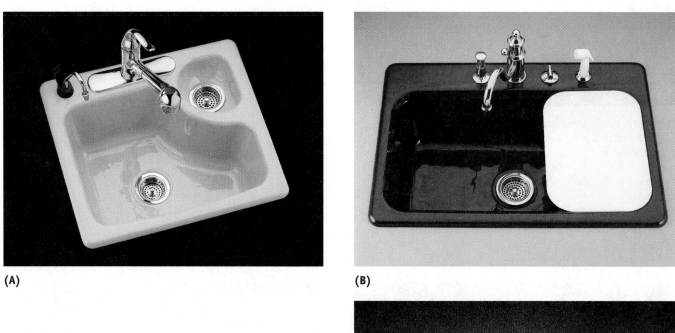

Figure 6–19 Specialty sinks give a unique look to a kitchen. *Courtesy of Kohler.*

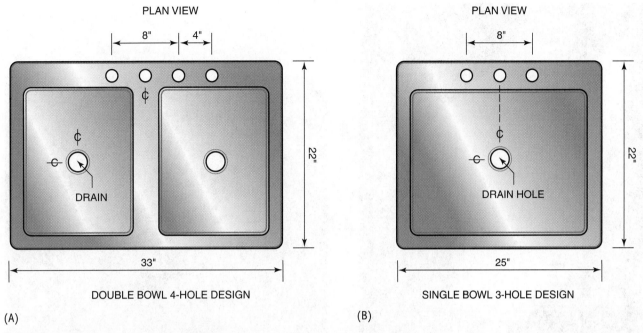

Figure 6–20 Kitchen sinks are available with numerous faucet hole design options.

utility sink in a garage or workshop area. The name laundry tray is a carryover from the days when clothes were hand-washed, and many laundry sinks had a countertop (tray) on one or both sides of the bowl. Today's laundry sink is a deep single bowl that allows clothes to be immersed for soaking. Many codes allow a washing machine to discharge directly into a laundry sink because of its deep bowl design. The drain opening of a laundry sink is similar to that of a bar sink; some manufacturers mold the drain into the bowl and do not require a separate basket strainer. Fiberglass is the most common material for manufacturing a laundry sink, but other plastic materials provide a less-expensive option. Laundry sinks are wall hung or floor mounted. As with any wall-mounted fixture, wood support must be installed by a plumber during the rough-in phase of construction. Figure 6-21 shows a typical floor-mounted laundry sink.

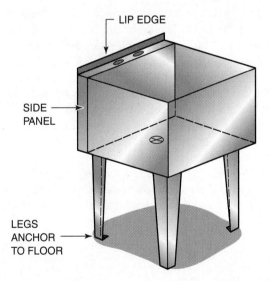

Figure 6–21 A laundry sink is installed in a washing machine area or a utility area.

FROM EXPERIENCE

A floor-mounted laundry sink is anchored to the floor, but the top of the sink must also be secured to a wall. A wall-mounted laundry sink must be securely anchored to a wood structure so it is not pulled from the wall when filled with water.

Bidets

A bidet is a specialty fixture that is not installed in most homes. It is a common fixture in many bathrooms in Europe and is growing in popularity in America. **Custom** homes typically install a bidet in a master bathroom. It is a personal hygiene fixture adjacent to a toilet. Bidets are available in

(A) (B)

Figure 6–22 A bidet is a personal hygiene fixture typically installed in a master bathroom. *Courtesy of Kohler.*

styles that match an adjacent toilet and are usually sold as a set with the toilet. The drain and water connection sizes are like those of a lavatory sink. They anchor to the floor rather than to a connecting flange, such as a closet flange that connects to a toilet. The bidet has a faucet located on the top of the fixture and a small cleansing spray head within the bowl. The faucet hole spread of a bidet varies, but is similar to a lavatory sink. Many styles have a seat similar to the adjacent toilet seat; others do not have a seat. Because the cleansing sprayer is located below the flood level rim of the bidet, the water supply must be protected against backflow with a vacuum breaker assembly. A bidet faucet is sold with the required vacuum breaker. Figure 6–22 shows a typical bidet in a master bathroom.

Summary

- The rough-in phase of construction is before the wall finishes are installed.
- The trim-out phase of construction is the final installation phase.
- Most fixtures are provided with a manufacturer rough-in sheet and installation instructions.
- A toilet has a maximum of 1.6 gallons per flush.
- The most common distance between a toilet and a rear wall is 12″.
- The two toilet bowl designs are round and elongated.
- A toilet seat must be the same shape as the bowl design.
- A bathroom sink is known as a lavatory.
- The most common faucet hole spread of a lavatory is 4″.
- A bathtub is available in numerous designs.
- A tub and shower unit is a one-piece fixture.
- Most bathtubs are installed before the wall finish is complete.
- A drop-in tub is typically installed on a platform constructed by a carpenter.
- A kitchen sink is installed during the trim-out phase of construction.
- The two most common laundry sink designs are wall hung and floor mounted.

Review Questions

1. The two bathtub designs indicating where the drain is located are known as
   a. Left hand and right hand
   b. Left head and right head
   c. Front wall and back wall
   d. None of the above are correct

2. A pre-molded shower base and a tiled shower base
   a. Can have different floor slopes by code
   b. Use different drain assemblies
   c. Have different drain sizes
   d. Are purchased at a plumbing wholesale outlet

3. A kitchen sink and a bar sink have
   a. The same size basket strainers
   b. Different size basket strainers
   c. The capability of accepting a garbage disposal
   d. Integral p-trap features

4. The most common flushing action of a residential toilet uses
   a. A blow-out design
   b. A siphon-jet design
   c. A deluge design
   d. An external p-trap

5. A bidet is a specialty fixture for
   a. Personal hygiene
   b. Bathing
   c. Clothes washing
   d. Hand washing

6. The most common hole design of a lavatory sink is
   a. A 4" spread
   b. A 6" spread
   c. A 4 hole
   d. A single hole

7. A handicap fixture must comply with the
   a. ADA
   b. DDA
   c. AAD
   d. DAA

8. A laundry sink is also used in residential applications as a
   a. Kitchen sink
   b. Utility sink
   c. Lavatory sink
   d. Bathing fixture

9. A one-piece tub and shower unit must be installed
   a. During the trim-out phase of construction
   b. After the doors and windows are installed
   c. During the rough-in phase of construction
   d. Before the plywood floor is installed

10. The two most common faucet hole layout designs for a kitchen sink are a three hole and a
    a. four hole
    b. two hole
    c. one hole
    d. five hole

11. **A bidet has a cleansing sprayer within the bowl, so the water supply piping requires**
    a. A gate valve
    b. A vacuum breaker
    c. Insulation
    d. Ball valve

12. **The most common bathtub is**
    a. 4' long
    b. 5' long
    c. 6' long
    d. 3' long

13. **Code usually requires a tile shower base installed on a wooden floor to have**
    a. A safety pan
    b. A special permit
    c. 6" square tiles
    d. 4" square tiles

14. **The common location of a toilet handle is on the**
    a. Right side of the tank
    b. Top of the tank
    c. Left side of the tank
    d. None of the above are correct

15. **A large bathtub design that does not have whirlpool features is commonly called**
    a. An oversized bathtub
    b. A garden tub
    c. A drop-in tub
    d. A hot tub

Chapter 7: Faucets and Drain Assemblies

You have learned about the types of fixtures commonly installed in residential construction. This chapter focuses on faucets and drain assemblies. Faucets are the outlet for the water we use, and drain assemblies connect plumbing fixtures to drainage systems with specific connections. Most fixtures have unique drain outlet connections, and drain assemblies are manufactured to accommodate each type. The drain outlet size varies with the particular fixture, and even though the drainage and vent piping serving the fixtures may be the same size, the drainage piping must be adapted to the fixture outlet size.

Faucets have evolved as decorative features and may create a theme in a bathroom. They are made in a variety of styles and finishes and with several basic operating features. Faucet selection often determines the type of fixture selected, but not all faucets are compatible with all fixtures. A manufacturer makes faucets that are compatible with certain fixtures, thereby creating categories of faucet designs. Kitchen and bathroom faucets are categories that are relevant to residential construction. Your knowledge of plumbing fixtures will help you learn about the correct faucet designs for each fixture category.

OBJECTIVES

Upon completion of this chapter, the student should be able to:
- understand the differences in basic faucet designs.
- recognize various faucet styles and finishes.
- identify the variations in fixture outlets and drain assemblies.
- order the correct faucet and drain assembly for a particular fixture.

Glossary of Terms

aerator removable threaded housing for a screen attached to a faucet spout that creates a uniform flow stream

BW&O abbreviation for bath waste and overflow drain assembly used on bathtubs

custom home a house built with fixtures upgraded from the basic fixtures installed in a spec home

diverter a device that routes the water from a tub spout to a showerhead, or from a showerhead to a handheld shower unit

escutcheon flange installed around a pipe to conceal pipe penetrations through a wall, floor, or ceiling

finish the color or polish of a faucet, drain assembly, or other fixture trim item

knock-out a manufactured portion of a sink or garbage disposer designed to be removed by an installer to install a faucet or dishwasher drain hose

pop-up drain assembly for lavatory sinks and bidets; abbreviated as PO

port opening an opening in a fixture, such as a drain or overflow hole that receives drain assemblies, to connect the fixture to the drain system

spec home a house built using average construction quality and product selection

T&S faucet abbreviation for a combination type faucet serving a tub and shower unit

trim refers to items that have chrome or other finishes

Faucets

There are so many faucet styles and faucet manufacturers that a plumber must constantly be updated on new styles and finishes. Each manufacturer publishes a product catalog to display its unique designs. Selecting which manufacturer's product to install is based on cost, quality, and preferred faucet design. A spec home may have basic faucet styles with mostly chrome finishes, but a custom home typically upgrades the fixture selection to include more popular finishes and more expensive faucet styles. Most master bathrooms and guest bathrooms have more expensive faucets than other bathrooms in a house. The faucet finish dictates what will be used for drain assemblies and bathroom accessories to create a color theme. Many faucets are available with different handle designs to help customize a bathroom. The exterior appearance of faucets can create the illusion that they operate differently, but manufacturers often use the same internal operating design for many faucets with different appearances. Although electronic operating features are available for faucets, the residential industry uses manually controlled faucets.

Faucets are categorized by the intended fixtures served and include a variety of styles for each fixture. Table 7-1 lists the faucet categories and the common handle configurations for each used in a residential home. It is helpful when manufacturers identify bathroom faucets based on their style, which allows a plumber to install the same style of faucet throughout a bathroom. A faucet designed for a kitchen sink would not be suitable for a shower and would not be offered as a bathroom faucet because it is not located in the same room.

A bathtub and a lavatory faucet would be purchased in the same style and finish. Some faucets operate with a single handle and others use separate handles for the hot and cold water. It is not uncommon to install a single-handle shower faucet and a two-handle lavatory faucet, but a whirlpool or garden (roman) tub faucet handle is usually the same style as the lavatory. Metal and acrylic are often used to manufacture faucet handles. Chrome, polished brass, and white are three popular finishes for faucets and drain assemblies. A combination of chrome and polished brass are sometimes used together in one faucet. Figure 7-1 shows three single-handle lavatory faucets with different handles and finishes.

FROM EXPERIENCE

A faucet installed on a sink through a countertop or through a tub platform is considered a deck-mounted faucet.

All faucets must be designed to prevent backflow of wastewater into the water distribution system. All codes dictate that a faucet must have an air gap or be protected with a vacuum breaker or approved check valve. The air gap codes cover both water distribution and drainage systems. This chapter discusses water distribution, and the drainage aspects are covered in the relevant portions of this book. The popularity of pull-out spray faucet spouts and handheld shower attachments has increased the threat of contamination to a potable water system. The backflow prevention accessories vary with each faucet and are explained with the relevant faucets throughout this chapter. Figure 7-2 is an illustration of a faucet air gap. Most faucets are manufactured to comply with water-distribution air gap codes.

FROM EXPERIENCE

The only completely effective form of backflow prevention is through the installation of an air gap. Many faucet accessories that attach to an aerator violate the air gap code because they lower the air gap below an allowable distance.

Water piping connects to faucets in a number of ways. Table 7-2 indicates the usual method for each faucet category. Some faucets connect to the water supply with male or female adapters and copper tube to create a soldered connection; others use a specially designed 3/8" OD supply tube that connects to male threads of a faucet. Table 7-2 lists the categories of residential faucets and their common forms of connections, which are determined by the manufacturer's design and the faucet style.

Table 7-1 Faucet Categories and Handle Options

Category	Single Handle	Two Handle	Three Handle
Lavatory	Yes	Yes	No
Bathtub[1]	Yes	Yes	No
Whirlpool or garden tub[2]	No	Yes	No
Shower	Yes	Yes	No
Tub and shower	Yes	Yes	Yes
Kitchen	Yes	Yes	No
Laundry	No[3]	Yes	No
Bidet	Yes	Yes	No

[1] Installed in wall
[2] Installed on tub
[3] A single handle lavatory faucet can be installed on a laundry tub

138 SECTION TWO *Fixtures and Equipment*

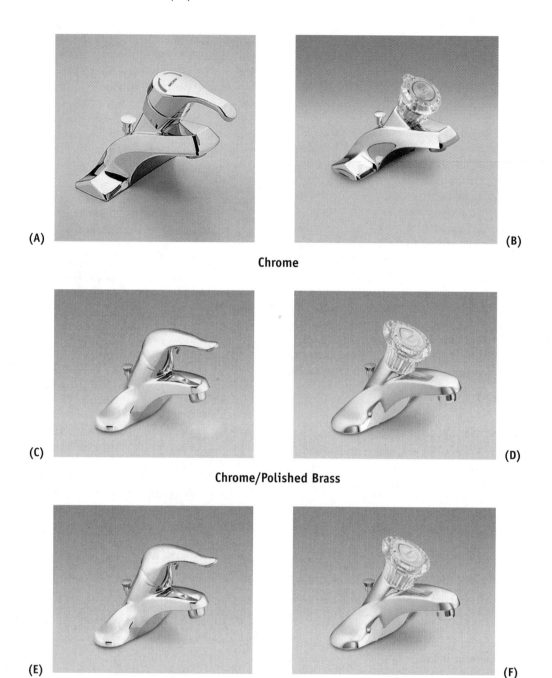

Chrome

Chrome/Polished Brass

Polished Brass

Figure 7-1 Faucet categories and handle options. *Courtesy of Moen, Incorporated.*

FROM EXPERIENCE

The connection to each faucet varies depending on the particular faucet, the type of water piping installed, and company preferences.

Lavatory Faucets

The most common lavatory faucets have a 4" spread between the hot and cold handles. Many lavatory faucets have a one-piece body design, and some have a three-piece design consisting of a hot and cold handle and a spout. Many three-piece faucets can be installed in 4", 6", and 8" spread lavatories, but others only work in 8" spread sinks. Single-handle and two-handle faucets are most popular. A single-handle faucet blends the hot and cold water using one handle. The

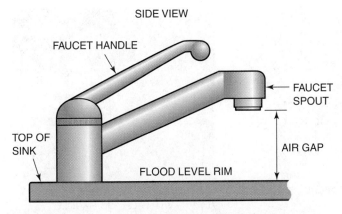

Figure 7-2 An air gap protects against backflow of wastewater into a potable water system through a faucet. *Courtesy of Moen, Incorporated.*

spout is located directly below the handle. With a two-handle design, the user must operate both handles to achieve a desired temperature. A single-handle faucet typically uses soft copper tubing to connect with the water-supply tube under the sink. Most 4″ spread two-handle faucets use a male-threaded extension to connect the hot and cold water supply tubing to the faucet. More expensive two-handle faucets can be installed in a 6″ and 8″ spread lavatory sink. The center hole of a lavatory sink is designated for the spout. Figure 7-3 shows several faucet styles.

 FROM EXPERIENCE

A 6″ and an 8″ spread faucet is often the same faucet, simply adjusted on a job site. A 6″ spread faucet is usually installed in a solid surface countertop that has had holes drilled on the job site.

A lavatory faucet installs onto a plumbing fixture in various ways depending on the type of faucet and the manufacturer's design. Most faucets use a tightening nut and flat washer to secure the faucet to a fixture. Installation instructions are included with each faucet, and a plumber should refer to this important information when installing an unfamiliar faucet design. Figure 7-4 illustrates a typical two-handle sink faucet connection to a fixture and the water piping connection.

 FROM EXPERIENCE

A faucet-connecting nut is tightened and loosened with a basin wrench. Many connecting nuts are plastic.

Bathtub and Shower Faucets

A bathtub faucet and shower faucet are two separate items with two different purposes. A tub faucet fills a tub, and a shower faucet serves a showerhead. A tub faucet is either deck mounted or installed in a wall. Deck-mounted faucets are used for large-capacity tubs, such as a garden (Roman) tub or whirlpool. A tub faucet installed in a wall usually also serves a showerhead, which makes it a combination tub and shower faucet. A faucet serving a tub or shower is often called a tub valve or shower valve even though, by definition, it is a faucet. Figure 7-5 is a single-handle shower faucet, but two-handle designs are also available.

 FROM EXPERIENCE

In a two-handle faucet, an internal washer usually stops the water flow; most single-handle designs are washerless.

Table 7-2 Common Faucet Connections

Category	Male Threads[1]	Female Threads[2]	Soldered[3]	3/8″ Tubing[4]
Lavatory	Yes	No	No	Yes
Bathtub	Yes	Yes	Yes	No
Whirlpool or garden tub	Yes	No	Yes	No
Shower	Yes	Yes	Yes	No
Tub and shower	Yes	Yes	Yes	No
Kitchen	Yes	No	No	Yes
Laundry	Yes	No	No	No
Bidet	Yes	No	No	Yes

[1] Connects with a female adapter or a specially designed supply nut
[2] Connects with a male adapter
[3] Receives copper tube
[4] Has 3/8″ OD tubing from factory

140 SECTION TWO *Fixtures and Equipment*

(A) TWO-HANDLE THREE PIECE

(B) TWO-HANDLE ONE PIECE

(C) SINGLE-HANDLE ONE PIECE WITH ACRYLIC HANDLE

(D) SINGLE-HANDLE ONE PIECE WITH HANDICAP LEVER HANDLE

(E) TWO-HANDLE THREE PIECE 8" SPREAD WITH LEVER HANDLES

(F) TWO-HANDLE THREE PIECE 8" SPREAD WITH CROSS HANDLES AND GOOSENECK SPOUT

(G) TWO-HANDLE THREE PIECE 8" SPREAD WITH LEVER HANDLES AND GOOSENECK SPOUT

Figure 7–3 Various styles of lavatory faucets allow a bathroom to be customized based on preference and cost.

A tub and shower **(T&S) faucet** provides water for bathing or showering by using a **diverter.** There are a number of ways to divert the water flowing through a tub spout to flow through a showerhead. When a T&S faucet is activated, the water flows to the tub spout, and the user must manually divert the water to the showerhead. A single-handle T&S faucet uses either a diverter-style tub spout or a push-button diverter located directly below the faucet handle. In a three-handle T&S faucet, the middle handle is the diverter. The body portion of a T&S faucet is installed during the rough-in phase of construction; the **trim** components are added after the wall finish is complete. Figure 7–6 shows a tub spout with a diverter feature. Figure 7–7 illustrates a tub and shower faucet in a wall. Figure 7–8 shows various types of T&S faucets.

CHAPTER 7 Faucets and Drain Assemblies   141

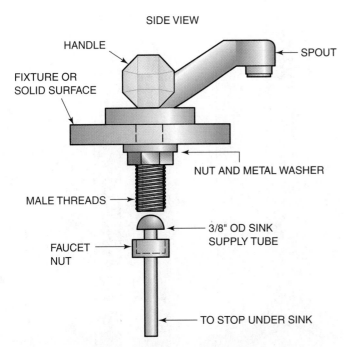

Figure 7-4 Faucets are tightened to a fixture with a nut and washer, and the water supply tubing connects to the faucet using various methods.

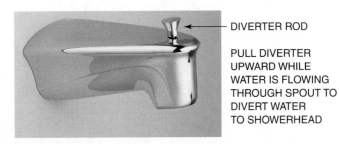

Figure 7-6 A tub spout can also divert water from the spout to a showerhead. *Courtesy of Moen, Incorporated.*

FROM EXPERIENCE

Some tub and shower faucets can be converted to only shower faucets by capping or plugging the outlet designated for the tub spout, but a shower valve cannot be converted to a tub and shower faucet.

Each shower faucet is sold with a wall **escutcheon**, a shower arm, and a basic showerhead. Showerheads can be purchased separately to provide a unique appearance or to create a desired spray pattern. Most showerheads are equipped with internal flow-regulators that only allow 1 gallon per minute (gpm) of water. Not all shower arms are compatible with every showerhead, but all have 1/2" threaded pipe nipples on the end that connect to the in-wall piping and are angled downward to direct the water flow toward the user. Most shower arms also have a 1/2" male thread on the end that connects to the showerhead, but some have pivot ends. The finish of the shower arm typically matches the finish of the faucet. Figure 7-9 shows two different showerhead designs with flow-pattern adjustment features.

FROM EXPERIENCE

A shower arm with no accessories to stop the flow of water through the showerhead is never subjected to high water pressure. Therefore, it does not have to be tightened as much as other threaded connections. A strap wrench can tighten the shower arm, so the finish is not damaged.

Many custom bathrooms or master bathrooms have more expensive faucet configurations, including handheld shower units. Most handicap codes dictate that a shower must be equipped with a handheld shower unit. These can be installed in a variety of ways. Some connect the piping system externally to the shower arm; others are installed with the

Figure 7-5 A single-handle shower faucet is a popular design. *Courtesy of Moen, Incorporated.*

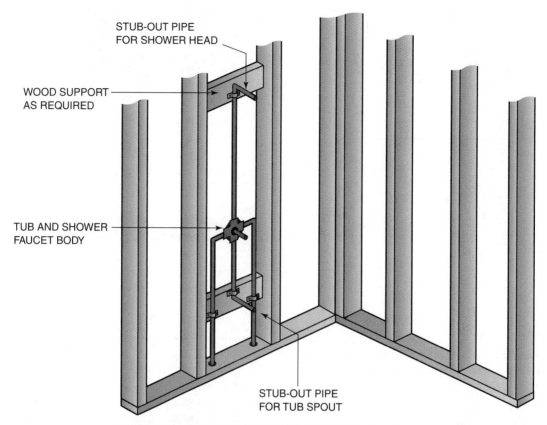

PIPING AND FAUCET ARE INSTALLED IN THE WALLS DURING THE ROUGH-IN PHASE OF CONSTRUCTION. FAUCET TRIM IS NOT INSTALLED UNTIL THE COMPLETION OF WALL FINISH, SUCH AS TILE AND GROUT.

Figure 7-7 A tub and shower faucet is installed in a wall, and the trim is installed after the walls are finished.

piping in the wall during the rough-in phase of construction. One piping configuration has a separate diverter valve that directs the flow of water from a shower head to a handheld shower. Most codes dictate that a vacuum breaker must be installed when a handheld shower is installed. Figure 7-10 shows a complex configuration of a shower faucet and several diverter valves that offer numerous showering options. Figure 7-11 is a typical vacuum breaker installed externally to protect against backflow if the handheld shower unit is placed in wastewater.

FROM EXPERIENCE

Some handheld shower units connect to the shower arm serving the shower head with a diverter tee, which eliminates the need for separate piping. However, this creates an aftermarket appearance and may not be satisfactory in some custom homes.

A framing contractor often erects a platform structure on a job site for a plumber to install large-capacity drop-in tubs. Most whirlpool or garden tubs are installed with a platform, but some have self-supporting structures. Most large-capacity tubs have a deck-mounted faucet placed so the user can fill the tub before entry. The solid surface of the tub must be ordered with the faucet holes in the correct location, or they must be drilled on the job site. A plumber might need to drill the faucet holes in a tub installed in a platform structure. If a platform is going to have a ceramic tile surface, a plumber must install the body of the faucet before the tile is placed and install the trim after the grout is applied between the tiles. The typical deck-mounted faucet has an 8" spread, and various styles are available. Figure 7-12 is an illustration of a typical drop-in tub on an erected platform that indicates a possible faucet location. Figure 7-13 compares some deck-mounted tub faucets.

FROM EXPERIENCE

Many faucets are sold with the rough-in and trim items in a single box, which creates a material-handling concern. The trim items should remain in the box and be placed in a secure location so they are not damaged or lost before the trim-out phase of construction.

CHAPTER 7 Faucets and Drain Assemblies 143

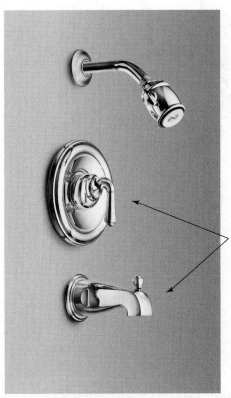

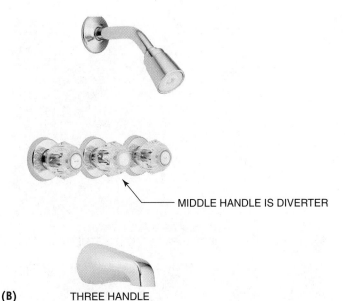

(B) THREE HANDLE

(C) TWO HANDLE WITH DIVERTER SPOUT

Figure 7–8 Various styles and types of tub and shower faucets are installed based on cost and preference. *Courtesy of Moen, Incorporated.*

Kitchen Sink Faucets

Cost and personal preference are usually the deciding factors in selecting a kitchen sink faucet. A kitchen faucet has a swivel spout, which allows the water flow to be used in both bowls of a kitchen sink. The styles and finish selections are vast. Chrome-plated faucets are installed in most spec homes, but many custom homes have color finishes. Spec homes tend to have more price-competitive faucets and custom homes typically install more expensive ones. Not all kitchen faucets fit every type of kitchen sink. A faucet selection can dictate a sink selection, but the sink choice can also determine the style of faucet that can be installed. The design of the faucet dictates the number of faucet holes required in a sink.

The most common type of kitchen faucet requires a sink with three faucet holes that are 4" apart or 8" from the hot and cold water supply connections to the faucet. Some single-handle faucets can be installed in a three-hole sink, because they are sold with a cover plate to conceal the unused hot and cold faucet holes. A competitively priced faucet with a spray unit requires a four-hole sink as shown in Figure 7–14.

 FROM EXPERIENCE

Sink hole covers, available in a variety of finishes, can be used to cover any unused holes in a sink. They are not desirable, however, because their presence indicates that an installation oversight occurred.

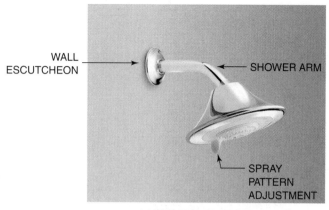

Figure 7-9 Various shower head designs are available to customize a bathroom and adjust the spray pattern. *Courtesy of Moen, Incorporated.*

Many manufacturers make faucet base plates with built-in spray units. This eliminates the need for a four-hole sink. Features like soap dispensers are sometimes added into the base plate of a faucet, eliminating the need for a more expensive sink. A plumber should remain knowledgeable about new designs and be creative in satisfying a customer's request. Figure 7-15 is a faucet with two features that can be installed in a three-hole sink.

 FROM EXPERIENCE

Soap dispensers that match the faucet finish have become very popular. The soap storage container is installed from below the sink and is removable from there. It is filled from above the sink.

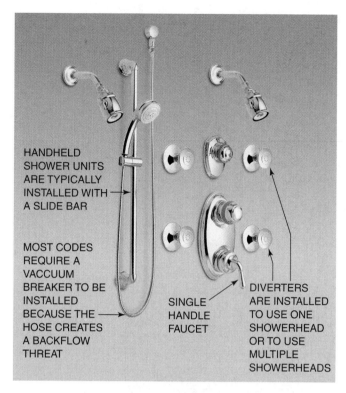

Figure 7-10 Hand-held shower units are required to satisfy most handicap codes and are popular in master bathrooms. *Courtesy of Moen, Incorporated.*

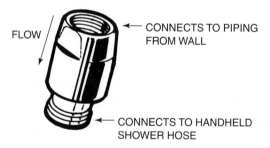

Figure 7-11 A vacuum breaker is required when installing a handheld shower because the hose can be immersed in the wastewater. *Courtesy of Moen, Incorporated.*

Faucets designed with a pull-out spout that is also the spray unit are very popular, and there is no need for a four-hole sink. The pull-out spout can be immersed in wastewater, which can allow backflow that threatens the potable water system. Most of these faucets are manufactured with an integral check valve to eliminate backflow, but many code officials require a second form of backflow prevention. A dual check valve will usually suffice as an additional means of backflow prevention. Figure 7-16 shows a pull-out spout faucet.

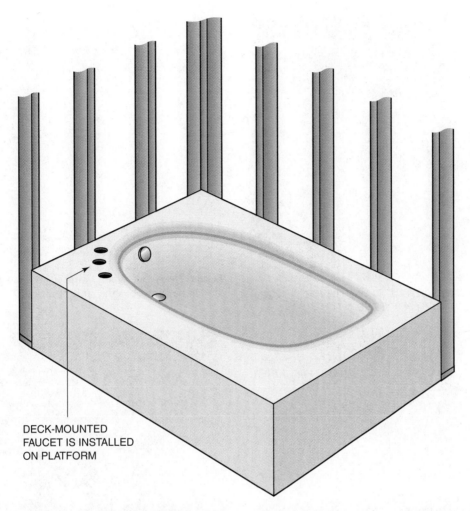

Figure 7–12 Large-capacity tubs are often installed in a platform structure and require a deck-mounted faucet.

CAUTION

CAUTION: Check local codes to be sure that a separate backflow prevention device is not required before you install a pull-out spray faucet. An installer can be held responsible for violating a code even while employed by a company. Not knowing a code is not an acceptable argument in court.

Most manufacturers make an economical faucet to remain competitive. A two-handle faucet with no spray unit but with a one-piece body is often the least expensive kind of kitchen faucet. A three-hole kitchen sink is required for most two-handle faucets with no spray. This faucet is often found in rental homes and commercial office kitchenettes because of its lower cost. Some versions of a two-handle faucet—depending on the type of handle, the spout, and the manufacturer—can be more expensive than other faucet designs. Some two-handle faucets that do not have a base plate still require a three-hole sink (Fig. 7–18). A two-handle faucet without a separate spray unit like the one shown in Figure 7–17 is ordered as "less spray."

 FROM EXPERIENCE

A competitively priced faucet is not necessarily inferior, but rather has less style and often fewer handle options than more expensive models.

Many kitchen faucets are designed for filling large pots with water. A gooseneck spout has a unique appearance and is typically two handled, but can also have a single handle. They can be purchased with or without separate spray units and are often more expensive than other faucet designs. Gooseneck faucets are often used for specialty sinks, such as bar sinks, and are used extensively in commercial applications. Other faucet designs can be used to customize a kitchen, many of which use either a one-hole or two-hole sink. Figure 7–18 shows two different gooseneck-style faucets.

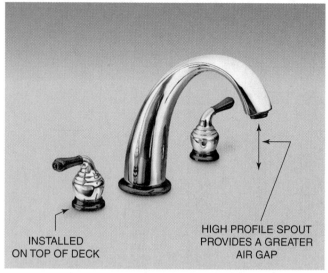

(A) DECK-MOUNTED TUB FAUCET WITH LEVER HANDLES AND HIGH PROFILE SPOUT

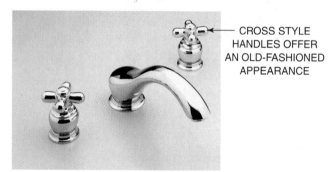

(B) DECK MOUNTED TUB FAUCET WITH CROSS HANDLES AND LOW PROFILE SPOUT

(C) DECK-MOUNTED TUB FAUCET WITH ACRYLIC HANDLES AND LOW PROFILE SPOUT

Figure 7–13 A deck-mounted tub faucet serves a whirlpool, and other large-capacity tubs are offered in a variety of styles. *Courtesy of Moen, Incorporated.*

 FROM EXPERIENCE

Most gooseneck spouts swivel from side-to-side so they can be used on a double bowl sink. A plumber should know that a rigid spout can only serve a single-bowl sink.

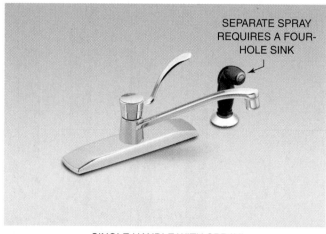

SINGLE HANDLE WITH SPRAY

Figure 7–14 One type of kitchen sink faucet using a four-hole sink has a separate spray unit. *Courtesy of Moen, Incorporated.*

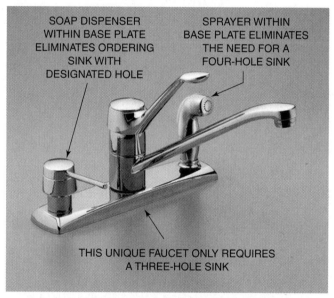

SINGLE HANDLE WITH SPRAY AND SOAP DISPENSER

Figure 7–15 A faucet design with a soap dispenser and a spray unit within the faucet base plate allows a three-hole sink. *Courtesy of Moen, Incorporated.*

Laundry Sink Faucets

Laundry tub faucets have a 4" spread and various spouts. Many laundry sink faucets have a swivel spout that does not get in the way when large items are placed in the sink. Some of these faucets have a hose thread on the outlet portion of the spout so a garden hose can be connected. To be legally installed, a hose-end spout must have a vacuum breaker to eliminate backflow into the potable water supply. Most residential-style laundry faucets are sold with **aerators** to eliminate a hose connection. A plumber must recognize the important differences between these designs.

CHAPTER 7 *Faucets and Drain Assemblies* **147**

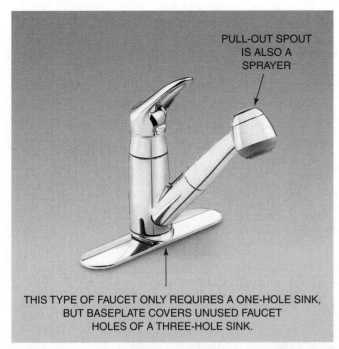

Figure 7–16 Pull-out spout faucets are popular, but they can jeopardize the potable water system if they are immersed in wastewater. *Courtesy of Moen, Incorporated.*

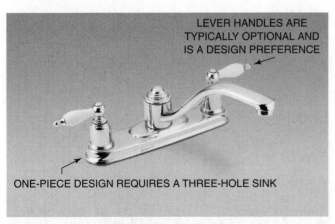

Figure 7–17 Two-handle faucets that do not have a spray unit offer an inexpensive option and can be installed in a three-hole sink. *Courtesy of Moen, Incorporated.*

A residential laundry sink has a ledge where the faucet is installed, adhering to air gap codes. Figure 7–19 shows a faucet with a hose-end spout that would need to have a backflow prevention device installed in the piping system or attached to the spout. Figure 7–20 shows two faucets with different spout ends; a typical laundry sink faucet has a swivel spout. Figure 7–21 is a laundry sink faucet with an aerator attached to the end of the spout; it does not require a backflow prevention device.

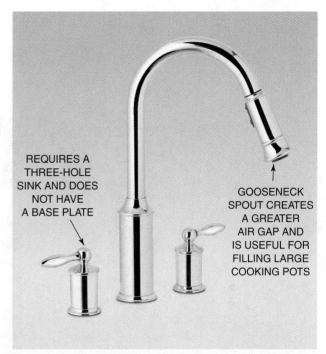

Figure 7–18 A gooseneck spout is a popular feature for filling large pots with water and creates a unique appearance. *Courtesy of Moen, Incorporated.*

CAUTION: A plumber should focus on all backflow prevention devices, but a laundry sink faucet with a hose-connection spout is a more serious threat than other faucet designs because a garden hose can be connected. View this type of faucet like a hose faucet on the exterior of a building.

148 SECTION TWO *Fixtures and Equipment*

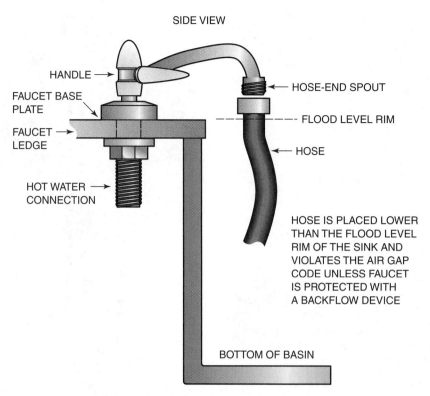

Figure 7–19 A hose-end spout of a laundry sink faucet that is not protected against backflow is illegal.

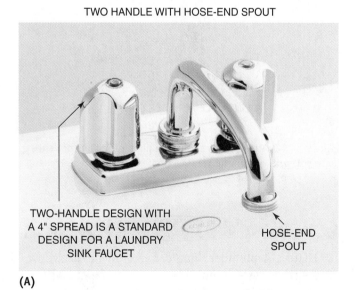

(A)

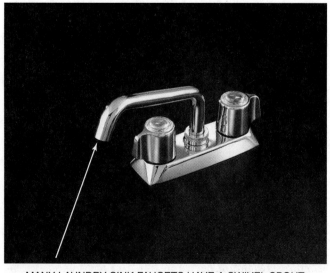

(B)

Figure 7–20 Laundry sink faucets are available with various spout designs, most of which swivel. *Courtesy of Kohler.*

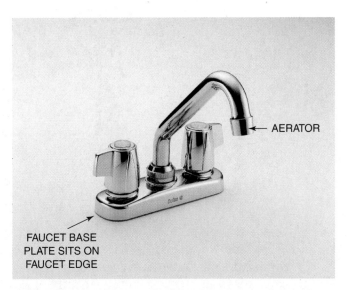

Figure 7-21 An aerator spout of a laundry sink faucet does not require a backflow prevention device. *Courtesy of Delta.*

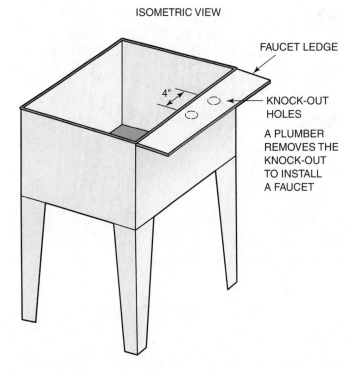

Figure 7-22 A laundry sink has a faucet ledge, so a faucet can be installed that complies with air gap codes.

 FROM EXPERIENCE

Laundry sink faucets typically have an internal washer to stop the flow of water. Most have a chrome finish.

Most residential laundry sinks are made from fiberglass or other soft materials and have **knock-out** holes, so the plumber can select the hole that matches the faucet being used. This allows a plumber to easily install either a kitchen or lavatory faucet on the sink ledge. A plumber can drill faucet holes in the ledge area if the faucet has a spread greater than the typical 4". Figure 7-22 is an illustration of a faucet ledge on a laundry sink.

Bidet Faucets

A bidet faucet must be compatible with the faucet holes and the vacuum breaker, if required. A vacuum breaker is required by code if the hygiene sprayer is located in the bowl area of a bidet, because it is below the flood level rim. The hygiene sprayer can be located in several different places, and a plumber must be sure that the correct faucet is purchased for the specific bidet. The style and finish of a bidet faucet is usually the same as the other faucets in the bathroom. Most have a two-handle design with no base plate, and a spread ranging from 4" to 8". Some are similar to a lavatory faucet, and others are single handled. Figure 7-23 shows several bidet faucet styles and designs.

 FROM EXPERIENCE

Though the term *knock-out* indicates that a plumber can forcibly remove material from the faucet hole locations, care must be taken not to damage the sink. A razor knife can be used to score the sink before tapping the knock-out area. A hole saw and drill can also be used to safely remove the knock-out areas.

 FROM EXPERIENCE

If a bidet requires a vacuum breaker, but does not have a hole for it, the rough-in water piping must be installed to accommodate the vacuum breaker. Review the fixture installation information during the rough-in phase of construction.

(A) TWO-HANDLE DESIGN WITH CROSS HANDLES

(B) TWO-HANDLE DESIGN WITH LEVER HANDLES

(C) SINGLE-HANDLE DESIGN

Figure 7–23 Bidet faucets are available in various styles and designs and must be compatible with the specific fixture.
A & B: Courtesy of Moen, Incorporated; C Courtesy of Kohler.

Drain Assemblies

Drain assemblies are chosen based on the fixtures they will serve. Codes dictate minimum drain sizes, and all fixtures and drain assemblies are manufactured according to those requirements. Most drain assemblies operate mechanically so the water level in the fixture can be controlled. Shower drains are not controlled mechanically and are considered to be floor drains. Most codes dictate that the largest foreign object that can enter a drainage system be 1/2″ diameter. Therefore, sink, shower, bidet, and bathtub drains must have straining capabilities. Most drain assembly connections must be sealed with putty or caulk to prevent leaks.

To identify a drain assembly, a plumber must know what fixture is being served and the manufacturer's designation for it. Drain assemblies vary depending on the manufacturer. The information provided here is generalized to show intent and is not specific to a particular manufacturer.

Lavatory Drain Assemblies

The drain assembly for a lavatory faucet is known as a **pop-up** (PO). Drains and faucets are usually sold together. The finish on the exposed portions of a pop-up matches the faucet. The PO assembly consists of several different elements that function as one unit. A PO rod, located within the faucet spout, is connected to a link assembly under the sink that operates the plunger. The rod is pulled upward to fill the

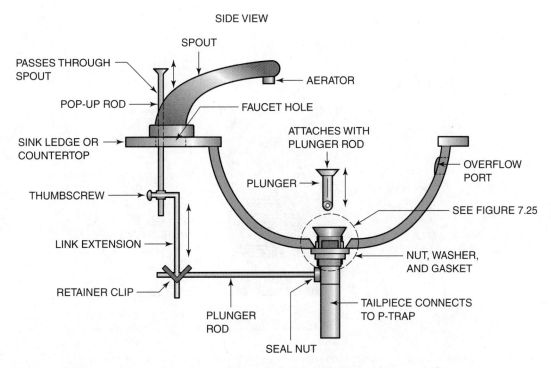

Figure 7-24 A pop-up drain assembly consists of several items linked together to work as one unit.

sink and pushed downward to drain it. The PO assembly also connects the fixture to the drainage system. All lavatory PO assemblies are 1-1/4" tubular size, and most are made of brass; however, some less expensive ones are plastic. The tubular tailpiece is not compatible with DWV pipe, so a trap adapter connects the two pipes. Many lavatories utilize tubular p-traps that connect with the drainage system. Figure 7-24 illustrates a PO assembly and how it operates.

 **FROM EXPERIENCE**

Some industry professionals think the abbreviation PO refers to **port-opening** device; others believe it stands for the first two letters in pop-up.

Most lavatories have an overflow port so water cannot rise above the rim of the fixture. The flanged drain portion of a PO assembly has a slot to receive water that enters the overflow port of a sink. The water then enters the drainage system. If a lavatory sink does not have an overflow feature, the slotted feature is irrelevant but can still be installed. In a lavatory overflow, the underside of the sink creates a hollow channel to drain high water levels into the PO flange body. There are many different designs, but most have a multiple-sectioned flange body. Figure 7-25 is a typical PO flange body design with an overflow slotted area.

 FROM EXPERIENCE

Most commercial sinks have a 1-1/2" grid strainer assembly, which does not have a PO feature but does have a mesh grid so objects cannot enter the drain system. Because there is no PO assembly, the sink cannot be used to retain water.

Bathtub Drain Assemblies

A bathtub drain assembly is termed a bath waste and overflow (**BW&O**). Several designs are available; the finish is purchased to match the faucet finish. A bathtub has an overflow port hole and a drain port that are always aligned vertically with the overflow hole above the drain hole. In most bathtubs, the holes are on the left or right end, but some tubs have holes located in the center. Large-capacity whirlpool tubs are more likely than standard tubs to have the holes in a variety of locations. If a tub is filled with water to a level above the bottom of the overflow hole, water enters the BW&O to discharge the water into the drainage system. The drain hole of a tub holds the drain assembly portion of the BW&O.

Trip-lever BW&O designs use an internal link assembly to control the water level in the tub. The user controls the internal operation manually with a lever located on the cover

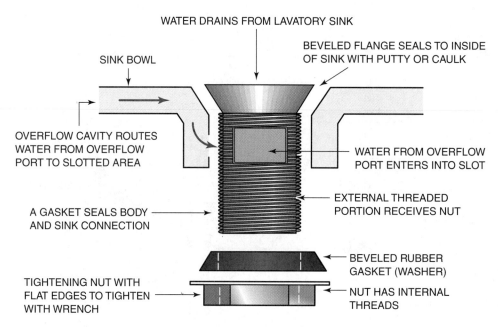

Figure 7-25 A pop-up flange body has a slotted area to receive water that drains through fixture overflow.

plate. When the lever is activated, the internal assembly raises and lowers a lift assembly. The internal link assembly is adjustable to satisfy different bathtub heights. A more competitively priced drain assembly is the touch-toe BW&O. It has no internal operating features; the user can fill and drain the tub by simply depressing the spring-loaded drain seal plug. Another popular style is the push-pull BW&O, which also does not have internal operating features. The push-pull design operates much like a touch-toe design. Figure 7-26 illustrates two different bath waste and overflow assemblies. Figure 7-27 illustrates the relation of a tub and a bath waste to an overflow assembly. Figure 7-28 is a trip-lever BW&O. Figure 7-29 is a push-pull BW&O that requires a plumber to install piping according to specifications for a specific bathtub.

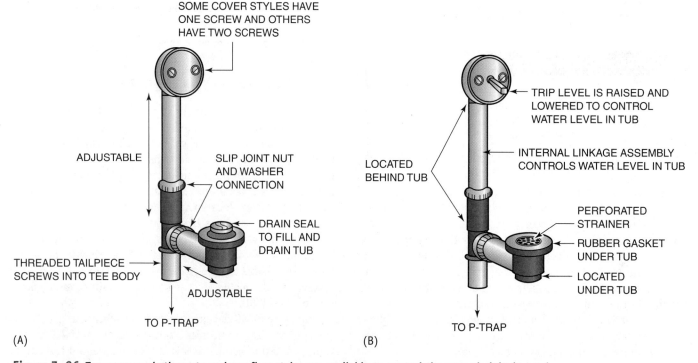

Figure 7-26 Two common bath waste and overflow styles are available to control the water height in a tub.

CHAPTER 7 *Faucets and Drain Assemblies* **153**

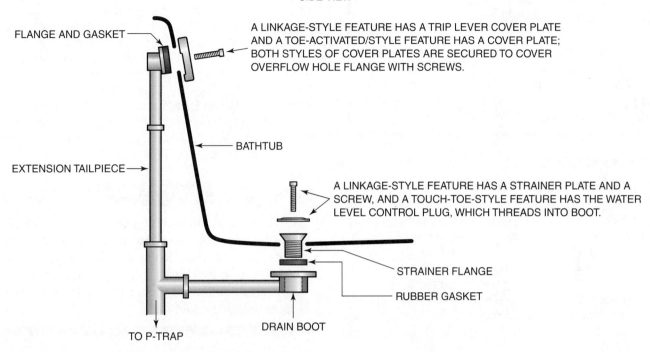

Figure 7–27 A bath waste and overflow assembly connects to the overflow outlet and drain outlet of a tub.

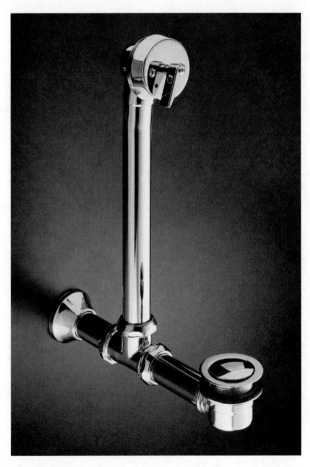

Figure 7–28 A trip-lever type of bath waste and overflow is adjustable in various tube heights. *Courtesy of Kohler.*

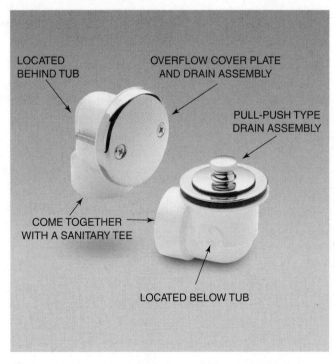

Figure 7–29 Many bath waste and overflow designs require a plumber to install piping to the requirements of a specific bathtub. *Courtesy of Moen, Incorporated.*

154 SECTION TWO *Fixtures and Equipment*

FROM EXPERIENCE

A plumber can manually adjust a trip-lever BW&O by removing the cover plate and the attached internal linkage assembly. Follow the manufacturer's instructions for the correct adjustment, and remember that not all kinds of trip levers are suitable for large-capacity bathtubs.

Shower Drains

There are several kinds of shower drains; their specific use depends on the type of shower base installed. A ceramic tile shower base requires a safety pan to protect the underlying wood floor and structure. Most safety pans are polyvinyl chloride (PVC) liners installed by plumbers during the rough-in phase of construction. A three-piece shower drain is needed to make sure that water does not seep around the drain and onto the wood floor. The three pieces have specific purposes. The threaded top portion is adjustable to allow for varying tile thickness; this is where the perforated strainer and screws are secured. The middle portion receives the top threaded portion and is bolted to the bottom portion. It secures the safety pan and has weep slots to allow any water that penetrates the tile shower base to drain into the floor drain. The bottom portion (body) rests flush with the wood floor and connects the piping to the p-trap. The minimum-size shower drain dictated by most codes is 2″ diameter. Shower pan installation is explained in the fixture installation section of this book. Figure 7–30 illustrates a three-piece shower drain and its relation to the wood floor and safety pan. Figure 7–31 shows two typical plastic three-piece shower drains.

CAUTION: Installing a three-piece shower drain incorrectly can result in serious property damage and can cause mold from leaks. All shower pan liners must be tested and inspected, and a plumber should always retain records of the inspection for any future legal action that could arise.

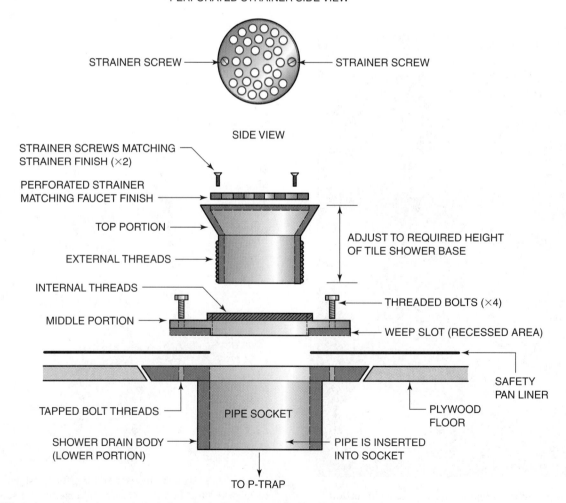

Figure 7–30 A three-piece shower drain is installed when a ceramic tile shower base is constructed.

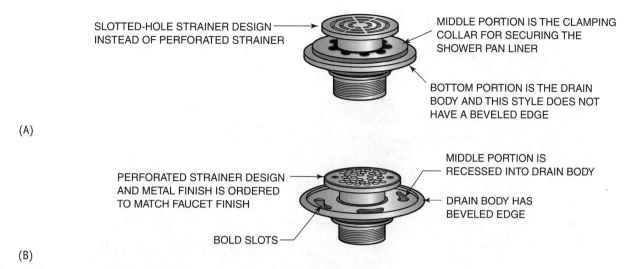

Figure 7-31 A three-piece shower drain is also considered to be a floor drain and offers a plumber a range of adjustment to compensate for various floor thicknesses.

If a tile shower is installed on a concrete floor, code typically does not require a safety pan; a one-piece shower drain can be installed there. A pre-molded shower base has a multi-piece shower drain that is installed by securing it to the base. A one-piece shower unit uses the same type of drain as a pre-molded base. Figure 7-32 is a side-view illustration of a one-piece shower drain often used with a tile base on a concrete floor. Figure 7-33 illustrates a one-piece shower drain used with a pre-molded shower base or one-piece shower unit.

FROM EXPERIENCE

A shower drain in a one-piece shower unit or shower base is often inaccessible after installation. Always tighten and seal the drain correctly and test for leaks before progressing to the final stages of construction.

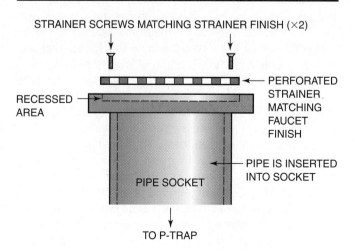

Figure 7-32 A one-piece shower drain is installed when a tile shower base is constructed on a concrete floor.

Kitchen Sink Basket Strainer

The drainage system serving a kitchen sink is connected to the fixture with a basket strainer. Regardless of the type of kitchen sink, the connection of the drainage system is the same. All have 1-1/2" drain connections. Basket strainers vary in the depth of the body portion and in the finish, which matches the color of the sink or the finish of a faucet. Colored sinks often use the same color basket strainer, and stainless steel sinks use a stainless steel strainer. A basket strainer can be installed either before or after a sink installation.

A plumber uses either putty or caulk to seal a basket strainer to the inside of a sink bowl. A rubber gasket is placed over the basket strainer from under the sink, and a fiber (cardboard) gasket, known as a friction washer, is placed between the tightening nut and rubber gasket to prevent the rubber from being deformed during the tightening process. Figure 7-34 shows a typical stainless steel kitchen sink basket strainer. Figure 7-35 is a side view of a deep basket strainer; shallow types have a threaded body, a larger tightening nut, and no spacer. A 1-1/2" flanged tailpiece is connected to a basket strainer. It is not illustrated but is explained in the fixture installation section of this book.

FROM EXPERIENCE

A kitchen sink basket strainer is accessible after a sink is installed. A plumber should not overtighten it during installation. Because the threads are shallow, and many of the securing nuts are aluminum, overtightening can cause damage to the basket strainer. In addition, overtightening can cause the rubber gasket to become deformed, thereby lessening its sealing capabilities.

156 SECTION TWO *Fixtures and Equipment*

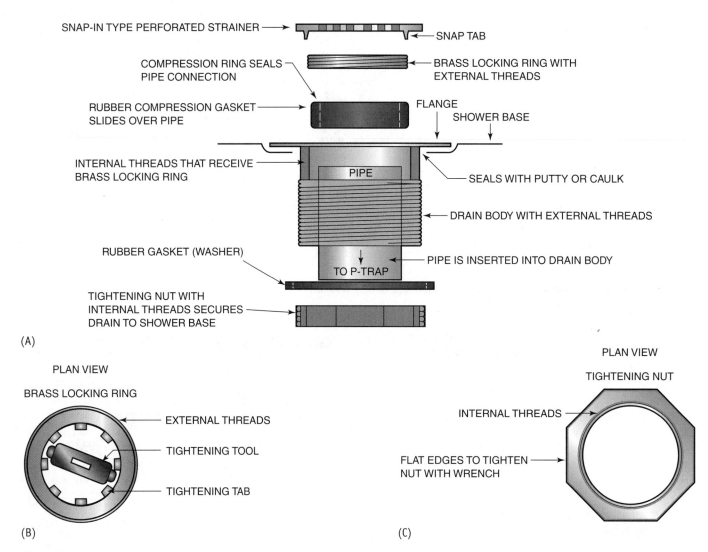

Figure 7-33 A multi-piece shower drain is installed for a pre-molded shower base and a one-piece shower unit.

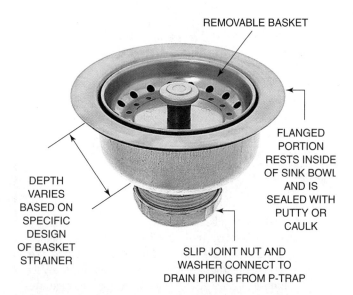

Figure 7-34 Kitchen sink basket strainers are available in various depths and color finishes. *Courtesy of Dearborn Brass.*

Laundry Sink Basket Strainer

Some laundry sinks are manufactured with a drain connection that, instead of a basket strainer, has a rubber stopper to allow the sink to be filled and drained. A basket strainer designed to connect the drain to a fixture that does not have a built-in drain connection is often called a junior basket strainer. The designation of junior refers to its being smaller than a kitchen sink basket strainer. The minimum size drain allowed by code to serve a laundry sink is 1-1/2", so a junior basket strainer is typically only available in that size. The drain connection uses a 1-1/2" flanged tailpiece in conjunction with a slip joint nut and washer, the same method used for a kitchen sink connection.

Most good-quality laundry sink basket strainers have a removable strainer to stop objects from entering a drainage system. Because some codes allow a laundry sink to receive discharge from a washing machine, it is important to have a removable strainer to catch the lint discharging with the wastewater. Laundry sinks are often used as utility sinks in

CHAPTER 7 Faucets and Drain Assemblies 157

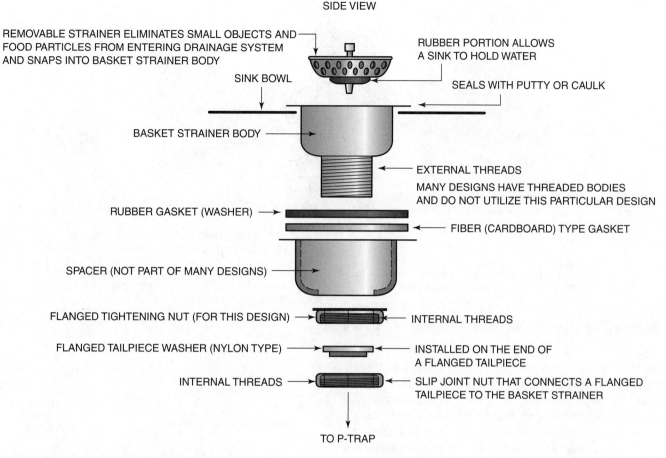

Figure 7-35 A kitchen sink basket strainer connects a drainage pipe to a kitchen sink.

garages and workshops, where it is also important to have a strainer installed. Figure 7-36 shows a common type of removable strainer. A strainer also has a rubber seal to allow a laundry sink to retain water. Figure 7-37 is an illustration of a laundry sink basket strainer and its relation to a sink bowl.

 FROM EXPERIENCE

The rubber stopper and chain assembly used to seal a laundry sink drain opening usually does not have a long enough chain to reach the faucet. Extra chain and a connecting link can be purchased at most hardware stores to ensure that the user will not have to reach into the wastewater to drain the laundry sink.

Bidet Drain Assembly

A bidet drain assembly is similar to a lavatory PO assembly. A linkage assembly allows a user to control the plunger with a lift rod located near the faucet. The drain

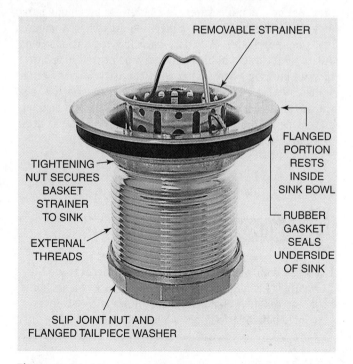

Figure 7-36 A laundry sink basket strainer is often called a junior basket strainer because it is smaller than a kitchen sink basket strainer. *Courtesy of Dearborn Brass.*

158 SECTION TWO *Fixtures and Equipment*

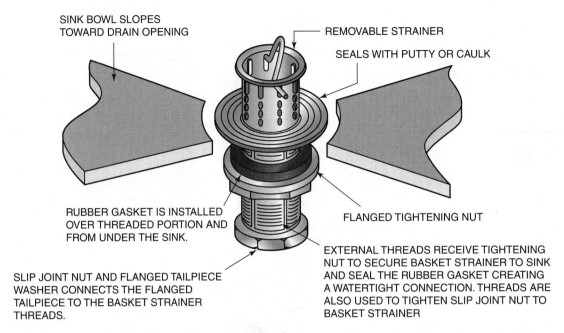

Figure 7–37 A basket strainer connects a laundry sink to the drain.

assembly is sold with the faucet and matches its finish. There are various styles of bidet faucets, and the fixture must be compatible with the faucet. Because most bidet faucets do not have a spout, the lift rod is typically installed in a designated hole in the fixture. The linkage assembly is hidden behind the bidet and remains accessible for adjustment and replacement. Figure 7–38 is a plan view illustration of a typical bidet fixture showing the holes provided by a manufacturer. This bidet has a hygiene spray located in the bowl area, which requires that a vacuum breaker be installed to satisfy backflow regulations. Most bidets with a vacuum breaker have a dedicated hole in the fixture, but some are served with the backflow device installed within the piping system. The illustration also shows the dedicated hole provided for the lift rod that operates the PO drain. Figure 7–39 is a side-view illustration of a bidet drain assembly in relation to the fixture.

FROM EXPERIENCE

A bidet drain assembly can be difficult to access once the fixture is installed due to its proximity to a wall and adjacent toilet. A plumber typically installs the drain assembly before installing the bidet.

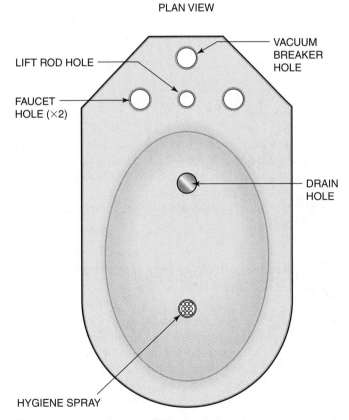

Figure 7–38 A bidet design may include a designated hole to install a lift rod to operate the pop-up drain located in the bidet bowl drain hole.

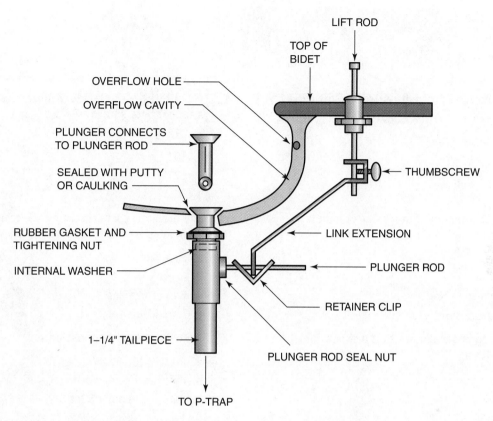

Figure 7-39 A bidet drain assembly is similar to a lavatory pop-up assembly and uses a linkage assembly to operate the drain from above the fixture.

Summary

- Numerous faucet designs are available for the various plumbing fixtures.
- A bathtub and shower faucet is usually installed during the rough-in phase of construction.
- A bath waste and overflow (BW&O) is the drain assembly for a bathtub.
- A lavatory faucet is typically sold with a pop-up (PO) drain assembly.
- The spout on some kitchen sink faucets is also a pull-out spray.
- A pull-out kitchen faucet requires a backflow device.
- The metallic finish of a faucet is typically coordinated with other fixture trim finishes in a bathroom.
- A typical sink faucet is either a one- or two-handle design.
- A tub and shower faucet is a one-, two-, or three-handle design.
- Not all faucets are compatible with all sink designs.
- Kitchen sinks require a basket strainer to connect the p-trap to the sink opening.
- A bidet faucet and drain assembly are sold based on the fixture design.
- Many bidets require a backflow prevention device.

Review Questions

1. **A kitchen faucet with a pull-out spray unit must be protected against**
   a. Backflow of wastewater
   b. Low water pressure
   c. Vandalism
   d. Excessive use

2. **The two most common lavatory faucet handle designs are single handle and**
   a. Three handle
   b. Two handle
   c. Four handle
   d. None of the above are correct

3. **Most handicap codes dictate that a shower must be equipped with a**
   a. Massaging shower head
   b. Two-handle faucet
   c. Handheld shower unit
   d. Three-handle faucet

4. **Water flow is routed from a tub spout to a shower head by activating the**
   a. Trip lever of a BW&O
   b. Diverter
   c. Hot and cold faucet handles
   d. Showerhead

5. **Most large-capacity tubs installed in a platform have a faucet that is**
   a. Deck mounted
   b. Wall mounted
   c. Single handle
   d. Three handle

6. **Many three-piece lavatory faucets can be installed in a faucet hole spread of 4", 6", and**
   a. 8"
   b. 10"
   c. 12"
   d. 3"

7. **The drainage system serving a kitchen sink is connected to the fixture with a**
   a. Lock nut
   b. Check valve
   c. Basket strainer
   d. Trap adapter

8. **A BW&O is the abbreviation for**
   a. Black, white, and optional
   b. Bolt, washer, and o-ring
   c. Bath waste and outlet
   d. None of the above are correct

9. **The minimum size shower drain dictated by most codes is**
   a. 1-1/4"
   b. 1-1/2"
   c. 2"
   d. 3"

10. **The minimum size of a lavatory drain assembly is**
    a. 1-1/4"
    b. 1-1/2"
    c. 2"
    d. 3"

11. **The minimum size of a kitchen sink basket strainer is**
    a. 1-1/4"
    b. 1-1/2"
    c. 2"
    d. 3"

12. **A basket strainer for a laundry sink is often called a**
    a. Miniature basket strainer
    b. Standard basket strainer
    c. Junior basket strainer
    d. Jumbo basket strainer

13 **A bidet and lavatory drain is typically controlled with a**

a. Linkage style drain assembly
b. Rubber stopper with a chain
c. BW&O
d. None of the above are correct

14 **Each shower faucet is sold with a wall escutcheon, shower arm, and**

a. Basic showerhead
b. Tub spout
c. BW&O
d. Lever-type handles

15 **A three-piece shower drain is required when installing a**

a. Pre-molded shower base
b. Shower pan liner
c. Three-piece shower unit
d. One-piece shower unit

16 **A kitchen faucet is compatible with**

a. Only certain types of kitchen sinks
b. All types of kitchen sinks
c. Lavatory sinks
d. None of the above are correct

17 **A standard laundry sink faucet typically has a**

a. 4" hole spread
b. 6" hole spread
c. 8" hole spread
d. 3" hole spread

18 **A backflow prevention device must be installed when a faucet outlet is below**

a. The soap dispenser
b. The top of the faucet handles
c. The approved air gap allowance
d. The handheld spray head

19 **The middle handle of a three-handle tub and shower faucet is a**

a. Warm-water regulator
b. Flow regulator
c. Pressure-balancing regulator
d. None of the above are correct

20 **A removable strainer in a basket strainer keeps objects from entering the drain and also**

a. Allows a sink to be filled with water
b. Seals water from leaking around the flange
c. Serves as a bathtub drain
d. Keeps water from backing up into the sink

Chapter 8: Plumbing Equipment and Appliances

Residential homes use plumbing equipment and appliances that connect to the water distribution and drainage systems. This chapter discusses plumbing equipment and how to determine specific types and variations. Some pieces of equipment, such as water heaters, are required by code for every home; others are preferred by homebuilders or homeowners. Some appliances also require plumbing systems. Garbage disposals, dishwashers, washing machines, and refrigerator icemakers are some examples.

OBJECTIVES

Upon completion of this chapter, the student should be able to:
- explain the differences in water heater designs.
- understand the basic principles of heating water.
- explain the variations in equipment connections.
- correctly order plumbing equipment.

Glossary of Terms

anode rod device installed in a water heater to protect the inside of a storage tank from corrosion

burner the main flame assembly that externally heats water in a gas water heater

collector heats water in a solar water heating system; also called a panel

element electrical heating device that internally heats water in an electric water heater

expansion tank device installed in a cold water piping system to absorb the expansion caused by heating water

flue the entire pipe system exhausting fumes from a gas water heater

high limit safety device on all water heaters to protect from overheating water

instantaneous a water heater that does not use a storage tank; also called tankless

pilot flame of a gas water heater that ignites gas entering a burner assembly

point-of-use water heater a small capacity water heater installed close to the fixture utilizing the water heater

pressure relief valve a reactionary device that protects a water heater against excessive pressure

tankless a water heater that does not store water in a storage tank; also called instantaneous

terminal where an electrical wire connects to a device; also referred to as a post

thermocouple heat-sensing device to ensure that a gas water heater pilot flame is ignited

thermostat a regulating device to control the temperature of a water heater

Appliance Connections

Appliances are a vital part of a functioning residential dwelling, and some of these appliances connect to plumbing systems. A plumber installs the water piping and drainage piping for several appliances. For some, the plumber installs the piping systems during the rough-in phase of construction, so the final connection can be completed by others. Other appliances do not require any special piping but are installed by a plumber during the trim-out phase of construction. Appliances served by plumbing systems and the system provided by a plumber for each are listed in Table 8-1.

Garbage Disposers

A garbage disposer is also known as a food waste disposer or a garbage disposal. This very simple appliance has been popular for decades. It is connected to a kitchen sink where the basket strainer would normally be installed. A garbage disposer is a motorized appliance that is activated manually with electrical current. Food waste is inserted into the garbage disposer, and, once the electrical supply energizes the motor, an internal rotating flywheel shreds food waste, which is discharged into the drainage system with water from the kitchen sink. The horsepower (hp) of the motor determines the capabilities of the garbage disposer; the greater the hp, the more capable the unit is for shredding food waste. The most common hp sizes for residential applications range from 1/3 to 3/4. Figure 8-1 is a standard residential garbage disposer.

FROM EXPERIENCE

A competitively priced stainless steel kitchen sink may vibrate excessively if the hp of the garbage disposer is too great.

A plumber installs a garbage disposer onto a kitchen sink with a specially designed mounting assembly that is unique to each manufacturer. This multi-piece assembly consists of a sink flange that is inserted and sealed into the sink drain outlet where a basket strainer would normally be installed.

Table 8-1 Appliances Requiring Plumbing Connections

Appliance	Hot	Cold	Drain
Garbage disposal	No	No	Yes
Dishwasher	Yes	No	Yes
Washing machine	Yes	Yes	Yes
Icemaker	No	Yes	No

Figure 8-1 A garbage disposer is a motorized food waste appliance connected to a kitchen sink. *Badger 5 Courtesy of In-Sink-Erator.*

A plumber completes the watertight connection from the underside of the sink. The disposer connects to the mounting assembly with a rubber gasket, which also serves to reduce noise. The discharge outlet has a 90° flanged tailpiece that connects to the p-trap under the sink. The sink flange is available in a variety of colors and finishes to match the color or finish of a sink or faucet. A standard garbage disposer is sold with a stainless steel sink flange. Figure 8-2 is a side view of a residential garbage disposer. Figure 8-3 shows several sink flanges and stopper colors and finishes. Figure 8-4 is a multi-piece mounting assembly similar to the one in the breakdown view in Figure 8-2. Figure 8-5 is a dual-purpose rubber gasket that seals the disposer to the mounting assembly and also reduces the noise created by the food-shredding operation.

FROM EXPERIENCE

The drain connection to a disposer with a 90° tailpiece has a specifically designed rubber washer that is not used in all disposers.

CHAPTER 8 Plumbing Equipment and Appliances

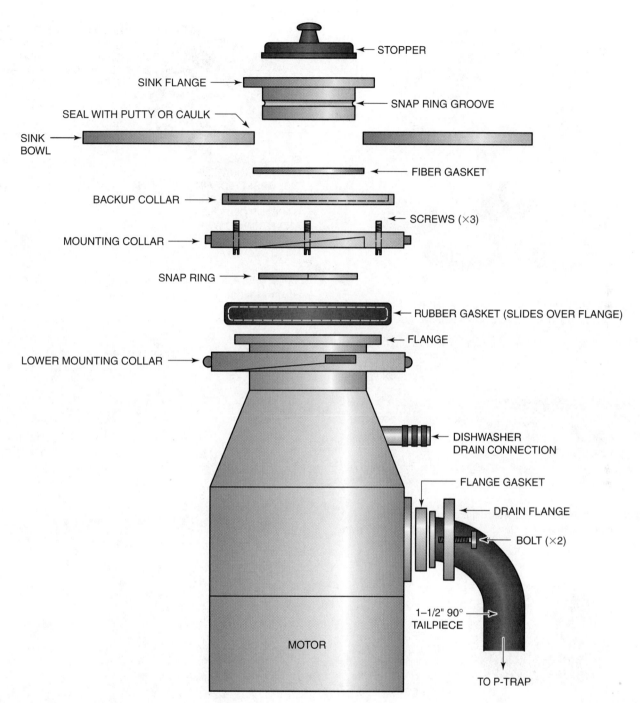

Figure 8-2 A garbage disposer connects to a kitchen sink and discharges food waste into the drainage system. It is also capable of receiving discharge from a dishwasher.

Most garbage disposers receive discharge from a dishwasher. A designated dishwasher connection is located on the side of a disposer, and a plumber must remove a knockout plug to connect the two appliances. The knock-out plug is removed from the disposer with a hammer and chisel before the disposer is installed. In many dishwashers, one end of the discharge hose is compatible with the connection to the garbage disposer. If a hose end is not compatible with the garbage disposer, a rubber drain connecter known as a boot is used for the connection. A rubber dishwasher boot requires that a small piece of copper be inserted into the boot and the dishwasher hose, and then all connections are sealed with hose clamps. A detailed explanation of a dishwasher installation is found in the fixture installation chapter of this book. Figure 8-6 shows a dishwasher boot with several types of hose clamps. Attempting to shred too much food or to shred items not intended for a disposer can cause the disposer to stop rotating (jam). A specially designed tool is provided with

Figure 8-3 Sink flanges are offered in a variety of colors and finishes that match the sink color or finish. *Courtesy of In-Sink-Erator.*

Figure 8-4 A multi-piece mounting assembly connects a garbage disposer to a sink. *Quick-Lock Sink Mount Courtesy of In-Sink-Erator.*

Figure 8-5 A rubber gasket creates a watertight seal and silences the shredding noise of a garbage disposer. *Courtesy of In-Sink-Erator.*

Figure 8-6 A rubber connector "boot" is installed to connect a dishwasher hose to a garbage disposer. *Courtesy of In-Sink-Erator.*

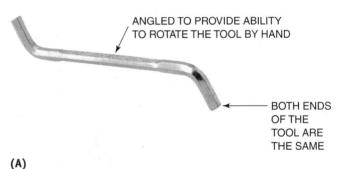

Figure 8-7 An angled tool is shipped with most garbage disposers and used to manually rotate the motor shaft in case the unit becomes jammed. *Jam Buster Wrench Courtesy of In-Sink-Erator.*

most garbage disposers, which the plumber should give to the customer. The tool is inserted into a compatible socket located under the electric motor portion of the disposer and manually rotated. This also rotates the motor shaft and flywheel portions to free the jammed garbage disposer. Figure 8-7 shows a garbage disposer tool and its use.

FROM EXPERIENCE

A plumber is not allowed by most codes to connect the electrical wiring to a garbage disposer. Never connect any wiring to an appliance without proper training and without possessing the required electrical license.

For many years, a home with a septic tank system could not use a garbage disposer due to the negative effects of food particles settling on the bottom of the septic tank. Now biodegradable solutions can be added to a septic tank to stimulate the decomposition of food waste within the septic tank. If the biodegradable solution is not added, the food waste will settle and have to be removed by a septic tank cleaning company. Some manufacturers make garbage disposers that inject small amounts of biodegradable solution with each use of the disposer. The solution is also good for the overall health of the septic tank, because it breaks down soaps, grease, and other waste items. Figure 8–8 is a garbage disposer designed for use with a septic system.

FROM EXPERIENCE

Some septic tank systems cannot receive discharge from a garbage disposer. A plumber must know whether the system can handle food waste and whether local codes allow food waste to be discharged into a septic tank system.

Dishwashers

A plumber installs a dishwasher during the trim-out phase of construction, usually when installing the sink and garbage disposer. The homebuilder or homeowner typically provides the dishwasher, and the delivery should coincide with the sink installation to increase productivity. Most dishwashers installed in a residential home are adjacent to the kitchen sink and located under the countertop. Figure 8–9 is a view of a dishwasher located within the cabinetry of a kitchen.

FROM EXPERIENCE

The color of the dishwasher is coordinated with other items in the kitchen and matches other appliances.

A dishwasher receives hot water from the same water source that serves the kitchen sink. The drain hose from the dishwasher is routed to the garbage disposer or to a tailpiece designed for that purpose. Many codes dictate that a dishwasher drain hose be routed through an air gap device so wastewater does not flow from the sink into the dishwasher. Backflow is a concern because the water connection to the dishwasher is below the flood level rim of the sink. The drain hose connection to the dishwasher is installed by the manufacturer, and the plumber routes the drain hose to the sink area. The water supply piping is typically 3/8" OD tubing routed from the sink area to the connection point under the dishwasher. The drain hose is routed through a hole drilled by a plumber into the side of the sink base

Figure 8–8 Special types of garbage disposers are installed for homes using a septic tank system. *Septic Disposer Courtesy of In-Sink-Erator.*

Figure 8–9 A typical residential dishwasher is installed adjacent to a kitchen sink and into a designated space under a countertop. *Courtesy of General Electric (GE Appliance).*

168 SECTION TWO *Fixtures and Equipment*

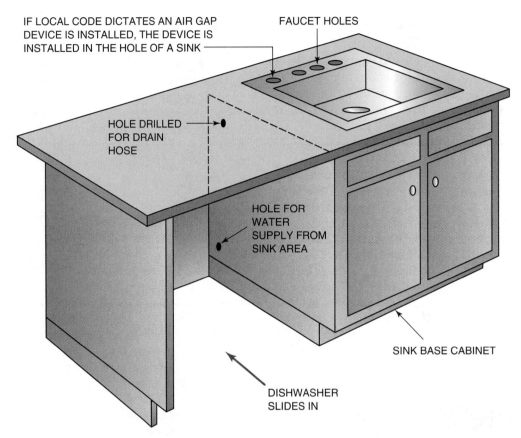

Figure 8-10 A plumber typically installs a dishwasher during the same phase of construction as the sink installation.

cabinet. Additional installation information is found in the fixture installation section of this book. Figure 8-10 is an illustration of a dishwasher in relation to the plumbing systems serving a sink.

 FROM EXPERIENCE

A licensed electrician connects the electrical wiring to a dishwasher. A plumber should never complete this task without being properly trained and in possession of an electrical license.

Washing Machine Box

A plumber does not install a washing machine but does provide the water and drain for it during the rough-in phase of construction. A washing machine box provides hot water, cold water, and a drain connection in one central location. A typical residential box is manufactured with a hub to receive 2" plastic pipe, because most codes dictate that the minimum size drain that can serve a washing machine is 2".

Most codes also dictate that the smallest water supply that can serve a washing machine is 1/2" (5/8" OD). The type of valve used to isolate the washing machine varies, but a boiler drain and a ball valve are the most common. The washing machine box is installed between two vertical wall studs. Because most residential boxes are plastic, they might not be allowable in multi-family dwellings. Plastic materials cannot be used in certain types of fire-rated walls; most codes dictate that metal washing machine boxes be used instead. A separate wall trim is usually sold with the rough-in box. It must be stored in a secure place until it is needed for the trim-out phase of construction. Figure 8-11 illustrates a typical residential washing machine box and its relation to the piping systems and wood structure. The installation of the box is explained in more detail in the installation sections of this book.

 FROM EXPERIENCE

A plastic washing machine box can be wrapped with drywall or other flame retardant materials to increase its fire rating, if allowed by local codes.

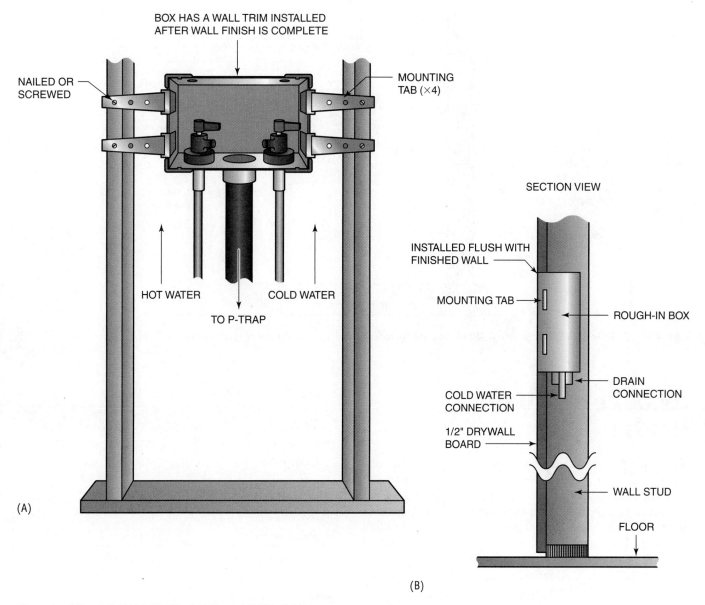

Figure 8–11 A washing machine box is installed during the rough-in phase of construction.

Icemaker Box

Most newer residential refrigerators are equipped with an icemaker and water dispenser. A plumber routes the cold water piping to the refrigerator area and installs an icemaker box at the termination point. The box is installed near the floor between two vertical wood studs during the rough-in phase of construction. The installation methods are explained in the water distribution section of this book. Most residential icemaker boxes are plastic, but metal ones are available for use in fire-rated walls. A plumber installs 1/2" pipe to the icemaker box and connects the piping to the angle valve provided with the box. The outlet of the angle valve has a 1/4" OD compression connection to allow the compatible tubing of the refrigerator to connect with the icemaker valve. Separate wall trim sold with the rough-in box must be stored in a secure place until it is required during the trim-out phase of construction. Figure 8–12 is an illustration of a typical icemaker box located between two vertical studs.

 FROM EXPERIENCE

If an icemaker box is not used, a 1/4" OD water supply is typically routed from under the kitchen sink to behind the refrigerator.

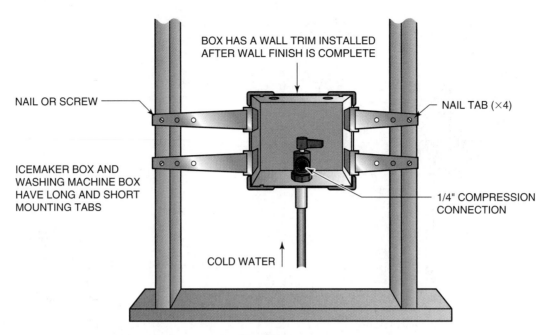

Figure 8-12 An icemaker box is installed behind a refrigerator to supply potable water to an icemaker.

Residential Water Heaters

Less demand is placed on residential water heaters than on commercial water heaters. Most residential water heaters have a storage tank that is heated by either electricity or gas. An adjustable thermostat senses the water temperature and controls the heating cycle. Once the cold water is heated to a desired temperature, the heating source automatically shuts off. When the **thermostat** senses that the water temperature is below the desired temperature, the heating source automatically begins the heating cycle again. Many homeowners are now purchasing **tankless** water heaters, also known as **instantaneous** water heaters. With a tankless water heater, a hot water faucet is opened and the water is heated as it flows through the heater. A point-of-use instantaneous water heater serves a single low-flow fixture, such as a lavatory sink. Water heaters come in various heights, diameters, and gallon capacities. Most residential types are offered in three basic sizes. The most common one has a 40 or 50 gallon storage tank capacity. The height and diameter of a water heater varies with the gallon capacity. Most manufacturers offer a tall, a standard, and a short version of residential water heater. The short version is called a low-boy, and the tall one, which is a slim version for use in tight confines is simply called tall. You must specify low-boy or tall when you order. If you do not, you will receive a standard size for the gallon capacity you order. The system should be filled with water and all trapped air should be removed before activating the gas or electricity. The types of heaters discussed in this chapter are listed in Table 8-2.

Table 8-2 Types of Residential Water Heater

Type	Gas	Electric
Instantaneous	Yes	Yes
Storage tank	Yes	Yes
Under counter	No	Yes
Atmospheric vented	Yes	N/A
Direct vented	Yes	N/A
Flame arrestor	Yes	N/A

N/A: Non-applicable

 FROM EXPERIENCE

The residential water heater capacity needed is based on the total number and types of fixtures installed in a home.

Gas Water Heaters

Natural gas and propane are the two types of gas used for heating water. In rural areas, where natural gas is not provided by a municipal gas company, a liquid petroleum gas called propane is used. A water heater designed for natural gas cannot be used with propane because the internal gas regulating orifice is different for each gas type, but many gas regulators (or gas valves) can be converted by a certified technician. A plumber should never attempt this procedure without possessing the required certification. Atmospheric-vented water heaters are the most common for residential applications. They must terminate in specific locations through a roof, so the water heater must be placed accordingly. The exhaust fumes from a gas water heater contain carbon monoxide, which can enter the occupied space of a home if the vent termination is improperly installed. A direct vent design allows the gas water heater to be installed in a more desirable location close to an exterior wall. The direct vent terminates through an exterior wall, and, like the atmospheric vent design, cannot be close to any opening into an occupied space.

The installation of water heaters is explained elsewhere in this book; this chapter focuses on the types of water heaters and their variations. Storage tank gas water heaters are the most popular because of their availability and cost. Instantaneous gas water heaters are slowly becoming more popular and less expensive than in the past. A less-common gas water heater is a power-vent type, which forces the exhaust from the water heater to the exterior of the house. If electricity is disconnected from a power-vented water heater, it cannot operate. This important safety feature is also why the power-vent water heater is not popular. Most gas water heaters have minimum clearances that they can be installed from combustible materials. These clearances are listed on the side of every gas water heater on a label known as a boiler plate. Other information pertaining to the specific water heater is also listed. Figure 8–13 shows several common gas water heaters for residential homes and a typical boiler plate.

FROM EXPERIENCE

A typical residential gas water heater is capable of producing more hot water per hour than most residential electric water heaters.

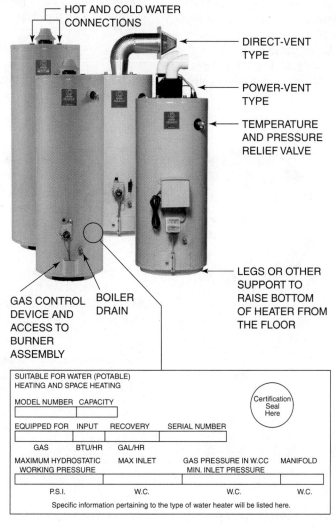

Figure 8–13 Gas water heaters are available in a wide variety of types for different installation locations. *Courtesy of State Water Heaters.*

Gas water heaters are rated by the gallons of hot water they can produce. A British thermal unit (BTU) is used to measure heat. The BTU rating is one thing to consider when ordering a gas water heater. One BTU is the amount of heat required to raise one pound of water one degree Fahrenheit (F). One gallon of water weighs 8.33 pounds, and 8.33 BTUs are needed to raise the temperature of one gallon of water one degree F. A specific length of time is used to determine the capabilities of a gas water heater; gallons per hour (gph) is the one used most often. Figure 8–14 illustrates the basic theory of heating water based on a BTU.

FROM EXPERIENCE

The incoming cold water replacing the hot water from a water heater is calculated by the weight per gallon when determining the heating capability. This is known as the recovery rate.

Temperature rise is the difference between the temperature of the incoming cold water and the temperature expected from the water heater. If the water entering a water heater is 50 degrees F and the thermostat is set at 120 degrees F, the temperature rise is 70 degrees F. Temperature

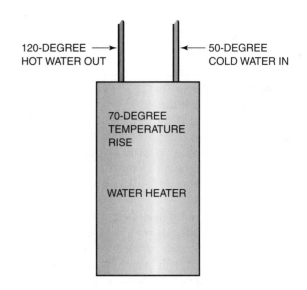

Figure 8–15 **Temperature rise is the difference between the incoming water and outgoing water in a water heater.**

rise determines the capabilities of a gas water heater with a certain BTU rating. Figure 8–15 illustrates the temperature rise of a water heater.

Selecting a gas water heater based on its capabilities requires manufacturer's data. Each manufacturer provides charts showing how many gph a specific model of gas water heater can produce based on its BTU rating. A common error a plumber makes when first using a chart is to go by the first-hour listing in the chart. The first-hour listing only indicates how many gallons a gas water heater can heat when no water is being removed. The recovery rate is the most important element in determining whether a specific water heater is suitable for a specific home. Because cold water is entering at the same time the hot water is exiting, the water in the tank is being cooled. The chart will typically list the first hour capabilities and several other temperature rise capabilities. The lower the temperature rise, the more gallons of hot water a water heater can produce in a specific time frame. Table 8–3 is a sample chart indicating a model number, the BTU rating, and the gph for several temperature-rise situations.

FROM EXPERIENCE

If the temperature rise is not known when sizing a water heater, sizing calculations typically use 100° temperature rise.

The two connection options for the cold water piping on a storage-tank water heater is either on the top or the bottom side of the heater. These are recognized as top fed

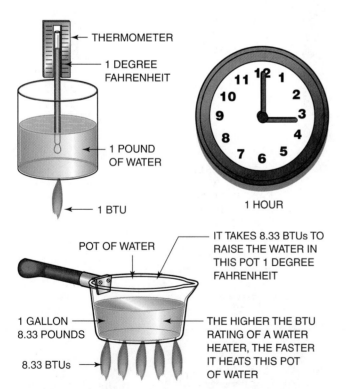

Figure 8–14 **A British thermal unit is used to measure heat. In a gas water heater, it is used to determine the length of time it takes to heat one gallon of water one degree Fahrenheit.**

Table 8-3 Gas Water Heater Selection*

Model Number	BTU	No Flow First Hour	Gallons per Hour Temperature Rise		
			100°	90°	60°
1234	35,000	60	40	45	55
2345	40,000	75	55	60	70
3456	45,000	80	50	65	75
4567	50,000	85	60	70	80
5678	55,000	95	70	80	90
6789	60,000	100	80	85	95

*Values are examples and are not real values

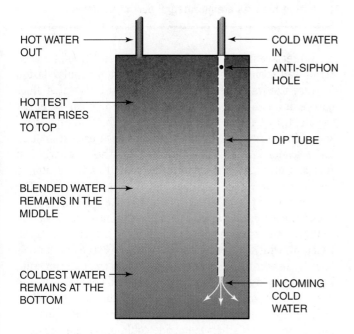

Figure 8-16 Cold water is routed to the bottom of a storage tank with a dip tube, and the hottest water rises to the top.

or bottom fed. A bottom-fed water heater needs a vacuum relief valve to prevent backflow from the heater into the cold water piping. Most residential storage-tank water heaters have both the hot and the cold water pipe connections on the top of the heater. A manufacturer-installed device known as a dip tube in the cold water connection of a top-fed water heater routes the incoming cold water to the bottom of the heater. Without a dip tube, the incoming cold water would mix with the hot water at the top of the water heater. With it, the hottest water will always rise to the top of the storage tank, and the coldest water will remain in the bottom of the tank. Figure 8-16 illustrates the natural separation of water based on temperature and how a dip tube routes the incoming water to the bottom of the tank.

 FROM EXPERIENCE

A dip tube is typically a thin-wall clear plastic tube with a flared end to keep it from falling into the water heater storage tank.

Every gas water heater must have a vent that terminates to open air to exhaust the carbon monoxide fumes created by heating the water. Carbon monoxide fumes are odorless and can kill occupants of a home. Conventional water heaters are vented atmospherically and must have adequate space around the heater for the fumes to be exhausted to the exterior. If there is not enough space, the exhaust fumes could enter the occupied area. The **burner** assembly is where the flames are that heat the water in a storage tank. The fumes rise through a pipe in the center of the tank with a spiral baffle that is inserted at the factory. The pipe is known as a baffle tube, but many other trade names are used. The sheet-metal vent pipe, known as a **flue** pipe, connects to a draft hood and is routed to a safe termination point above the roof. Many codes require a heating and ventilation license to install the flue pipe. Figure 8-17 is a typical residential atmospheric-vented water heater. Figure 8-18 illustrates the water heater's drafting process.

 FROM EXPERIENCE

The size of a flue pipe connection to a draft hood is dictated by the water heater design and determined by the manufacturer. A larger flue pipe might be needed on a particular installation. That is determined by the contractor based on local codes.

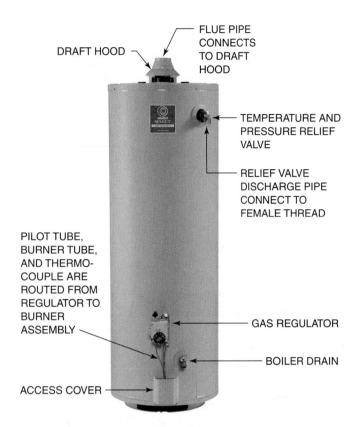

Figure 8–17 A conventional gas water heater is a natural draft venting design that relies on atmospheric conditions to function properly. *Courtesy of State Water Heaters.*

A gas regulator is an automated device that controls the gas flow to a burner assembly. The gas supply pipe is connected to the gas regulator, whose design is based on safety. Most codes do not allow anyone to disassemble a gas regulator for repair who is not certified to repair it. A residential gas regulator has two water temperature sensing probes that are immersed in the water. One probe senses the low water temperature; the other senses the high water temperature. When the water is cooler than the desired temperature, the gas regulator allows gas to flow to the burner assembly. When the water reaches the desired temperature, the regulator closes internally to stop the flow of gas to the burner assembly. If a malfunction occurs and the high temperature probe senses excessively hot water, the regulator will close the gas supply to the burner assembly.

The regulator also controls the gas flow to a **pilot** flame, which then ignites the burner flame so the heating cycle can begin. A safety device known as a **thermocouple** has two distinct ends. One end is connected to the gas regulator, and the other has a sensing probe that is immersed in the pilot flame. The sensing probe lets the regulator know that a pilot flame is present so gas sent to the burner assembly will be ignited by the pilot flame. If the thermocouple does not sense a pilot flame, the regulator remains in the closed position and does not send any gas to the burner assembly. A red, manual-override button is depressed to allow gas to flow through the pilot tube to light a pilot flame. The button remains depressed for at least one minute after lighting the pilot flame, so the thermocouple can sense the heat of the flame. The thermocouple then keeps the gas port open when the override button is no longer depressed, which keeps the gas flowing through the pilot tube. Figure 8–19 illustrates a gas regulator, and Figure 8–20 is a burner assembly for a residential water heater.

FROM EXPERIENCE

A residential gas regulator is different from a commercial gas regulator because the temperature sensing and regulating features are an integral part of its design.

A direct-vent gas water heater is vented through an exterior wall rather than the roof. This is a very popular design because the water heater can be installed in the first floor garage of a two-story house and the flue pipe can be easily installed. In a direct-vent water heater, the flue pipe is inserted into a fresh-air intake pipe. This ensures that enough air is available for combustion and for the evacuation of fumes. Unlike an atmospherically vented water heater, a direct-vent water heater does not have a draft hood. The connection of the fresh-air intake and the flue pipe to the top of the water heater is sealed. The distance from a wall to a direct-vent water heater is dictated by the manufacturer, based on which water heater is installed. Extension kits allow the heater to be located farther from the exterior wall, but typically not more than 1′ from the side of the heater to the exterior wall. The termination through the wall is strictly regulated and cannot be too close to doors, windows, and gas meters. Carbon monoxide could enter the occupied space through a window or door, and gas fumes from a leaking gas meter could enter the fresh air intake. Figure 8–21 illustrates a direct-vent water heater.

FROM EXPERIENCE

A direct-vent water heater has a sealed access panel and does not use inside air as a source of combustion.

July 1, 2003, marked the beginning of a transition period for a new water heater design. The use of a flame arrestor to minimize the possibility of an explosion when flammable liquids are stored in the same space as a gas water heater was mandated for certain water heaters. This design uses a filtering media to keep fumes from stored materials from entering the burner assembly area. Table 8–4 shows the tran-

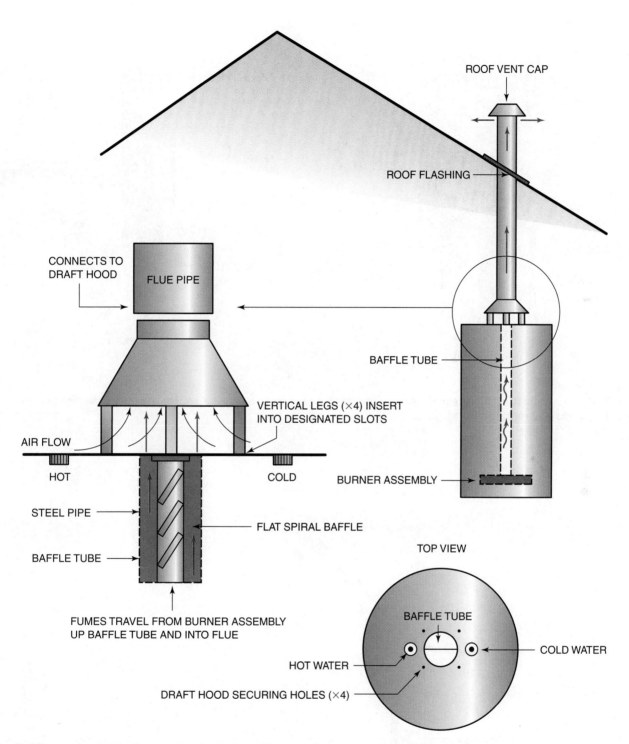

Figure 8-18 An atmospherically vented water heater relies on inside air to aid in evacuating fumes.

sition schedule and types of water heaters affected. Manufactured homes are excluded from this mandate. When the inventory of atmospheric residential water heaters is depleted in the country, all gas water heaters that are 75,000 BTU or less and installed inside a residential home will have to comply with this regulation. Figure 8-22 shows the unique features of a flame arrestor design.

 FROM EXPERIENCE

A flame arrestor does not create a safe environment for storage of flammable liquids near a water heater; it only minimizes the risk.

176 SECTION TWO *Fixtures and Equipment*

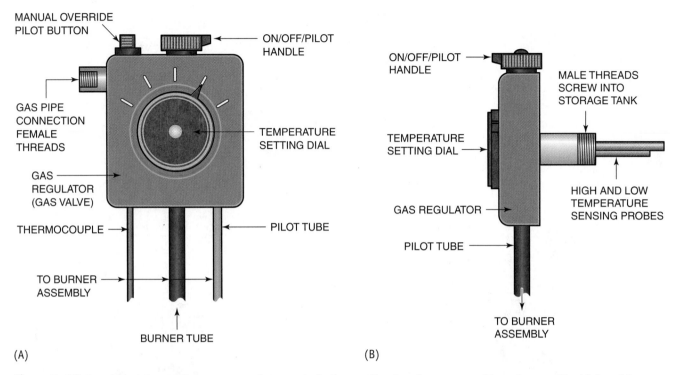

Figure 8–19 A residential water heater gas regulator controls the gas flow to a burner assembly and senses the high and low temperature of the water within a storage tank.

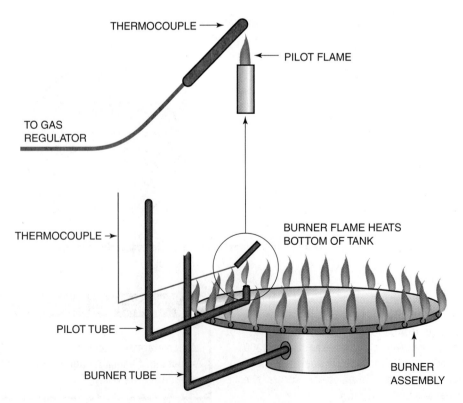

Figure 8–20 A burner assembly heats the external bottom of a storage tank.

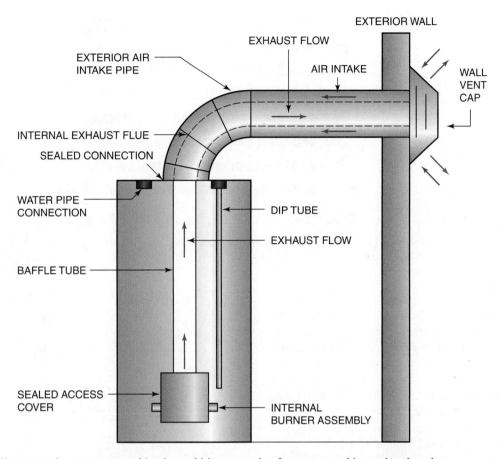

Figure 8–21 A direct-vent heater uses outside air to aid in evacuating fumes to provide combustion air.

Table 8–4 Flame Arrestor Water Heater Transition Schedule

Date	Water Heater Type
July 1, 2003 *	30, 40, and 50 gallon atmospheric-vent models
July 1, 2004 *	30, 40, and 50 gallon power-vented models
July 1, 2005	All gas-fired models of 75,000 BTU or less

*Excludes manufactured homes

A tankless heater design is suitable for many residential applications. Also known as an instantaneous water heater, most tankless heaters use gas, but electric ones are also available. They are becoming more popular as designs are improved, and the cost savings from not maintaining a storage tank of heated water is attractive to many homeowners. Tankless heaters are more expensive than tank-type water heaters, but the cost savings over several years outweighs the initial cost of installation. One disadvantage of tankless heaters is that they often cannot provide an adequate volume of hot water for high-demand homes. Water flow is regulated to ensure that the desired temperature leaves the heater. Water flows through a coil that is heated by the burner, and, if the water flows too fast through the coil, it

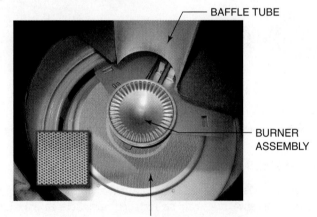

(A)

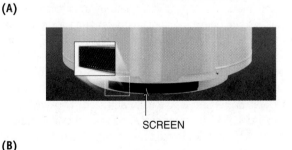

(B)

Figure 8–22 A flame arrestor design minimizes the risk of explosion when flammable liquids are stored near a water heater. *Courtesy of State Water Heaters.*

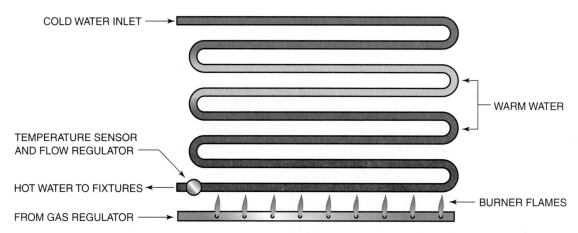

Figure 8-23 Gas tankless water heaters have a coil that is heated by a burner. The water flow is regulated to remain in the coil until it is heated to a desired temperature.

will not be heated to the desired temperature. Two or more tankless heaters can be installed in a piping configuration known as a series. In a series configuration, one heater preheats the water and then routes it to the cold water pipe of another heater to be heated to the desired temperature. This process lowers the temperature rise of each heater. The water does not have to remain in the coil as long, so more gph are produced. The heater is mounted to a wall, and the flue is terminated through an exterior wall similar to a direct-vent water heater. Figure 8-23 illustrates a coil design for heating water in a tankless water heater.

 FROM EXPERIENCE

A tankless water heater can be a reliable source of adequate hot water, but the flow rate is diminished if it is improperly sized.

Electric Water Heaters

Electric water heaters are more common than gas water heaters because they can be installed anywhere in a home. They do not require venting or gas piping and are less expensive to purchase. The negative aspect of an electric water heater is that the operating cost is greater than for a gas water heater. Residential electric water heaters are offered in a variety of sizes and are not as heavy as gas water heaters. Water is heated by a heating **element,** and the flow of electrical current is regulated by a safety device known as a **high limit.** The temperature setting is manually adjusted. The most common residential electric water heater has two thermostats and two elements. Unlike the immersed thermostat on a gas water heater, the thermostats on an electric water heater are in contact with the external portion of the storage tank. These are surface mounted thermostats. Figure 8-24 shows the exterior appearance and different sizes of residential water heaters.

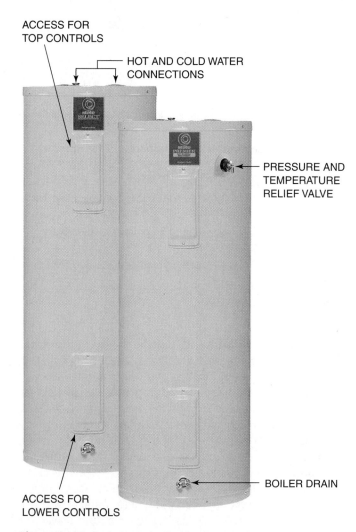

Figure 8-24 Various sizes of electric water heaters are available. *Courtesy of State Water Heaters.*

FROM EXPERIENCE

An electric water heater is typically a better selection than gas if the installation location is limited to the middle portions of a home.

A standard residential electric water heater is classified as a 240-volt, 4500-watt, non-simultaneous water heater. The heat of an electric water heater is measured in wattage of a heating element rather than BTU rating. One thousand watts is one kilowatt (kW), and 1 kW equals 3412 BTUs. This conversion process is used to compare the two energy sources, so the correct size water heater can be installed when converting to a different type. A 4500-watt element is also a 4.5-kW element, and, when 4.5 is multiplied by 3412, it indicates that the 4500-watt element is also a 15,354 BTU element. This provides fewer BTUs than most residential gas water heaters and is a leading reason why a gas water heater provides more hot water than a 4500-watt electric water heater. Additional electrical information is presented in the troubleshooting section of this book. Figure 8–25 illustrates the basic conversion from wattage to BTUs.

FROM EXPERIENCE

The average residential new-construction plumber might not be required to size a water heater, but knowing the sizing process is crucial for advancement in a plumbing career.

There are several different wattage ratings of heating elements. The higher the wattage rating, the faster it can heat water. The length of time it takes to heat water in an electric water heater is referred to as the recovery rate. In a non-simultaneous operation, two elements are immersed in the water of a storage tank, but only one element is heating at a time. The two elements and two thermostats are referred to as upper and lower because of their physical location on the exterior of the tank. Three wires are routed by a licensed electrician from the circuit breaker panel to the water heater. The ground wire is green and, because it is not energized, it can be a bare copper wire. The other two wires are often referred to as line voltage one (L1) and line voltage two (L2). Each serves a particular purpose and both work together to provide the 240 volts needed for an element to operate. L1 and L2 are connected separately to the high-limit device, which is a safety device that interrupts electrical current if an unsafe condition is present. The high-limit device has a red manual reset button and is typically combined with the upper thermostat as a single device. The electrical current from L2 is internally routed (relayed) through the high-limit switch to a **terminal** (post) identified on the high-limit device as #4. Because there are two heating elements, two wires are routed from the #4 post of the high-limit switch. One is designated for each element; both are energized with 120 volts. These two wires provide a constant 120 volts to each element and wait for L1 to provide the other 120 volts, so a heating cycle can begin.

If the high-limit device does not sense any unsafe conditions, L1 is internally relayed to the #2 post of the high-limit device, which has an external metal connector (jumper) that routes L1 externally to the #1 post of the upper thermostat. At this point, L1 is internally routed either to the #2 post of the upper thermostat, which energizes the upper element, or to the #4 post of the upper thermostat. Its routing depends on the internal water temperature of the storage tank. When the internal routing is to the #4 post of the upper thermostat, the #1 post of the lower thermostat is energized. Because cold water is routed to the bottom of the tank through the dip tube, the lower thermostat senses the difference in water temperature when hot water is used, and reacts by internally routing L1 to the #2 post of the lower thermostat. This energizes the lower element with the additional 120 volts required for the heating element to operate. If hot water is withdrawn from the tank faster than the lower element can recover, the upper thermostat senses the lowering of water temperature near the top of the tank.

When the upper thermostat senses this difference in water temperature, the L1 voltage is internally relayed from the #4 post of the upper thermostat, thereby removing L1 from the lower thermostat and routing it back to the #2 post of the upper thermostat, which is routed to the upper element. This energizes the upper element with 240 volts to begin its heating cycle. When the upper element completes its heating cycle, the upper thermostat senses the increased water temperature and relays the electrical current back to the lower thermostat via the #4 post of the upper thermostat. L1 remains inactive at the #1 post of the lower thermostat until the water temperature decreases again. Figure 8–26 shows a common wiring configuration of a 240-volt, non-simultaneous residential electric

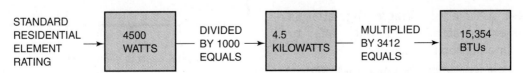

Figure 8–25 One thousand watts equals one kilowatt, and one kilowatt equals 3412 BTUs.

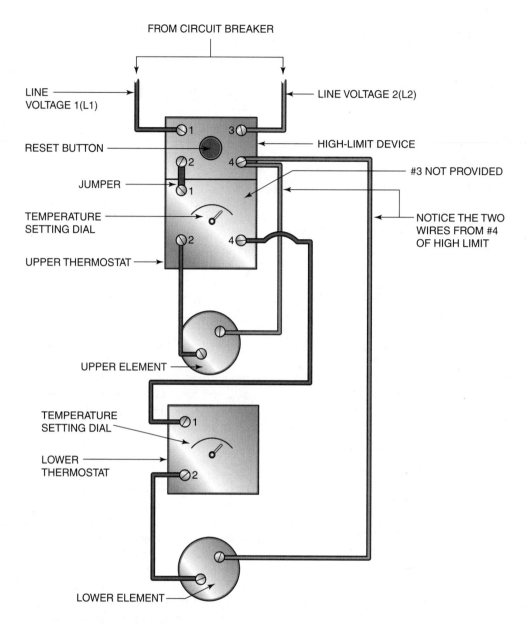

Figure 8-26 Most residential electric water heaters have a 240-volt, non-simultaneous operating configuration.

water heater. Other variations exist, so a plumber must review the exact wiring schematic for a particular water heater. This schematic is only an example and is not intended for use on a job site.

CAUTION: Never assume the wiring configuration is the same for every water heater. Always refer to manufacturer information for each one.

FROM EXPERIENCE

The thermostats, elements, and internal wiring are installed at the factory and the electrical system and connection to the water heater is performed by an electrician.

A residential heating element is offered in various designs. The most common is a simple U-shape, which is screwed into

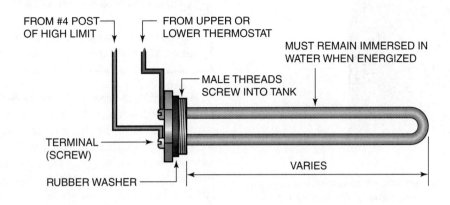

Figure 8-27 A residential heating element is typically a 240-volt, 4500-watt type.

a designated female threaded connection in the water heater tank. A rubber washer is placed over the male threads of an element before installing it into the tank. A socket-type tool tightens the element into the tank. Two connector screws (posts) secure the wires to the element. They are not dedicated so either wire can connect to either screw. Figure 8-27 illustrates a typical residential heating element.

FROM EXPERIENCE

The voltage and wattage ratings of an element are listed on the exterior of an element, so the wrong element is not installed.

Many homeowners are seeking ways to conserve both energy and water. Using smaller electric water heaters that serve certain fixtures has become a popular option. If a water heater is located too far away from a particular fixture, several gallons of water can be wasted waiting for the hot water to flow from a faucet. For every gallon of hot water that is wasted, an equal amount of cold water is entering the water heater and must be heated, which consumes electricity or gas. A small-capacity electric water heater can be installed under a counter-type sink to provide immediate hot water. Most small-capacity water heaters store two to five gallons of water, which may not be enough. However, using creative piping configurations, a small capacity water heater can provide immediate hot water flow from a faucet. By the time that hot water is used up, the hot water from the remote water heater has entered the small water heater. Figure 8-28 shows a small-capacity water heater that uses 120 volts and can be installed under a countertop sink.

FROM EXPERIENCE

Most under-counter water heaters have an electrical plug that is compatible with a 120-volt receptacle, but local electrical codes may require a direct electrical connection to the water heater.

Instantaneous electric water heaters are also a popular choice for conserving water. Most are considered

Figure 8–28 Under-counter electric water heaters are used for remote sinks or other low-volume fixtures. *Courtesy of In-Sink-Erator.*

point-of-use water heaters. They are not designed to provide large volumes of water, but rather to serve low-volume fixtures such as lavatory sinks. They are available in 120 and 240 volts. Most mount on a wall and are very compact. Figure 8–29 shows a point-of-use instantaneous water heater located under a wall-hung lavatory sink, and a view of the heater's internal features.

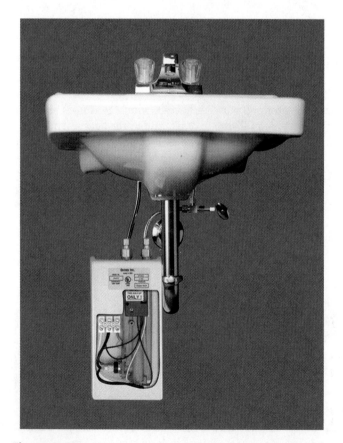

Figure 8–29 Point-of-use instantaneous electric water heaters are installed directly below a fixture. Internal wiring and component design are unique to each manufacturer. *Courtesy of Eemax.*

FROM EXPERIENCE

The higher the voltage rating of an instantaneous water heater, the more gpm of hot water it provides.

Many point-of-use electric water heaters have special water outlets (dispensers) for preparing hot beverages. The dispensers are available in various designs and finishes to match existing kitchen faucet trim. Many office kitchenettes have used these dispensers for decades, and now many custom homes also include them in their kitchens. Figure 8–30 shows several instantaneous hot water outlets for preparing hot beverages.

FROM EXPERIENCE

A sink may have to be ordered with an additional hole to accommodate a hot water dispenser.

Solar Water Heater

Solar water heating is popular in many regions of the country, but it has never fully replaced gas or electric. The initial installation cost of a solar water heating system may deter many homeowners; however, there are federal tax credits for installing alternate energy systems. There are various solar water-heating designs, but the roof-mounted version is the most popular. The location and angle of the solar panel is important for optimal efficiency of the entire system. Most solar systems are connected directly to the gas or electric water heating system to provide adequate hot water during non-solar heating periods, such as during the night and on cloudy days. Because the solar **collector** (panel) is located away from the storage tank, a pump is needed to circulate the water through the system. Unheated water enters the collector and remains there until it is heated to a desired temperature. A system controller made up of temperature sensors activates the pump to circulate the heated water to the storage tank. This process continues until all the water in the storage tank is heated. Figure 8–31 is an illustration of the basic concept of a solar water-heating system. Figure 8–32 is a cut-away view of a solar collector.

FROM EXPERIENCE

A solar water-heating system used for heating potable water must comply with all codes regulating the entire potable water system.

CHAPTER 8 Plumbing Equipment and Appliances 183

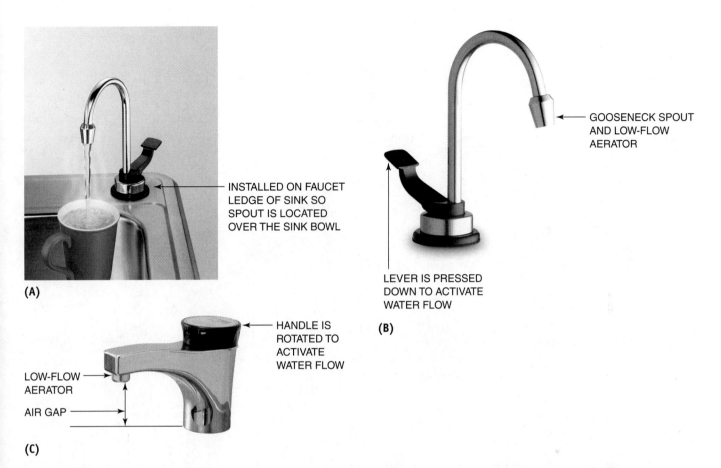

Figure 8–30 Various styles of point-of-use water outlets are available and are popular for preparing hot beverages. *GN Series Courtesy of In-Sink-Erator.*

The pump used to circulate potable hot water has a nylon, stainless steel, or bronze impeller to comply with code regulations. All water-heating systems must have safety features such as temperature and **pressure relief valves**, and backflow prevention devices. Cold regions must have freeze protection devices and manual draining capability, such as a boiler drain. The system control will vary based on the complexity of the system. The system's manufacturer will provide adequate information about purchasing compatible operating accessories. In addition, local codes will dictate any specific installation requirements. Figure 8–33 is a typical circulating pump for a small domestic hot water system. Figure 8–34 is a schematic illustration of a solar water-heating system with typical control, safety, and operating features.

 FROM EXPERIENCE

Pumps are selected based on the system they serve. The height of a solar collector above the storage tank is a determining factor in selecting the correct pump.

System Protection

Every water-heating system must be protected against dangers that can occur when water is heated. During a heating cycle, water expands and can cause the pressure relief valve to drip, which, in turn, can cause piping systems to leak. **Expansion tanks** are installed to absorb the expansion of a system. The connection from the piping system to the water heater can corrode, which is a cause for concern regarding the longevity of a water heater. Another concern is preventing rust from entering the piping system. An internal accessory known as an **anode rod** is installed by the manufacturer to protect the inside of the water heater.

You have learned about electrolysis and the use of dielectric unions, temperature and pressure relief valves, and backflow prevention. Now we will focus on specific parts of water heating systems. For example, earthquake-prone regions of the country require that securing straps and accessories be installed to protect the water heater and associated piping.

Lined Pipe Nipples

The pipe nipples on many water heaters are lined with a corrosion-resistant material such as PEX. The connection to

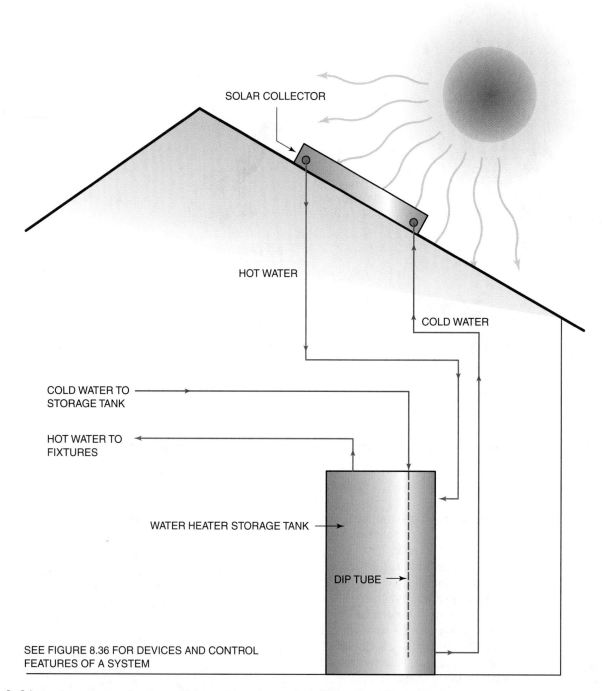

Figure 8–31 A solar collector absorbs sunlight, and heat is transferred from the collector to the water.

copper pipe is still a dissimilar connection, so electrolysis protection is required, but the lining eliminates internal corrosion of the nipple. Corrosion of a pipe nipple allows rust to settle in the bottom of a storage tank; it can also enter the piping system. Rust can obstruct the flow of water in a faucet aerator, create abrasion within a piping system, and cause premature failure of valves and devices. If a water heater has female threaded connections, lined nipples can be ordered separately. A plumber must take precautions when connecting copper tube to a lined nipple. Heating the lining of the pipe nipple directly with a torch or connecting a fitting that has been soldered and not allowed to cool will melt the internal lining of the nipple. Figure 8–35 shows a lined pipe nipple that connects the piping system to a water heater.

CHAPTER 8 *Plumbing Equipment and Appliances* **185**

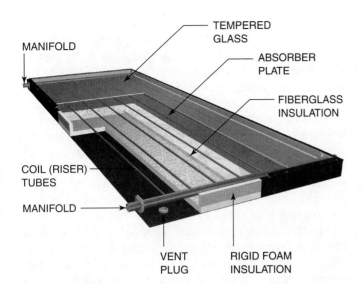

Figure 8-32 A popular solar collector design uses a manifold and coil to heat the water quickly and evenly. *Courtesy of Sun Earth, Inc.*

Figure 8-33 A circulating pump moves water through a piping system. Low horsepower pumps are inexpensive to operate. *Courtesy of Taco, Inc.*

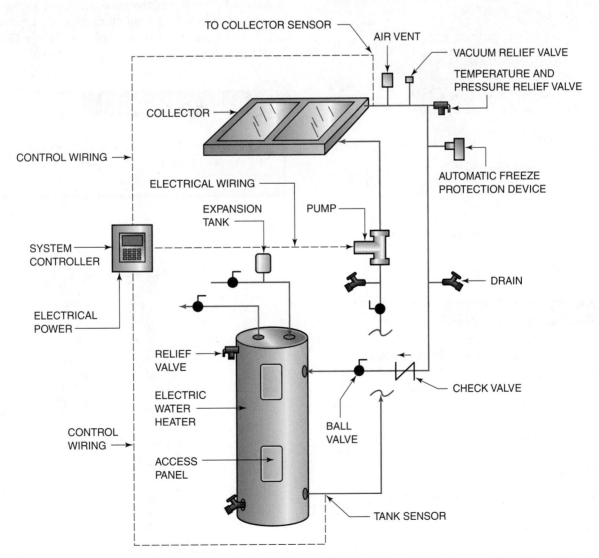

Figure 8-34 A complete solar water heating system has a complex series of controls, safety devices, and isolating valves.

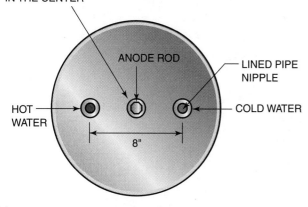

Figure 8–35 Most manufacturers line the pipe nipples that connect to a water heater with non-corrosive materials. *(A) Courtesy of State Water Heaters.*

 FROM EXPERIENCE

If pipe nipples are not included with a water heater and lined nipples are not available, it is best to install brass nipples. Galvanized nipples are legal, but may corrode, and installing black steel nipples violates plumbing codes.

Anode Rod

The storage tank for a residential water heater is manufactured from carbon steel. Most residential water heaters have glass linings to prevent the carbon steel tank from corroding. The lining is a very thin coat of porcelain enamel designed to fill every internal crevice created during the manufacture of the tank. However, the shipping and handling process can cause small fractures in the glass lining. An anode rod is installed by the manufacturer to combat glass-lining imperfections or minor damage. The anode rod is a sacrificial magnesium-based device that dissolves (corrodes) over a period of time. As the magnesium dissolves, electrons are created and are attracted to the exposed steel portions of the storage tank. Eventually, the anode rod dissolves to a point that it becomes ineffective, resulting in corrosion of the steel tank's interior. Most manufacturer warrantees dictate that the tank must be drained at least annually and that the anode rod must be inspected. Homeowners rarely do this. Not following this requirement is the leading cause of premature failure of a water heater. If the water heater is more than five years old, it is likely that the anode rod is depleted and no longer protecting the storage tank from corrosion. The water quality, the amount of water used, the temperature of the water, and the condition of the glass lining are major factors in the life expectancy of an anode rod. The anode rod in most residential water heaters is in the top of the tank, and is removable with a socket tool. Figure 8–36 is an illustration of a typical anode rod. It compares the new one and the old one that is no longer effective.

 FROM EXPERIENCE

Anode rods are not typically available locally and might have to be specially ordered from the water-heater manufacturer.

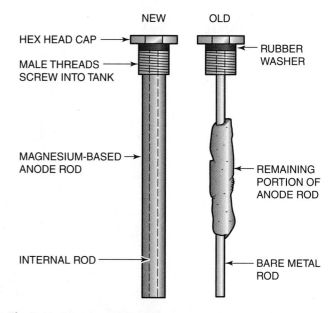

Figure 8–36 An anode rod is designed to dissolve as it protects the storage tank from corrosion.

Expansion Tanks

Most codes dictate that an expansion tank must be installed near a water heater to protect the piping system from the high pressure caused by a heating cycle. Many leaks in piping systems occur when no water is being used. Homeowners sometimes return from vacation to find that a leak has occurred and do not realize that, because the water heater has been heating while they were away, excessive pressure has built up within their piping system. A temperature and pressure relief valve is installed on every water heater. However, many leaks occur in a piping system when the pressure is less than the amount needed for the relief valve to perform its protective duties. Many water meters have a check valve so water will not flow back into the municipal potable water system. Those water meters create a closed system when no water is flowing, thereby leaving no room for expansion of the heated water. Most residential water heaters use small expansion tanks, but several different types and sizes are available. Regardless of size, all expansion tanks operate basically the same. Figure 8-37 shows a typical expansion tank.

An expansion tank for a potable hot water system has an internal rubber membrane called a bladder. Because water cannot be compressed, air is injected on one side of the bladder; water then enters and exits the expansion tank as the system pressures dictate. The air pressure in the expansion tank should be equal to the cold-water supply pressure of the system. It must be checked by a plumber before the water heater is put into service. If the incoming cold water is 60 pounds per square inch (psi), then a plumber must make sure that the air pressure in the expansion tank is also 60 psi. When the heating cycle increases system pressure and causes water to enter the tank, the air is compressed within the expansion tank. By allowing for expansion, the piping system is not subjected to the pressures that would be caused by expansion. The expansion tank should be close to the water heater and must be on the downstream side of a check valve. It is installed in the cold-water piping system serving the water heater. Figure 8-38 illustrates the operating features of an expansion tank serving a residential water heater. Figure 8-39 illustrates the typical location of the expansion tank in relation to the water heater and piping configuration.

> **FROM EXPERIENCE**
>
> An expansion tank can prolong the life of a water heater and the piping system.

Figure 8-37 An expansion tank is a safety device that protects a hot water system from excessive pressure caused by expansion of water during a heating cycle. *Courtesy of State Water Heaters.*

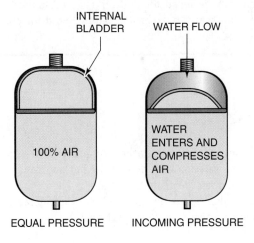

Figure 8-38 An expansion tank receives water as a hot water system expands during a heating cycle and discharges water as a system returns to normal pressure.

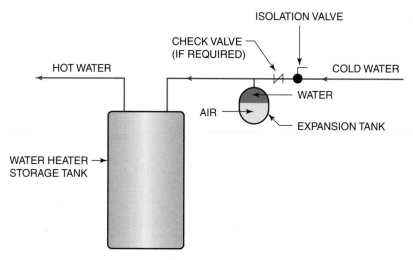

Figure 8-39 An expansion tank is installed in the cold water piping to a water heater and downstream of a check valve.

 FROM EXPERIENCE

An expansion tank for a potable hot water system is specifically designed for that purpose. A plumber must make sure that the expansion tank is rated for potable water use.

Summary

- Washing machine installations typically have a specially designed wall box that houses the hot and cold water supply connections and a 2" drain.
- An icemaker box is installed behind a refrigerator to provide a water supply connection.
- A dishwasher is typically installed during the kitchen sink installation phase of construction.
- A residential water heater is usually installed during the trim-out phase of construction.
- Most residential electric water heaters are 240-volt and non-simultaneous.
- A gas water heater has strict installation locations.
- One British thermal unit (BTU) is the amount of heat required to raise one pound of water one degree F.
- Residential water heaters with a BTU rating of 75,000 or less must have a flame arrestor design.
- According to most codes, an expansion tank must be installed with every water heater.
- An anode rod is a factory-installed sacrificial rod designed to protect the inside of a storage tank.
- A thermocouple must sense a pilot flame before allowing gas to flow through a gas regulator.
- The termination location of a gas water heater flue pipe is strictly regulated by code.

Review Questions

1. A dishwasher can be connected to a branch tailpiece or to a designated connector of a
   a. Garbage disposer
   b. Basket strainer
   c. P-trap
   d. Vent

2. A typical residential washing machine box has a hot and cold water connection and a(n)
   a. Overflow hole
   b. Pop-up assembly
   c. Drain connection
   d. None of the above are correct

3. A typical residential icemaker box is installed during the phase of construction known as the
   a. Trim-out phase
   b. Rough-in phase
   c. Appliance installation phase
   d. Underground phase

4. A typical residential water heater expansion tank has an
   a. Internal bladder
   b. External water inlet
   c. Internal check valve
   d. Exterior bladder

5. Every gas water heater must have its exhaust flue pipe terminated
   a. Through the exterior wall
   b. Through the roof
   c. To open air
   d. Inside the room it occupies

6. To route incoming cold water to the bottom of a hot-water storage tank, a device is installed known as
   a. A dip tube
   b. A funnel pipe
   c. An anode rod
   d. A vacuum relief valve

7. An electric water heater is heated with one or more
   a. Solar collectors (panels)
   b. Burner assemblies
   c. Elements
   d. Regulators

8. A British thermal unit (BTU) is used to
   a. Measure heat
   b. Heat water
   c. Size a circulating pump
   d. Capture heat

9. A direct-vent gas water heater is vented through the
   a. Exterior wall
   b. Roof
   c. Interior wall
   d. None of the above are correct

10. A tankless gas water heater has a
    a. Storage tank
    b. Coil design
    c. Heating element
    d. Ball valve

11. An electric instantaneous water heater has a
    a. Storage tank
    b. Heating element
    c. Ball valve
    d. Gate valve

12. Heated water from a solar collector (panel) is circulated with
    a. A flow regulator
    b. Air pressure
    c. Water pressure
    d. None of the above are correct

13 A water-heater pipe nipple designed to minimize corrosion is

a. Lined
b. Plastic
c. Black steel
d. None of the above are correct

14 An anode rod installed in a water heater to protect the internal steel storage tank is designed to

a. Be sacrificed
b. Be permanent
c. Eliminate draining the tank
d. Be reenergized

15 One BTU will raise the temperature of one pound of water

a. 1 degree Fahrenheit
b. 8.33 degrees Fahrenheit
c. 10 degrees Fahrenheit
d. 83.3 degrees Fahrenheit

16 A residential water heater gas regulator (or gas valve) controls the flow of gas to a

a. Thermocouple
b. Burner assembly
c. Flue pipe
d. Chimney

17 A typical residential electric water-heater design is a 240-volt

a. Simultaneous type
b. Non-simultaneous type
c. Instantaneous type
d. Three-phase type

18 A thermocouple senses the flame from a

a. Burner assembly
b. Pilot
c. BTU
d. Flue pipe

19 Every water heater must have a thermostat and a

a. Check valve
b. High-limit device
c. Ball valve
d. Flue pipe

20 The carbon monoxide fumes created by a gas water heater are deadly and

a. Odorless
b. Have a distinct odor
c. Have a distinct taste
d. Have a distinct color

SECTION THREE

Layout and Installation

SECTION THREE
LAYOUT AND INSTALLATION

- Chapter 9:
  Blueprint Reading and Drafting

- Chapter 10:
  Material Organization and Layout

- Chapter 11:
  Water Service Installation

- Chapter 12:
  Water Distribution Installations

- Chapter 13:
  Drainage, Waste, and Vent Segments and Sizing

- Chapter 14:
  Drainage, Waste, and Vent Installation

- Chapter 15:
  Fixture and Equipment Installation

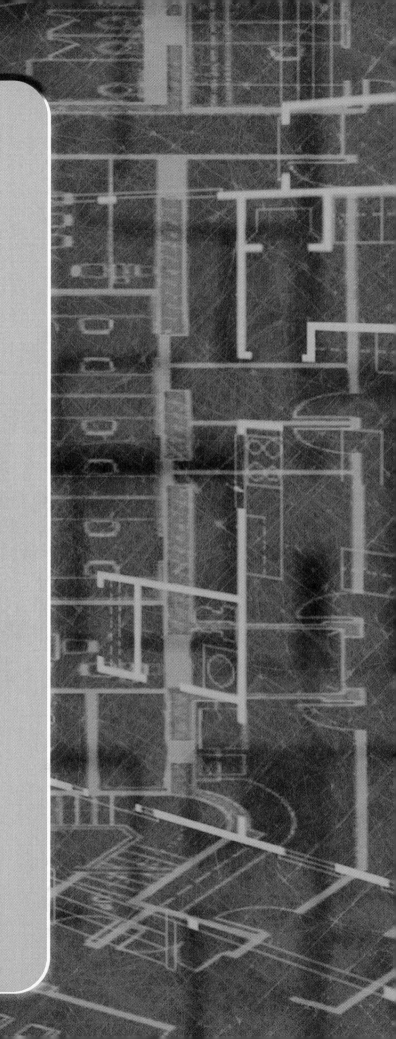

Chapter 9 | Blueprint Reading and Drafting

You have learned about the safe use of tools and equipment and how to order materials. Now we will discuss the installation of plumbing systems. Blueprint reading and drafting are valuable skills for interpreting and communicating designs on a construction site. Single-family residential blueprints are not as comprehensive as multi-family and commercial blueprints, but a basic understanding of abbreviations and symbols is necessary to progress in a plumbing career. A typical single-family residential blueprint shows the location of plumbing fixtures in relation to the structure. A plumber then routes piping systems based on particular job conditions. The fixture locations are part of the design intent. Plumbing codes and the types of fixtures being installed also help determine the size and routes of piping systems. A residential plumber must have thorough code knowledge and be able to use a blueprint for intent and layout purposes. However, a plumber arrives on a single-family residential construction site when most or all of the framing is complete, which minimizes the need to use blueprints to install the piping systems.

OBJECTIVES

Upon completion of this chapter, the student should be able to:
- understand basic plumbing symbols and abbreviations.
- interpret basic residential architectural blueprints.
- create simple sketches of piping systems.
- understand the different illustrated views of a piping system.

Glossary of Terms

drafting triangle a drafting tool used to illustrate straight or angled lines

drawings blueprints

isometric view a three-dimensional view of a piping system indicating the scope of work

joist horizontal board to support a structure; floor joists and ceiling joists are the two most common types

load-bearing describes a wall supporting a portion of the structure above

plan view a view of a design from the top; also known as a bird's-eye view

riser diagram an isometric or side view of a large portion or detailed area of a piping system and typically utilized to reflect several stories of a building

section view a view, usually detailed, of a design from the side; also known as a side view

side view a view of a design from the side; also known as a section view

sketch an illustration focusing on a certain portion of an area or piping system

stud vertical board to erect walls; 2" × 4" and 2" × 6" are the two most common sizes in residential construction

tee a fitting having three connections and offered in a variety of sizes

Plumbing Symbols

Symbols are used throughout the construction industry to illustrate devices, equipment, fixtures, piping systems, and other related items. Many projects require engineers and architects to create new symbols relating to job conditions. Knowledge of common symbols is essential in reading blueprints, regardless of the type of construction. A residential plumber is typically only exposed to common symbols relating to plumbing fixtures. Symbols are fairly standard, but architects may have varying styles. Templates are used to manually illustrate symbols, but most architects use Computer Aided Drafting (CAD) programs.

The following information covers common blueprint symbols a plumber might be exposed to during a plumbing career. Understanding plumbing symbols allows a plumber to advance into multi-family and commercial plumbing employment without learning on a job site. It is important to recognize that the purpose of a training program is to help build a career with progressive opportunities—one that does not limit you to a single area of the plumbing industry.

A piping system is illustrated from three different views. A **plan view** looks down onto a design and is also known as a bird's-eye view. A side or **section view** illustrates a piping system or portion of a design from the side. An **isometric view** is a three-dimensional illustration exposing large portions of a system. It is not a detailed view, but it shows the intent of the system, known as the scope of work.

90-Degree Offsets

A 90-degree offset fitting is one of the most common in the plumbing industry. A basic symbol with unique variations is used to illustrate the position of a 90-degree offset that turns up and turns down. A side-view symbol for a 90-degree toward is the same as for a plan-view 90-degree up. A side-view 90-degree away is illustrated like a plan-view down. A circle indicates a pipe that is either facing toward you or away from you. If a line that represents a pipe is connected to a circle, that symbol indicates that the 90 degree is facing toward you. If the line penetrates the circle, the 90 degree is away from you. Figure 9–1 compares photographs of actual piping offsets with the illustrated symbols of a plan-view 90-degree down and up.

FROM EXPERIENCE

Think of the circle for a 90-degree up as being the same as looking inside a fitting or pipe with no visible obstruction.

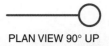
PLAN VIEW 90° UP

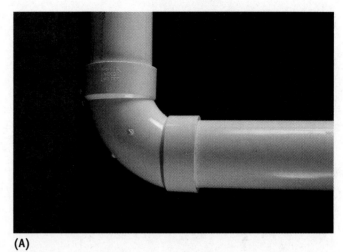
(A)

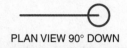
PLAN VIEW 90° DOWN

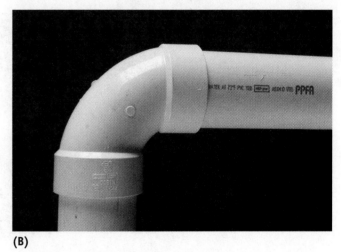
(B)

Figure 9–1 A 90-degree offset is one of the most common fittings in the plumbing trade.

45-Degree Offsets

A 45-degree offset is useful in many piping situations. Most architects do not use the symbols for a 45-degree offset, but instead leave the piping details to the plumber at the job site. A detailed blueprint created by a contractor is known as a shop drawing. Its intended purpose is to create a more specific installation blueprint based on the design intent of the architect. There are 45-degree offset symbols to indicate three different piping configurations. A 45-degree up symbol is a full circle; for a 45-degree down, the line representing the connecting pipe penetrates the circle. An in-line 45-degree offset in a piping system is often used to

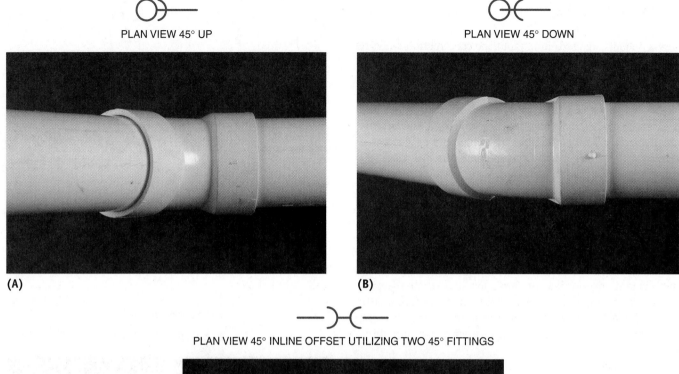

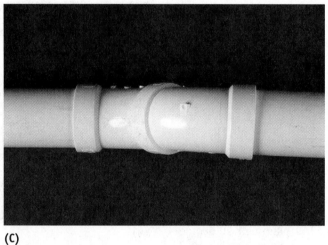

Figure 9–2 A 45-degree symbol indicates that half the fitting hub is visible on a plan view.

offset over or under another pipe or building structure. The symbol illustrates the unique features of a 45-degree fitting. Figure 9–2 compares photographs of the actual piping offsets with the symbols for a plan-view 45-degree fitting down, up, and creating an inline offset.

FROM EXPERIENCE

The line through the circle is an obvious sign that the fitting is turning down, but notice the difference in the semicircle for the up and down positions.

Tees

Tees have three connecting pipes and are illustrated much like a 90-degree offset. A full circle with no line penetrating it represents a tee turning up on a plan view and a tee turning toward you on a side view. A continuous line through a circle represents a tee turning down on a plan view and a tee away on a side view. Numerous piping configurations can be made with tees to achieve design intent. Detailed illustrations are needed to indicate the exact pipe route, but understanding basic symbols is the first step in learning blueprint reading. Figure 9–3 compares photographs of actual piping offsets with illustrated symbols of a plan-view tee down and up.

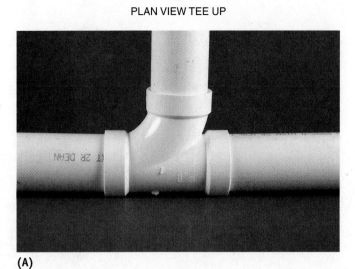

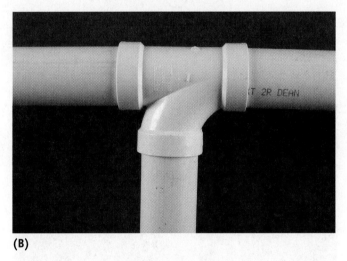

Figure 9–3 A tee has three pipes connecting into one fitting.

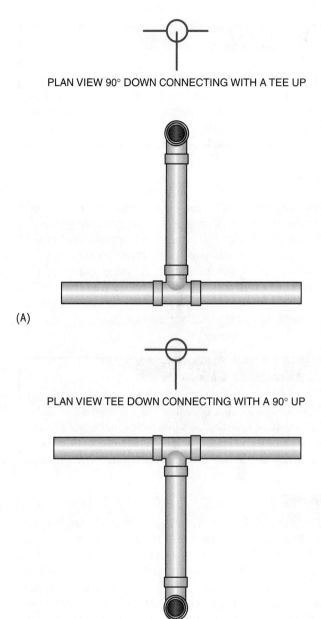

Figure 9–4 A perpendicular fitting connection has symbols that could represent numerous piping configurations.

A tee up and down is similar to a 90-degree symbol; the circles are the feature that determines the position.

Perpendicular Tee Configuration

A perpendicular tee configuration often connects a branch pipe to a main piping system. Two basic symbols indicate a branch pipe that connects to the bottom or to the top of the main pipe. The same symbol that indicates a tee turning down represents a perpendicular branch that connects to the bottom of the main pipe. When a perpendicular branch pipe connects the top of a main pipe, the 90-degree down symbol is visible and a line represents the main pipe, but the tee up is hidden. Figure 9–4 compares photographs of the actual piping offsets with the illustrated symbols for perpendicular tee configurations.

198 **SECTION THREE** *Layout and Installation*

FROM EXPERIENCE

Perpendicular is considered 90 degrees from the connecting pipe. Many design features can be hidden under this symbol, so further investigation of design intent is usually required.

P-trap

The plan-view symbol for a p-trap is a combination of a 90-degree up and a 90-degree down symbol. A p-trap is a U-shaped fitting that holds water and can only be installed in one position, so only one symbol is required. Many p-trap designs can swivel. A variation of the symbol illustrates that the inlet of the p-trap should be offset from the outlet. Figure 9–5 compares photographs of the actual p-traps and the symbols for p-trap configurations.

FROM EXPERIENCE

A p-trap symbol that appears to be swiveled often indicates that a dimension on a blueprint is not exact or that it can be field verified by a plumber.

Piping

The types of piping systems in a residential plumbing system are limited compared with the types of systems that could be installed on a commercial and industrial job site. The common residential systems include cold water, hot water, hot water return, drain, and vent. Stating that a symbol is used to identify piping can be misleading because the entire illustrated piping system is identified by unique lines. Dots are inserted to break the solid lines representing the three different water distribution piping systems. A vent is indicated by a continuous dotted line, and a drain by a solid line. The use of a solid line to identify other piping systems is acceptable. Inserting the type or the abbreviation for a particular system is similar to inserting dots. Figure 9–6 illustrates the common piping systems pertaining to a residential plumbing system.

FROM EXPERIENCE

Hot-water return systems are not common in single-family residential construction, but knowing the symbol is important.

P-TRAP

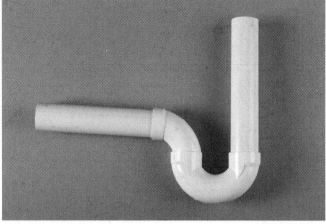
(A)

P-TRAP SWIVELED

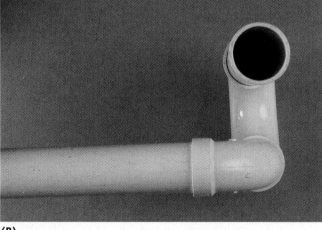
(B)

Figure 9–5 A p-trap symbol is a combination of 90-degree offset symbols.

Cap, Reducer, and Plugs

A plumbing system is typically installed in phases, and a plumber must cap or plug a system to test for leaks. Cleanouts are installed in drainage systems to provide access for clearing obstructions. A cleanout plug is illustrated in the same manner as plugs used for other purposes. A cleanout installed in a floor has a cover, and a circle with the abbreviation CO indicates that to a plumber. Some cleanout symbols may use abbreviations that specify the location of the cleanout, such as WCO for wall cleanout. A cap is typically a temporary item and may not be indicated by an ar-

(A) DRAIN IS A SOLID LINE

(B) VENT IS DASHED LINE

(C) COLD WATER HAS A SINGLE DOT

(D) HOT WATER HAS DOUBLE DOTS

(E) HOT WATER RETURN HAS TRIPLE DOTS

IRRIGATION
(F) CUSTOMIZED SYMBOL

Figure 9–6 Piping systems are illustrated to indicate the specific type of system or they can be customized.

chitect, but it is often illustrated on a shop drawing to ensure that it is ordered. A reducer is used to transition between two different pipe sizes and can be confused with a flow-direction arrow. When a symbol is used to indicate a reduction in pipe size, notification of the reduced pipe size is placed near the symbol. Figure 9–7 illustrates common symbols for a cap, reducer, and plug.

FROM EXPERIENCE

FCO is the abbreviation for floor cleanout used by many architects. ECO can represent end cleanout or exterior cleanout; its physical location in a design indicates its abbreviated meaning.

Valves and Devices

Numerous symbols indicate types of valves within a system. Custom symbols or industry standard symbols can be

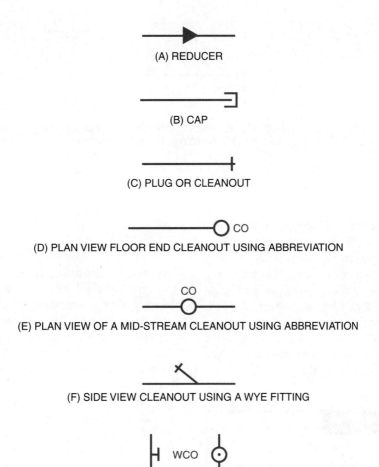

Figure 9–7 Cap, reducer, and plug symbols indicate specific installation instructions within a piping system.

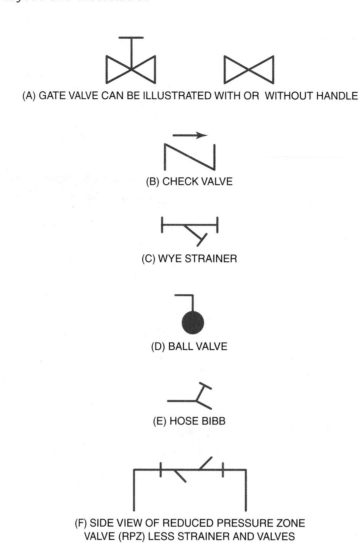

Figure 9–8 Valve and device symbols indicate an installation location within a piping system.

used to identify a valve. Some valve symbols indicate a handle, but the actual symbols used depend on the drafting style of an individual architect. Symbols and their meanings for a particular job are listed by an architect on a blueprint page for a plumber's review. Some symbols can actually indicate the type of connection to a valve, such as threaded, flanged, or welded. A check-valve symbol is standard and must include an arrow indicating a direction of flow to make sure it is installed correctly. Figure 9–8 illustrates numerous valves and devices often found on a residential blueprint.

FROM EXPERIENCE

RPZ usually indicates the water service piping entering a building, or it can mean irrigation systems.

Fixtures

Symbols identifying specific fixtures installed in residential construction are basic. Variations may occur on a particular job, but the intent is easily understood based on their location within a building. Symbols do not indicate the actual fixture but do illustrate design intent. A variation on a sink symbol might indicate the number of bowls or a corner sink. A bathtub symbol will indicate the drain location, so a plumber knows where to install the rough-in piping, and the correct tub to order. Three basic bathtub symbols are common. Most corner or garden style tubs are found in a master bathroom. A toilet symbol is typical for all jobs, and a bidet symbol is similar to a toilet symbol but without the indication of a toilet tank. A shower symbol typically has crossed lines indicating that the floor is sloped toward the drain. The drain is located in the center of the shower symbol, so a plumber needs to understand where to install the drain. The

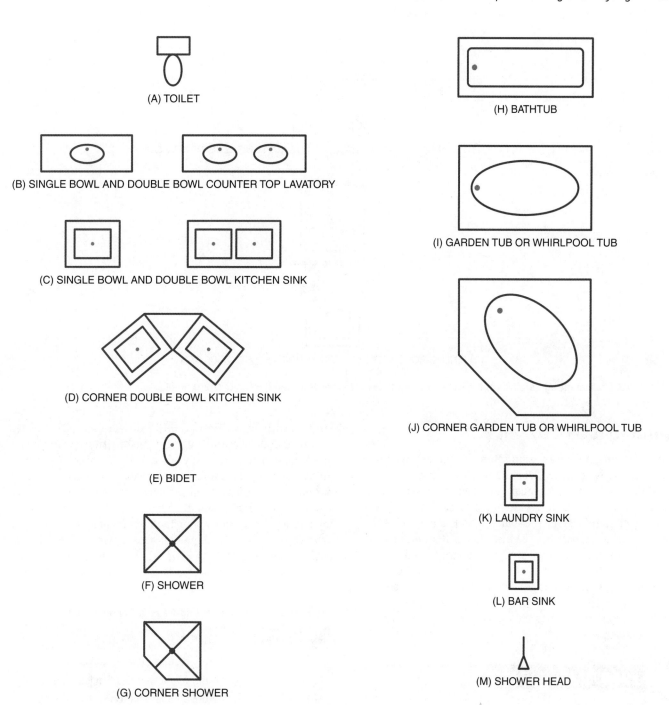

Figure 9-9 Plumbing fixtures are illustrated with unique symbols.

specific fixture, not the symbol, dictates the drain location. The same sink symbols sometimes identify different types of sinks; only the size of the symbol might vary. If the same symbol is used to identify different types of fixtures, an abbreviation or other identifying mark must clarify the design intent. Figure 9-9 illustrates common residential plumbing fixtures and their unique symbols.

 FROM EXPERIENCE

Angle shower and bathtub symbols vary based on the corner of the room in which they are located.

(A) WATER HEATER

(B) FLOOR DRAIN CAN BE ILLUSTRATED USING MANY DIFFERENT METHODS

(C) DISHWASHER

(D) SINGLE-BOWL KITCHEN SINK WITH GARBAGE DISPOSER

Figure 9–10 Equipment and drains are identified using abbreviations within or adjacent to the symbol.

Equipment and Drains

Abbreviations indicate design intent for plumbing equipment and drains, such as floor drains. A circle with WH is typical for a water heater symbol. A garbage disposer may be identified with GD within the kitchen sink symbol, and, if a double bowl sink is used, the abbreviation is placed in the bowl where the disposer is to be installed. A floor drain symbol can be a simple circle or a circle placed inside a square and will be identified as FD, but many different methods are used. A dishwasher, washing machine, and dryer are shown as squares with their abbreviations to indicate their location within a room. Figure 9-10 shows some common symbols and their abbreviations.

Basement and garage floors are the most common areas for floor drains in residential homes.

Abbreviations

Abbreviations are required on blueprints to eliminate clutter within a design. Common abbreviations are industry standard, and unique abbreviations often describe a specific system, item, or term. An abbreviation legend is created by an architect and listed on a blueprint page. Some abbreviations can describe several different items, so a plumber must not automatically assume the abbreviation stands for a particular item without verification. WH can represent either water heater or wall hydrant, but their installation locations are different and usually easily identified. As your career progresses, you will recognize common abbreviations and might not have to review a legend to interpret them. One of the first steps in reviewing a blueprint for the first time is to search for abbreviations and understand their meaning. You may have to interpret non-plumbing abbreviations when coordinating your installations with other trades. Table 9-1 lists abbreviations that you might encounter during your career.

Table 9–1 Common Plumbing-Related Abbreviations

Abbreviation	Stands for
ABS	Acrylonitrile Butadiene Styrene
ADA	Americans with Disabilities Act
AFF	Above Finished Floor
AG	Above ground
AGA	American Gas Association
AISI	American Iron and Steel Institute
API	American Petroleum Institute
ASA	American Standards Association
ASME	American Society of Mechanical Engineers
ASTM	American Society for Testing Materials
AW	Acid Waste
AWWA	American Water Works Association
B & S	Bell and Spigot
B to B or B - B	Back to Back

Table 9–1 Continued

Abbreviation	Stands for
BD	Building Drain
BFP	Backflow Preventer
BID	Bidet
BLK	Black
BM	Bench Mark
BOP	Bottom of Pipe
BOT	Bottom
BS	Building Sewer
BT	Bathtub
BTU	British Thermal Unit
BV	Ball Valve or Butterfly Valve or Branch Vent
C2H2	Acetylene
C4H10	Butane
°C	Celsius or Degrees Centigrade
C	Centigrade or Hundred or Center
CC	Cubic Centimeter
C to C or C - C	Center to Center
C × C	Copper by Copper
C × C × C	Copper by Copper by Copper
C to F	Center to Face
C × F	Copper by Female
C × M	Copper by Male
CF or CU FT	Cubic Foot/Feet
CFM	Cubic Feet Per Minute
CFS	Cubic Feet Per Second
CHW	Chilled Water
CHWR	Chilled Water Return
CHWS	Chilled Water Supply
CI or C. I.	Cast Iron
CI or CU IN	Cubic Inch
CM	Centimeter
CMU	Concrete Masonry Unit
CO	Cleanout or Carbon Monoxide
CO2	Carbon Dioxide
COMP	Compression or Companion
CP	Chrome Plated or Control Point
CPVC	Chlorinated Polyvinyl Chloride
CS	Cast Steel or Carbon Steel
CV	Circuit Vent
CV or CK V	Check Valve
CY or CU YD	Cubic Yard
CW	Cold Water
D	Diameter
DEG or °	Degree

Table 9–1 Continued

Abbreviation	Stands for
DF	Drinking Fountain
DFU	Drainage Fixture Unit
DH	Double Hub
DIAM	Diameter
DR	Drain or Drainage
DS	Downspout
DW	Dishwasher
DWG	Drawing
DWV	Drainage, Waste, and Vent
E to C or E - C	End to Center
E to E or E - E	End to End
ECO	End Cleanout
EWC	Electric Water Cooler
EWH	Electric Water Heater
F to F or F - F	Face to Face
°F	Degrees Fahrenheit
FCO	Floor Cleanout
FD	Floor Drain and Fixture Drain
FF	Finished Floor
FIG	Figure
FIP	Female Iron Pipe
FLG	Flange
FLGD	Flanged
FLR	Floor
FP	Full Port
FPS	Feet Per Second
FS	Floor Sink
FT or ′	Foot or Feet
FTG	Footing or Fitting
FU	Fixture Unit
FV	Flush Valve
G	Gas
GAL	Gallon
GALV	Galvanized
GI	Galvanized Iron
GND	Ground
GP	Gauge Pressure
GPF	Gallons Per Flush
GPH	Gallons Per Hour
GPM	Gallons Per Minute
GPS	Gallons Per Second
GV	Gate Valve
HB	Hose Bibb
HD	Heavy Duty or Hub Drain
HG	Mercury

Table 9-1 Continued

Abbreviation	Stands for
HGT	Height
HHW	Heating Hot Water
HHWR	Heating Hot Water Return
HHWS	Heating Hot Water Supply
HOR	Horizontal
HP	High Point or High Pressure or Horse Power
HTG	Heating
HTR	Heater
HW	Hot Water
HWH	Hot Water Heater
HWR	Hot Water Return
ID	Inside Diameter
IE	Invert Elevation
IN or "	Inch
INC	Increaser
INV	Invert
IPC	International Plumbing Code
IPS	Iron Pipe Size
IW	Indirect Waste
JS	Janitor Sink
K	Kelvin
K or KIP	Kilopound
KG	Kilogram
KM	Kilometer
KO	Knock Out
KS	Kitchen Sink
KW	Kilowatt
L or LGTH	Length
LAV	Lavatory
LB	Pound
LH	Left Hand
LIQ	Liquid
LP	Low Pressure
LPG	Liquid Petroleum Gas
LV	Loop Vent or Low Voltage
LW	Light Weight
M	Motor or Thousand
MAINT	Maintenance
MALL	Malleable
MATL	Material
MAX	Maximum
MECH	Mechanical
MED	Medium
MFG	Manufacturing

Table 9-1 Continued

Abbreviation	Stands for
MFR	Manufacturer
MH	Manhole
MI	Malleable Iron or Mile
MIN	Minimum
MIP	Male Iron Pipe
MISC	Miscellaneous
MM	Millimeter
MR	Mop Receptor
MS	Mop Sink
MSS	Manufacturer's Standardization Society
N	North
N2	Nitrogen
NG	Natural Gas
NC	Normally Closed
NFWH	Non Freeze Wall Hydrant
NH	No Hub
NIC	Not In Contract
NIP	Nipple
NO	Normally Open or Number
NOM	Nominal
NPS	National Pipe Size
NTS	Not To Scale
O	Offset
O2	Oxygen
OD	Outside Diameter
OS&Y	Outside Screw and Yoke
OZ	Ounce
P&T	Pressure and Temperature
PB	Lead
PC	Plumbing Contractor or Pre-cast Concrete
PCF	Pounds Per Cubic Foot
PE	Plain End or Polyethylene
PEX	Cross Linked Polyethylene
PI	Pressure Indicator
PG	Pressure Gauge
PLG	Plumbing
PLMG	Plumbing
PRES	Pressure
PRV	Pressure Reducing Valve
PSF	Pound Per Square Foot
PSI	Pounds Per Square Inch
PSIA	Pounds Per Square Inch Absolute
PSIG	Pounds Per Square Inch Gauge

Table 9-1 Continued

Abbreviation	Stands for
PT	Pint
PVC	Polyvinyl Chloride
QTY	Quantity
R	Radius
R & L	Right and Left
RCP	Reinforced Concrete Pipe
RD	Roof Drain
RED	Reducing or Reducer
RF	Roof Flashing
RGH	Rough
RH	Right Hand
RI	Rough-in
RL	Roof Leader
RM	Room
RPM	Revolutions Per Minute
RPS	Revolutions Per Second
RPZ	Reduced Pressure Zone Valve
RV	Relief Vent or Relief Valve
RWL	Rain Water Leader
S	Sink or Sewer
SA	Shock Absorber
SAN	Sanitary
SCD	Screwed
SCHED	Schedule
SCM	Square Centimeter
SD	Storm Drain
SEC	Second
SF	Square Foot or Square Feet
SH	Single Hub
SHR	Shower
SHWR	Shower
SIN	Square Inch
SK	Sketch or Sink
SM	Square Meter
SO	Side Outlet
SP	Soil Pipe
SPEC	Specifications
SQ	Square
SQ FT	Square Foot
SQ IN	Square Inch
SQ YD	Square Yard
SS	Stainless Steel or Sanitary Sewer or Soil Stack or Service Sink
STD	Standard
STL	Steel

Table 9-1 Continued

Abbreviation	Stands for
SUPT	Superintendent
SV	Stack Vent or Safety Valve or Service Weight Pipe
SY	Square Yard
T	Travel
T&P	Temperature and Pressure
TBM	Temporary Bench Mark
TD	Trench Drain
TEMP	Temperature
TG	Temperature Gauge
TH	Thermostat (see TSTAT)
THD	Threaded
THK	Thick
TI	Temperature Indicator
TLT	Toilet
TOC	Top of Concrete
TP	Trap Primer
TSTAT	Thermostat (see TH)
TYP	Typical
UG	Underground
UH	Unit Heater
UL	Underwriter's Laboratories
UNO	Unless Noted Otherwise
UR	Urinal
USS	United States Standard
V	Volt or Vent or Valve
VAC	Vacuum
VB	Vacuum Breaker
VCP	Vitrified Clay Pipe
VCT	Vinyl Composite Tile
VERT	Vertical
VIF	Verify In Field
VOL	Volume
VS	Vent Stack
VTR	Vent Through Roof
W	Width
W & D	Washer and Dryer
WB	Washer Box
WC	Water Closet
WCO	Wall Cleanout
WH	Wall Hydrant or Water Heater
WI	Wrought Iron
WM	Washing Machine or Water Meter
WP	Water Pump
WS	Waste Stack

Table 9-1 Continued

Abbreviation	Stands for
WSFU	Water Supply Fixture Unit
WWP	Water Working Pressure
XH	Extra Heavy
XHVY	Extra Heavy
XS	Extra Strong
XXH	Double Extra Heavy
XXS	Double Extra Strong
YD	Yard
YLW	Yellow
YR	Year

Architectural Blueprints

An architectural blueprint is the master plan of an entire project. Single-family residential construction blueprints show the construction of the home but often do not show the piping. It is the responsibility of a plumber to interpret the framing construction based on the architectural design intent. An architect provides a carpenter with detailed information to ensure that the structural load of the building is constructed as designed. The architect provides numerous detailed views of areas that require clarification. The fixture locations and the direction of floor **joists** are two areas of a design that a plumber focuses on initially when designing a pipe route. The drain termination serving a toilet, shower, and bathtub are crucial and do not leave much room for error. It is not uncommon for a plumber to request a design change when a portion of the building conflicts with the piping system. An architect may indicate that a fixture be installed in a certain location, but plumbing codes might not allow a pipe route there. Conflicts with wood **studs** and joists are common, and a plumber can request that the joist or stud design be altered to accommodate the plumbing system.

Architectural blueprints are illustrated on a small scale to represent a full-sized structure. Industry-standard scales are used, the most common being that 1/4" represents one linear foot, but other scaling options are also used. An architect provides dimensions for the total length, width, and height of the building in feet and inches. A scale or scale ruler (see Figure 9–21) determines dimensions that are not provided by an architect. To avoid errors, it is usually not recommended that a scale ruler be used to lay out exact dimensions. A plumber should coordinate unknown dimensions with a carpenter or some other person who is responsible for the location of walls. Doors, windows, and other fixed items dictate wall locations; plumbing fixtures are then installed to coordinate with wall positions. Any deviation from original design intent impacts the installation of other items in the house.

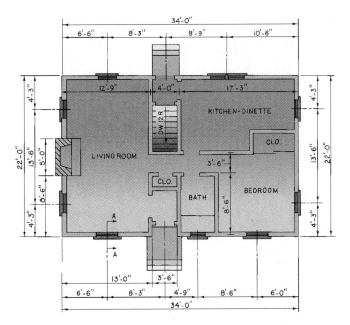

Figure 9–11 A floor plan provides dimensions to critical construction areas and illustrates major components of a design.

A plan view of a single-family residential blueprint is known as a floor plan. This illustrated view indicates which floor of the building it represents. In addition, all blueprint pages have an alphabetical and numerical identification, as well as a text description. For example, the first-floor blueprint might be identified as A1, and the second-floor plan as A2. The actual identification used depends on the project and the architect's style. A floor plan includes major design features such as stairs, doors, windows, plumbing fixtures, and dimensions. The floor plan may not show all dimensions of every interior wall or partitions. This is not a problem because the framing is typically complete before a plumber begins the installation on a single-family home. Figure 9–11 shows a typical single-family residential floor plan.

 FROM EXPERIENCE

Plumbing blueprints are often indicated by the letter *P*, such as P1, P2, or P3; electrical is *E*, and heating is *H*.

The location and direction of floor joists are indicated on a blueprint known as a foundation plan, but the joist directions are typically shown on all floor plans. Other contractors use this plan when they install the foundation of a building, and a plumber uses it to coordinate the pipe routes that serve fixtures. A plumber cannot simply drill holes or cut notches where desired because codes regulate the size and location of holes and notches in wood boards. A plumber often must install piping in the same direction as a joist to

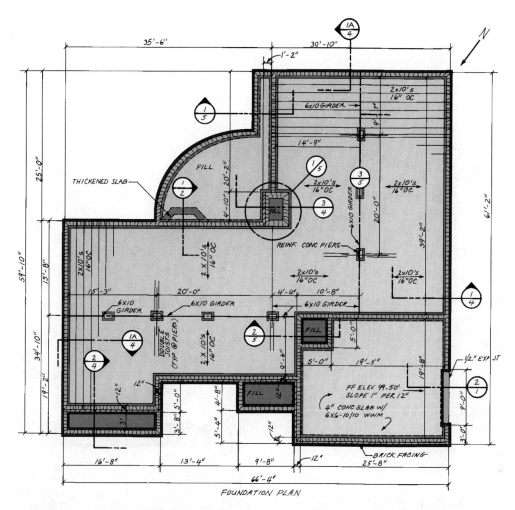

Figure 9–12 A foundation plan indicates locations of structural supports and direction of floor joists. *Courtesy of PTEC-Clearwater-Architectural Drafting Department.*

provide a productive pipe installation and minimize construction errors and code violations. Knowing abbreviations pertaining to framing or general construction will increase your plumbing knowledge also. The abbreviation OC means on center. It can describe how far apart the joists are being installed. Floor joists are larger than wood studs; the most common sizes for a single-family residential home are 2″ × 8″ and 2″ × 10″. A wood board is not the exact size of the dimensions it is identified by. For example, the actual dimension of a 2″ × 4″ board is 1-1/2″ × 3-1/2″. Figure 9–12 shows a typical residential foundation plan.

 FROM EXPERIENCE

The location of underground plumbing pipes that will continue through the floors above should be coordinated with the direction of the floor joists to increase productivity and avoid drilling and notching the joists.

Architectural Symbols

There are numerous architectural symbols. A plumber should be familiar with relevant ones that show details within a blueprint. Review Figure 9–13 and notice the circular symbols. A plumber must know two common symbols that will help find larger views of specific construction areas. A circle is segmented in half. An arrow on one side of the circle indicates the view direction, and a tail indicates the scope of the view. The bottom half of the circle contains the blueprint page information, and the top half has the detail number to view once you are on the correct blueprint. The detailed illustration will have a circular symbol without an arrow or tail to indicate that the detail you are viewing is the one referenced on another blueprint. The identifying letters and numbers within the circle are the same as those on the blueprint page that the detail clarifies. Figure 9–13 shows two detail symbols to direct a plumber to a larger, detailed construction area. Figure 9–14 illustrates various ways to indicate a detailed area is available for review.

208 SECTION THREE *Layout and Installation*

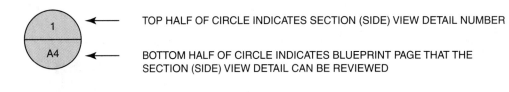
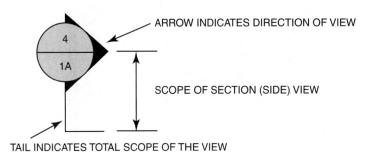

Figure 9-13 Detail symbols direct a plumber to a larger detailed view of a construction area.

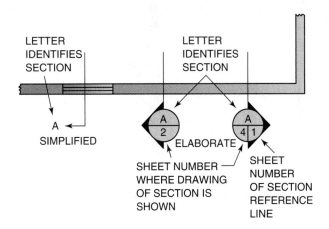

Figure 9-14 Various methods indicate that a detailed view of an area is available for review.

Details are typically illustrated to scale and are identified by the detail symbol on the blueprint page.

Ceiling joists and other framing elements are shown on a blueprint known as a roof-framing plan. A plumber installs vent pipes through a roof, attempting to make all vent penetrations where they are not visible from the front of the home. The roof is complete when a plumber installs the vent system in the attic, but this plan is not widely used by a plumber. The peaks and valleys of the roof, the distance of the roof trusses, and the direction of the trusses are indicated on a roof-framing plan. As with the foundation and floor plan, the roof-framing plan shows crucial dimensions.

Most blueprints indicate the direction of North to orient you when reading a blueprint. Figure 9-15 is a typical residential roof-framing plan with a direction arrow indicating North abbreviated as N.

 FROM EXPERIENCE

A plumber should locate a vent-pipe penetration through a roof based on the safest installation of a roof flashing around the pipe.

An elevation plan shows the exterior design of all sides of a building. The detailed information provided by an architect varies by project. An elevation plan exposes the differences and similarities of all sides; which sides include brick or siding; and the location of doors, windows, chimney, and other features on each side of the home. Most elevation-plan blueprints indicate the finished ground (grade) level in relation to the finished floor (FF). Any indication of a finished grade is from mean sea level. If a numerical value of 101.50 feet is indicated as the finished grade, it represents the height above the sea when the water is between high and low tide. Figure 9-16 is a typical residential elevation plan, and Figure 9-17 illustrates the height of a building above mean sea level.

 FROM EXPERIENCE

An elevation plan is used for all areas of construction, not just the exterior. A side-view illustration of a plumbing rough-in could also be called an elevation plan.

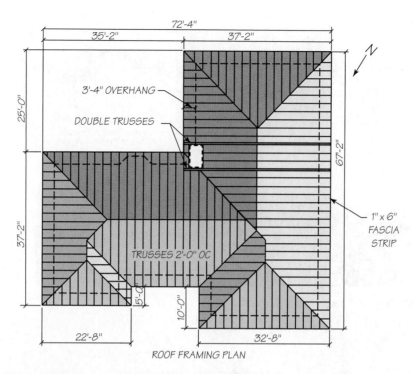

Figure 9–15 A roof framing plan illustrates the roof truss directions and peaks and valleys of a roof design. *Courtesy of PTEC-Clearwater-Architectural Drafting Department.*

Many detail symbols used by a plumber represent interior features such as kitchen cabinets. The location and dimensions of a sink or refrigerator are determined from detailed drawings of a kitchen area. When the location of a specific detail is indicated on a plan view, an architect will often provide more detailed information about that area and use typical detail symbols to indicate that further information is available. Figure 9-18 is an example of a detailed view of a kitchen area. Within that detail, symbols (A and B) indicate that a side view of each portion of the cabinets is available. The wall cabinet widths are identified; for example, W20 is a 20" wide wall cabinet. B36 means a 36" base cabinet.

 FROM EXPERIENCE

A plumber does not install cabinets, but must coordinate the piping locations serving a kitchen sink and an ice-maker box.

A plumber often routes piping in an exterior wall and must know the width of the wall and its relation to other parts of the exterior of the building. A section view of an exterior wall is often in the area of a window or door. The section view includes detailed information for all contractors to use in coordinating ceiling heights, exterior finishes, or window heights. A plumber needs to make sure that a hose faucet does not conflict with a window sill or the height of the ceiling structure. A plumber must locate windows when installing underground piping to avoid turning up under a window. Exterior **load-bearing** walls have strict drilling regulations, so pipe might have to be relocated under the floor if the wall studs cannot be drilled to offset the pipe around the window. Figure 9-19 is an example of a sectional view of a window area.

 FROM EXPERIENCE

Always review window and door locations during an underground piping phase of construction to avoid conflicts.

A plumber installs underground piping serving drains, sewers, and water piping. A foundation wall rests on a support structure known as a footing. A plumber might have to install piping below the footing depth, so a detail of the footing might be needed to clarify the depth below the floor. A plumber might also need a footing detail to coordinate with other installations, such as when a drain is needed to keep water from settling near the footing area. This is known as a French Drain. Figure 9-20 shows a footing detail with information that could be used by a plumber.

210 SECTION THREE *Layout and Installation*

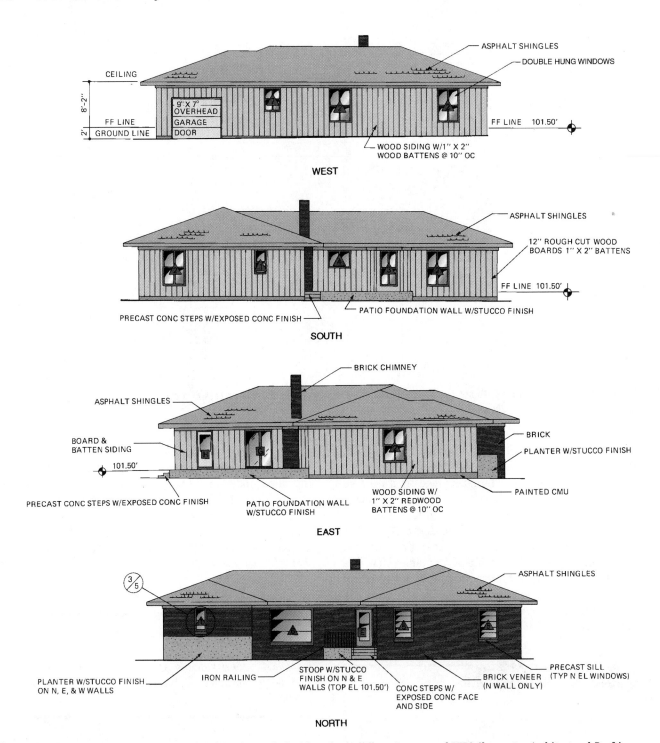

Figure 9–16 An elevation plan illustrates side views of all sides of a building. *Courtesy of PTEC-Clearwater-Architectural Drafting Department.*

 **FROM EXPERIENCE**

Architectural blueprints typically have footing details to provide accurate information on the footing. A home with a basement requires more foundation work than a home with a crawl space.

A plumber must know what finished products will be used for floors, ceilings, and walls to coordinate the installation of piping during the rough-in phase with the fixtures the piping serves. During the rough-in phase, the floor of a residential house is typically plywood, known as the subfloor. If the finished floor will be tiled, the piping serving the toilet must be installed to accommodate the tile. Therefore, a plumber must know what the finished floor height above the

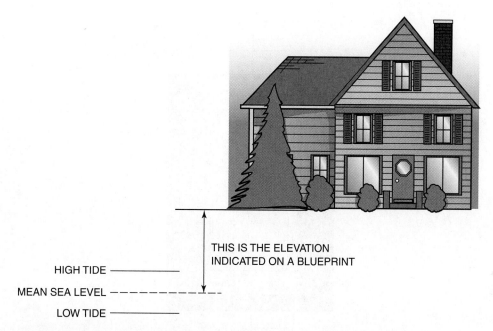

Figure 9–17 Mean sea level is determined from the average water level between high and low tide.

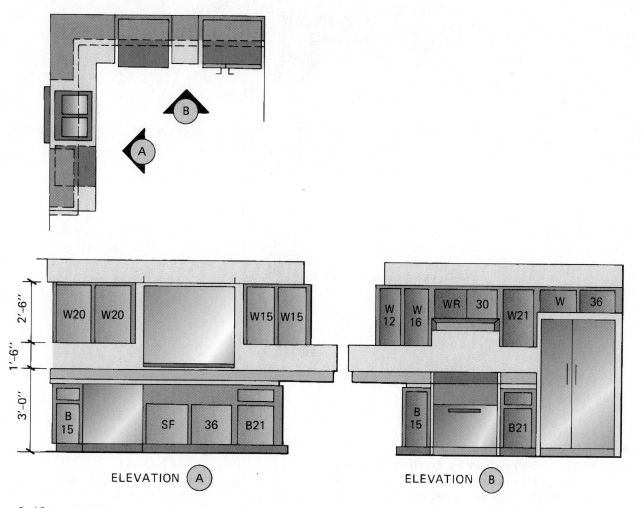

Figure 9–18 Details illustrate a specific area of construction such as kitchen cabinets. *Courtesy of PTEC-Clearwater-Architectural Drafting Department.*

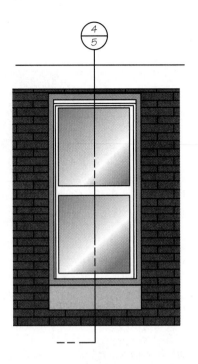
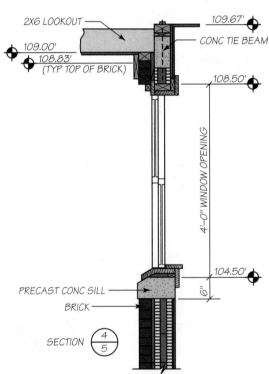

Figure 9–19 Window details are an area of construction that requires a detailed view. *Courtesy of PTEC-Clearwater-Architectural Drafting Department.*

plywood subfloor will be. The wall finish in a residential house is typically 1/2" drywall with either paint or wallpaper. If tile is used, a plumber must install piping to accommodate the wall finish. An architect provides a room finish schedule indicating the finishes to be used in each room. Table 9-2 is an example of a room finish schedule.

Drafting

Reading blueprints means interpreting a design; drafting is a way to communicate a design. A plumber on a residential project typically installs plumbing systems based on a visual inspection of the job. Making a simple **sketch** on pa-

Table 9-2 A Typical Finish Schedule

			Finish Schedule				
Room	**Walls**	**Paint Colors**	**Base**	**Floor**	**Ceiling**	**Cornice**	**Remarks**
LIV. RM.	DRY WALL	BONE	WOOD	OAK	PLASTER	WOOD	BOOKCASE
DIN. RM.	"	"	"	"	"	PICT. MLDG	CLIPBD.
KITCHEN	"	EGG SHELL	TILE	VINYL	"	—	—
HALL	"	"	WOOD	OAK	"	WOOD	SEE DTL.
ENTRY	"	"	"	"	"	—	—

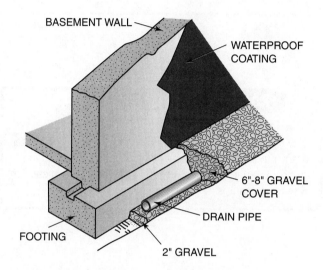

Figure 9-20 Installation of certain aspects of construction can be clarified with a detail.

per or even a piece of cardboard box to indicate design intent is a form of drafting. More formal presentation of design intent is done on drafting paper with special drafting tools. To clearly illustrate a design, a person needs basic drafting skills and the ability to communicate the design effectively. A plan view illustrates piping systems that relate to other areas of construction such as walls or fixtures. A section or side view shows installations such as fixture rough-ins to indicate elevations of water and drain pipes above a floor or the distance from a wall or other piping. An isometric view is extremely useful when a piping system has numerous offsets and the design intent requires clarification. Having a knowledge of drafting tools helps in creating sketches that are clear and precise. A correctly illustrated sketch eliminates communication errors and creates a productive job site.

Drafting Tools

Drafting tools are specialty items that can be purchased at an office supply store. A plumber does not need expensive drafting tools to create sketches on a job site. If the sketches are for distribution to other employees, contractors, or an ar-

chitect, drafting tools should be used to create professional-looking sketches. Many tools can be used to create sketches and blueprints; as your drafting skills increase, you will be able to draw more effective freehand sketches on job sites. Drafting tables and boards provide flat surfaces on which drafting paper can be held in place with specially designed drafting tape.

Scale Ruler

A scale ruler is used to determine dimensions of a blueprint, but it also contains a standard 12″ ruler. The two different types of scale rulers are architectural and engineering. A plumber working with blueprints for the interior of a building uses an architectural scale. If the work area is located on the exterior of the building, such as in a parking lot, an engineering scale ruler is used. A typical architectural scale ruler is 13″ long, 1″ wide, and triangular in shape. The scaling options range from 3/32″ to 1″ representing one linear foot on a blueprint. The smaller the scale, the more feet it can represent on a blueprint. The two most common scales for plumbing blueprints are 1/8″ and 1/4″. Two scaling options are used along the same row of dimensions; one is read left to right and the other right to left. A portion-of-a-foot option is available to determine linear inches. Figure 9-21

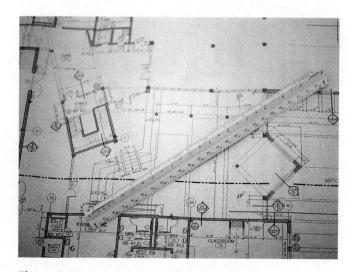

Figure 9-21 A scale ruler is often called a scale and is used to dimension a blueprint.

214 SECTION THREE Layout and Installation

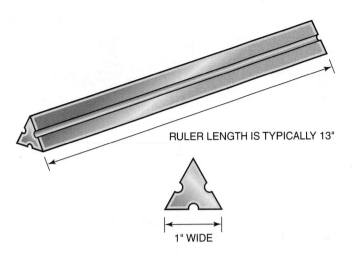

Figure 9-22 The physical dimensions of a typical scale ruler are standard, but many variations exist.

Table 9-3 Scale Options and Dimensions

Scale	Dimensions To
3/32"	124 feet
1/8"	92 feet
3/16"	62 feet
1/4"	46 feet
3/8"	28 feet
1/2"	20 feet
3/4"	14 feet
1"	10 feet

shows a scale ruler for determining blueprint dimensions. Figure 9-22 illustrates the size and shape of a triangular scale ruler. Table 9-3 lists the scaling options of an architectural scale ruler. Figure 9-23 is a detailed view of two scaling options of an architectural scale ruler.

 FROM EXPERIENCE

A 6" flat scale ruler is helpful on a job site and fits into a shirt pocket, but most only provide two scaling options. A plumber often uses a standard tape measure when the scale is 1/8" or 1/4".

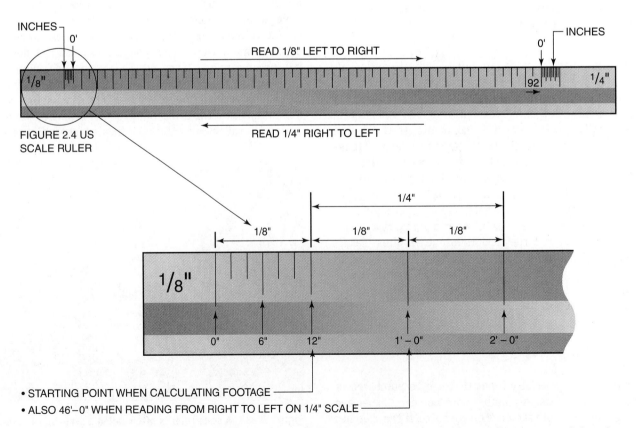

Figure 9-23 An architectural scale ruler has several scaling options and has two different scales on the same row of the ruler.

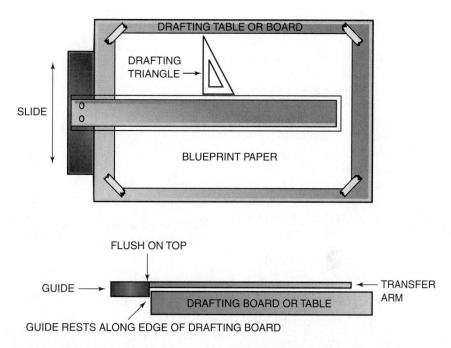

Figure 9-24 A slide square is used to align other drafting tools and illustrations on a blueprint page.

Drafting Triangles

A slide square is placed on a drafting table to provide a true horizontal plane with which to draw horizontal lines or to guide a drafting square to create other uniform angles. A **drafting triangle** has a 45-degree angle and is used to create plan- and side-view illustrations. A dual-purpose triangle, with 30- and 60-degree angles, is used for isometric sketches. Drafting triangles are available in various sizes. Small ones are most common for creating sketches on a job site. Drafting techniques for isometric sketches are the most difficult to learn. When you have learned the basic plan-view symbols, your blueprint-reading skills will also be strengthened by learning drafting skills. Figure 9-24 shows a slide square and a drafting triangle with a drafting table or board. Figure 9-25 shows an isometric drafting triangle.

FROM EXPERIENCE

Circular pieces of drafting tape dispensed from a roll are called drafting dots. They are sold in most office supply stores that sell drafting supplies.

It takes practice to become comfortable drawing a piping system in three-dimensions using an isometric drafting triangle. One of the first steps is learning to create an isometric horizontal line. The slide square in Figure 9-24 created a true horizontal line. When the longest edge of an isometric drafting triangle is placed on the slide square or in the true horizontal position, the 30-degree angle of the drafting

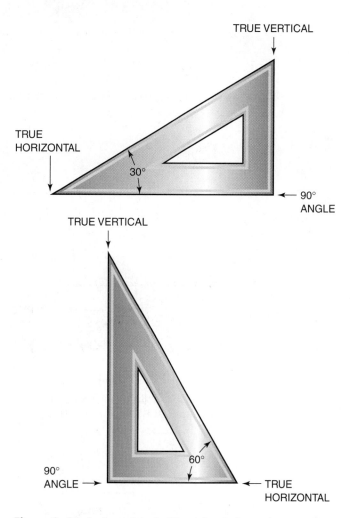

Figure 9-25 An isometric drafting triangle is used to create isometric blueprints and isometric sketches.

triangle creates a line that represents isometric horizontal. Once isometric horizontal is established, all lines representing connecting piping are drawn as if the 30-degree angle is horizontal. True horizontal is 90 degrees from true vertical so isometric horizontal relates to both known positions. Multiple isometric offsets within a single design must be aligned with isometric horizontal to remain consistent. Proper placement of the isometric drafting triangle is essential to create a professional design. Figure 9-26 illustrates that isometric horizontal is 30 degrees from true horizontal and 60 degrees from true vertical. Figure 9-27 illustrates various offsets using an isometric drafting triangle. Figure 9-28 shows numerous offsets; it can be used as a practice example to develop isometric drafting skills.

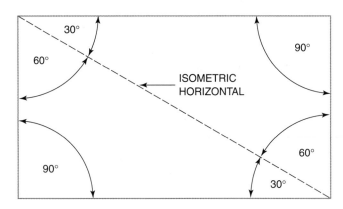

Figure 9-26 Isometric horizontal is 30 degrees from true horizontal and 60 degrees from true vertical.

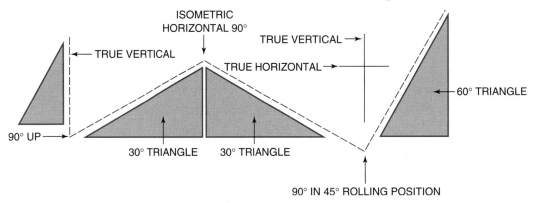

Figure 9-27 An isometric drafting triangle is used in various positions to illustrate different offsets.

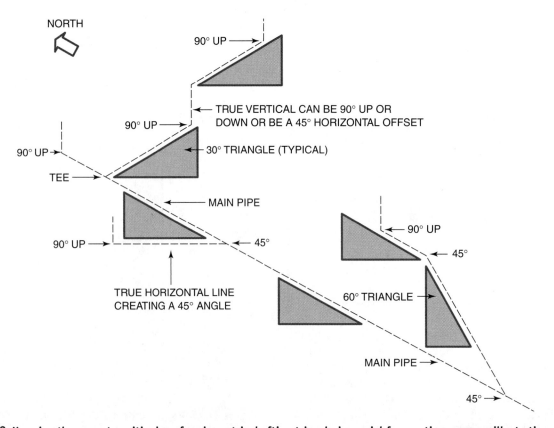

Figure 9-28 Knowing the correct positioning of an isometric drafting triangle is crucial for creating a proper illustration.

CHAPTER 9 Blueprint Reading and Drafting 217

FROM EXPERIENCE

Practice drawing non–plumbing-related objects if you have difficulty recognizing the different angles of the pipe. This gives you practice with isometric angles while illustrating familiar objects.

Symbol Templates

Symbol templates are used to create industry-standard or custom symbols, or to draw features of a building or piping system. Multi-purpose templates have basic shapes such as squares, triangles, and circles; others contain more specific shapes. Circular templates like that shown in Figure 9–29 enable you to draw a plan view, 90-degree symbols, or larger circular objects such as storage tanks. Identification templates include options such as arrows, room identification symbols, revision symbols, and N for North (Fig. 9–30). A plumbing template includes common plumbing fixtures and shapes to

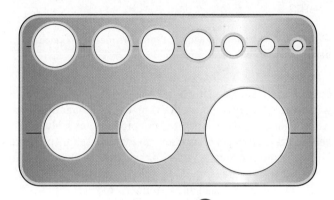

USED TO CREATE PLAN VIEW 90-DEGREE
UP OR SIDE VIEW 90-DEGREE TOWARD

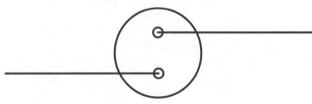

USED TO ILLUSTRATE A PLAN VIEW OF A STORAGE TANK
WITH 90-DEGREE DOWN PIPING

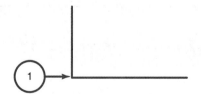

USED TO INDICATE SPECIFIC ACTIVITIES AND
TYPES OF FITTINGS OR EQUIPMENT

Figure 9–29 A circular template creates numerous symbols.

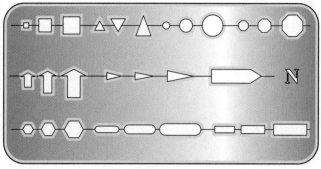

USED TO IDENTIFY A REVISION HAS OCCURRED ON
A BLUEPRINT, THIS ILLUSTRATES REVISION ONE

USED TO INDICATE FIXTURE TYPE OR OTHER
SPECIFIC EQUIPMENT TYPE OR ACTIVITY

USED TO IDENTIFY ROOM NUMBERS
OR OTHER UNIQUE AREAS

USED TO INDICATE NORTH

Figure 9–30 An identification enclosure template creates numerous identifying symbols.

create plumbing-related symbols (Fig. 9–31). Examples of how each template can be used are included in the figures.

FROM EXPERIENCE

Most templates have various arrow symbols. You can select an arrow style for pointing at objects within a sketch to establish your own drafting style.

Drafting Paper

There are many different kinds of drafting paper. Most graph paper, which is used to learn basic drafting skills, has square segments to guide you in maintaining true horizontal and vertical when sketching. Graph paper is usually 8-1/2" × 11", but larger sizes are available. Graph paper with squares is not used for isometric sketches because isometric horizontal is at a 30-degree angle (Fig. 9–32). When a square is segmented from corner to corner, it creates a 45-degree

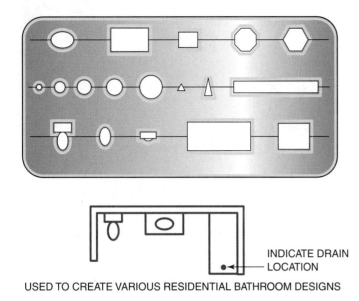

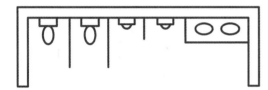

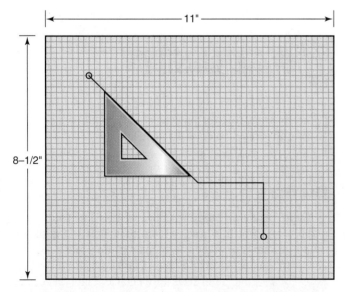

Figure 9–31 A plumbing template has various fixture symbols and other shapes to create numerous symbols.

Figure 9–32 Graph paper can be used to create side-view and plan-view drawings, but the square graphs are not designed for isometric sketches.

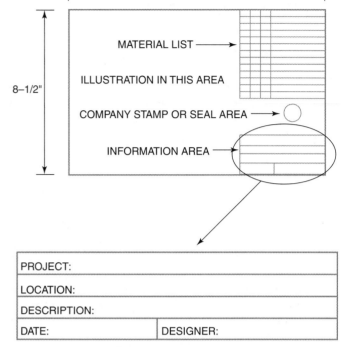

Figure 9–33 A shop drawing is typically created on smaller paper to use as an installation guide.

angle, which is more suitable for a plan- or side-view sketch. A more expensive type of drafting paper is manufactured with vellum and is designed to be reproduced as a blueprint. Vellum drafting paper is available in large sizes for full-sized blueprints and in smaller sizes for sketches. A small sketch, known as a shop drawing, is typically created on 11″ × 17″, so it can be reproduced in a standard copy machine (Fig. 9–33). Vellum drafting paper is blank and can be used for plan, side, or isometric sketches. A material list is often made on a shop drawing (Fig. 9–34) where an information area has been created (Table 9-4). An information area can include the date, the location of an installation, and the name of the person who created the sketch. If the sketch is submitted to another contractor or to an architect, the company might add their corporate seal to give the sketch a professional, legal appearance. A material list describes the material required for the specific sketch on the shop drawing. An illustration can include alphabetical or numerical symbols that are explained in the material list. This keeps the illustration from being cluttered with text and keeps the focus on the design intent.

 FROM EXPERIENCE

Graph paper is helpful for beginners, but you should attempt to illustrate on blank paper as soon as you are comfortable doing so to strengthen your drafting skills.

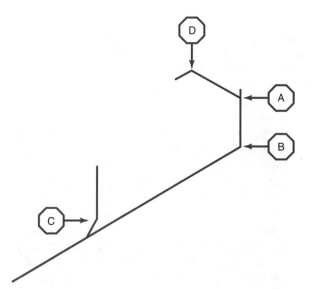

Figure 9–34 A shop drawing typically includes symbols indicating specific material required.

Table 9–4 A Materials List for a Sketch Indicates the Materials Required to Complete the Installation

Item	Qty.	Size	Description
A	1	3″ × 1-1/2″	PVC Sanitary Tee
B	1	3″	PVC Long Sweep 90° Elbow
C	1	3″ × 2″	PVC Combo Wye & 1/8th Bend
D	1	1-1/2″	PVC 90° Elbow
N/A	6	1-1/2″	Feet of PVC Pipe
N/A	4	2″	Feet of PVC Pipe
N/A	18	3″	Feet of PVC Pipe
N/A	1	1-1/2″	Plastic Test Cap
N/A	1	2″	Plastic Test Cap
N/A	1	3″	Plastic Test Cap

Isometric Drafting

This chapter has exposed you to basic isometric drafting techniques and an isometric drafting triangle. Comparing isometric sketches to plan-view and side-view sketches of the same design demonstrates the importance of an isometric sketch. An isometric view of a piping system clarifies piping configurations that are hidden from a plan and a side view. A scale is typically not used on an isometric sketch because it is a representation of a plan- or side-view sketch. Dimensions can be included in an isometric sketch to show fabrication intent or distances from specific walls or columns. A plumber who wants to advance from single-

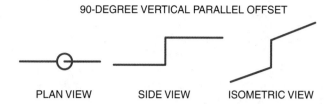

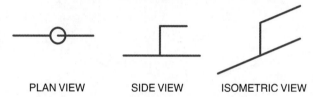

Figure 9–35 Plan-view comparisons with a side view and isometric view indicates that design intent is not clear on a plan view.

family construction to multi-family or commercial plumbing will be faced with using an isometric sketch of a system. Figure 9–35 provides illustrations comparing the same plan view with side and isometric views. Figure 9–36 can only be illustrated in an isometric view to clarify the design intent and the importance of using the North symbol. Figure 9–37 compares several plan, side, and isometric views and shows that all the offsets are only visible with an isometric view.

 FROM EXPERIENCE

A plumber who can draw isometric sketches can communicate a design effectively on a job site with other employees and contractors. This drafting skill is impressive to an employer and demonstrates the employee's dedication to the plumbing industry.

The view angles used to illustrate an isometric drawing are northeast, southeast, northwest, and southwest. Because an isometric illustration is three-dimensional, the view angles are tilted from the navigational directions, such as north. A view angle is where you would be standing if you

220 SECTION THREE *Layout and Installation*

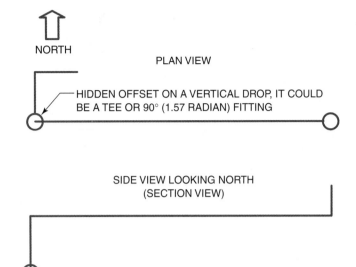

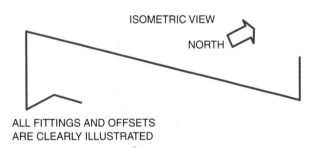

Figure 9-36 Plan-view comparisons with a side view and isometric view indicates that design intent is not clear on a plan view.

were looking at the piping system on the job site. An isometric sketch can become confusing when lines cross and appear to connect. The view angle that you choose should have a minimum number of lines crossing to avoid misinterpretation of design intent. The angle you choose to illustrate an isometric sketch should indicate navigational directions. Using north, south, east, and west helps someone else interpret your design. Figure 9-38 demonstrates the isometric view angles. Figure 9-39 shows the same piping configurations from various views.

 FROM EXPERIENCE

Imagine placing yourself at a different elevation from the piping system while drawing an isometric sketch. This may help you determine the isometric view angles.

Isometric sketches often need clarification. If many lines seem to connect or the actual offset is not clear, detail boxes are used to strengthen the design intent. A detail box is a dashed three-dimensional drawing at an isometric angle. A 45-degree piping offset is often difficult to illustrate because, depending on the actual illustration and isometric angle, true vertical lines can represent a 45-degree angle. A rolling offset is when the piping system is routed along different paths before and after the offset. A detail box can show that the offset is rolling to the side. A detail box can also indicate the dimension of an offset. A 45-degree angle is made by simply segmenting a square in half at an angle, so a 45-degree offset that is not installed rolling to one side has the same vertical dimension as its horizontal dimension. When the offset is rolling to one side, you must include the distance from the pipe to the route of the bottom and top horizontal pipes. Figure 9-40 shows two examples of detail boxes that clarify isometric 45-degree offsets.

A detail box should be drawn with dashed lines when the pipe is illustrated with a solid line, so the features do not appear to be merged.

Riser Diagrams

Vertical piping passing through several floors cannot be illustrated on a plan-view blueprint. A side-view sketch can illustrate piping that does not have connections in different directions, but an isometric sketch can fully illustrate a vertical piping system. The isometric illustration of piping through several floors is known as a **riser diagram.** Some people may find a riser diagram difficult to interpret at first glance, but knowing the drafting techniques used helps in interpreting it. To avoid becoming overwhelmed by the complexity of drafting a riser diagram remember that it begins as a simple isometric sketch. When you practice, use colored pencils to show lines that intersect, to help remember that the piping does not connect. Using a navigational direction, which is typically required for a riser diagram, will keep you orientated to the design intent and how the diagram relates to the building structure or plan view of each floor. Figure 9-41 is an example of a riser diagram beginning as a simple sketch. Figure 9-42 is a continuation of the sketch that becomes a piping system rising several floors within a building.

 FROM EXPERIENCE

A riser diagram is not intended to illustrate every fitting or offset, but rather is an overview of design intent based on the number of fixtures, valves, equipment, and other major elements of a system.

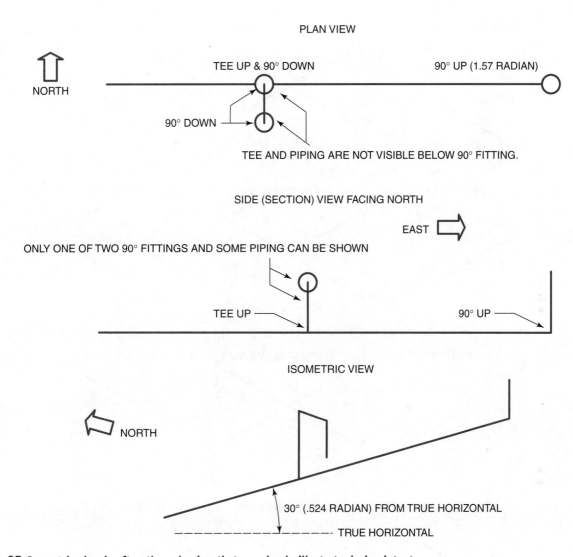

Figure 9–37 Isometric view is often the only view that can clearly illustrate design intent.

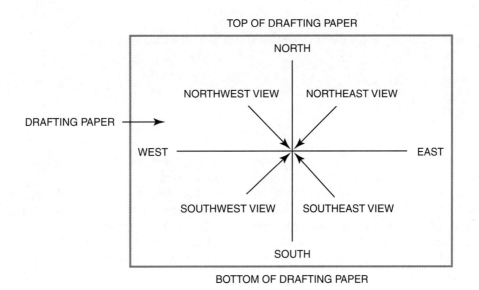

Figure 9–38 An isometric-view angle is determined before drafting an isometric illustration.

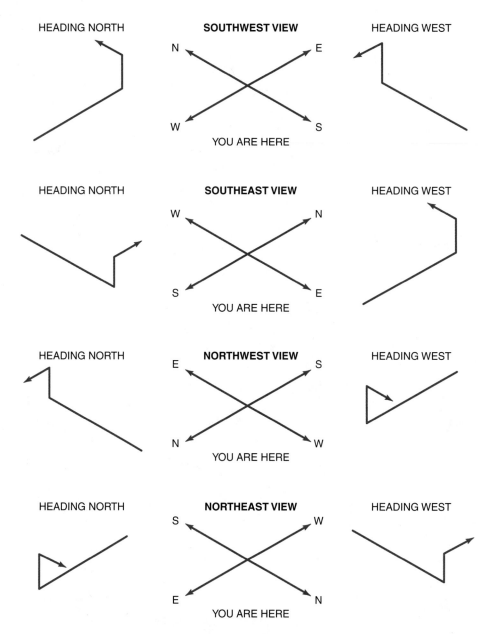

Figure 9–39 Where you are standing as you view an isometric sketch determines the view angle.

The specific piping system being illustrated dictates what information is provided on a riser diagram. Sizing information is provided mainly at major intersection points or when a change of size has occurred. Valve locations and important connections must be included to ensure design intent is clear and the system is installed correctly. Reference to areas such as columns or floors is vital so an installer can relate a riser diagram to a certain place in a building. Abbreviations are used to identify fixtures, and a legend is created to indicate their meaning.

When fixture identifications are similar, such as WC for toilet, a numerical system is included with the abbreviation. If two different types of toilets are illustrated, the numerical system indicates that difference to a plumber. The numerical system is also used when an abbreviation represents two completely different items; for example, WH means wa-

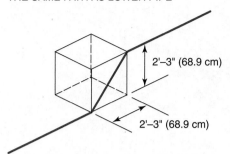

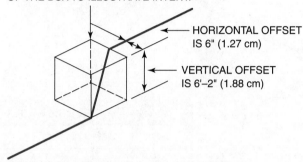

Figure 9-40 A detail box is used to clarify design intent of a piping offset.

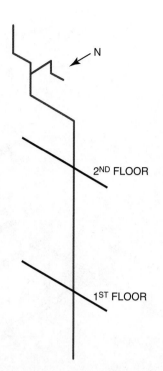

Figure 9-42 A riser diagram continues from a simple isometric sketch to a piping system rising through several floors.

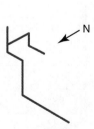

Figure 9-41 A riser diagram begins with a simple isometric sketch and then develops to represent several floors of piping.

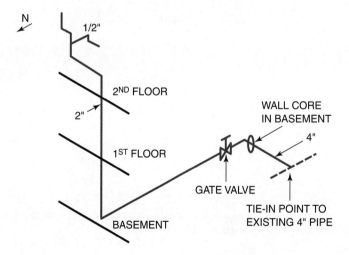

Figure 9-43 Valve locations, sizes, and other unique characteristics are added to illustrate design intent.

ter heater and wall hydrant. If fixtures illustrated in a small area are the same, the word *typical* indicates that. This method avoids clutter and maintains a clean illustration; too much information can be confusing to a plumber. Figure 9-43 is a continuation of the riser diagram illustrated in Figure 9-42 and includes more detailed information. Figure 9-44 expands the riser diagram into a more complex illustration with abbreviations and a legend.

224 SECTION THREE *Layout and Installation*

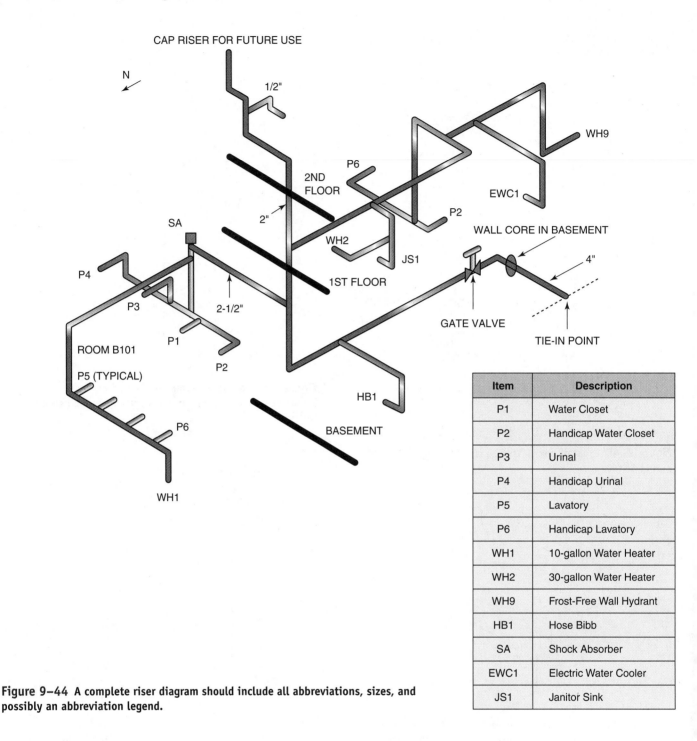

Figure 9-44 A complete riser diagram should include all abbreviations, sizes, and possibly an abbreviation legend.

Item	Description
P1	Water Closet
P2	Handicap Water Closet
P3	Urinal
P4	Handicap Urinal
P5	Lavatory
P6	Handicap Lavatory
WH1	10-gallon Water Heater
WH2	30-gallon Water Heater
WH9	Frost-Free Wall Hydrant
HB1	Hose Bibb
SA	Shock Absorber
EWC1	Electric Water Cooler
JS1	Janitor Sink

Summary

- Common plumbing symbols indicate the intent of a design.
- Abbreviations are used on a blueprint to eliminate clutter and are usually listed on a legend with their meanings.
- Most residential blueprints do not illustrate the pipe routes for plumbing systems.
- An architect typically only illustrates the fixture locations on a residential blueprint.
- Drafting skills allow a plumber to effectively communicate a design to co-workers and other tradespersons.
- A plan view is also known as a birds-eye view.
- A side view illustrates an area of construction from the side.
- An isometric view is a three-dimensional view of a piping system.
- A riser diagram is an isometric view of a large portion of the entire piping system.

Review Questions

1. **WC is the abbreviation for a**
   a. Toilet
   b. Water connector
   c. Waste collector
   d. Water collector

2. **A drafting triangle with 30- and 60-degree angles is designed for drawing**
   a. Plan-view sketches
   b. Side-view sketches
   c. Isometric-view sketches
   d. Sectional-view sketches

3. **An architectural blueprint illustrates**
   a. Every fitting required
   b. Design intent
   c. The entire project's requirements
   d. None of the above are correct

4. **A circle with a line halfway through it is a symbol that represents**
   a. 90-degree down
   b. 90-degree up
   c. Isometric 90 degree
   d. Tee down

5. **A cold-water pipe is illustrated in a plan view with**
   a. A solid line with a single dot
   b. A solid line with a double dot
   c. A dashed line
   d. A solid line with triple dots

6. **The abbreviation for a cleanout is**
   a. CL
   b. CNT
   c. CT
   d. None of the above are correct

7. **Correctly identify the following symbols.**

   ANSWER: _____

   ANSWER: _____

   ANSWER: _____

   ANSWER: _____

8. **The average sea level height used to determine the elevation of a job site is**
   a. High tide
   b. Low tide
   c. Ebb tide
   d. Mean sea level

9. **To determine the type of floors, walls, and ceiling to be used in a building, a plumber consults a**
   a. Room finish schedule
   b. Specification book
   c. Manufacturer data catalog
   d. Booklet from the previous job

10. **On a blueprint illustrated with a 1/4" scale every inch represents**
    a. 2 feet
    b. 4 feet
    c. 8 feet
    d. 1/4 of one foot

11. **A sketch made by a contractor to install a piping system is known as a**
    a. Shop drawing
    b. Architectural blueprint
    c. As-built sketch
    d. Post-project sketch

12. **To clarify design intent of a rolling 45-degree offset, the drafting technique used is a**
    a. 45-degree drafting triangle
    b. Side-view illustration
    c. Detail box
    d. Sectional-view illustration

13. **An isometric sketch that illustrates several floors of piping is called a**
    a. Riser diagram
    b. Riser sketch
    c. Riser blueprint
    d. Plan-view drawing

14. **Drafting paper used for plan-view and side-view sketches that has a square grid pattern is known as**
    a. Copy paper
    b. Graph paper
    c. Vellum paper
    d. Printer paper

15. **The four isometric angles used to illustrate a piping design are**
    a. N, S, E, and W
    b. N, SE, SW, and NW
    c. NE, SE, NW, and SW
    d. NE, SE, E, and W

16. **WH is the abbreviation for wall hydrant and**
    a. Wall hose
    b. Water heater
    c. Water hose
    d. Wash hydrant

17. **A vent pipe illustrated on a plan view is a**
    a. Solid line with a single dot
    b. Solid line with a double dot
    c. Solid line with a triple dot
    d. None of the above are correct

18. **AFF is the abbreviation for**
    a. After floor finish
    b. After final finish
    c. Above finished floor
    d. Above fixture finish

19. **An architectural blueprint indicating the locations for plumbing fixtures is a**
    a. Location plan
    b. Floor plan
    c. Foundation plan
    d. Shop drawing

20. **The actual size of a 2" × 4" board is**
    a. 2" × 4"
    b. 1-1/2" × 4"
    c. 2" × 3-1/2"
    d. None of the above are correct

Chapter 10: Material Organization and Layout

Material organization skills are essential for increasing job site productivity. Organizing material in list form eliminates material procurement errors and minimizes quantity overruns. Your blueprint reading and drafting skills will allow you to create a shop drawing and organize your material according to design intent. You will not simply order arbitrarily based on unknown circumstances. Knowing exact quantities and installation phases creates a systematic and productive work atmosphere. Organizing installation data from manufacturers and creating shop drawings are two important ways to increase productivity. Proper ordering techniques can decrease labor costs by minimizing material handling, thereby allocating more time for installations. By coordinating with other trades and organizing materials, the layout of piping systems can be expedited. This chapter discusses proven methods for creating a productive job site and stimulating creative installation ideas. Apprentices who learn and implement correct organizational skills become adept plumbers.

OBJECTIVES

Upon completion of this chapter, the student should be able to:

- respect material organization methods to increase productivity.
- demonstrate safe material-handling techniques.
- use manufacturer installation data properly.
- demonstrate proper system layout.
- Know that the layout and installation of a plumbing system differs for each project.

Glossary of Terms

carpenter individual who installs the wood framing or other woodwork

joist horizontal board used for structural integrity

load-bearing the portion of a structure that bears the weight of the structure, such as a load-bearing wall

procure the process of receiving material through ordering or gathering

slab concrete floor that defines a building design that does not have a crawlspace or basement

stud vertical board used to erect a wall

trench an installation area below ground created by excavating soil; also known as a ditch

Communication

For a construction site to operate effectively, each trade must provide effective written and oral communication. Written communication is the most reliable, but, if information is required immediately, oral can be more productive. A plumber reviewing a blueprint is receiving communication about a design from the architect. When a plumber provides an interpretation of a blueprint to a coworker, an oral communication is continuing from the initial written communication. A cycle of errors occurs if the oral or written communication is misunderstood. An electrician who tells a plumber where wiring will be installed so the plumber can install piping in the same area is communicating design intent orally. If the electrician provides incorrect information, and a plumber installs piping in the wrong location, an error has occurred and someone is blamed. Regardless of who is to blame for an error caused by poor communication, a loss of time and money results. Common oral and written methods of communication used in the construction industry are listed in Table 10-1.

Written Communication

The most effective form of communication on a job site is written communication. Safety training programs typically include weekly job site meetings with written safety lessons. Every employer must have a Hazardous Communication Program that distributes information concerning dangers and safety procedures to employees. A Material Safety Data Sheet (MSDS) is a written document required by the Occupational Safety and Health Administration (OSHA) that must be available upon request for every product containing dangerous or hazardous materials. Each job site must catalog information for specific products, including gases used to solder copper tubing. The MSDS indicates the dangers and the medical treatment necessary in case of exposure. A physician might need this information. Always review safety-related information before using a product. A list created to order material is a form of written communication,

Table 10-1 Communication Methods

Method	Oral	Written
Two-way radio	Yes	No
Mobile phone	Yes	Yes
Telephone	Yes	No
Facsimile	No	Yes
Blueprint	No	Yes
Tape recorder	Yes	No
Computer / E-mail	No	Yes
Meetings	Yes	Yes
Safety	Yes	Yes

Table 10-2 Typical Written Communications

Method
Blueprint
Shop Drawing
Rough-in Sheet
Building Permit
Material List
Letter of Transmittal
Extra Work Order
Web Site
Facsimile (Fax)
Electronic Mail (E-mail)
Letter
Memo
Material Safety Data Sheets

and an incomplete material list is an example of ineffective communication. The initial source of written communication must be accurate for any continuing written communication to be effective. An error on a blueprint can create a chain of errors by everyone involved. The architect is responsible for correcting the errors. If a misinterpretation of a blueprint causes the errors, the person or company who misinterpreted is responsible. A shop drawing created by a contractor is based on an architect's design intent. If the contractor creates an incorrect shop drawing, the drawing provides incorrect written communication. It is extremely important for a plumber to thoroughly understand design intent when reviewing all forms of written communication. If the information appears to be wrong, it can be questioned and corrected before performing an installation. You will recognize information that might be incorrect as your career progresses. Table 10-2 lists several forms of written communication.

 FROM EXPERIENCE

Always request design changes in writing for legal reasons. A verbal design change is easily disputed by other parties and typically does not have merit in a court of law.

Oral Communication

One of the leading causes of construction errors is incorrect oral communication. An error is often caused unknowingly by repeating incorrect information received from another source on a construction site. On a typical single-family residential construction site, a plumber and an apprentice install the entire plumbing system. The architectural blueprints typically do not show the piping systems, which are determined based on design intent, codes and job-site coordination. If a floor joist interferes with a pipe route, a plumber can ask the carpenter to alter the joist layout. This common and acceptable process involves oral communication and often written documentation is never established.

With technological advancements such as mobile phones and two-way radios, a plumber has immediate contact from a job site to a supervisor. Providing information immediately rather than creating a paper trail results in increased productivity. Interpreting code and organizing material are two reasons why oral communication is vital for a productive job site. Many mobile phones also have two-way radio features and some models are able to communicate nationally. Smaller construction sites may use a two-way radio for verbal communication within a smaller radius. Co-workers on a job site can communicate questions and information without leaving their work areas. Most two-way radios have multiple channels to communicate with other contractors who are using specific channels. A crane operator typically has a dedicated channel so crucial communications are not interrupted by those not involved in the hoisting procedures. Several variations of two-way radios use a home-base accessory to create a private two-way system. A two-way radio typically uses a public communication tower, so contractors can all communicate freely with each other on a job site.

FROM EXPERIENCE

Two-way radios and mobile phones are only effective when the batteries are charged. Plumbers should treat all communication devices as important tools.

Material Organization

Material organization skills are essential for completing a task in a productive manner. If an item is not ordered, a task cannot be completed. The time spent procuring an overlooked item could be used to perform another task. This wasted time can be a leading cause of a project losing money. Labor rates increase, but material prices fluctuate; if a low-cost item stops the progress of a task, the cost of the item increases. To organize material, a plumber must be knowledgeable about the task. Because a plumber must determine the pipe route for most residential projects, plumbing companies usually provide extra material for a job. This eliminates the need to stop work to procure a particular item. This overstocking method may seem costly, but, if labor costs are decreased, the residential material costs are usually justifiable.

For large projects, materials might be warehoused in a central location. Materials for small projects might be kept

Table 10–3 Material Organization Methods

Method
Creating Materials Lists
Palletizing Material
Bagging and Tagging
Warehousing On or Off Site
Procuring Material per Job Site or Installation

on a vehicle. For a single-family home, a plumber might only need to work in one location for a short period, so on-site material storage is not necessary. One way to save costs on any project is to make a list of needed materials before beginning a task. Regardless of where the material is procured, having a list lets you know if all material is available before an installation begins. All material should be organized and delivered to a job site based on a particular project. Often plumbing companies organize material in storage bins on a truck, but the plumber should take required material into the house to avoid numerous trips to a truck to retrieve material. Theft can occur on construction sites, so material should be stored in a secure area. Table 10-3 lists several material organization methods.

Palletizing

A pallet is a wooden platform that is raised from the floor to accommodate a fork truck or pallet jack. Larger projects, where work is performed in numerous areas, can use pallets to store material. Most small residential construction projects cannot organize to a point where palletizing material is effective. Often material is placed in one area of a job site and then moved several times before it is installed. Constant movement of material requires manual labor, which increases the labor cost of a project. Palletizing material allows it to be relocated with a pallet jack, which requires less labor. Figure 10-1 shows a pallet with material organized for relocation to another area of a job site. Figure 10-2 is a pallet jack used to relocate a pallet.

FROM EXPERIENCE

The handle of a pallet jack is pumped up and down to raise the forks from the ground and a lever located in the handle is triggered to lower the forks.

Bagging and Tagging

Bagging and tagging is a trade name for organizing material according to activity or installation. Heavy objects like cast-iron fittings can be placed in a burlap bag, and plastic

Figure 10–1 Organizing material on a pallet to minimize relocating with extensive labor.

materials can be stored in a cardboard box. Whether a bag or box is used, the organizational process remains the same. A material list provides a format for keeping track of the material needed for an installation. Organizing those materials in a box or bag ensures that they will be available when needed. Identifying the organized material is crucial for locating the correct, stored items. A tag is used for a burlap bag. A label is used for boxes, or the installation location is simply written on the side of the box.

Material that is not available when the organizational process begins should be noted on a tag or label. A material tracking chart or book can be used to remind a plumber that an order is incomplete and to ensure that all missing material is received before the installation begins. Some items, such as a washing machine box, icemaker box, and shower faucet, are installed during two different phases of construction. In that case, the trim plates must be stored and protected, even removed from the job site until they are required for the trim-out phase of construction. Table 10–4 is an example of a material tracking chart that can be created on a job site. Figure 10–3 shows two ways to organize material using a tag or a label.

Figure 10–2 A pallet jack easily relocates material stored on a pallet and minimizes labor costs.

 FROM EXPERIENCE

Storing material in a box or bag without identifying the contents can increase labor cost by forcing you to search for material.

Table 10–4 Material Tracking Chart

Qty.	Size	Description	Ordered	Received	Notes
3	1/2"	Copper Cap	Yes	3	P.O. #12761
2	3/4"	Copper Cap	Yes	2	
2	3/4"	Copper 90°	Yes	1	Reordered 5/27
1	1/2"	Copper 90°	Yes	0	Reordered 5/27
1	3/4"	Copper Tee	Yes	0	Reordered 5/27
1	1/2"	Copper Tee	Yes	1	
2	3/4" × 1/2"	Copper Tee	Yes	1	Reordered 5/27

232 SECTION THREE *Layout and Installation*

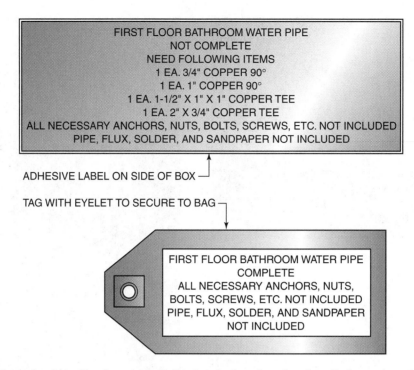

Figure 10-3 A tag or label is used to identify a bag or box indicating the location of an installation and any items required to complete a material order.

Material Handling

As previously discussed, organized material procurement, storage, and handling is vital for a productive installation. Plumbing systems are manually installed. By distributing material to the proper work areas the installation process is expedited, because the plumber does not have to look for material in remote locations. Most residential materials are lightweight except for equipment like water heaters and some bathtubs. Because single-family homes are fairly small, equipment is not needed to move materials from place to place on the job site. Carts, dollies, and other wheeled pieces of equipment are available to move heavy objects. Machinery such as a fork lift is often needed to place one-piece tub and shower units onto upper floors. More than one person is needed to carry heavy objects like cast-iron bathtubs to designated locations. Many heavy objects must be carried up stairs; using a dolly prevents back injuries. A support to ease lower-back strain is often supplied by an employer, but it is the employee's responsibility to wear it when needed. Figure 10-4 is a photo of a pipe cart specifically designed to carry heavy pipe on a job site. Figure 10-5 is a typical back support for lifting heavy objects.

 FROM EXPERIENCE

Back support is not common on many job sites. It is more common for material handlers in warehouses or for truck drivers.

(A)

(B)

Figure 10-4 A pipe cart eliminates the need to carry heavy pipe and can be operated by one person. *Courtesy of Vestil Manufacturing.*

CHAPTER 10 Material Organization and Layout   233

Figure 10-5 Back support is required to minimize the risk of lower back injury.

(A)

(B)

Figure 10-6 A ladder rack is also used to deliver pipe to a construction site, so the weight load of the rack must be known. *Courtesy of Werner Ladder.*

Vehicle Racks

Ladder racks and pipe racks are installed on pick-up trucks and vans. Vehicle racks are designed for specific vehicle types. Most piping is sold in 10′ and 20′ lengths, which means a rack is needed to deliver pipe to a job site. Ladders and pipe are usually carried on the same rack and must be secured, often with flexible securing straps. They must be tightened well to ensure that the piping and ladders remain in place while driving. Numerous styles of racks are available, many of which are custom made by local fabrication shops. A plumber must be aware of the weight capacity of the rack and the vehicle. A heavy-duty rack may be able to handle several tons of weight, but the truck might not. When heavy loads are placed high above a truck, the vehicle cannot make turns as quickly. Abrupt stopping or accelerating can cause the material on the rack to become airborne if the straps break or are not adequately secured. The rack system connections to the vehicle must be inspected to make sure the anchoring bolts are secured.

State highway regulations dictate that objects protruding from the front or rear of a vehicle must have a bright-colored flag attached or have some other means of indicating an unsafe condition to other drivers. Most states dictate that an object protruding more than 36″ from the vehicle must be identified with a flag, but a plumber must know local regulations to avoid violating them. Pipe carriers can be purchased that hold small-diameter pipe, so it does not have to be secured to the pipe rack. Pipe carriers can be custom made out of PVC pipe with a cleanout on each end. Flexible tubing used for residential water distribution systems cannot be safely strapped to a pipe rack, so use of a pipe carrier is common. Most pipe carriers are secured to the pipe rack with U-bolts, which are simply metal rods bent in a U-shape and threaded on each end. Figure 10-6 shows several different types of ladder and pipe racks. Figure 10-7 is a photo of a ladder secured to a pipe rack. Figure 10-8 is an illustration of a custom-made pipe carrier.

CAUTION: Rubber straps can break and cause injury, especially in cold climates. Always wear eye protection.

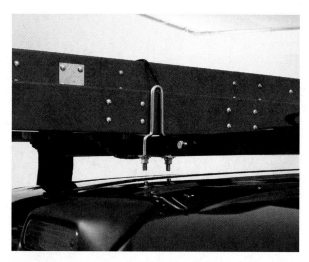

Figure 10-7 A ladder and pipe is secured to a truck rack with flexible straps. *Courtesy of Werner Ladder.*

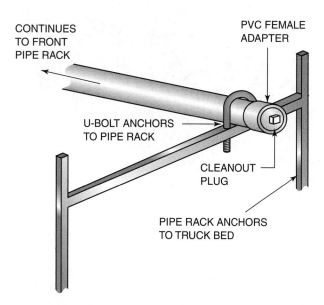

Figure 10-8 Pipe carriers can be made or purchased to hold small-diameter pipe on a truck rack.

FROM EXPERIENCE

Always inspect straps for deterioration, and discard them when abrasion or cracking is visible.

Layout

When all material is procured and organized, the installation process can be completed. Layout can be done at the same time as material procurement if an adequate labor force is available. A typical residential single-family home layout is within a wood structure, and all the walls, floor joists, ceil-

Table 10-5 Three Home Designs

Type	Above ground	Underground
Basement	Yes	Yes
Crawl space	Yes	No
Slab	Yes	Yes

ing joists, and roof are constructed before a plumber arrives on the job site. Plumbing systems are usually installed before electrical and HVAC systems because drainage systems need to be more exact. They often also occupy more space within a wall or ceiling cavity. Drainage piping is larger than most other residential piping; a 3" pipe consumes most of the width between 2" × 4" wall studs. Because the actual dimension of a 2" × 4" stud is 1-1/2" × 3-1/2", a plumber must be accurate when laying out a hole for a 3" pipe in a wall. There are three basic home designs. The layout process is dictated by the design. A home with a concrete floor and basement is referred to as a **slab** design. Some piping is installed below ground for a slab. Some designs have a basement and some a crawl space. All homes require above-ground piping, with the variations in design dictating the above-ground piping layout. Table 10-5 indicates the three basic home designs according to the foundation used.

A building drain is the lowest horizontal drainage pipe. It receives all other drainage pipes and conveys waste and wastewater to a building sewer. A building sewer connects to the building drain and conveys waste and wastewater to the septic tank or a municipal (public) sewer. The minimum depth below ground that the building drain can exit the building is determined by code; the maximum depth is determined by the sewer connection to the septic tank or public sewer. Codes, sizing, and additional system identification are explained in Chapter 13. When a house is built on a concrete slab, the underground plumbing system must be installed before the house can be built. A house with a basement does not necessarily have a drainage system installed below the floor. In a house with a crawl space, there are more piping design options, and a plumber installs the building drain portion of a drainage system above ground. Figure 10-9 illustrates a building drain exiting a house built on a concrete slab. Figure 10-10 shows two possible exit points of a building drain from a house with a basement. Figure 10-11 illustrates a building drain exiting a house built with a crawl space.

FROM EXPERIENCE

Different regions of the country use different foundation designs. In many southern regions, slab or crawl space designs are common; northern regions have more homes with basements.

CHAPTER 10 *Material Organization and Layout* **235**

Underground Layout

Underground layout of a piping system depends on the design intent of the building and codes. The layout phase of an underground piping system is crucial and errors can have serious consequences. If an incorrect installation is not discovered until after the concrete floor is placed, an area of the floor may have to be removed to correct the error. There are similarities in the layout process for every building as to wall locations, wall widths, and techniques to ensure accuracy of pipe installations. Material-saving methods can also determine pipe routes while adhering to code regulations and design intent. Using equipment for excavating a pipe **trench** must be considered while laying out a piping system to eliminate extensive hand excavation. Moving previously excavated soil by hand decreases productivity. The location of excavated soil must be determined during the layout process, so it does not interfere with other pipe trenches and wall layout locations. The person excavating the soil must know where to place the soil.

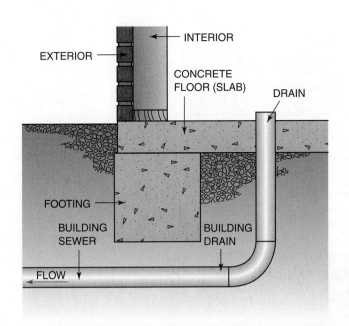

Figure 10-9 A house built on a concrete slab requires that the building drain be installed below ground.

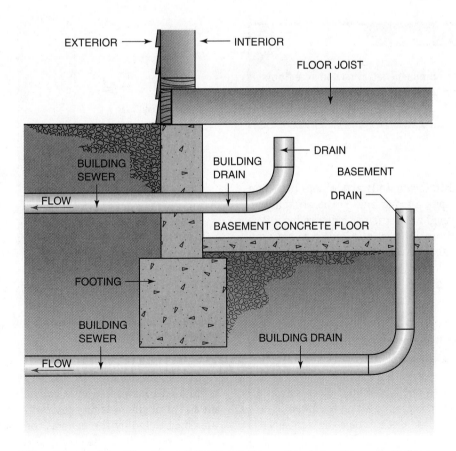

Figure 10-10 Piping exiting a basement is either through the foundation wall or below the concrete floor.

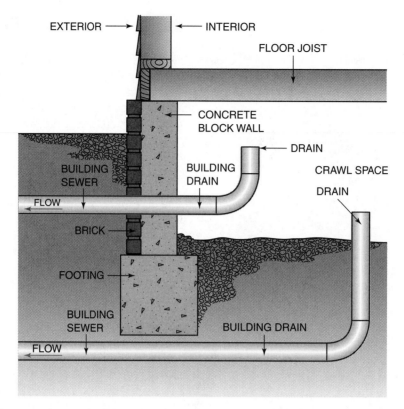

Figure 10–11 Piping exiting a crawl space is either through the concrete block wall or below the concrete footing.

FROM EXPERIENCE

Many codes require all piping penetrating concrete floors to be sleeved with a protective pipe or other approved sheathing to avoid damage to the piping systems.

Wall Layout

A blueprint is used to locate walls and adhere to design intent. The phrase to "pull a string line" means stretching a string from an established wall location to determine a definite installation location. The string is known as masonry string and is more commonly used by a brick mason to install brick or concrete blocks. Figure 10–12 illustrates a pipe installed between two strings that represent where a wall will be installed.

FROM EXPERIENCE

String is sold as a roll. A reel-type string holder is available to roll up the previously used string for future use. A thin wood board or small piece of pipe can also serve as a string holder.

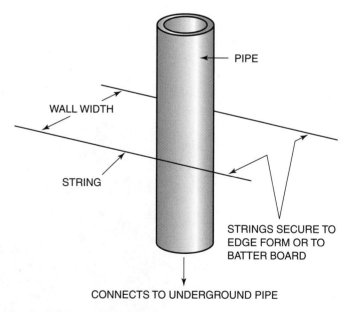

Figure 10–12 Strings are used to simulate a wall to install underground piping.

Edge forms are used to create the outline (footprint) of a building constructed on a concrete slab. The edge forms on a typical single-family residential project are wood and are constructed by the general contractor. A plumber uses nails to secure (tie) the string to the edge form. If a metal edge form is used, a wooden structure built by a plumber,

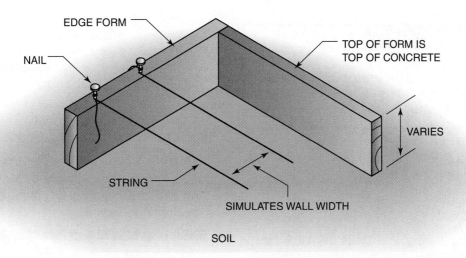

Figure 10–13 A wooden edge form creates an outline of a building constructed on a concrete slab.

known as a batter board, is used to pull the string to simulate a wall. Most residential construction projects do not require strings to be pulled (stretched) as far as some commercial projects. If a string is pulled more than 100′, it is less accurate than at shorter distances and is more likely to conflict with other construction activities. Many batter boards are fabricated and installed by the general contractor to indicate specific design features. Figure 10–13 illustrates a typical wooden edge form used to create a footprint of a building constructed on a concrete slab. Figure 10–14 shows a batter board used as a securing point for a string to indicate placement of piping turning up into a future wall. The knot used to tie the string to the nail is based on preference, but the tightness is crucial to ensure the wall width is accurate. A cinch knot is often used as the final securing point of a string. See page 246 for step-by-step instructions for laying out an edge form and for tying a cinch knot.

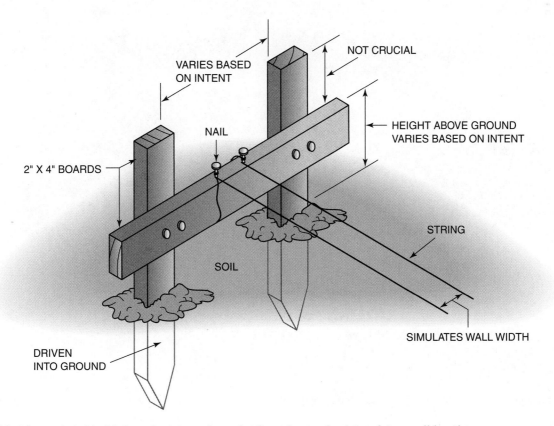

Figure 10–14 A batter board is fabricated as a securing point for string to simulate a future wall location.

FROM EXPERIENCE

The average width of a wall layout batter board is 24", allowing room to adjust the wall location on the horizontal portion of the batter board.

For step-by-step instructions on edge-form layout, see the Procedures section on pages 246 – 247.

Trench Layout

The pipe route can be established once the wall locations are determined. The strings must be removed for the excavation phase and installed again for the pipe installation. A loose piece of string can also be placed on the soil to identify the pipe route and spray paint applied over the string to mark the route. A powder, such as lime, similar to what is applied to athletic fields, can be used as well. It is available in bag form at local building supply stores. The pipe route should include the main trench and all branch trenches for connecting branch pipes to the main pipe. After the layout is complete, excavation can begin. If the process takes several days, perhaps because of inclement weather, the lines may have to be drawn again. Figure 10–15 illustrates string placed on soil to lay out a pipe trench. Figure 10–16 shows an excavated trench and the importance of not placing excavated soil in dimension areas.

CAUTION

CAUTION: When applying spray paint or powder, be sure to wear a respirator or dust mask if required by the MSDS information. Eye protection might also be required on windy days.

FROM EXPERIENCE

A beverage bottle or cup can be used to apply a powder product onto the soil for laying out a pipe route. If a plastic bottle is used, cut the spout portion off to create a larger opening in the bottle.

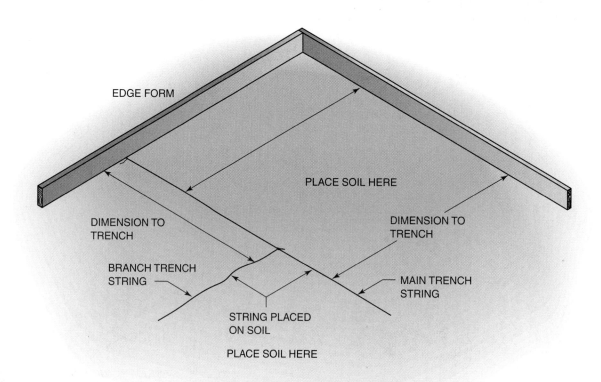

Figure 10–15 Loose string is placed onto the soil and spray paint or some other method is used to identify where a trench is required.

CHAPTER 10 *Material Organization and Layout* 239

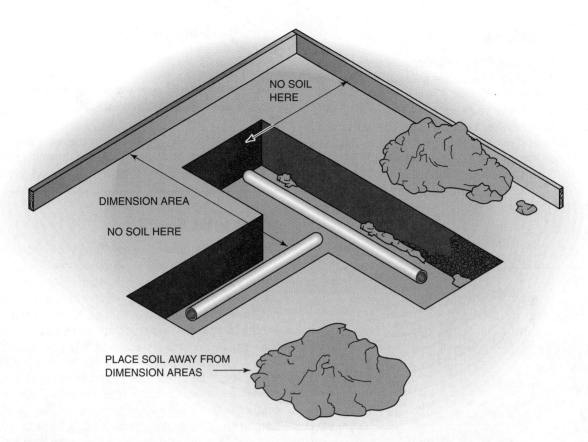

Figure 10-16 Location of excavated soil should not conflict with dimension areas to install a pipe in a trench.

Above-ground Layout

Because piping installed above ground does not require excavation, it can be relocated more easily than piping installed below ground. Residential construction typically requires laying out pipe routes through wood **studs, joists,** and floors. Piping systems installed in crawl spaces, attics, and most basements are exposed. The fixture locations are determined from a blueprint, and the type of fixture dictates a portion of the piping layout. Drainage and venting codes are important in choosing the size and route of the piping. Codes regulating drilling and notching of wall studs and joists must be followed to avoid structural failure and code violation. The entire route of a pipe must be determined to accommodate unique challenges in an installation. A plumber is essentially creating a map of a pipe layout route, including the approved fittings for the installation. Numerous other considerations that are specific to each job site determine the actual layout procedure. These include the hole location, physical size of a drill bit, the type of drill, and so forth. Table 10-6 lists common layout considerations that must be addressed before drilling any holes for above-ground piping.

Table 10-6 Above-ground Layout Considerations

Consideration	Notes
Fixture location	Actual location within a room and in relation to other fixtures.
Type of fixture	This dictates minimum pipe sizes.
Codes	All codes dictate the pipe size, fitting installed, distances of drains from a vent, drilling of wood studs and joists, and other information.
Location of wood studs and joists	Some studs and joists can be relocated or altered by a carpenter at the request of a plumber.
Physical size of drill and drill bits	Drill and bits must be able to fit within the space between studs and joist to be used safely.
Coordination with other trades	A plumber must recognize the basic locations of certain items installed by other trades.

Fixture Locations

An architect may locate fixtures to accommodate a design preference or to satisfy a customer and not consider the challenges it creates for pipe installation. A plumber typically determines the pipe route upon arriving at a job site. With the framing complete, a plumber uses the blueprint to lay out the fixtures. The plumber either drills a small hole or drives nails part way through the floor where the pipe will penetrate the floor to show the location from the floor below. Figure 10-17 illustrates the fixture location of a typical bathroom layout. Figure 10-18 illustrates the process of using a nail to indicate where to install piping for a fixture or group of fixtures.

FROM EXPERIENCE

Drilling holes indiscriminately creates an unprofessional appearance; a pipe route must be thoroughly investigated to avoid wasting time drilling unnecessary holes.

Floor Joist Conflicts

Possible pipe routes and any conflicts are determined by viewing the location of the holes from the floor below a fixture location. Conflicts with floor joists are common; drainage systems connecting to specific fixtures often cannot be relocated around a floor joist. Drainage piping can usually be relocated easily around wall studs, and often wall studs can be relocated to accommodate pipe installations. A floor joist cannot be moved easily, because the floor and walls are resting on it; it is a **load-bearing** structural feature of the house. A **carpenter** should be told about any conflict with a floor joist. The carpenter may have a solution, or the coordination process might have to involve the architect. A plumber should never alter the location of structural boards or cut a floor joist without knowing all structural codes and getting written permission. A carpenter might be able to cut a floor joist and add other means of support to allow for pipe installations. Most codes allow a small portion of the top and bottom of a floor joist to be slightly

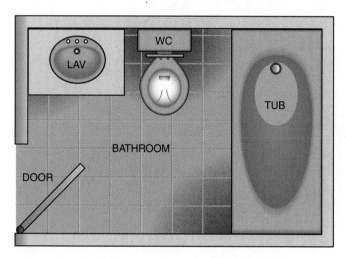

Figure 10-17 Fixture location is established to determine pipe route, which exposes possible conflicts.

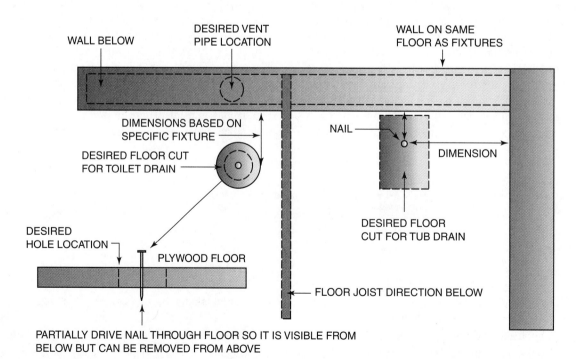

Figure 10-18 Partially driving nails into the floor allows fixture locations to be viewed from below the floor.

CHAPTER 10 *Material Organization and Layout* **241**

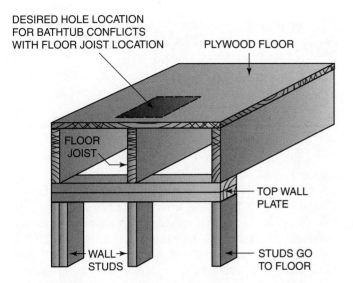

Figure 10–19 Floor joists often conflict with fixture drain locations.

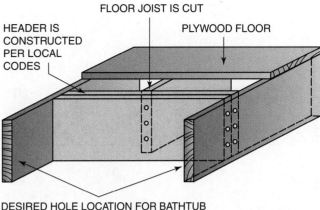

Figure 10–20 A carpenter can cut a floor joist and install a header per local code to provide an opening for a tub drain location.

notched to install piping that partially conflicts, but a plumber must know local codes before proceeding. Figure 10-19 illustrates a common conflict involving a bathtub drain. Figure 10-20 shows a common solution to a floor joist conflict installed by a carpenter.

CAUTION: Cutting, drilling, or notching a floor joist can eventually cause structural failure of a house. A plumber or plumbing company can be held responsible for damage even years after the violation occurred.

 FROM EXPERIENCE

Attempt to identify conflicts immediately upon arriving on a job site to expedite corrective action. This process may take several days and can cause a delay in installation.

Wall Layout

Once all conflicts are eliminated, the next step is to determine the wall layout to install the piping serving the fixtures. Many codes exist concerning drilling through walls. The size and location of a hole varies based on the load-carrying responsibility of the wall stud. Not all wall studs have the same purpose. Some are considered partitions while others are load bearing; that is, they carry the load of the structure above. Codes vary in different regions of the country. A plumber must

know local requirements to avoid code violation and expensive errors. When the pipe route is determined, the walls are marked with a pencil or ink marker to indicate the center and the size of the hole, or the drill bit to use. Most plumbers have an apprentice drill the holes. Being specific during the layout process eliminates errors in drilling.

Because every fixture needs a drain and a water supply, the holes for both are typically laid out and drilled at the same time. Laying out both systems at the fixture location exposes possible conflicts between the two systems. Because the drainage piping is larger in diameter, it takes precedence during the layout phase and is usually installed before the water piping. Drilling all necessary holes in a particular work area at the same time increases productivity. When laying out a wall for a pipe that is routed from one floor to another, a plumb-bob tool is useful for transferring a line from the bottom plate and top plate of the wall. However, using a plumb-bob can decrease productivity and is rarely needed on residential construction because the piping is usually plastic or flexible. The vertical wood studs are typically installed level; transferring the dimensions using a tape measure is adequate and more productive. Figure 10-21 illustrates one method of identifying the hole location and size. Figure 10-22 shows a wall layout of a pipe passing through a bottom and top plate. Figure 10-23 illustrates a wall layout for a pipe installed horizontally through wall studs.

CAUTION: Some codes do not allow large-diameter holes to be drilled in exterior wall studs to avoid structural failure during a hurricane, earthquake, or tornado.

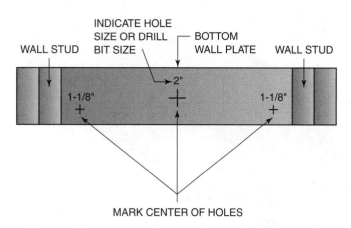

Figure 10-21 A thorough layout process eliminates drilling errors and increases installation productivity.

FROM EXPERIENCE

Locating piping directly against a wall stud eliminates the need for a tape measure, and the piping can be supported directly to the wall stud.

A drain, vent, and water piping system often requires horizontal routing within the wall and poses challenges to accommodate the pipes within one wall space. Because a 2″ × 4″ wall is only 3-1/2″ wide, a plumber must consider all of the piping that must be installed within a wall when laying out the pipe routes. If a 3″ pipe is installed in a 2″ × 4″, there is not enough room for water piping to cross over the 3″ pipe in the same wall. The water piping routes are usually less important than the drainage pipe route because the fixture connection can usually be performed with flexible tubing during the trim-out phase. Many residential homes use flexible tubing during the rough-in phase of an installation, which allows offsets to be created around drainage piping in a wall. A plumber should lay out piping systems with the least amount

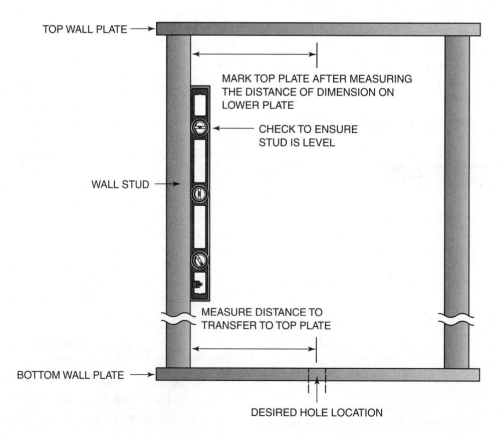

Figure 10-22 A hole layout for a pipe installed vertically in a wall can usually be performed with only a tape measure.

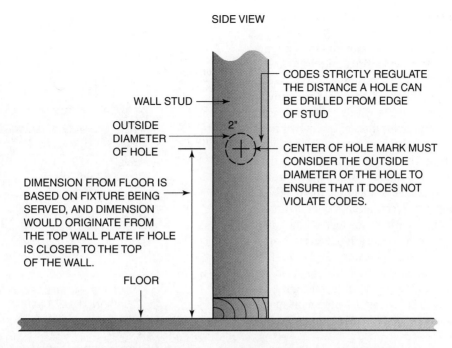

Figure 10-23 A wall-stud layout must be based on local codes pertaining to hole size and location.

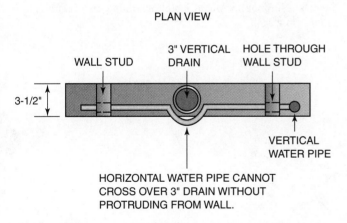

Figure 10-24 A 3″ pipe installed in a 2″ × 4″ wall does not leave enough room for a horizontal water pipe.

Figure 10-25 A 2″ pipe in a 2″ × 4″ wall must be installed toward one side of the wall to accommodate a horizontal water line.

of fittings and offsets possible to expedite installation. Figure 10-24 illustrates a 3″ pipe in a 2″ × 4″ wall that is obstructing the route of a water pipe. Figure 10-25 illustrates a 2″ pipe that accommodates a water pipe in the same wall.

FROM EXPERIENCE

Remain creative when laying out the pipe and, while focusing on code regulations, also focus on cost-saving installation methods.

Manufacturer Rough-in Sheet

A fixture manufacturer provides installation requirements for each fixture on a manufacturer rough-in sheet. Some provide installation data in book form, which can be obtained by request. The information varies depending on the fixture being installed. A plumber uses the rough-in sheet to determine the drain, hot and cold water locations, heights above the floor, and the distance between the drain and water piping where it stubs out the wall. Wall-hung fixtures require support, either from a manufactured support assembly known as a fixture carrier or from wood backing installed by the plumber to anchor the hanger brackets used to install the fixture during the trim-out phase. A rough-in sheet indicates the height and spread (distance apart) of the hanger brackets in relation to the drain and water piping. A plumber must verify the dimensions provided in a rough-in sheet to make sure the overflow rim of the fixture conforms to local codes or installation requirements. A manufacturer provides information based on national standards, but the dimensions might vary based on your local codes and whether the fixture is being used for handicap purposes. If the fixture flood level rim varies from the rough-in sheet, then all dimensions listed on the rough-in sheet will adjust based on that change. If the manufacturer states that the flood level rim of a sink is 32", but your installation requirements dictate 36", then all vertical dimensions provided on the rough-in sheet will be increased by 4". Figure 10–26 is an example of a rough-in sheet for a wall-hung lavatory. Figure 10–27 illustrates a rough-in of a wall-hung sink that requires support. It also shows some typical rough-in dimensions. These might vary from an actual installation depending on specific manufacturer requirements, job conditions, or company preference.

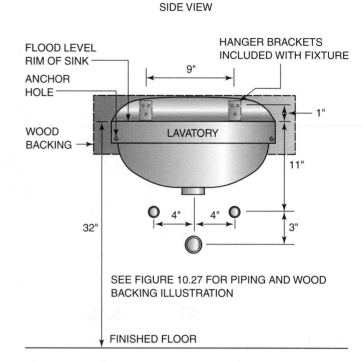

Figure 10–26 A manufacturer rough-in sheet provides important information about installing piping and fixture support in a wall.

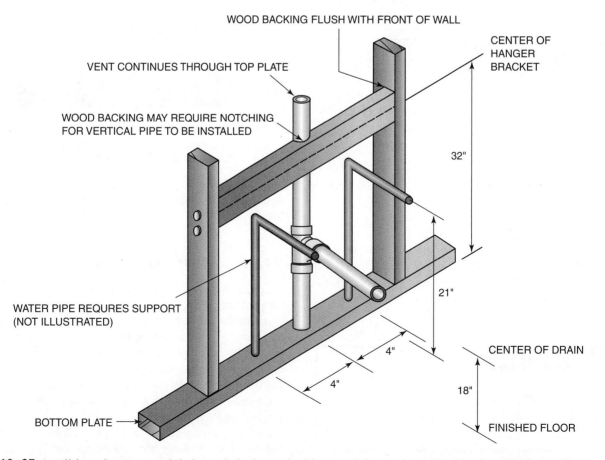

Figure 10–27 A wall-hung lavatory rough-in has a drain, hot and cold water piping, and wood backing installed in a wall.

FROM EXPERIENCE

Most residential fixtures are basically the same regardless of manufacturer; a plumber usually installs the rough-in piping without referring to rough-in sheets. Fixtures such as one-piece toilets, bidets, and large-capacity bathtubs generally require a rough-in sheet.

Summary

- Good written and verbal communication skills are required for effective communication.
- The best form of communication is typically in writing.
- Verbal communication can increase productivity.
- Mobile phones and two-way radios are two forms of verbal communication.
- Organizing materials eliminates material shortages.
- Bagging and tagging is a method used to organize material for specific tasks.
- Vehicle racks have maximum weight loads.
- Securing materials and ladders to a vehicle rack must be performed with proper techniques so items do not fall while driving.
- Layout of a piping system must include fixture types and various plumbing codes.
- Drilling and notching codes must be addressed during an above-ground layout process.
- Excavated soil placement must be considered during a layout process for underground installations.
- Fixture rough-in requirements and fixture clearance codes must be known during the layout process.

Procedures

Edge Form Layout Procedure

A A wall dimension provided by an architect on a blueprint will state either the rough-wall or the finish-wall dimension. If a finish-wall dimension is provided, deduct the thickness of the drywall or other finish material to find the rough-wall width. Locate the wall from a specific reference point using a blueprint or shop drawing. Measure and mark the center of the wall from the appropriate area, such as the perpendicular edge form. Mark the wall width on both sides of the center mark, which is 1-3/4" on each side if the wall is a 2" × 4" width. With a hammer, pound two nails into the edge form to create the wall width.

B If the distance from the edge form to another location, such as another edge form, exceeds 100', a batter board may be required. Drive the batter board assembly into the ground to show that the horizontal portion is aligned squarely with the nails in the edge form. Repeat the wall layout process from step A. At this point in the layout process, four nails are installed to secure the string. Tie one end of the string to any of the four nails using a knot that will remain secure with no pressure applied to the string. Route the string to the nails located in the opposite edge form or batter board and then back to the remaining nail adjacent to the initial knot location, which creates a rectangle. Tie a cinch knot, explained in step C.

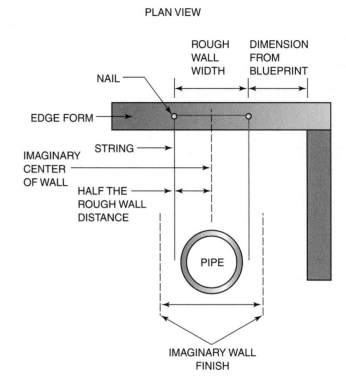

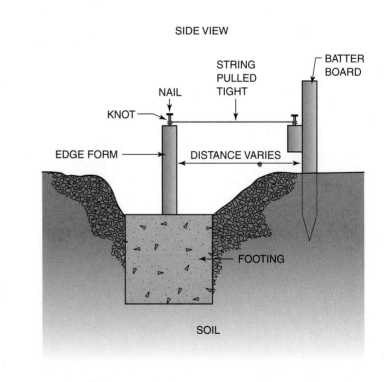

C A cinch knot allows a plumber to tie string quickly and remove it without untying the knot or cutting the string. With the string in both hands, loop it around the nail. Wrap the loose end of the string over and under the main string several times. Pull the string with both hands in opposite directions to tighten the string. Once it is taut, continue pulling on the main line and insert the loose end through the loop created around the nail. Release the main string line, and the pressure of the taut string will cinch the loose end to complete the knot.

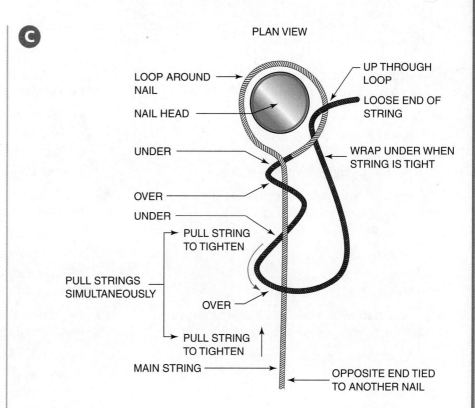

D To loosen the cinch knot, pull the main string toward the knot; with your other hand, pull the cinch knot away from the nail; and remove the loop from the nail. Reroll the string for future use.

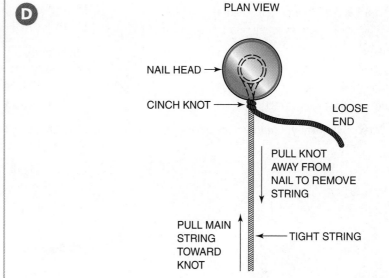

Review Questions

1. Material organization techniques that identify and store material within a box or bag are called
   a. Bagging and tagging
   b. Bagging and identifying
   c. Boxing and bagging
   d. Boxing and identifying

2. Underground layout often requires the simulation of wall locations and widths using
   a. Concrete blocks
   b. Wall studs
   c. String
   d. Trenches

3. The manufacturer's instructions for installing the piping that serves a fixture is provided on a(n)
   a. Trim-out sheet
   b. Rough-in sheet
   c. MSDS sheet
   d. Architectural blueprint

4. Pipe protruding too far from the back of a vehicle must be
   a. Painted
   b. Plastic pipe
   c. Steel pipe
   d. None of the above are correct

5. The most effective form of communication is
   a. Written
   b. Oral
   c. Voice mail system
   d. Telephone

6. The three basic single-family home designs are slab, basement, and
   a. Two story
   b. One story
   c. Crawl space
   d. Three story

7. To minimize nonproductive handling of stored material, it can be placed directly
   a. On the floor
   b. On a pallet
   c. In the back of a truck
   d. In the attic

8. Two main wall layout considerations are design intent and
   a. Codes
   b. Soil type
   c. Concrete thickness
   d. Wall type

9. A wall-hung lavatory must be supported with a fixture carrier or
   a. Drywall
   b. Plastic anchors
   c. Wood backing
   d. Concrete anchors

10. A plumber can determine a pipe route on the floor below the fixture location by
    a. Partially driving nails into the floor
    b. Asking the carpenter
    c. Using rough-in sheets
    d. Completely removing the plywood

11. MSDS is the abbreviation for
    a. Material Safety Data Sheet
    b. Mechanical Sizing Distribution Standard
    c. Material Sizing and Design Standard
    d. Master Size and Design Sheet

12 To communicate directly with a crane operator typically requires a dedicated channel of a

a. Two-way radio
b. Mobile phone
c. Transistor radio
d. Three-way radio

13 A floor joist conflicting with a pipe route should be altered by a

a. Carpenter
b. Plumber
c. Mechanical engineer
d. None of the above are correct

14 Layout of a pipe trench can be performed by placing string on the ground and using

a. Spray paint
b. An ink marker
c. Pencil
d. Crayon

15 If an edge form is not feasible for wall layout to install underground piping, a plumber can

a. Excavate additional soil
b. Use a batter board to pull strings
c. Install the pipe close to its design intent
d. Use the excavated soil to secure strings

16 A plumber must be aware of safety concerns when using a pipe rack and know that the rack has

a. Maximum load ratings
b. Unlimited load rating
c. Minimum load ratings
d. Minimum and maximum ratings

17 To protect the structural integrity of wall studs and floor joists, they are regulated by

a. A plumbing engineer
b. Drilling and notching codes
c. Drainage codes
d. Architectural blueprints

18 A tool used to transfer the center of a hole from the bottom plate of a wall to the top plate is called a

a. Transfer tool
b. Plate hole aligner
c. Plumb-bob
d. Hole aligner

19 To expedite the drilling process during the wall layout process, a plumber can

a. Indicate the hole or drill bit size on the wood
b. Indicate the type of pipe that will be inserted through the hole
c. Indicate the tool to use for drilling the hole
d. Drill each hole anywhere

20 The largest diameter drainage pipe that can fit in a 2" × 4" wall is

a. 2"
b. 3"
c. 4"
d. None of the above are correct

Chapter 11 | Water Service Installation

You have learned about the products used to install a water supply system. This chapter discusses water sources and the distribution and installation of potable water to a house. A water service installation includes the piping that runs from a water source to an exterior connection of a house's water distribution piping. Plumbing codes dictate that a home can only be considered habitable if potable water is provided. The main sources of potable water are a municipality, a community, or a private source. A well can provide water to a single home or it can provide a community water supply system for a subdivision. A municipal water system is considered a public supply source that provides potable water to its customers. The two sources of water are below ground or on the surface, and both are subject to pollution. Public water supply systems typically use surface water, and private systems tend to use groundwater. The Environmental Protection Agency (EPA) regulates the potable water quality throughout the United States. The importance of protecting our drinking water supply and the negative impact of pollution are discussed in this chapter.

OBJECTIVES

Upon completion of this chapter, the student should be able to:
- understand correct installation techniques for installing a water service.
- understand the basics of municipal and private water systems.
- understand basic codes pertaining to burial depths and locations.
- understand and respect water quality issues and regulations.

Glossary of Terms

aquifers geologic formations containing water

backfill loose soil placed into an excavated area; also called fill

brackish lowland water close to the ocean; has high salt content

branch a pipe installed laterally from a main pipe

compacting the process of compressing loose soil placed back in a trench; also known as tamping

filter an accessory that removes particulates from water, but does not purify water

purification a process to cleanse the water to ensure it is considered potable

trench excavated pocket of soil to install piping; also known as a ditch

water distribution system though an entire system is distributing water, this refers to the piping inside a house

water service piping from a water meter or well to a building; connects to the water distribution system

Water Source

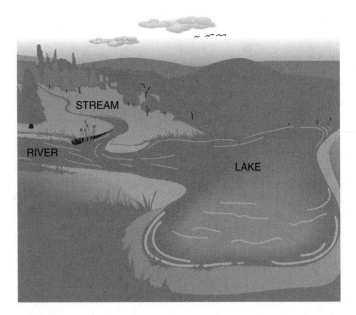

Figure 11-1 Surface water is a primary source of water used by many municipalities.

Potable water is considered safe for human consumption and is often referred to as drinking water or domestic water. It is easy to take our clean water for granted. Communities throughout the country were established based on the availability of adequate surface water or groundwater. Lakes, streams, and rivers are three primary sources of surface water. Groundwater (**aquifer**) is the natural water table below the surface. It is the primary source for homes using a well pump system. Rainwater and natural underground springs (veins) replenish the available water above and below ground. Both sources of water can become polluted from environmental neglect, and not all pollutants can be removed from water to make it safe for drinking. Municipalities must remove harmful pollutants and purify the water to very strict EPA standards. One benefit of a municipal water supply is the addition of fluoride, which has decreased tooth decay. Private well systems must also adhere to EPA standards, but inspection for water quality is the responsibility of the owner of the well. It is important to respect the environment to minimize pollution that contaminates our water sources. A plumber is on the front line in protecting the water supply, which must be a top priority when installing potable water systems. Figure 11-1 shows the primary sources of surface water. Figure 11-2 illustrates the primary source of groundwater. Figure 11-3 shows the two primary ways that surface and groundwater is replenished.

Figure 11-2 Groundwater is a primary source for private well systems.

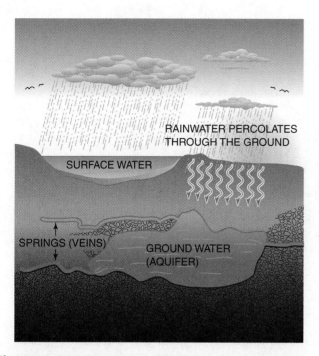

Figure 11-3 Underground springs and rainwater replenish available water sources.

Groundwater is typically cleaner than surface water because it is filtered as it percolates through the soil. Storm water runoff can be a primary source of surface water pollution.

Public Water System

The EPA defines a public water system as one that is in service for a minimum of 60 days per year and that provides water to at least 15 connections or 25 individuals. Public water system designs vary based on the location of their water source, the number of customers, and the technological advancements adopted by each community. The Safe Drinking Water Act (SDWA) is a federal law regulating public water systems; it creates guidelines for local municipalities to define what is considered a public water system. Many jurisdictions that are not annexed within a town or city are served by a community water system that is considered a public water system by EPA standards. A homeowners' association or a private company maintains most community water supply systems. A campground, a factory, or an isolated residential subdivision that is not annexed into a city are places where community water systems are used. A city might have to use several pumping stations and possibly a water tower to have adequate pressure to deliver water to its customers. A water tower provides pressure within a piping system. One vertical foot of water exerts 0.433 pounds per square inch (psi) of pressure. A 100-foot head of water will create 43.3 psi of water when it is not flowing through the piping system, which is considered static water pressure. High-rise buildings sometimes need to install booster pumps to take the water to a height that the city water pressure is not capable of delivering. Most community water supply systems use groundwater as a primary source along with a submersible pump located below ground. Figure 11-4 illustrates a municipal water system providing water from a surface water source. Figure 11-5 shows a community water supply system that is considered to be a public system, but uses a groundwater source.

Some codes do not allow a plumber to work on a public water supply system without a utility contractor license.

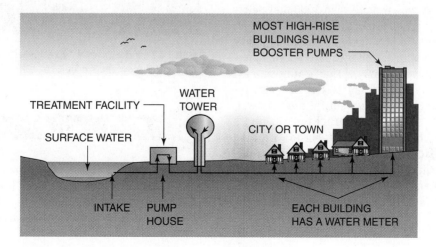

Figure 11-4 A municipal water supply system is owned by a city or town.

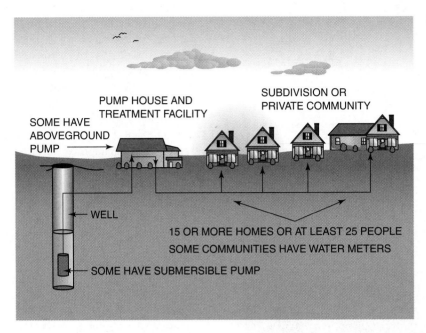

Figure 11-5 A homeowner association or private company can own a community water supply system that is considered a public system.

Private Water System

Water systems serving fewer than the EPA criteria for defining a public water system are considered private water systems. A water source used by an individual or by a few homes is considered a private water system. The usual source of water for private systems is groundwater, but cottages located on lakes or rivers might use the surface water for drinking water. A well is typically drilled or bored into the ground, so a pump can extract water from the water table. The nature of soil, rock, or sand is a factor in the availability of water and in the health of the earth, which dictates the quality of groundwater. Deep wells and shallow wells are the two basic classifications, both of which are considered standard designs. A shallow well is less than 50 feet deep, and a deep well is more than 1000 feet below ground. Deep wells typically provide cleaner water than shallow wells. The region of a country often dictates how deep a well must be drilled to access adequate water. Mountainous regions may have natural spring water trapped close to the surface between rock layers, but flat, arid regions do not. Oceanic regions have plentiful groundwater, but the water close to the ocean is **brackish** (having a high salt content) and typically not suitable for drinking without intense **purification.** Figure 11-6 compares different geographic regions and where water might be located in relation to the surface of each. Figure 11-7 illustrates a well that is used to extract water from the ground.

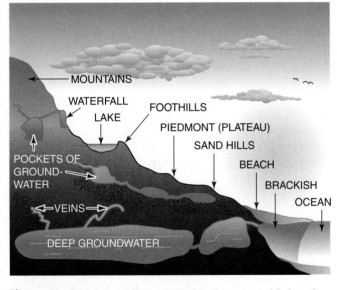

Figure 11-6 Water quality varies with the geographic location and depth below the surface.

 **FROM EXPERIENCE**

A steel casing (pipe) is installed and grouted in place to stabilize the soil near the top of the well. The top of the well casing must be capped.

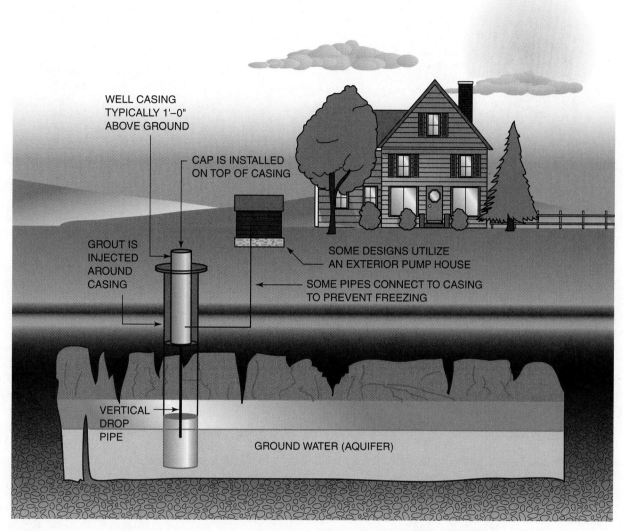

Figure 11-7 A well is a cavity created in the earth to extract water from the water table below ground.

EPA Standards

Water, regardless of the source, is typically tested for pureness by local health administrations and regulated in each state or county. Certain minerals in the water are acceptable; high mineral levels can be lowered or removed with water purification systems; and sediment can be filtered from the water. The presence of arsenic is a major concern with private water systems. Municipal systems are constantly inspected for dangerous levels of arsenic or toxic pollutants, but some private systems are not inspected on a regular basis. Development of rural areas can change the quality of the groundwater and slowly pollute a once-clean source of water. Animal farms, buried fuel tanks, chemical plants, or other possible sources of pollution are suspected when groundwater becomes polluted. A septic tank can also be a source of pollution to groundwater if toxic substances enter the drainage system.

Water Quality

Water clarity is not the same as water quality; even crystal-clear water can be contaminated with deadly toxins or other contaminants. Water purification and filtration systems are very popular with private well users, and municipal water systems are purified before being routed to users. A variety of water treatment applications can increase water quality, with the end result being potable, or safe, water based on EPA standards. The United States operates under a Safe Drinking Water Act (SDWA), which was passed in 1974. The Act has been amended several times since its adoption

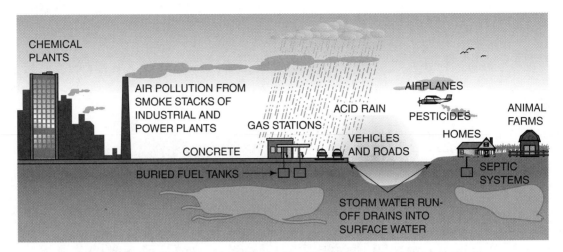

Figure 11-8 Environmental threats to our drinking water system are numerous and are regulated by law.

to deal more effectively with the ever-changing environmental threat to the nation's water sources. The SDWA gives the EPA authority to dictate drinking water standards and to delegate its regulating authority to state or local agencies. Figure 11-8 illustrates some possible threats to our nation's drinking water.

> **FROM EXPERIENCE**
>
> A plumber must always install safe piping systems to protect the health of America. This includes using lead-free products and installing piping systems per code.

There are two levels of EPA drinking water standards. The National Primary Drinking Water Regulation (NPDWR) is enforceable by law. The National Secondary Drinking Water Regulation (NSDWR) is a recommended standard but it is not enforceable by law. Protecting drinking water from contaminants that have adverse effects on our health is the main intent of the NPDWR. Regulations pertaining to the maximum level of contaminants allowed and the approved water treatment processes are strictly enforced under the NPDWR. The NSDWR is concerned with the discoloration of consumers' skin or teeth and the taste, odor, or color of water. State or local authorities can adopt the secondary standards as part of their primary standards, but the EPA does not enforce the secondary standards. It is important for a homeowner with a private water source to have it tested on a regular basis, especially if massive development occurs in their neighborhood.

Water Filtration

Water filtration and purification systems are used to clean and treat every gallon of water distributed by a municipal water system. Water **filters** remove sediment and improve taste, odor, and color. Contaminants are removed, and chlorine, fluoride, and other necessary chemicals added during a purification process to disinfect the raw source of water. After the water is treated, it is distributed to a city, community, or individual home, but the piping system that carries the water can lower its quality. The distribution route can expose the potable water to entirely different contaminants. To protect the drinking water, codes regulate the products and installation techniques used, including the solvents or solder that connect the piping system. Backflow preventers, air gaps, and other forms of anti-siphon devices are included in all **water distribution systems.** Water filtration devices and individual water filters are common for water dispensers, icemakers, and other point-of-use locations. Many kinds of water filters are available; some focus on specific tasks such as improving taste, removing odor, or clarifying the color of the water. A consumer or plumber selects the proper filtration method based on the water quality improvement needed at the point of use. Nitrate and coliform bacteria are two of the leading threats to water quality for private water systems. Radon and pesticides can have more devastating health effects, and private water systems should be tested if specific health problems arise. Figure 11-9 illustrates a whole-house water filter design. Figure 11-10 illustrates a point-of-use water filtration system.

> **CAUTION**
>
> **CAUTION:** A plumber should never attempt to regulate water quality using a water softener system without being qualified. The misuse of salts can increase a person's sodium levels and cause severe health problems.

CHAPTER 11 *Water Service Installation* 257

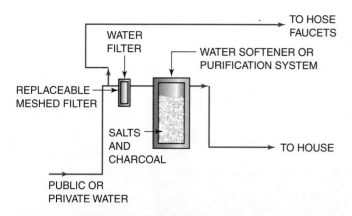

Figure 11-9 A whole-house filtration system serves the entire system and can include water treatment features.

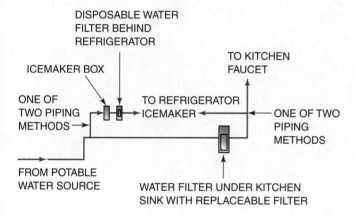

Figure 11-10 A point-of-use filter can improve water quality for specific locations of a water piping system.

 FROM EXPERIENCE

Water sample kits are available at most plumbing supply stores or through your state or local health agency. A water sample can be submitted for testing; basic testing is often free, but testing for some contaminants can be expensive.

Water Service Installation

A **water service** system is made up of the piping leading from a water main or well to a building. It is important to understand the relevance of this segment of a piping system in relation to codes. Piping allowed for water service installations can differ from the piping used in the internal water distribution system of a home. Polyethylene (PE) and polyvinylchloride (PVC) are allowable by code for a water service system, but not for a water distribution system within a building. Most codes dictate that the smallest allowable water service is 3/4". Public or community water systems use water meters to record gallons of water used. This allows the authorities to issue an invoice for payment. A private water system uses a well and does not record the gallons of water used. The termination of a water service is the entry point into a building where it connects to the water distribution system. Figure 11-11 illustrates a water service pipe served by a municipal or community water system. Figure 11-12 shows a water service pipe from a private

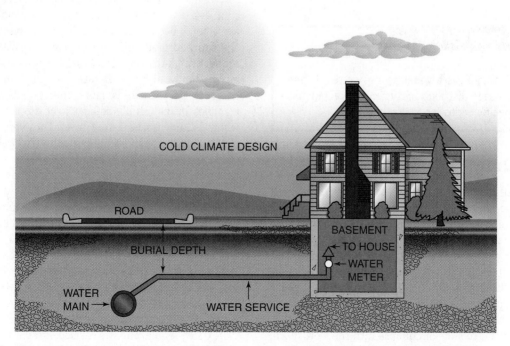

Figure 11-11 A water service pipe connecting to a municipal or community water system uses a water meter.

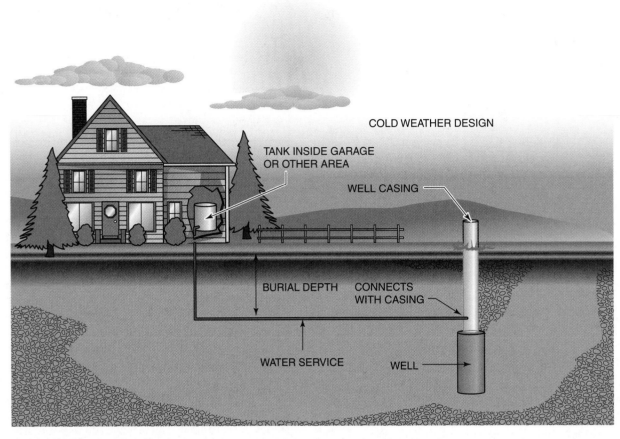

Figure 11–12 A private water service connects from a well to the house.

water system; various designs are possible with some tanks being in a dedicated pump house or within a home.

A code book categorizes a water service system differently from a water distribution system, so a plumber must know specific codes for each system.

Trench Safety

An employee working in and around an excavated *trench* must be trained to determine the stability of the soil. OSHA classifies soil in several different categories, and an employer is responsible for training all employees regarding current safety standards. One qualified individual on the job site must be designated as the person responsible for determining soil stability. Weather can cause changes in the soil, and different areas within a trench may be classified differently. Rain can soften the soil and cause the sides of a trench to slide or collapse into the workspace. Cold weather can cause the excavated area to freeze, but it can thaw if the following day is above freezing; therefore, soil conditions may be classified differently during the course of an installation. In some soil conditions, a trench must be sloped when the workspace is more than 4′ below the finished grade of the soil. Local regulations might set higher standards than those set forth by OSHA. Some regulations might allow or dictate a stepped design to be excavated if the trench is more than 4′ deep. The sides of trenches in less stable soil must be sloped at a 45° angle, which is known as a 1-to-1 ratio. For every foot of vertical depth, one foot must be provided horizontally on each side of the excavated area. For large excavated areas or trenches that are subject to collapse, specially engineered trench boxes or shoring may be required. OSHA and local regulations may require trench boxes and shoring for all excavations in earthquake-prone areas of the country.

An extension ladder is often needed as a point of entry and escape when people are working in trenches. A safety harness with a lifeline is recommended, even though OSHA may not require it. Many pipe trenches are considered to be a confined space by OSHA, which requires special training, a confined-space attendant, and an air-quality monitoring device. OSHA regulations are revised based on an annual review

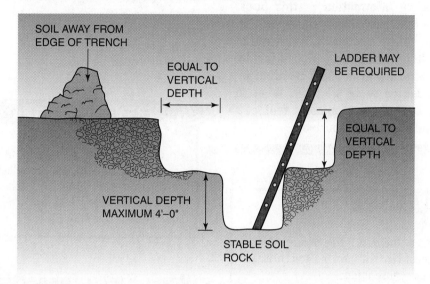

Figure 11–13 A stepped excavated trench may be allowable or required by OSHA or local codes.

of safe workplaces. Do not assume any information meets regulations unless it is from a current OSHA publication or local guidelines. Excavated soil placed too close to the edge of a trench applies weight to the soil and can cause the trench to collapse. Always place the excavated soil a safe distance from the open trench area. Review OSHA and local regulations to determine the safe distance for soils in your area. Figure 11–13 illustrates a trench with both sides excavated with steps. Figure 11–14 illustrates a trench that is sloped at a 45-degree angle, or a 1-to-1 ratio.

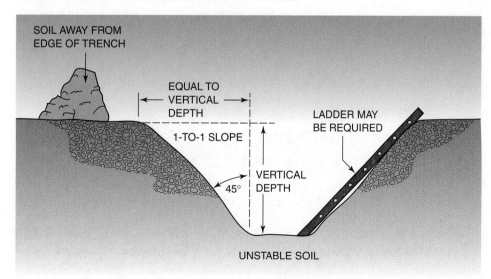

Figure 11–14 Unstable soil requires a trench to be sloped at a 45° angle, and other regulations may require that all trenches be excavated in this manner.

CAUTION: Trench safety information in this book does not necessarily represent OSHA regulations. It is provided to create safety awareness. Never enter an excavated area without being properly trained in OSHA standards. A person who is OSHA certified in trench safety is needed on some job sites. When a trench collapses, it can sweep a person's feet out from under them, and they can be buried.

Always keep the edges of the trench free of loose debris and rocks. Small objects can fall into the trench while you are working and can cause injury.

Burial-Depth Requirements

Codes that dictate the burial-depth requirements of a water service pipe vary based on the frost levels in the area. Extremely cold regions may have to bury the water service as deep as 5′. Regions that rarely experience freezing temperatures may only require a burial depth of 1′. A municipal water system is buried in the street or utility easement, and a branch pipe is routed from a water main into the property being served. Homes in cold climates often have the water meter installed in the house because it would freeze if installed outside. In warm regions, the water meter is usually installed near the street in a box or vault designed for that purpose. If the placement of the water meter box causes the piping connection to be above the required burial depth allowed by code, the piping must offset downward to an approved depth immediately after the connection with the meter. Figure 11-15 illustrates how burial depths are determined for codes.

An inspector usually measures from the ground elevation at the time of the installation and not the expected finished level. A plumber must know the expected finished grade, not only the elevation at the time of an installation.

Water Meter Connection

The water meter connection to the water service depends on the type of meter provided by the local water authority. The water meter connection is outside and below ground if an exterior meter is installed in an approved box or vault. If the connection is located inside the building, it typically is in a basement or some other location where the temperature remains above freezing. Every water meter has a direction of flow and an arrow indicating which connection is incoming and which is outgoing. The water service is routed to the incoming side of an interior meter and the outgoing side of an exterior meter. The municipality or community typically installs the water meter and box or vault, but may only provide the meter to a plumber when the water meter is in-

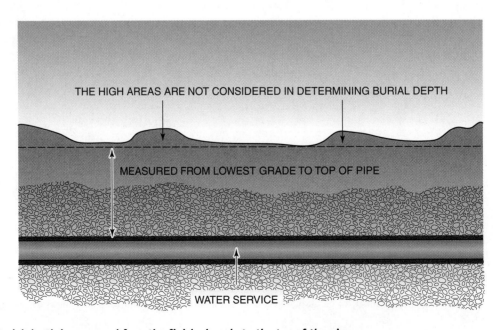

Figure 11-15 Burial depth is measured from the finished grade to the top of the pipe.

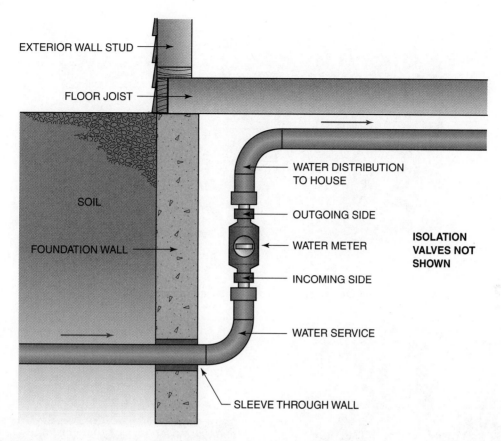

Figure 11-16 The water service connects on the incoming side of an interior water meter.

stalled inside. For interior meter installations, a plumber routes the water service from the connection provided by the municipality or community. Some municipalities route the water service to the inside of a house, and the plumber connects to the outgoing side of the meter. Most codes dictate that a licensed utility contractor must install the main water piping in the street. For water meters inside a building, the contractor then installs a valve known as a curb cock for a plumber to connect the water service. Some codes require a means of backflow prevention on the customer side of the meter, such as a double-check valve assembly. Figure 11-16 is an illustration of an interior water meter connection to the water service. Figure 11-17 shows an exterior water meter connection to the water service.

FROM EXPERIENCE

A municipality or other water authority owns the water meter; a plumber cannot alter, remove, or repair it.

Most water meter connections require an adapter or pipe nipple for the specific type of pipe being installed on each side of the meter. Most water meters have a rubber washer similar to a garden hose washer and a union connection that allows the meter to be removed from service without cutting the water piping. Some meters have an isolation cock (valve) that requires a special tool to isolate the water flow through the meter. Most interior water meters do not have an isolation valve and require a plumber to install a valve. Isolation valves on both sides of a meter are recommended to completely isolate the system if a meter is removed from service. Figure 11-18 is a typical interior water meter measuring in gallons of water. Figure 11-19 shows an interior water meter with isolation valves.

FROM EXPERIENCE

When no water is being used from a faucet, toilet, or other point-of-use location, a trickle indicator on a water meter indicates whether a leak is present in the piping system.

Well Connection

A water service connects to a well in one of two ways depending on the regional climate. The well connection with a steel well casing for cold-climate installations is below ground. A two-piece fitting known as a pitless adapter allows

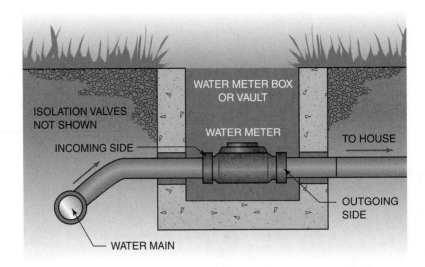

Figure 11–17 The water service connects on the outgoing side of an exterior water meter.

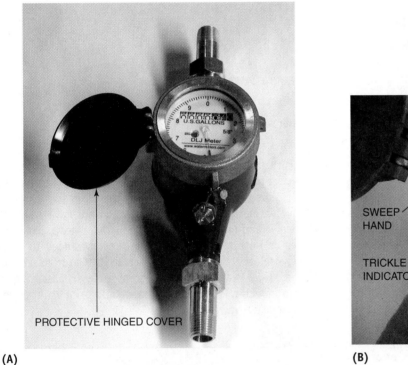

Figure 11–18 A water meter connects to the water supply system and records gallons of water that flow through the meter. *Courtesy of watermeters.com.*

Figure 11–19 Connections to an interior water meter typically have an isolation valve on both sides. *Courtesy of watermeters.com.*

the vertical piping in a well to be removed without accessing the below-ground water service piping. The pitless adapter is removed from the well by first removing the well cap or plug, and then inserting a 1″ diameter threaded pipe—either purchased or fabricated in a T-handle shape. The pipe is threaded into the pitless adapter and pulled upward to remove the vertical piping from the well. In warm climates, a pitless adapter is rarely used. The piping from the well is often routed through the top of the well casing, and a specially designed seal is used to plug it. Figure 11–20

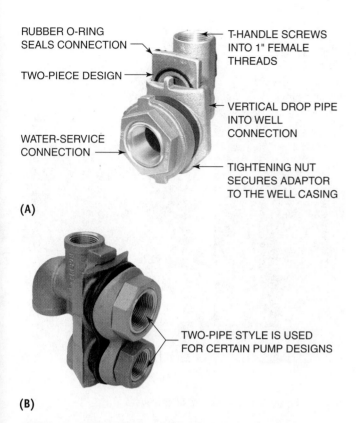

Figure 11-20 A variety of pitless adapter designs are available to customize water service well connections. *Courtesy of Simmons Manufacturing Company.*

compares several pitless adapter designs. Figure 11-21 shows how a pitless adapter connects the water service to a well. Figure 11-22 illustrates the use of a T-handle pipe to remove half of the pitless adapter.

House Connection

Depending on whether the site uses a municipal or a private source, the water service connects to the water distribution system differently. For a private well system, accessories such as a storage tank and pressure switch must be installed to regulate the pressures and the volume capacity of the system. For a municipal or community system with a water meter, the pressure coming into a house varies depending on the design of the system. Codes often dictate that a pressure-regulating device be installed when a water service connects to the water distribution system of a building. Code dictates that a municipal or community water system must have a pressure-reducing valve if the incoming pressure exceeds 80 pounds per square inch (psi). A well system is controlled by a pressure switch, which typically does not allow water pressure to exceed 60 psi; therefore, some codes do not require a pressure-reducing valve. A well system uses a storage tank at the entry of the water service and typically has all the regulating accessories, including the required isolation valve, in a central location. Every connection between a water service and a water distribution system

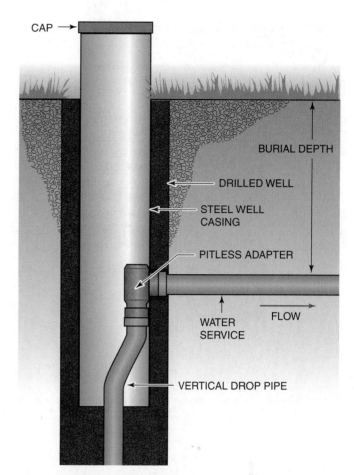

Figure 11-21 A pitless adapter is installed connecting a water service to a well casing.

must have an isolation valve that is readily accessible. Most codes dictate that a valve in a crawl space cannot be installed farther than 3' from a crawl space door. Some codes allow the main isolation valve to be installed outside if it is located in a specially designed valve box and is no more than 2' from the exterior wall. PVC and PE piping are not allowed to enter a home according to many codes and must be converted to an approved material such as copper. Figure 11-23 illustrates the typical house connection between a municipal water service and a distribution system. Figure 11-24 shows a typical house connection between a private water service and a distribution system.

Same Trench with Sewer

If a water service is installed in the same trench as a drain or sewer, backflow could occur. If the drain or sewer developed a leak, the surrounding soil could become contaminated with sewage. If the water service developed a leak and the water system was shut off, the sewage could backflow into the non-pressurized water service through the leaking area. To protect against backflow, codes regulate the installation of a water service in the same trench as drainage piping.

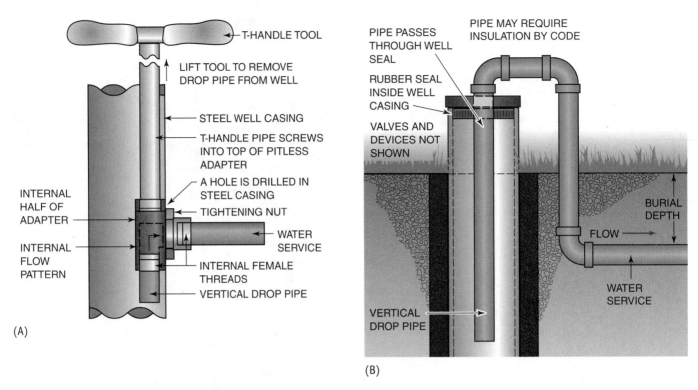

Figure 11–22 The water service from a well either originates from a pitless adapter or from the top of the well.

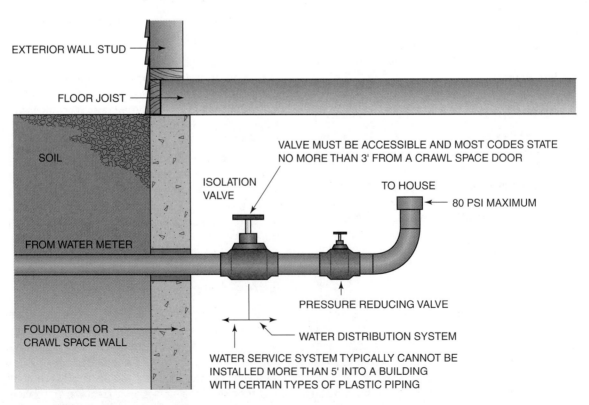

Figure 11–23 A connection from a municipal water service to a distribution system requires a valve and possibly a pressure-reducing valve.

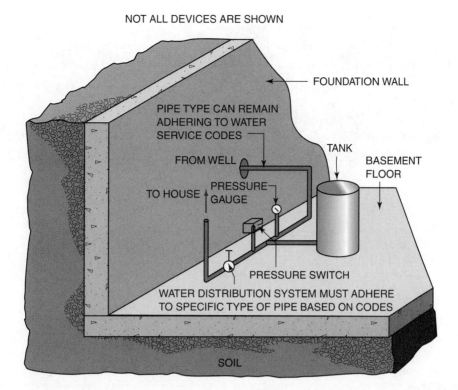

Figure 11-24 A connection from a private well water service to a distribution system requires pressure- and volume-regulating accessories and an isolation valve.

When placing soil back in a trench, the soil must be compacted or tamped. The **compacting** process is achieved with a manually operated and motorized piece of equipment known as a tamper, which applies quick jolting actions to a base-plate to compact the soil in the trench.

CAUTION: Using compacting equipment can cause a trench to collapse. When working in a trench, always have a co-worker outside the trench.

Same Elevation

Most codes dictate that a water service installed at the same elevation as a sewer must be separated from the sewer by at least 5′ of undisturbed or compacted soil. These codes often dictate that a separate trench must be excavated, which might not be possible on some job sites. If a separate trench approach is possible, the installation costs increase. Excavating a separate trench for a water service in a cold climate can mean excavating two trenches, each up to 5′ deep, to adhere to burial-depth codes. A single trench can be used,

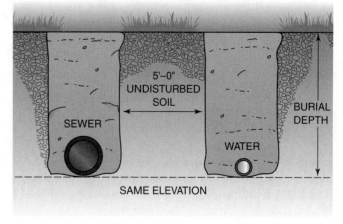

Figure 11-25 A water service installed at the same elevation may be required by local code to be 5′ away from the sewer.

but the 5′ of earth separating the two pipes would have to be compacted, which would also be very expensive. Figure 11-25 illustrates a side view of a water service installed at the same elevation as a sewer.

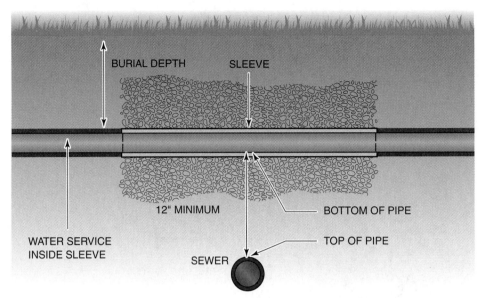

Figure 11–26 A water service must be installed in a sleeve when crossing a sewer trench.

Different Elevation

To avoid excavating a separate trench, some codes allow the water service to be installed in the same trench, but at a higher elevation than the sewer. If using the same trench is allowed by code, the bottom of the water service must be at least 1' higher than the top of the sewer, measured from the highest elevation of the sewer. Because a sewer is sloped, a plumber must install the water service relative to the sewer's highest point, where it exits the house. The best way to ensure code compliance is to excavate a trench with a ledge on which to install the water service. The piping used for the sewer must be approved for this exception, which usually means cast iron, PVC, ABS, or copper must be used.

 FROM EXPERIENCE

A code book lists approved material for water service and sewer piping installed in the same trench. Not all materials approved for other types of installations are approved for installations in the same trench.

See page 267 for a step-by-step procedure on installing a water service in the same trench as a sewer.

Perpendicular Installation

True perpendicular is a pipe that is installed at a right angle or 90° from another pipe. Many codes require interpretation of intent as opposed to specific definitions of terms. A water service that crosses a sewer trench at any angle is considered perpendicular in this scenario, because the regulating code is concerned with the safety of the water supply system. To ensure that the water service pipe is protected against damage from possible excavation to access the sewer, most codes require that pipe crossing a sewer trench be installed in an oversized piece of pipe known as a sleeve. Figure 11–26 shows a water service pipe installed in a sleeve crossing a sewer trench.

Summary

- Protection of water supply sources must be a primary concern of a plumber.
- Lakes, streams, and rivers are three primary sources of surface water.
- Groundwater (aquifer) is the natural water table below the surface.
- The Environmental Protection Agency (EPA) regulates the quality of municipal water sources.
- Water clarity is not the same as water quality; even crystal-clear water can be contaminated with deadly toxins or other contaminants.
- A water service system is the piping from a water main or well to a building.
- An employee working in and around an excavated trench must be trained to determine the stability of the soil.
- Codes dictate the burial depth requirements of a water service pipe and vary based on the frost levels of the particular region.
- There are strict code regulations concerning the installation of a water service and sewer in the same trench.

Procedures

Installing a Water Service in the Same Trench as a Sewer

A Excavate a trench with a ledge that is at least 1' higher than the top of the sewer when it exits the house. Most codes dictate that the minimum size sewer is 4", and the OD of a 4" pipe is roughly 4-1/2"; therefore, the ledge must be at least 16-1/2" above the bottom of the sewer trench. The sewer slopes downward from the house toward the municipal sewer connection or septic tank. The initial layout is based on the burial depth of the water service, which is typically dictated by code to be deeper than the sewer to protect against freezing. The sewer connection at the street or septic tank may be too high for the trench to accommodate this design.

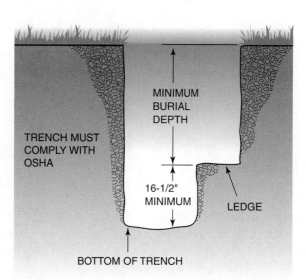

B Install the sewer and water pipe according to job conditions and local codes. The sewer might have to be installed before the water service because it is lower in the trench. A local plumbing inspector usually inspects both piping systems at the same time, so you will not be able to place soil over the sewer pipe until after an approved inspection. If any portion of the ledge collapses as you are working in the trench, you will have to compact the soil to create a firm installation bedding beneath the water service.

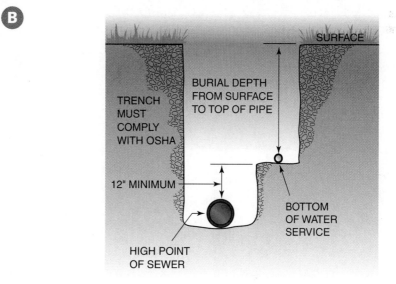

Procedures

Installing a Water Service in the Same Trench as a Sewer (continued)

C Placing soil back into the trench is known as **backfilling.** When backfilling within the footprint of a building or in trenches that will have sidewalks or driveways installed above them, the soil must be compacted more thoroughly than in trenches that will be under a grassy landscape. Many codes dictate that the first 1' of soil placed on top of the piping must be compacted, and that no more than 6" vertical layers can be placed and compacted at a time. The final backfill soil is placed to meet the local standards pertaining to vertical layers between compacting. It is best to crown the soil slightly where the trench is located, as it will settle over a period of time.

C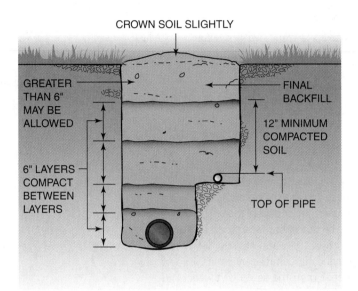

Review Questions

1. **EPA is the abbreviation for**
   a. Environmental Protection Agency
   b. Earth Protection Agency
   c. Environmental Protection Association
   d. Early Pollution Action

2. **Water serving a private water system is**
   a. Provided by a municipal water supply
   b. Provided by a community water supply
   c. Extracted from a well
   d. Always considered safe to consume

3. **A water service can be described as the piping connecting a water source to**
   a. The water distribution system in a building
   b. A municipal water supply system
   c. A well system
   d. A specific fixture in a building

4. **The burial depth of a water service is measured from the finished grade of the soil to**
   a. The bottom of the pipe
   b. The top of the pipe
   c. The center of the pipe
   d. A water meter

5. **Some codes dictate that a water service installed perpendicular to a sewer must be**
   a. Installed above ground
   b. Installed 5′ away horizontally
   c. Sleeved
   d. Below the sewer

6. **The two main sources of potable water are**
   a. Groundwater and surface water
   b. Salt water and brackish water
   c. Leach fields and reclaimed water
   d. Polluted and contaminated water

7. **OSHA defines many excavated trenches as**
   a. Confined spaces
   b. Unregulated workspaces
   c. Non-threatening
   d. Above-ground spaces

8. **A 1-to-1 slope of a trench creates a(n)**
   a. Step-design trench
   b. 45° sloping trench
   c. Vertical wall trench
   d. 11° slope

9. **A water service connects to a well casing below ground with a**
   a. Water service well connector
   b. Male adapter
   c. Pitless adapter
   d. Female adapter

10. **Some codes allow a water service to be installed in the same trench as a sewer if it is a minimum of**
    a. 12″ above the top of the sewer
    b. 12″ below the bottom of the sewer
    c. 12″ to either side of the sewer
    d. 24″ below the sewer

11. **The minimum size water service allowable by most codes is**
    a. 12″
    b. 3/4″
    c. 1″
    d. 1-1/4″

12. **Water quality can be improved by purification and**
    a. Filtration
    b. Bottling
    c. Circulation
    d. None of the above are correct

13. **Three common non–health-related problems with some well water that can be improved by a filter are taste,**
    a. Color, and arsenic
    b. Toxins, and arsenic
    c. Odor, and color
    d. Color, and toxins

14. **The abbreviation SDWA stands for**
    a. Sanitation Department and Water Authority
    b. Sanitary Domestic Water Action
    c. Safe Drinking Water Act
    d. Safe Domestic Water Act

15. **Some codes dictate that a pipe trench must be backfilled and compacted with**
    a. 6" layers until 12" above the pipe
    b. 12" layers until 6" above the pipe
    c. 12" layers until 12" above the pipe
    d. 6" layers until 6" above the pipe

16. **A water service installed at the same elevation below ground as a sewer must be**
    a. Installed in a sleeve
    b. Separated from the sewer by 5' of compacted or undisturbed soil
    c. A minimum of 12" from the sewer
    d. Surrounded by concrete

17. **A municipality sells water to customers, and the number of gallons of water used is recorded with a**
    a. Water meter
    b. Gallons recorder
    c. Pressure regulator
    d. Gallons gauge

18. **Most codes dictate that a water distribution system in a home cannot be more than**
    a. 12" above the ground
    b. 80 psi
    c. 3' from the floor joist
    d. 180 psi

19. **Most codes dictate that a main isolation valve in a crawl space be**
    a. No farther than 3' from the crawl space door
    b. No closer than 3' from the ground
    c. No closer than 2' from the crawl space door
    d. No farther than 6" from the ground

20. **To provide access in and out of a pipe trench, OSHA may require**
    a. An extension ladder
    b. A scaffold
    c. A backhoe
    d. A rope

Chapter 12: Water Distribution Installation

You have learned about water service piping and how water is routed into a building. In this chapter, you will learn how to install a water system in a residential house. The type of material installed can dictate your layout approach, but the design intent of a building will be the same regardless of material type. An architect indicates where the plumbing fixtures are located, and a plumber provides hot and cold water to them. The type of fixture dictates where to install the water supply that connects to faucets, and plumbing codes dictate the minimum size of the piping and the routing through walls and ceilings. This chapter discusses installing a piping system and the basic sizing needed based on fixture requirements. Your plumbing career begins with manually installing piping and then might progress into designing piping systems. You will be introduced to the hands-on aspect of the plumbing trade and be reminded of the importance of focusing on safety while performing any task. Installing a water distribution system requires working from a ladder, using power tools and chemicals, and soldering copper pipe with a torch. Remember that safety is more important than productivity; an injured worker cannot be productive. The importance of water quality has already been covered. Recognize that you are responsible for protecting the health of anyone drinking water from piping you install.

OBJECTIVES

Upon completion of this chapter, the student should be able to:

- know correct techniques for installing a water distribution system.
- respect that a plumber installs water piping per code and protects water quality.
- know drilling and notching codes to ensure structural safety of a building.
- understand the rough-in aspects of a water supply to specific fixtures.

Glossary of Terms

brazing welding process used without flux to weld copper tube; also known as silver soldering

flux chemical paste to solder copper tube

joint fitting connection to a pipe

joist horizontal board that is part of a complete framing system; categorized as floor joist and ceiling joist

partition a non–load-bearing wall designed to separate rooms, not to support a structural load

silver solder metal filler used to braze copper tube; it has high silver content and is in stick form

solder metal filler supplied in roll form to solder copper tube; it cannot contain more than 2% lead

soldering process of welding copper tube using flux and solder

stud vertical board to create a wall; categorized as load-bearing and non–load-bearing

torch tool that is ignited and creates a flame to solder or braze copper tube

Layout and Sizing

The layout of a water distribution system varies based on the job site and the fixtures being served. Basic considerations and techniques discussed in this chapter demonstrate that creativity is essential in installing a water distribution system. Water piping is typically installed after the drainage and vent piping because the water piping system consumes less space in walls and ceilings and the code regulations are less strict. The layout process begins with locating where each fixture is installed, determining what the requirements are for each fixture, finding out where the water service enters the building, and determining the route of the main piping system. Once the main route is known, you can decide how the smaller branch pipes will connect or be routed. The rough-in and termination of the water piping is based on the specific fixture and often on company preference to increase productivity. In previous chapters, you learned about rough-in sheets to install piping based on the manufacturer's requirements. Most residential fixtures allow for the water piping to be installed without using rough-in sheets because most residential fixtures typically allow for minor deviations from or use of company preferences of heights and distances of the stub-out locations. Codes state that the hot water must be installed on the left side of a fixture unless otherwise directed by the fixture manufacturer.

Pipe Sizing

The sizing of a system depends on the quantity and type of fixtures being served. You will not be expected to know theoretical pipe sizing at this stage in your career, but you should know proper code book use and basic pipe-sizing principles. A code book indicates the minimum size pipe that can be routed to each type of fixture; when all calculations are finished, you will determine the actual pipe size to install. This process is simplified by plumbing code officials dictating the maximum number of fixtures that can be served on a specific pipe size. This removes the design aspect from the piping process on a job site. The design criteria for sizing the piping system is based on the maximum number of gallons per minute or per flushing cycle of a particular fixture and then calculated for all the fixtures combined. All residential fixtures are not used at the same time, so a residential piping system is designed differently than a commercial water distribution system. You must first understand the difference in the segments of a water distribution system before using the sizing charts in a code book. The three main segments are the main, branch main, and individual supply. An individual supply is a pipe serving one fixture; this is where a design approach is initiated. Calculating the requirements of a single fixture and then totaling the requirements for all fixtures served determines the flow requirements of the entire water distribution system. Figure 12–1 illustrates the three main segments of a water distri-

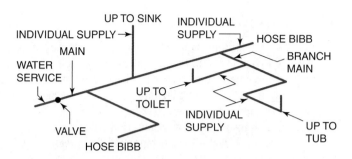

Figure 12–1 Three main segments of a water distribution piping system are the main, branch main, and individual supply.

Table 12–1 Common Individual Fixture Pipe Sizes

Sizes are based on International Plumbing Code

Fixture	Smallest Size
Bathtub[1]	1/2" (3/8")
Large Capacity Bathtub	1/2" (3/8")
Bidet	3/8"
Dishwasher[1]	1/2"
Hose Faucet	1/2"
Kitchen Sink[1]	1/2" (3/8")
Laundry Sink[1]	1/2" (3/8")
Lavatory Sink	3/8"
Shower with One Head[1]	1/2" (3/8)
Toilet[1]	1/2" (3/8")

[1] Sizes can be one pipe size smaller based on distance the pipe travels, pressure within the piping system, and whether it is installed as a manifold system in a parallel manner.

bution system. The sizes of individual fixture supply pipes are found in charts in a code book. Table 12–1 lists common residential fixtures and the minimum individual pipe size that can serve each one. There are many exceptions, so you must always read footnotes listed with the chart or table. If a fixture is not listed in a code book, use the manufacturer's recommendation for minimum pipe size. If a manufacturer dictates a larger minimum pipe size for a specific fixture or faucet than what is listed in a code book, the larger size is considered to be the minimum requirement.

 FROM EXPERIENCE

The sizes indicated in a code book are minimum standards; a plumber typically installs 1/2" pipe in walls even if a code book allows 3/8". The sizes indicated in a codebook that are smaller than 1/2" are for the final connection to a fixture.

Sizing Theory

Theory—determining the available water of a certain pipe size based on volume and pressure—plays a major role in pipe sizing. The outcome of the calculation is the number of gallons that can be delivered to adequately supply a specific fixture. A water supply fixture unit (wsfu) is the term used to describe the end result of a flow calculation. The pipe size determines how much water is in a pipe, and the pressure determines how fast that water can be delivered to a fixture. The theoretical basis of a wsfu is that 1 gallon per minute (gpm) equals 1 wsfu. The metric system is often used to describe water flow because it does not use fractions, which simplifies the mathematical process. One gallon equals 3.785 liters and one gpm is .0631 liters per second (3.785 ÷ 60 seconds).

Knowing theory is important when taking a plumbing exam or as your career develops into a management position. Although an apprentice and a plumber installing piping systems on a job site are not expected to know the theory of a water piping system, most training programs include some basic sizing information. The rate the water flows through the piping is known as velocity, which is measured in feet per second. The number of feet per second is how the gpm requirement is determined, but if the same pressure is exerted in a 1/2" pipe and a 1" pipe, the larger pipe will produce more gallons of water per minute. Doubling a pipe size does not provide twice the amount of water. Two 1/2" pipes do not have the same capacity as one 1", even though the mathematic approach indicates that would be true. A 1" pipe has the same capacity as four 1/2" pipes. The theoretic explanation derives from knowing that the square area of a circle is less than the square area of an equal size square. The circle only consumes 78-1/2-percent of that square. Pi (π) is the ratio of the circumference of a circle to its diameter. In mathematical formulas, its value is 3.1416. The 0.7854 (78-1/2%) used in the formula is derived by dividing pi by 4 (3.1416 ÷ 4 = 0.7854). Figure 12-2 illustrates the square area of a circle compared with that of a square. Another widely used term in the pipe-sizing process is the cross-sectional area of a pipe. The term *cross sectional* refers to diameter because a pipe is circular and does not have a length and a width. Cross-sectional information is used more often when a code refers to an effective opening of a drain. This will be discussed later in the book when it is relevant.

FROM EXPERIENCE

Theory is important for taking a plumbing exam, but most plumbers on a job site use codes to size pipe and do not perform sizing calculations. The pressure of the piping system, the type of pipe used, and the total length of pipe and number of fittings installed play a major role in the actual gpm that flow through a pipe. The velocity of the water determines the acceptable gpm per code. A code book lists the minimum gpm requirements of specific fixtures. If this theoretic information intrigues you, and you would like to learn more about the engineering process, refer to a code book or professional sizing manual for in-depth explanations of the complex procedures involved.

See page 305 for a step-by-step procedure to determine the volume of different pipe sizes.

Job Site Sizing

A plumber must know the minimum pipe size that can serve a specific fixture based on codes, manufacturer recommendations, and company preference. However, the pipe actually installed on a job may be larger than the minimum standards to increase productivity. Another factor is the type of connections to valves located beneath a fixture. Both a toilet and a sink use an isolation valve known as a stop. The most common stop connects 1/2" (5/8" OD) pipe to 3/8" OD tubing. The 1/2" pipe stubs out the wall or through a floor, and the 3/8" portion of the stop connects to the fixture with supply tubing. The size of the stop to be installed is often more important than code in determining the size of the rough-in water pipe. Some codes allow two fixtures, each with a 1/2" or less individual supply pipe size requirement, to be served by a 1/2" branch main; other codes allow three fixtures on a 1/2" pipe. Many companies install 3/4" pipe up the last branch connecting two different fixtures to eliminate any potential volume problems. Figure 12-3 illustrates common piping installations for residential plumbing fixtures.

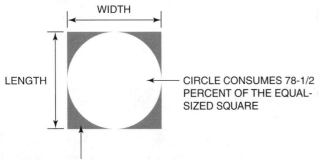

Figure 12-2 A circle consumes 78-1/2 percent of an equal sized square.

FROM EXPERIENCE

A code book indicates minimum sizing, but company preference often dictates the installation of larger piping.

CHAPTER 12 Water Distribution Installation

COLD WATER DISTRIBUTION SYSTEM TO FULL BATHROOM

COLD WATER SUPPLY TO HALF BATHROOM

(A)

(B)

Figure 12-3 Job-site pipe sizing varies based on local codes and company preference.

The specific fixture dictates the termination point of an individual supply pipe. A faucet serving a shower and tub terminates with its connection to the faucet. A sink, bidet, and toilet have exposed termination points with stops installed below the fixture. A washing machine and icemaker terminate to a dedicated outlet box, and the appliance connects to the isolation valves within the box. A dishwasher usually uses the same water supply that serves the kitchen sink; a 3/8" OD tubing is routed under the sink to the dishwasher. Most codes dictate that the individual supply piping must be routed to a maximum of 30" from the fixture it serves to ensure adequate water flow to the fixture. The pipe size for a toilet and sink is reduced from 1/2" to 3/8" at the isolation stop below the fixture, which explains why the 30" maximum distance is important. Figure 12-4 illustrates a typical kitchen sink water supply system serving a kitchen faucet and dishwasher.

FROM EXPERIENCE

The actual piping arrangement below a kitchen sink is determined on a job site based on conditions and fixture layout.

Wall Layout

The wall rough-in for a water distribution system is determined by the fixture type and the job site. There are standard dimensions for typical fixtures, but company preferences vary. A plumber makes sure the piping that stubs out a wall accommodates the fixture being served. A sink installed in a countertop offers more flexibility as to where

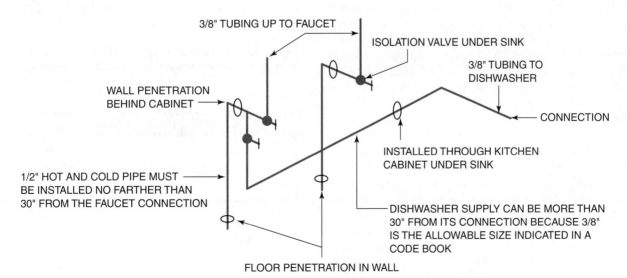

Figure 12-4 A dishwasher water supply typically originates below a kitchen sink.

Table 12-2 Common Fixture Water Pipe Requirements

Fixture	Hot	Cold	Stub Out
Bathtub[1]	Yes	Yes	No[1]
Icemaker	No	Yes	No
Kitchen Sink	Yes	Yes	Yes
Laundry Sink	Yes	Yes	Yes
Lavatory	Yes	Yes	Yes
Shower	Yes	Yes	No[2]
Toilet	No	Yes	Yes
Washing Machine	Yes	Yes	No

[1] Large capacity and whirlpool tubs often use deck-mounted faucets and do not use in-wall piping

[2] Bathtub and shower have a shower head and tub spout stub out, but water is connected to faucet in wall

a pipe terminates than does a stub out serving an exposed fixture connection. Because water flows through the piping system under pressure, the piping can be routed in any position to accommodate specific job conditions. Drainage piping, on the other hand, must be installed with specific fittings and in certain positions to allow the water to flow by gravity. The water piping system is typically installed after the drainage and vent system because it has less critical installation requirements. Not all fixtures require a hot and cold water supply, nor do they all require stub out pipes. Table 12–2 lists common fixtures and their specific requirements.

Toilets

A plumber installs the in-wall piping for a toilet based on the location of the center of the toilet, with the standard dimension being 6" to the left of the toilet center. The stub-out location is standard for most toilets. If a one-piece toilet is being installed, the plumber must use a rough-in sheet provided by the manufacturer. Two-piece toilets with a separate tank and bowl allow for a slight variation as to where the stub out is installed. A company may want specific rough-in heights from a floor, but the stub-out location on the left side of a toilet is considered industry standard. A rough-in occurs before any flooring material is installed. If the floor is being tiled, the plumber must install the water piping higher to accommodate the tile. The wood trim known as base molding is higher in many custom homes than spec homes. If the plumber installs the stub out too low, the piping conflicts with the base molding. If the stub out is too high, the final connection to the toilet may be impossible. A water piping wall layout may be on a lower floor than the toilet if the water supply serving it is installed through the floor. This installation is rare in new construction, but is often used in renovations. Figure 12–5 illustrates a standard water and floor stub out for a two-piece toilet.

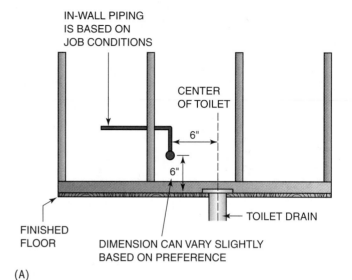

(A)

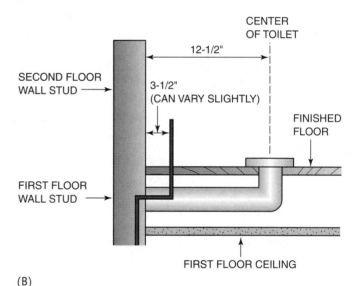

(B)

Figure 12–5 A two-piece toilet has a standard stub-out location and varies in height installed above the floor.

Stub-out pipes typically extend 6" from the wall or above the floor.

Lavatory

A lavatory installed in a countertop has different requirements than a pedestal lavatory. The area inside the sink base cabinet is not visible, so a plumber can install the stub

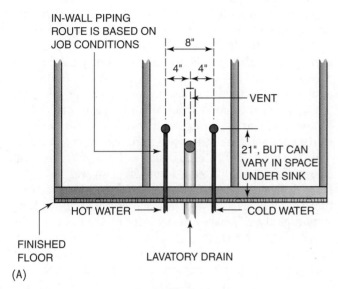

(A)

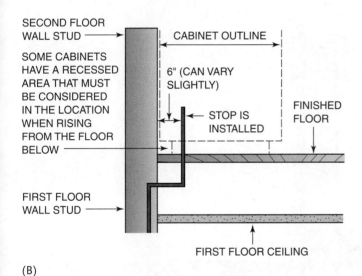

(B)

Figure 12–6 A lavatory installed in a countertop allows for more variable stub-out locations.

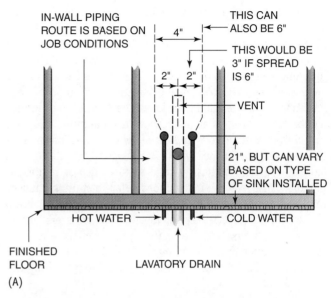

(A)

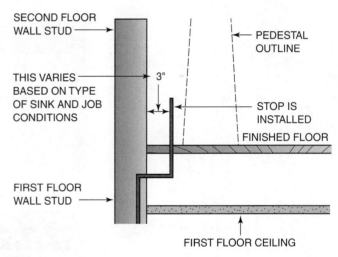

(B)

Figure 12–7 A pedestal sink requires that the stub outs are more exact to create a professional appearance.

out without being exact. A pedestal sink has visible piping below the sink. A plumber must install the stub outs at the same height and a specific distance apart from the center of the pedestal sink. The water stub out is typically centered with an 8″ spread for a cabinet-type sink. The stub out for a pedestal sink usually has a 4″ or 6″ spread, so the isolation stops and flexible faucet tubing are more visually appealing. The maximum height for most lavatory water piping regardless of the type is 21″ above the finished floor. As with many vertical dimensions, company preference and specific fixture types may vary. Figure 12–6 shows a wall and floor stub out for a countertop-type lavatory. Figure 12–7 illustrates a wall and floor stub out for a pedestal lavatory.

 FROM EXPERIENCE

The height above the floor that a stub out is installed is often determined by company preference. Many companies locate the water piping at about 14″ for a countertop lavatory. This requires longer supply tubes, but the isolation valves do not conflict with the drain assembly.

Bathtubs

Most bathtubs are used for bathing and showering. As shown in Figure 12-8, using a combination tub and shower (T&S) faucet does not change the sizing requirements for the individual fixture supply. The faucet and associated piping for a one-piece tub and shower unit as well as the unit itself are installed in the wall during the rough-in phase of construction. This lets a plumber know the installation requirements for a particular faucet. A manufacturer provides installation instructions indicating where the faucet is in relation to the finished wall. Many walls that surround a bathtub are tiled or have a solid surface such as marble. A plumber must read the instructions before installing a faucet in a wall during a rough-in phase to avoid having to access the wall for corrective action during the trim-out phase. Figure 12-10 illustrates a plastic protector with a wall-thickness gauge that is used on tub and shower faucets. The height of a shower head and tub spout is determined by the specific installation. Most companies use a standard rough-in height, but preference can determine the actual installation.

Water piping routed from a floor below can be within the wall or through the floor. An open space on the floor is created when a tub is installed. If the wall is located directly over a floor joist, the holes for the water piping can be drilled through the floor and the piping can be offset in the wall under the tub. A tub spout must be far enough above the height of the tub to avoid violating the air gap code. A shower head must be located so it does not conflict with the top layer of wall tile or the top of a one-piece tub and shower unit. Large-capacity bathtubs, such as a garden (roman) tub or a whirlpool tub, often do not use a shower head. The faucet is frequently installed on a platform that supports the tub; this is known as a deck-mounted installation. Many large-capacity tubs have a faucet ledge on the tub to receive a faucet. The tub and faucet are typically installed during the rough-in phase. If a large-capacity tub will be installed during the trim-out phase, the water supply is routed to the area

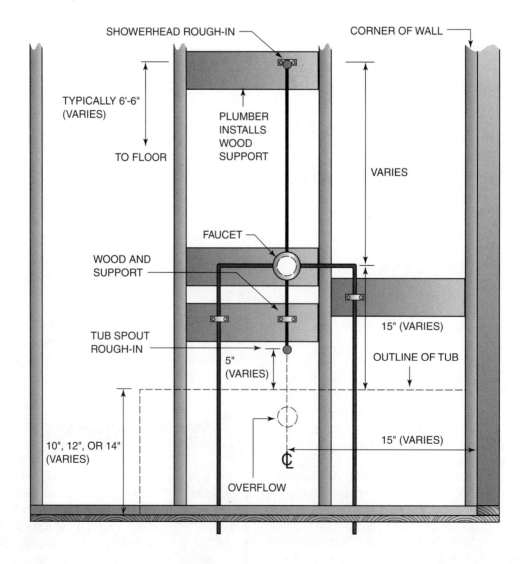

(A)

Figure 12-8 A tub and shower faucet installation varies based on fixture type and company preference.

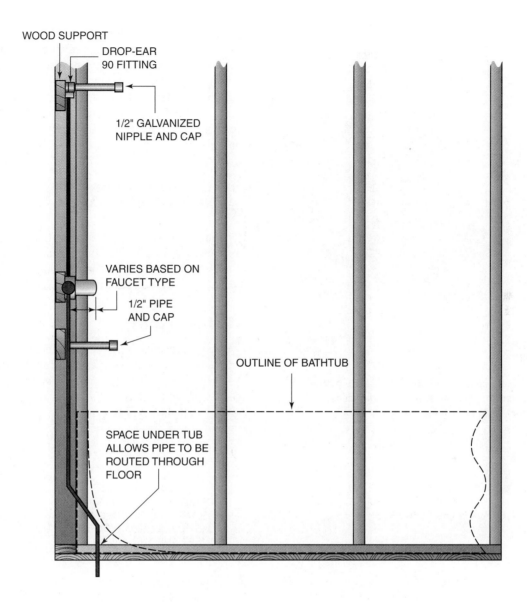

(B)

Figure 12–8 Continued

where the faucet will be installed. Handicap codes regulate specific heights and distances from walls and placement of handheld spray units. A plumber must refer to a code book for handicap installations. Figure 12–8 shows a tub and shower faucet installation. Figure 12–9 illustrates a water pipe routing through a floor serving a large-capacity tub.

 FROM EXPERIENCE

A tub and shower faucet is typically installed in the center of the bathtub, but master bathrooms often have customized designs. Handicap tub and shower faucets are not always centered; often they are located on the side wall opposite the drain location.

Showers

A shower faucet is similar to a tub and shower faucet but without the tub spout connection. A shower faucet installation is the same as the tub and shower faucet illustrated in Figure 12–8, with the faucet style determining the actual installation. The shower head height is usually 6'6" from the finished floor of the shower. The distance of the faucet above the floor varies depending on company preference, but generally ranges from 3' to 4' from the finished floor of the shower. A tiled shower requires a plumber to know the finished wall thickness from the wall **studs.** A faucet has a plastic protector and wall thickness gauge to protect the faucet finish and to indicate where to install the faucet based on various job-site conditions. The center of the faucet is usually centered in the shower. One-piece shower units may have a dedicated area to install the faucet. Handicap

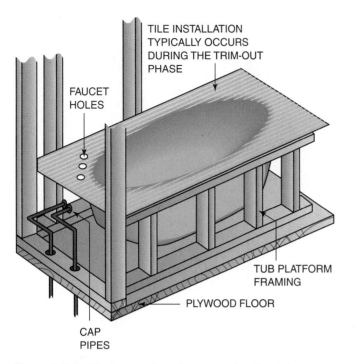

Figure 12-9 A large-capacity tub may require the piping system to be routed to a general area where the faucet installation occurs during a trim-out phase.

showers have factory-installed grab bars with the faucet placed in a dedicated area above them. Some showers have handheld spray units and multiple shower head features that dictate the placement of flow-control devices and valves. The actual installation is based on job-site conditions and manufacturer's instructions. Figure 12–10 shows a plastic protector and wall thickness gauge for a typical single-handle shower faucet.

 FROM EXPERIENCE

Handicap shower faucets are installed in different locations depending on the type of shower. A plumber must refer to a code book for specific requirements.

Kitchen Sinks

Water piping that serves a kitchen sink is similar to that in the countertop lavatory installation illustrated in Figure 12–6. The piping configuration under the sink depends on specific job conditions and whether or not a dishwasher is being installed. A typical configuration is illustrated in Fig-

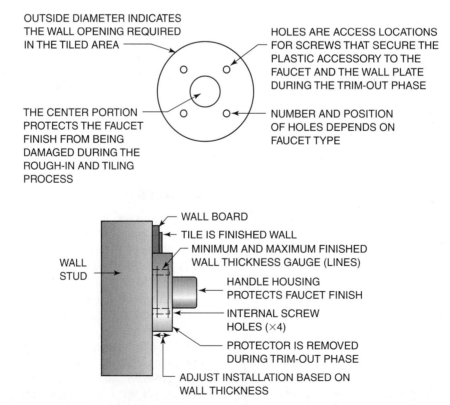

Figure 12-10 A plastic protector and wall-thickness gauge accessory protects the faucet finish and indicates the faucet installation in a wall.

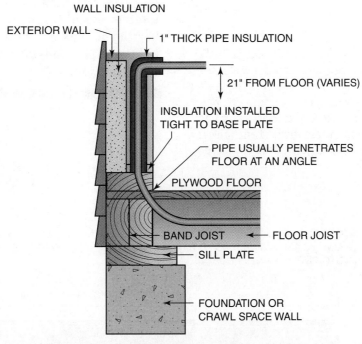

Figure 12–11 Water piping installed in an exterior wall must be protected against freezing.

ure 12-4. Most kitchen sinks are installed along an exterior wall and centered beneath a window. Floor penetrations can be used to route the water piping vertically into the cabinet if there is a crawl space or basement below the sink cabinet. The insulation requirements of exterior walls vary in different regions of the country, but most plumbing codes dictate that the water piping must be installed on the interior side of the wall insulation. At least 1" of insulation must be installed over the pipe according to most plumbing codes. Figure 12-11 illustrates a water pipe installation in an exterior wall.

 FROM EXPERIENCE

The outside diameter of an insulated pipe must remain inside the wall for the wallboard to be installed; therefore, a plumber must know how thick the insulation is.

Drilling and Notching

Drilling and notching codes vary extensively by region of the country and often by state or local area, depending on climate conditions. States set minimum standards, and local governments adopt ordinances to strengthen state codes. A plumber must know local regulations to avoid expensive replacement of wooden structural boards in a building. Drilling through wood boards can weaken them, and positioning or sizing a hole incorrectly can create an unsafe condition. The possibility of snowfall, hurricanes, tornados, and earthquakes are main reasons for variances in drilling and notching codes. Determining a pipe route depends, in part, on drilling and notching codes. A plumber selects the location and size of a hole based on local regulations. The information here is not intended to explain your local codes, but it will help you interpret them. It is essential to know whether structural wooden boards are load bearing to correctly determine the allowable sized holes that can be drilled. Notching is making a cut on the edge of a wooden board; the depth of the cut is strictly regulated and is determined by the area of the joist being notched. Holes and notches that place the outside diameter of a pipe close to the edge of a stud, joist, or plate must be protected against nails and screws. Most codes require that a nail plate (stud guard) be installed wherever a 1-1/2" nail or screw could penetrate the piping. A nail plate is a 1/16" thick steel plate that comes in different lengths and widths. Piping installed horizontally through a wall stud requires only a 1-1/2" wide nail plate, but pipes passing through a top or bottom plate vertically must be protected with a 4" wall plate. Some local codes specify the kind of nail plates that can be used. That information is listed in a local code book. It is important to understand how drilling and notching codes protect the inhabitants of a house. Plumbing and building inspectors focus on these areas in a job-site walkthrough, and violations are taken very seriously.

Walls

The two wall types commonly described in a code book are load-bearing and non–load-bearing. Exterior walls are load-bearing and are typically regulated like interior load-bearing walls. More stringent codes may have different regulations for exterior walls. Some interior walls are classified as **partitions** and are regulated differently and with fewer restrictions than load-bearing walls. Figure 12–12 shows the basic wall types.

CAUTION: The allowable span of a joist determines whether a wall installed within the span is load-bearing or non–load-bearing. Never assume that a wall stud is non–load-bearing without proper identification.

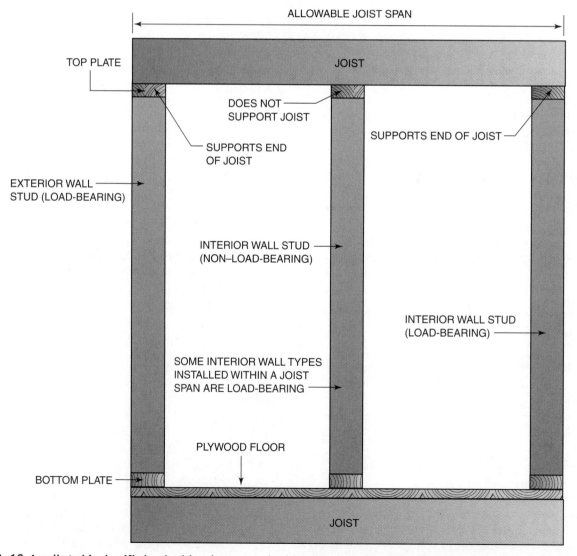

Figure 12–12 A wall stud is classified as load-bearing or non–load-bearing; each has different drill and notching regulations.

FROM EXPERIENCE

A main consideration in determining a layout is to create a safe working space in which to drill holes. The easiest pipe route may not be the safest one.

The two common sizes of wall studs are 2″ × 4″ and 2″ × 6″. The sizes used to indicate wall widths are nominal and not the exact dimensions. As with pipe, nominal sizes of wood show the intent of a design. A plumber must know the actual board sizes to determine the maximum size hole to drill or notch to cut based on the percentages that can be removed. The actual size of a 2″ × 4″ board is 1-1/2″ × 3-1/2″, and the actual size of a 2″ × 6″ board is 1-1/2″ × 5-1/2″. This is extremely important to understand to avoid creating an unsafe structure. The mathematical process that determines the drill bit or notch size usually results in numerical values in decimal form that do not match actual drill bit sizes. Changing these decimal values to useable values is known as rounding off. Most dimensions can be rounded up or down but, for safety reasons, these values must be rounded down to the nearest lower dimension to make sure safe drilling and notching procedures are followed. If a code dictates a certain percentage and the numerical value is 6-5/8″, a plumber must round the result to 6-1/2″, not 6-3/4″. Determining the percent of a hole diameter or depth of a cut requires basic conversion skills among percentages, fractions, and decimals. Table 12–3 lists mathematical comparisons of equivalents.

FROM EXPERIENCE

The width of a board that is installed vertically, such as a wall stud, is considered its depth when it is installed horizontally, such as a joist.

See page 307 for a step-by-step procedure for determining percentages of a hole diameter and notch.

The location of a drilled hole or notched area is regulated to protect the structural integrity of a wall stud. A notch and a hole cannot be located in the same area, and a hole cannot protrude into a 5/8″ protected area on either edge of a wall stud (Fig. 12–13). Some codes state that 40 percent of a load-bearing stud and 60 percent of a non–load-bearing stud can be drilled (Fig. 12–14 and Fig. 12–15). Other codes allow a load-bearing stud to be drilled 60 percent if the stud is double, but not more than two consecutive doubled studs can be drilled (Fig. 12–16). Still other codes state that a maximum of 25 percent of a load-bearing stud and 40 percent of a non–load-bearing stud can be notched (Fig. 12–17). Table 12–4 lists load-bearing and non–load-bearing 2″ × 4″

Table 12–3 Mathematical Equivalents

Percentage	Decimal	Fraction	Useable
10	0.10	0.8/8″	1/16″
20	0.20	3.2/16″	3/16″
25	0.25	1/4″	1/4″
30	0.30	4.8/16″	1/4″
40	0.40	6.4/16″	3/8″
50	0.50	1/2″	1/2″
60	0.60	4.8/8″	1/2″
70	0.70	11.2/16″	11/16″
75	0.75	3/4″	3/4″
80	0.80	12.8/16″	3/4″
90	0.90	14.4/16″	7/8″
100	1.00	4/4″	1″

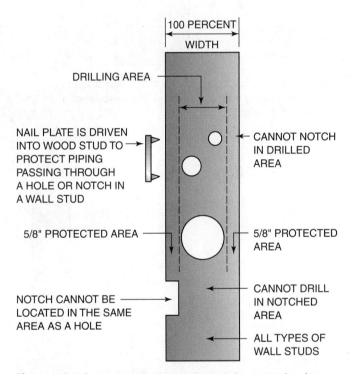

Figure 12–13 Most codes dictate that a hole cannot be closer than 5/8″ from the edge of a wall stud or located in the same area as a notch.

and 2″ × 6″ wall studs and the maximum allowable hole sizes based on information in this book.

CAUTION: The drilling and notching information in this book is not intended for use on a job site. Know your local codes; this information is only to help you interpret your local codes.

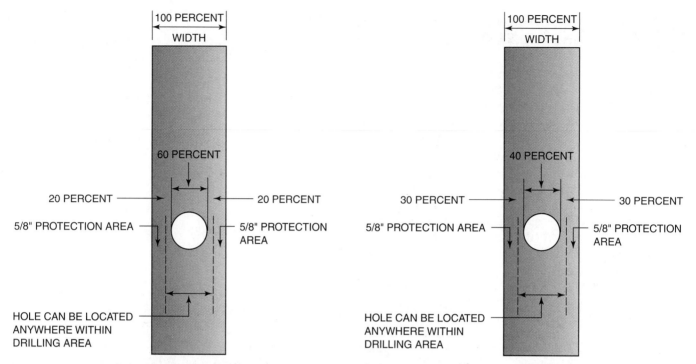

Figure 12-14 Most codes dictate that a hole drilled in a non–load-bearing wall stud can consume a maximum of 60 percent of the width.

Figure 12-15 Most codes dictate that a hole drilled in a load-bearing wall stud can consume a maximum of 40 percent of the width.

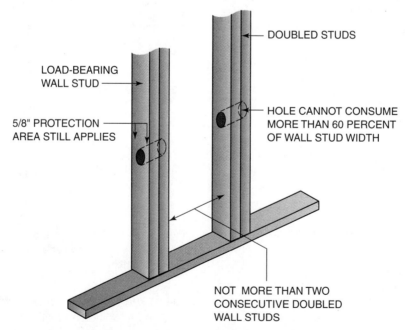

Figure 12-16 Most codes allow a hole drilled through a load-bearing stud to consume 60 percent of the width provided the stud is doubled and no more than two consecutive double studs are drilled.

Joists

A **joist** is a horizontal structural support classified as either a floor joist or a ceiling joist. The joists that separate different levels of a house are floor joists, and ceiling joists are the structural boards of the highest ceiling in a house. The size of a joist depends on the design intent and the structural load of a particular area. Two common sizes are 2″ × 8″ and 2″ × 10″, but a 2″ × 12″ is used for larger build-

Table 12-4 Maximum Hole Sizes Based on Wall Type and Width

CAUTION: These values are for lesson purposes only and are not intended to reflect your local codes.

Board Size	Stud Type	Maximum Hole	Decimal Value	Notes
2" × 4"	NLB Stud	2-1/16"	2.10"	Maximum of 60% of its width
2" × 4"	DLB Stud	2-1/16"	2.10"	Maximum of 60% of its width
2" × 4"	SLB Stud	1-3/8"	1.40"	Maximum of 40% of its width
2" × 6"	NLB Stud	3-1/4"	3.30"	Maximum of 60% of its width
2" × 6"	DLB Stud	3-1/4"	3.30"	Maximum of 60% of its width
2" × 6"	SLB Stud	2-1/8"	2.20"	Maximum of 40% of its width

Abbreviation List
NLB = Non–Load-Bearing Wall Stud
DLB = Doubled Load-Bearing Wall Stud
SLB = Single Load-Bearing Stud

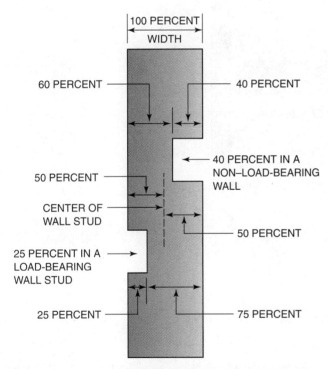

Figure 12-17 Most codes allow 25 percent of a load-bearing and 40 percent of a non–load-bearing stud to be notched.

ings or for greater span requirements. The actual dimension of a 2" × 8" is 1-1/2" × 7-1/2"; the actual dimension of a 2" × 10" is 1-1/2" × 9-1/2"; and the actual dimension of a 2" × 12" is 1-1/2" × 11-1/2".

Because a joist has a different structural purpose than a wall stud, it also has different drilling and notching code regulations. Some codes do not allow any holes to be drilled in the middle one-third of a span; other codes require a joist to be reinforced with exterior grade plywood if a large-diameter hole is drilled through it. Some codes do not allow a hole to be drilled in a floor joist in the area that has a load-bearing wall directly above it. Thorough code knowledge is required before drilling and notching. A joist can be difficult and expensive to replace, and an inspector can demand that a company provide a certification from a structural engineer for a joist that has been altered. The dimension of a board known as width for a wall stud is called depth for a joist. The length of a wall stud is known as the span of a joist. Figure 12-18 illustrates the difference between a floor joist and a ceiling joist and the defining terms used to relate codes to depth and span.

 FROM EXPERIENCE

A floor joist is a load-bearing structural board. Often in a residential house, several load-bearing floor joists are joined together to create a beam. Never drill or notch a beam without prior approval and careful code review.

A carpenter often notches the ends of a joist during the framing phase of construction. A plumber might also have to notch a joist to install piping that conflicts with the joist, but strict codes dictate the maximum depth of the notch. Some codes dictate that the end of a joist cannot be notched more than 25 percent of its depth. Other codes allow notching at the bottom and top of a joist as long as the notch does not remove more than 1/6 of the depth of the joist. Still other codes do not allow any notching in the middle one-third of a joist. According to some codes, holes cannot be drilled in a 2" protected area around the top and bottom edges of a joist. The 2" protected area also typically includes the area where a joist contacts the load-bearing wall or other vertical support. Where holes are allowable in a joist, some codes dictate that a hole cannot consume more than 1/3 of the joist depth. Figure 12-19 illustrates a joist with notching regulations. Figure 12-20 illustrates the protected areas of a joist pertaining to drilling holes.

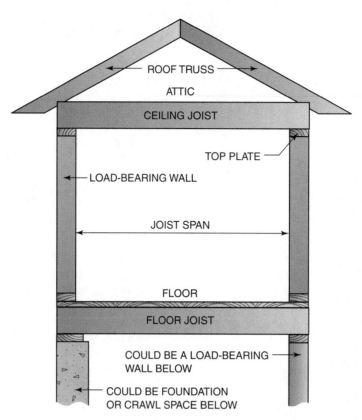

Figure 12–18 A joist depth is also known as the width of a board, and its span is known as its length.

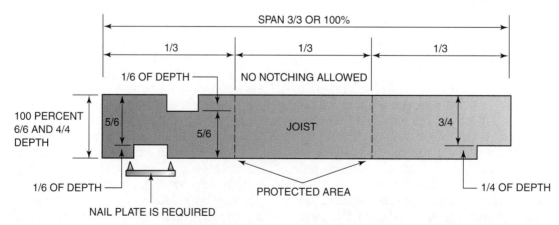

Figure 12–19 The depth of a joist is also the width if installed in the vertical position; the depth of a notched area is regulated and varies based on local codes.

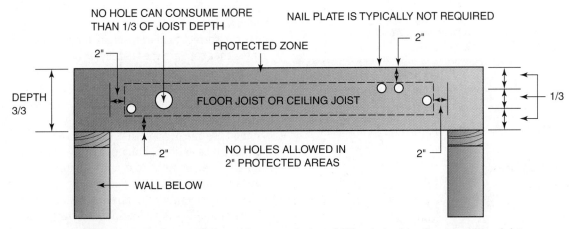

Figure 12-20 Some codes dictate that a 2" protected zone be created when drilling holes in a floor or ceiling joist.

Table 12-5 Maximum Hole Sizes Based on Joist Depth

CAUTION: These values are for lesson purposes only and are not intended to reflect your local codes.

Board Size	Stud or Joist	Maximum Hole	Decimal Value	Notes
2" × 6"	Joist	1-3/4"	1.83"	Maximum of 1/3 of its depth
2" × 8"	Joist	2-1/2"	2.50"	Maximum of 1/3 of its depth
2" × 10"	Joist	3-1/8"	3.17"	Maximum of 1/3 of its depth
2" × 12"	Joist	3-3/4"	3.83"	Maximum of 1/3 of its depth

Table 12-5 lists 2" × 8", 2" × 10", and 2" × 12" joists indicating the maximum allowable hole sizes based on information in this book.

 FROM EXPERIENCE

Calculators are available to convert decimals to fractions that recognize dimensions in feet and inches.

Hangers and Supports

Residential plumbing systems use very few types of hangers and supports compared with commercial piping installations. Supports are usually attached to wooden boards with nails and screws. Copper tubing uses copper-plated hangers and supports and flexible tubing such as PEX uses plastic materials. Plastic pipe products such as CPVC and PVC are more rigid than flexible tubing, and, if allowable by local codes, can typically withstand metal supports without being damaged. A plumber often installs additional wood boards so that hangers and supports can be installed where desired or where dictated by code. Maximum spacing between hangers and supports is also dictated by codes, and different materials have different spacing requirements. Pipe routes are often determined based on the most practical or productive method of installing the required number of supports. The location of piping can determine when support is needed. Water piping stub outs require a support bracket or other adequate means of support to eliminate the possibility of damage to the piping when wallboard or cabinets are installed. Earthquake-prone regions are regulated by seismic codes and may require special methods and spacing allowances to support piping systems. A plumber must be knowledgeable about specialized codes. Table 12-6 lists common types of residential water piping and the maximum spacing needed between hangers or support. Codes vary, and the information in Table 12-6 does not reflect spacing code for your local area.

Types

Residential hanger selection is based primarily on the piping material installed. Several hanger and support types can be used with all material types. Many manufacturers offer several hanger designs to accommodate plastic and metal

Table 12-6 Residential Water Distribution Piping Maximum Hanger Spacing

CAUTION: These values are for lesson purposes only and are not intended to reflect your local codes.

Pipe Type	Vertical Installations	Horizontal Installation
Brass	10'	4'
Copper 1-1/4" OD and smaller	10'	6'
CPVC 1" and smaller	10'[1]	3'
CPVC 1-1/4" and larger	10'[1]	4'
Galvanized Steel	15'	12'
PEX	10'[1]	32"
PVC	10'[1]	4'

[1] 2" and smaller pipe must have additional support half the distance indicated when installed vertically through a floor (mid-story support)

pipe. A clevis type and an adjustable swivel-ring type hanger are available copper plated for installing copper tubing and rubber coated for installing plastic tubing. It is important to use the correct hanger to avoid damage to the pipe. Code dictates that copper pipe that is not insulated must be installed with a copper plated hanger to prevent electrolysis (corrosion). Whether the pipe is horizontal or vertical is also a deciding factor in the type of hanger selected. Some hangers and supports can be used in both positions; others are designed for one or the other. Location is another determining factor for selecting a hanger. A split-ring hanger can be used for all pipe positions, but is used more often when a pipe is installed along a wall or floor.

Residential water distribution systems are mostly installed in walls and floors and suspended from joists. The walls and joists that the pipes pass through can serve as support provided the distance between joists and walls does not exceed the maximum distance allowable by code. Hangers that increase productivity are more common than ones that use accessories, such as threaded rod, concrete anchors, or an-

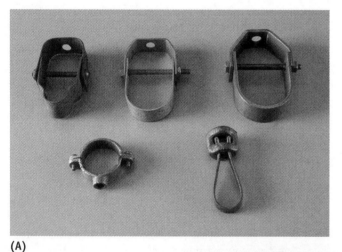

(A)

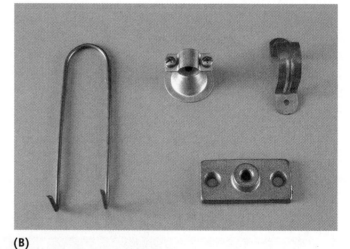

(B)

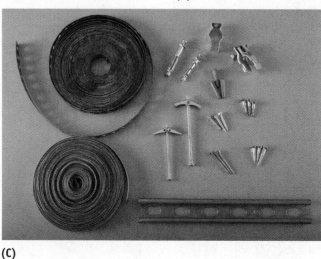

(C)

Figure 12-21 Various types of hangers, supports, and anchors are available for supporting piping systems.

choring plates. Vertical residential piping installed through each floor typically does not need a support known as a riser clamp, but vertical support is required based on hanger spacing codes. The most common hanger used by a residential plumber is a clip often referred to as a strap. Plastic tubing needs a plastic securing clip, and copper uses a copper-plated clip. The typical plastic clip has a single nail to drive into the wood with a hammer and provides a productive installation. A typical copper clip has two holes and requires screws or nails to secure the clip to the wood structure.

Company preference can be a deciding factor in hanger selection; some companies use wood to support piping. Wood can be used to secure wall stub-out piping. A copper-plated stub-out bracket is available to secure piping exiting a wall to serve a fixture. The copper tube that is inserted into the holes of a stub-out bracket is soldered to the bracket to provide pipe stability. The actual method used to support piping in various locations and positions depends on the specific installation. The most important aspects of supporting piping are adherence to codes, productive installations, and completing an installation in a safe and quality manner. Figure 12–21 shows numerous types of hangers and supports. Some are more common for residential construction than others, but it is important to recognize the various types. Figures 12–22 through Figure 12–33 illustrate various uses and installation methods for several of the hangers and supports from Figure 12–21.

FROM EXPERIENCE

Various hangers are used in the commercial and industrial area of the plumbing trade. A plumber can obtain manufacturer catalogs at local plumbing wholesale stores to become acquainted with the various types.

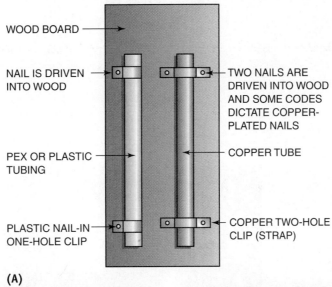

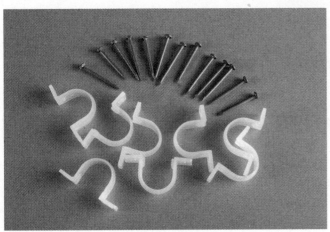

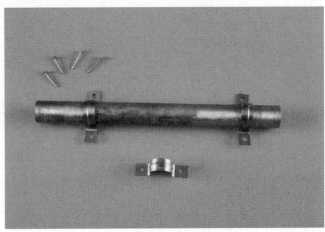

Figure 12–22 Clips and straps support lightweight piping to wood structure.

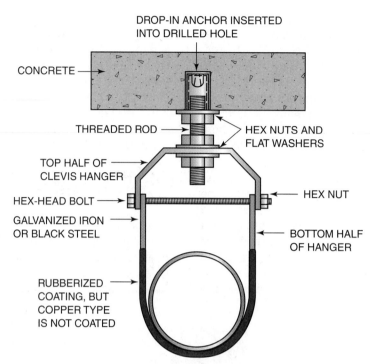

Figure 12-23 A concrete drop-in anchor is a common attachment method; a clevis hanger is a common type for installing horizontal piping.

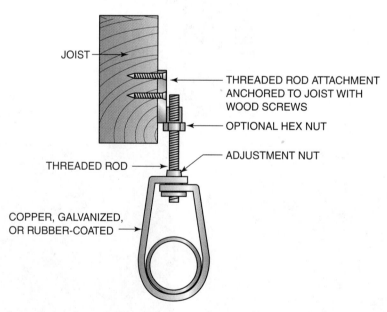

Figure 12-24 A threaded rod attachment is used for wood structures; an adjustable swivel ring hanger is another common type used for installing horizontal piping.

CHAPTER 12 Water Distribution Installation **291**

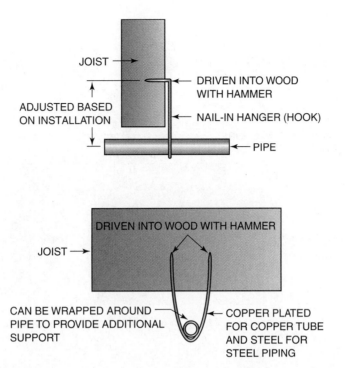

Figure 12–25 A nail-in hanger, also known as a hook, provides a productive installation for lightweight pipe.

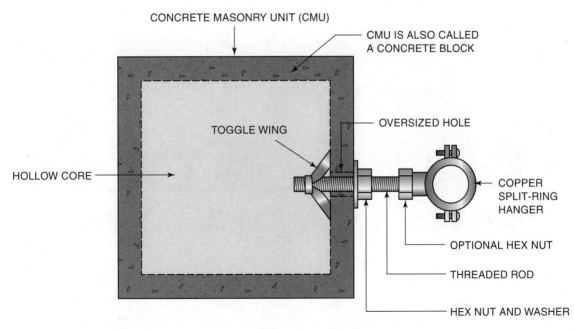

Figure 12–26 A toggle bolt and threaded rod assembly can be used to install a split-ring hanger to a concrete masonry unit.

292 SECTION THREE *Layout and Installation*

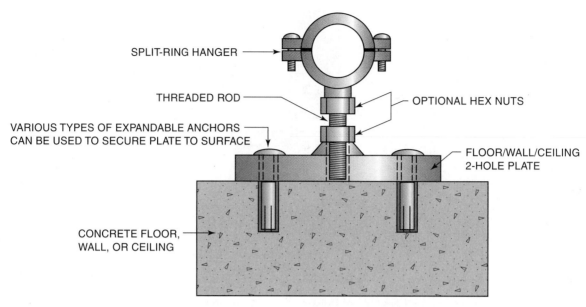

Figure 12-27 A 2-hole plate and various expandable anchors can be used to install piping along a floor, wall, or ceiling.

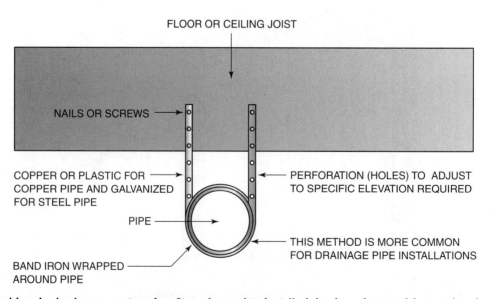

Figure 12-28 Band iron is also known as strapping. It can be used to install piping in various positions and anchors to the structure with screws or nails.

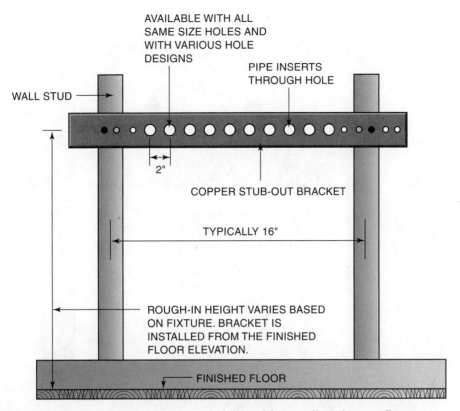

Figure 12-29 A copper stub-out bracket can secure horizontal piping exiting a wall serving as a fixture.

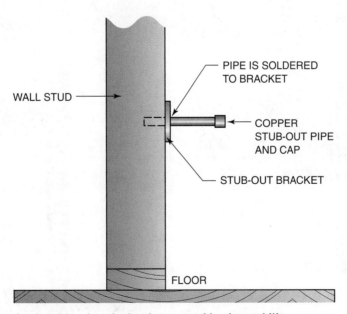

Figure 12-30 A copper stub-out pipe is soldered to the bracket to provide pipe stability.

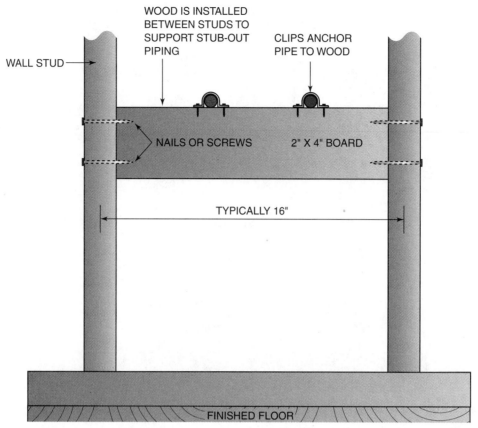

Figure 12–31 A 2″ × 4″ wood board can be used to support stub-out piping serving a fixture instead of a manufactured stub-out bracket.

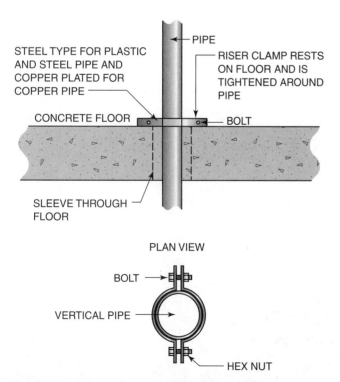

Figure 12–32 A riser clamp is installed to support a vertical pipe penetration through a floor but may not be required by local codes in residential construction.

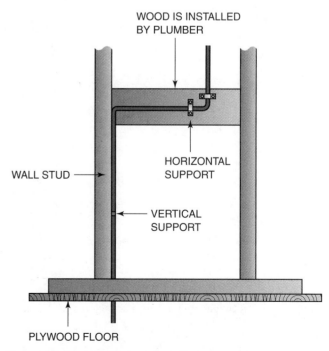

Figure 12–33 Wood is installed where required to support piping at necessary intervals as dictated by code or specific installation location.

Compounds and Sealants

Compounds and sealants are used extensively when connecting threaded and soldered pipe fittings to create a watertight joint. A compound used for threaded connections, known as pipe dope, is applied only to the external portion of a male thread before screwing it into a compatible female threaded fitting, A sealant known as Teflon tape that provides the same watertight connection as pipe dope is also used for threaded connections. **Flux** is a paste used for **soldering** copper tube. It is placed in a fitting socket and on the external portion of the tube, so the solder can flow into a fitting socket. **Solder** is used to weld a copper fitting connection. Its melting point varies based on the percentage of alloys used in the manufacturing process. Soft-soldering for welding is sold in roll form. Another welding process known as brazing or **silver soldering** is used for copper connections. It employs a solder that is in stick (rod) form that has a higher silver content than most types of soft-solder. **Brazing** is performed without flux and at higher temperatures. Soft copper tubing can be connected to another copper tube or to a piece of equipment with a flared connection. A flaring tool is used to fold or spread copper tubing to mate with a compatible manufactured flared fitting. The fitting design and flaring tool have compatible flare angles to ensure that a proper flare is created. Plastic piping uses a solvent-welding process that requires a cleaner, glue, and primer to create a permanent connection. Each of these basic connections is unique; a plumber can connect different types of piping with threaded connections. Pipe dope is used extensively with metal threads, and Teflon tape is used for plastic and metal threads.

Material Safety Data Sheets

A Material Safety Data Sheet (MSDS) must be available upon request from an employer or wholesaler. A person working with any product must know its dangers to ensure safe use and avoid possible health concerns. It is also important to know about emergency treatment in case it is needed. When soldering, fumes are expelled by the heated copper pipe, flux, and solder into the work area. PVC solvents present unique health and safety concerns and are extremely flammable and explosive. If your skin comes into contact with most chemicals, it presents health concerns. An MSDS identifies the dangers, precautions, safety recommendations, and first aid treatment for each of these concerns. A plumber should keep all MSDS information in a book or file in case of emergency and must know where the information is located at all times.

Sealants

Pipe dope is available in many mixtures depending on the manufacturer. Not all pipe dope is suitable for use with pipes carrying drinking water. A code book lists current standards for the lead content and types approved for your local area. The kinds usually used by plumbers are sold in a can with a brush attached to the inner portion of the cap. Some homeowner supply stores sell pipe dope in a squeeze tube, which is typically more suitable for a homeowner than a contractor. A plumber must know the proper use of all compounds. Pipe dope should never be applied to the inside of a female-threaded fitting. Using too much pipe dope or applying it inside a fitting can cause the compound to travel through the piping system and cause obstructions. A non-adhesive Teflon tape is used on male threads when it is preferred and when it is required by code. Many codes do not allow some pipe dope compounds on plastic threads because the pipe dope expands when it sets up (dries) and can crack the plastic fitting. Company preference usually dictates the pipe dope used. Figure 12–34 is a photo of a popular brand of pipe dope

Figure 12–34 A sealing compound known as pipe dope or Teflon tape is applied to male threads to seal the connection to female threads. *Courtesy of RectorSeal.*

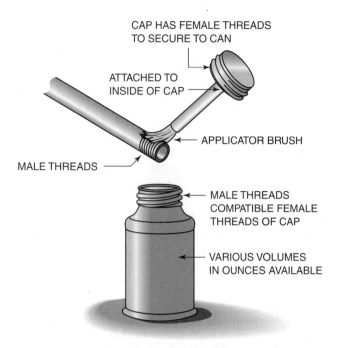

Figure 12-35 A brush applicator connects to the inside of the cap that seals a can of pipe dope.

used for numerous piping systems. Figure 12-35 shows the internal applicator brush connected to the cap that seals the can of pipe dope.

 FROM EXPERIENCE

Slightly loosening a threaded joint may cause the connection to leak. The fitting must then be removed from the pipe end and pipe dope or Teflon tape must be reapplied. If you slightly overtighten a threaded fitting, you might have to turn the fitting one more rotation to achieve the desired positioning of the fitting.

See page 309 for a step-by-step procedure for using Teflon tape.

Flux and Solder

Flux must be used with care; avoid contact with your eyes and mouth. A soft-solder connection needs flux to allow solder to flow into a fitting socket when the appropriate heat is applied to the external portion of a fitting. Some flux and solder products are not approved for use with potable water piping, a plumber must know all current codes. The maximum amount of lead that solder can contain is 2 percent, but, as with all codes, that could be reduced when new regulations are adopted. Solder is a combination of alloys, typically consisting of 95 percent tin and 2 percent lead with

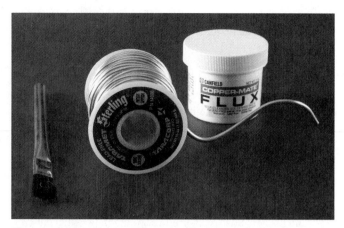

Figure 12-36 Flux and solder are used to solder a copper connection when heated with a torch.

the remaining percentage made up of silver or other alloys. Tin is the primary alloy used for solder to provide strength for the solder **joint** connection. The heat range for melting the solder varies with the particular solder used; some have melting points as low as 400 degrees F while others average 700 degrees F. Flux is a paste. Some are self-cleaning and some are sold with an applicator brush similar to that on a can of pipe dope. A flux brush, often called an acid brush, is purchased separately and used to apply the flux onto the pipe end and into the fitting socket. Too much flux causes the excess to enter the piping system, which can obstruct filters and screens on faucets and other devices. Too little flux will not allow the solder to be drawn into the fitting socket when heated. Figure 12-36 shows a can of flux, a flux brush, and a roll of solder for copper tube.

 **CAUTION**

CAUTION: Hot flux can cause serious burns and can splatter as it is heated. Wear protective clothing, gloves, and eye protection.

 FROM EXPERIENCE

Flux can cause serious eye irritation. Plumbers should wash their hands thoroughly after each use to avoid accidentally rubbing their eyes with residual flux on their hands.

Soldering

As mentioned previously in this chapter, soldering is a welding process commonly known as soft-soldering. It is

Table 12-7 Steps Required to Solder a Copper Connection

1	Measure pipe
2	Cut pipe
3	Ream pipe
4	Sand pipe end
5	Clean fitting socket
6	Apply flux to pipe end
7	Apply flux to fitting socket
8	Insert pipe into fitting socket
9	Light torch
10	Apply heat
11	Wipe melted flux as necessary
12	Apply solder as necessary
13	Wipe excess solder as necessary
14	Allow to cool

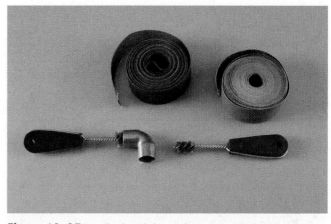

Figure 12-37 A wire brush is used to clean the inside of a fitting socket, and sand cloth is used to clean a pipe.

CAUTION: Severe burns can result from the torch flame and the molten flux and solder. Always remain aware of your surroundings, and keep an approved fire extinguisher in your work area.

Practice is required to be efficient at soldering, so focus on quality as you learn. To complete a quality solder joint, be sure to control the melting flux. Solder will follow a flux trail along the pipe. Always wipe excess flux from the pipe and fitting before applying solder.

designated soft-soldering to indicate that it requires flux and a roll of solder. Several steps (Table 12-7) are needed to complete a solder joint, which begins with measuring the pipe length desired and ends with cooling the solder joint. The cutting process is completed with a copper tubing cutter; a hacksaw should only be used if necessary, and never on medical gas systems. Cutting copper with a hacksaw creates small copper particles that can enter a piping system. Most standard tubing cutters have a reamer attachment on the side of the tool. It is inserted into the tubing to remove the ridge created while cutting the pipe. It is crucial that the pipe end and fitting socket be clean and oil-free to prevent contamination. Sand cloth, also known as emery cloth, can be used to clean the pipe end. It is sold in roll form, with some types available in a meshed design. A plumber simply tears off a piece of sand cloth in the desired length, typically about 6", when working with small-diameter copper tube. Copper tube and fittings are protected with a clear coating at the factory so the pipe does not tarnish (oxidize). This coating must be removed before applying flux. Special pipe cleaning tools are available for this, but most plumbers use sand cloth. The internal socket of a fitting also must be cleaned, and, because most residential copper fittings are 3/4" and 1/2", a wire brush is used. Larger copper fittings can be cleaned by inserting a piece of sand cloth into the fitting socket. A wire brush, often referred to as a fitting brush, is sold according to the fitting size it is intended to clean. Most wire brushes are only used in the clockwise direction to avoid breaking the bristles, which shortens the lifespan of the brush. Table 12-7 lists the basic steps in a soldering process. Figure 12-37 is a photograph of a wire brush and a roll of sand cloth.

See page 310 for the step-by-step procedure on how to solder a copper connection.

Brazing

The brazing process requires extreme heat but does not require the use of flux. Brazing is also known as silver soldering. The solder is sold in either flat or round rods, often called sticks. Silver solder has various percentages of actual silver with 2 percent, 5 percent, and 15 percent being the most common. The heat range to successfully complete a silver-soldered joint varies with the type of silver solder used. Temperatures exceeding 800 degrees F are typically needed to melt the silver solder. New copper pipe and fitting sockets usually do not need to be cleaned with sand cloth before brazing. Some codes dictate that copper pipe installed below ground cannot be soft soldered, but brazing is acceptable by all codes. The silver-soldering process requires that the pipe and fittings are dry fitted (no flux), free of dirt and oil, and heated until the pipe and fitting are cherry-red (glowing). The heating process begins on the pipe; then the flame is moved onto the fitting and continuously moved to maintain the temperature of the pipe and fitting. Once the materials are properly heated, the brazing rod is placed on the pipe where the fitting socket begins. Residential construction uses small-diameter pipe, and a small handheld **torch** can provide adequate heat, but piping larger than 1" requires a larger torch. The goal is to complete a silver-solder joint that has a uniform cap (bead) of solder along the ridge of the fitting. The position of a joint also dictates the approach for applying the silver solder. The three basic installation positions are horizontal, vertical, and upside-down vertical; the most difficult position is upside-down vertical. Figure 12–38 illustrates a completed silver-soldered joint that has a strong cap (bead) covering the ridge of the fitting. Figure 12–39 illustrates the three basic positions that a solder joint can be installed.

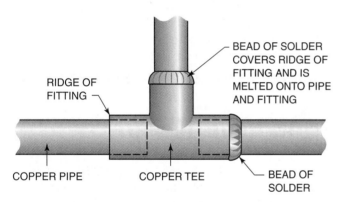

Figure 12–38 A brazed copper joint should have a uniform bead along the ridge of the fitting.

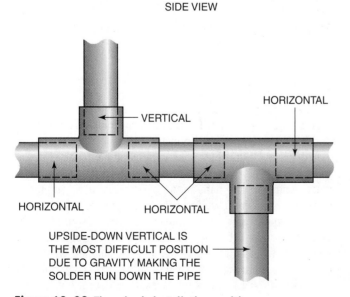

Figure 12–39 Three basic installation positions are common.

CAUTION: Any contact with the copper pipe during or after applying heat will cause severe burns. Always wear leather gloves and a long-sleeved shirt to avoid injury.

 FROM EXPERIENCE

Bending the first 2" of the silver-solder rod to a 45° angle may make the soldering process more comfortable. Applying a small amount of heat to the solder while pressing it against the cool portion of the pipe makes it easier to bend the rod.

See page 314 for a step-by-step brazing process.

Flaring

Flaring copper is the process of folding or spreading the pipe end with a flaring tool. Soft copper is used to make a flared pipe end that is mated with a compatible threaded flare fitting and flare nut to complete a connection. A flare nut, which is typically sold separately from the fitting, secures the pipe to the fitting. A threaded flare connection does not require pipe dope or Teflon tape and is considered a ground joint because it is a metal-to-metal connection. Buried copper water piping, such as a water service from a water meter and gas connections to equipment, uses flared connections. The greatest benefit of a flared connection is that it can be disturbed or moved while under pressure without leaking, and the connection can be loosened with wrenches. A solder joint is a permanent connection that requires a union to be installed to access the piping system, but a flare nut serves as a union-type connection. The flare fitting and nut are brass, which is acceptable for gas and water. Figure 12–40 shows a flared connection.

 FROM EXPERIENCE

Soft copper tubing often is not perfectly round, but a flaring tool can be used to make an out-of-round copper tube round. To avoid crushing soft copper tubing, a plumber should not forcibly cut it.

See page 316 for step-by-step procedure on creating a flare.

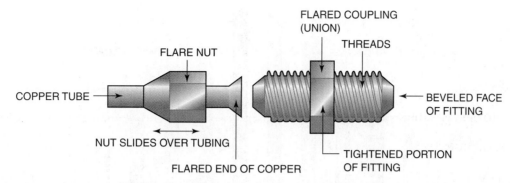

Figure 12-40 A flared connection is durable and used for soft copper tubing that is subjected to movement.

Working with Flexible Tubing

Flexible tubing is widely used in the residential plumbing industry to increase productivity and decrease material costs. Many regions of the country still use copper piping, and commercial water distribution systems primarily use copper piping. Cross-linked polyethylene (PEX) has become the most common flexible tubing for residential water distribution systems. Polyethylene (PE) was used widely for decades to install water service to a house, but other plastic piping such as polyvinyl chloride (PVC) is used more often today. PE is still a popular choice for installing the drop-pipe for a submersible pump in a well. Many residential water distribution systems use a manifold piping design that distributes water individually to each fixture, as opposed to how it has been discussed in this chapter. The stub out serving a fixture usually converts to copper to provide a quality installation, but most codes allow PEX to be used for stub-out purposes. Valves, male adapters, female adapters, washing machine boxes, and icemaker boxes are available that connect directly to PEX. Most codes dictate that PEX cannot connect directly to a water heater and that it must be protected against damage from the flue pipe of a gas water heater. The most common pipe conversion near a water heater is to copper. Figure 12-41 illustrates a PEX piping system converting to copper for a water heater connection.

 FROM EXPERIENCE

Connecting PEX directly to a water heater might cause the pipe to be subjected to extreme heat, which could void the warranty and cause leaks.

Manifold Systems

In a manifold system, the water service piping enters the house to a designated area where the manifold is installed; all piping serving fixtures originate from that location. A manifold system is often referred to as a parallel system. The size of each pipe connected to the manifold is based on the specific fixture requirement determined from a code book similar to Table 12-2. Some codes allow the sizes listed in a table to be reduced by one pipe size if the water pressure is greater than 35 psi and if the pipe length from the manifold to the fixture does not exceed 60′. Some codes dictate that all individual piping connected to a manifold system must have separate isolation valves and that they be identified as to the fixture they serve. All manifolds must be accessible. Hot and cold water manifolds are installed separately. Figure 12-42 illustrates a typical water distribution manifold serving a house. Figure 12-43 shows a manifold cold water distribution flow diagram.

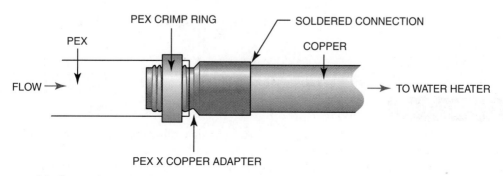

Figure 12-41 Some codes dictate that PEX cannot connect directly to a water heater, so it is usually converted to copper.

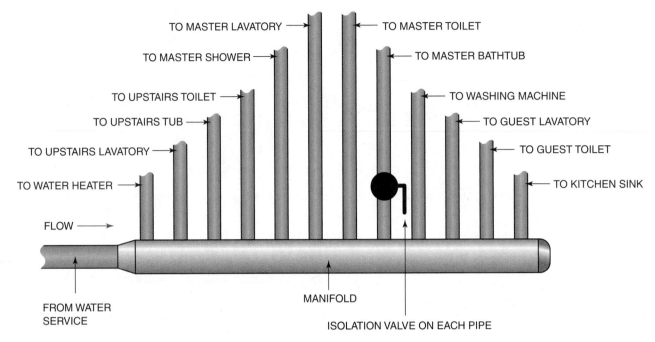

Figure 12-42 A manifold piping system has individual pipes serving each fixture.

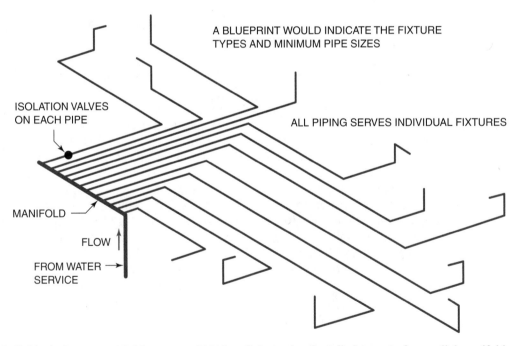

Figure 12-43 Individual pipes are routed from a manifold to a fixture when installed as part of a parallel manifold water distribution system.

FROM EXPERIENCE

A manifold system allows each fixture to be isolated for repair or to isolate a leak in a specific pipe.

Connection Types

Flexible tubing connections are based on product type. PE uses a ribbed (barbed) fitting that inserts into the tubing and is secured with stainless steel hose clamps. PEX uses one of two basic connections to secure the ribbed fitting into the tubing. The fittings used with PEX tubing are each compatible with a particular type of PEX tubing. One crimping process is very simple. A metal ring slides over the PEX before the fitting is inserted into the tubing. A crimping tool is then placed over the crimp ring and secured in place by closing the scissor-like crimping tool. The crimping tool must be manually calibrated (adjusted) periodically according to manufacturer's requirements. The other popular connection type is unique to certain manufacturers. A PEX ring slides over the tubing and a specially designed tool is inserted into the end of the tubing to expand the piping. When the tool is removed from the expanded pipe end, a fitting is inserted into the tubing and the PEX ring slides toward the tubing end to secure the fitting in place. The fitting is sealed into the tubing when the expanded tubing forcibly returns to its natural round shape. Regardless of the type of PEX system used, the process is simple, and a plumber only has to make these connections one time to be competent with either method.

Working with Plastic Pipe

Plastic water piping is used for water service, well pump, and water distribution systems. Most codes do not allow PVC for the water distribution system in a house; it can only be installed within 5' of where piping enters a house. Plastic hangers and supports are generally used to hang plastic piping to avoid damaging the exterior of the pipe. Water distribution systems use CPVC for plastic pipe installations. Fitting connections for both PVC and CPVC are solvent-welded (glued). CPVC can distribute hot water, but it must be installed according to specific manufacturer recommendations. CPVC expands and contracts when hot water is distributed, and expansion joints may be required on long horizontal installations. To avoid breakage of the pipe or fittings during expansion, CPVC should not be secured too tightly at offset areas. CPVC that is secured too tightly where it passes through wood floors, studs, and joists can cause a squeaking noise, thereby generating complaints from homeowners. Converting from one material, such as copper, PVC, or CPVC, to another is performed with specific adapters. A male and female adapter is the most common method. Valve, male, and female adapters are available that connect directly to CPVC with a solvent-weld. Some CPVC male and female adapters have plastic threads, but the brass-threaded types are more durable and provide a higher quality installation.

Cutting

Special shear-type cutting tools that cut plastic piping squarely should be used instead of a plastic pipe saw to keep shavings from entering the piping system. The cutting process is very simple but having a sharp cutting blade is crucial to create a smooth cut. The pipe is inserted between the cutting blade and the jaw of the tool, and the scissor action handle is squeezed (closed) to cut the pipe. If a saw is used, all shavings must be removed from the pipe end before proceeding with the solvent-welding process. A deburring tool or pocket knife removes any ridge created from cutting the pipe. Once a square cut is performed and all burrs are removed, the next step is to prepare the pipe and fitting to be solvent-welded.

Solvent-Welding

Solvent-welding uses chemicals to fuse a pipe and fitting together to create a watertight connection. Plumbers refer to the process as gluing a joint and rarely use the term *solvent-welded* to refer to the process. CPVC glue is only compatible with CPVC, and PVC glue with PVC pipe. An all-purpose glue is available but is not allowed by most codes. An improperly solvent-welded joint can cause severe water damage to a home; a contractor can be held responsible if a connection fails. Always read manufacturer information pertaining to the actual process, and certainly follow the manufacturer's recommendation for temperature of the work area where a connection is being completed. Some manufacturers offer various types of glue. A medium-bodied clear glue is the most common for PVC; CPVC glue is yellow. A blue glue is available with more tolerance for moist working conditions and, according to manufacturers, it has a shorter drying time. Extremely cold or hot weather has a negative effect on the curing times and bonding capabilities of the glue. Most codes dictate that a purple primer must be used before gluing the connection. Clear primer and cleaner are available, but code officials can visually confirm that purple primer was used because it permanently stains the piping. Most manufacturers dictate a three-step process. If all steps are not completed, the warranty is voided, and the solvent-weld is considered inadequate. Some manufacturers may not dictate that a cleaner and primer be used, but most codes require it. Codes often state that the regulations for specific installations are based on manufacturer's instructions. The most stringent procedures always rule over other regulations.

CAUTION: Working in closed spaces with solvents and glue can create explosive conditions and unsafe breathing conditions.

CAUTION: Never use an open flame near solvents and glue.

FROM EXPERIENCE

During the rough-in phase of construction, a plumber may not be concerned about spilling primers and glue. During the trim-out phase, a plumber must be more careful because primers, cleaners, and glues can permanently damage flooring surfaces.

See pages 320–322 for a step-by-step solvent-welding procedure.

Testing

Testing a water distribution system is mandatory; the pressure applied is dictated by code. Most codes state that test pressure must be one and one-half times the operating pressure, but other codes dictate a specific pressure such as 100 psi. The maximum operating pressure allowable by code for a water distribution system is 80 psi, and codes dictate that a pressure-reducing valve must be installed when incoming water pressure is more than 80 psi. If the one and one-half pressure regulation is dictated by your local code, the test pressure is 120 psi (80 + 40). The type of test used is based on contractor preference. The two methods allowable by code use air or water. Air tests use an air compressor, and water tests use a hydrostatic pump. Which method is chosen depends on the preference of a contractor and the connection to a particular portion of a piping system. When water is used, a garden hose is attached to a boiler drain somewhere in a piping system, such as a washing machine box, and potable water is supplied to the piping being tested. The hydrostatic pump is manually activated to pressurize the piping system. Most hydrostatic pumps used for residential construction are manually operated and do not require electricity. Most cold-weather regions use air tests during colder months so the piping system will not freeze. Air is injected into the piping system, typically through a 1/2" threaded connection such as the shower head. Air is the most widely used method because many construction sites do not have potable water available when the test occurs, but electricity is available to operate an air compressor. An advantage of using air pressure is that a piping system can be depressurized to repair a leak. There is no need to drain water from a system.

A water heater is typically installed during the trim-out phase of construction. The hot and cold water piping are usually connected to create a single piping system for testing purposes. If the water heater pipes are not connected, the hot and cold stub outs for another fixture can be connected. A tub and shower faucet can also be opened to simulate a warm-water flow condition that distributes the test pressure evenly between the two piping systems. The tub and shower faucets are installed during the rough-in phase and are under pressure during the test. The hose bibbs are usually installed after a pressure test so the pressure does not escape through them. All stub-out pipes serving fixtures are capped to allow the piping system to be pressurized. A pressure gauge must be installed somewhere in the piping system so the plumbing inspector can make sure the test is satisfactory. Most codes dictate the duration of a test, and, because an inspector is usually not required to remain on a residential job site very long, most tests are 15 to 20 minutes long. A testing accessory known as a test block has a pressure gauge and air inlet device. Figure 12–44 illustrates an air test connection to a piping system. Figure 12–45 shows a hydrostatic test connection to a piping system. Figure 12–46 illustrates a test block to air test a piping system.

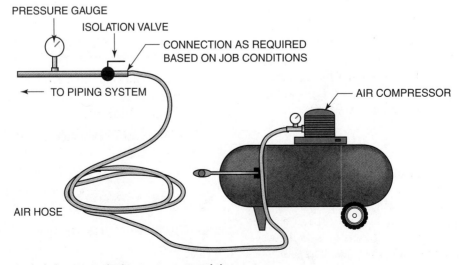

Figure 12–44 Compressed air is one method to test a water piping system.

CHAPTER 12 *Water Distribution Installation* **303**

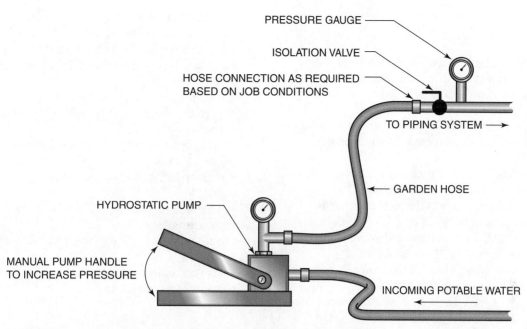

Figure 12-45 A hydrostatic water test is another method to test a water piping system.

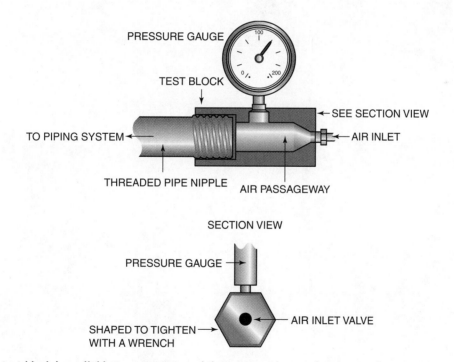

Figure 12-46 A test block is available to connect to a piping system when performing an air test.

FROM EXPERIENCE

The actual connection of the water piping system to the air compressor or hydrostatic pump is determined by jobsite conditions. A plumber typically has fabricated piping arrangements to use on all jobs.

Summary

- The layout of a water distribution system varies based on the actual job site and the fixtures being served.
- The sizing of an entire system is based on the quantity and type of fixtures being served.
- The three main segments of a water distribution piping system are the main, branch main, and individual supply.
- A code book lists common fixtures and minimum pipe sizes that can be installed to each fixture.
- A water supply fixture unit (wsfu) is the designated term to describe the end result of a flow calculation.
- One gallon is 3.785 liters and one gallon per minute is .0631 liters per second (3.785 ÷ 60 seconds).
- The rate the water flows through the piping is known as velocity and is measured in feet per second.
- The wall rough-in for a water distribution system is determined by the fixture type and job-site conditions.
- The standard water rough-in dimension for a tank-type toilet is 6" to the left of the toilet center.
- The maximum height for most lavatory sink water piping, regardless of the type, is 21" above the finished floor.
- A combination tub and shower (T&S) faucet does not change the sizing requirements of the individual fixture supply.
- Drilling and notching codes must be known to install piping in walls.
- Maximum hanger and support spacing is dictated by code.
- Hanger and support selection is based on the type of piping installed.
- Copper must be soldered with approved lead-free products.
- Follow manufacturer instructions when performing a solvent-weld connection.

Procedures

Determining the Volume of Different Pipe Sizes

A The first step in determining pipe size is to recognize that 78-1/2 percent is written as 0.7854, and that what was the length and width of a square is now the diameter (D), because it is circular. The formula to find the square area of a circle is D × D × 0.7854. Simple mathematics would lead you to believe that, because 1/2" × 3 = 1-1/2", three 1/2" pipes would provide the same amount of water as one 1-1/2" pipe, but that is incorrect.

- In this example, we are determining the area of 1-1/2" pipe and how many 1/2" pipes it would take to provide the same volume of water as the 1-1/2" pipe.

- in^2 represents square inches

- 1-1/2" pipe formula is 1.5 × 1.5 × 0.7854 = 1.768 in^2 (1.76715)

- 1/2" pipe formula is 0.5 × 0.5 × 0.7854 = 0.196 in^2 (.19635)

B The next step is to divide the square area of the larger pipe by the square area of the smaller pipe. The resulting answer is the number of 1/2" pipes that are required to equal one 1-1/2" pipe. This process works for all pipe sizes. There are other important design considerations in the complete pipe-sizing process, but this is a basic first step.

- 1.768 ÷ 0.196 = 9.02

- This demonstrates that nine 1/2" pipes provide the same volume of water as one 1-1/2" pipe.

A

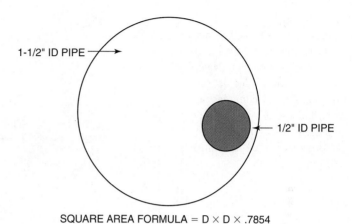

SQUARE AREA FORMULA = D × D × .7854

B

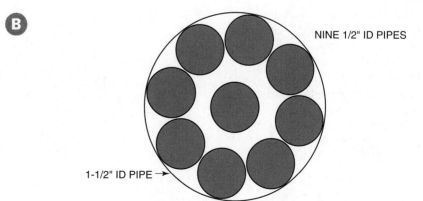

IT TAKES NINE 1/2" ID PIPES TO PROVIDE THE SAME VOLUME AS ONE 1-1/2" ID PIPE

Procedures

Determining the Volume of Different Pipe Sizes (continued)

C The next step is to add height or length to the square area to determine volume of a pipe over its installation route. One gallon of water equals 231 cubic inches of water. Using our 1-1/2 ID pipe, we proceed knowing that the square area of the 1-1/2" pipe is 1.768 in².

D The length of a pipe is multiplied by the square area and then divided by 231 to calculate the gallon capacity of the pipe. If the 1-1/2" pipe is 10' in length, we convert that to inches by multiplying 10' × 12", which results in 120 inches. The next step is to multiply the square area of 1.768 by 120, which results in 212.16 cubic inches (cu²). Dividing 212.16 by 231 determines the number of gallons the 10' piece of 1-1/2" pipe holds. Therefore, 212.16 ÷ 231 = 0.918 gallons or just a little less than one gallon of water. The entire formula discussed in this procedure is D × D × 0.7854 × L ÷ 231 = gallon capacity of a round pipe.

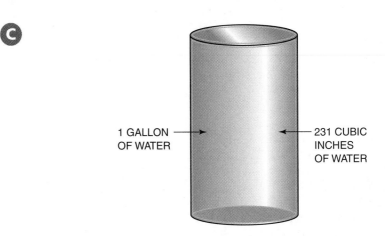

D × D × 0.7854 × L ÷ 231 = GALLON CAPACITY OF ROUND PIPE
1.5 × 1.5 × 0.7854 × 120 ÷ 231 = 0.918 GALLONS OF WATER

Procedures

Determining Percentages of a Hole Diameter and Notch

A The first step in determining the maximum allowable hole size to drill or notch to cut in a wood board is to find 100 percent of the width of a wall stud or depth of a joist. A 2″ × 4″ board is actually 3-1/2″ wide, or 3.5″ in decimal form. The next step is to identify the specific code stating the percent that can be drilled or notched.

A CAUTION: OTHER CODES APPLY TO DRILLING AND NOTCHING REGULATIONS. THIS PROCEDURE IS ONLY FOR DETERMINING PERCENTAGES OF A WALL WIDTH OR DEPTH

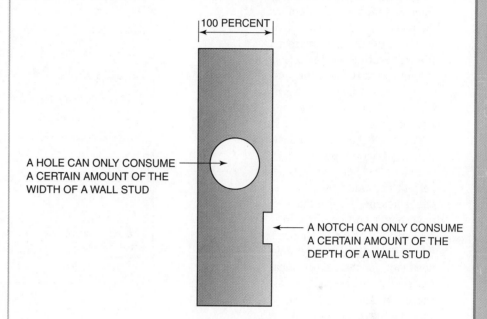

B Understanding the components of a fraction is crucial in converting decimals to fractions. The top half of a fraction is the numerator, and the bottom half is the denominator. To convert a fraction to a decimal, you must divide the numerator by the denominator. When converting decimals to fractions, you will choose a desired denominator during the mathematical process. The standard denominators are 2, 4, 8, 16, 32, and 64, all of which are even numbers. A fraction with an even-numbered numerator and denominator (2/16) can be reduced to a lower usable form; for example, 2/16″ is the same as 1/8″.

B A fraction has identifying descriptions

Numerator ⟶ $\frac{1}{2}$
Denominator ⟶

Numerator divided by denominator results in the decimal equivalent of the fraction

$$2\overline{)1.0}$$ = 0.5

Denominator of choice relates to the final mathematical determination desired, such as 2, 4, 8, 16, 32, or 64

$$\frac{1}{2} \quad \frac{1}{4} \quad \frac{1}{8} \quad \frac{1}{16} \quad \frac{1}{32} \quad \frac{1}{64}$$

A fraction is reduced to have an odd numerator (inches)

$$\frac{2}{16} = \frac{1}{8}$$

Procedures

Determining Percentages of a Hole Diameter and Notch (continued)

 You must then change the percentage to a decimal; for example 50 percent is written as 0.50. Then multiply the wall width by the percentage allowable by code—in this case, 50 percent. The wall width is written in decimal form, and the resulting answer is the maximum number in inches that can be removed by drilling or notching. The next step is to convert the resulting dimension to a fraction. Because 1.75" is a mix of a whole number and a percent of a whole number, the one represents 1", and the 0.75 represents a portion of 1". To convert the portion of an inch to a fraction, you must select a denominator for the mathematical process. In this example, 4 was chosen. Multiply the decimal value by the denominator of choice to find the numerator portion of the fraction. In this example, $4 \times 0.75 = 3$, so 3 is the numerator, resulting in 3/4". Add that value to the whole value that was not included in the process to achieve the entire value. The answer indicates that 50 percent of 3-1/2" is 1-3/4".

C

A percentage is converted to decimal form

$$50\% = 0.50$$

The stud width is multiplied by the decimal equivalent of a percentage

Wall width ⟶ ⟵ Maximum dimension that can be drilled

$$3.5" \times .50 = 1.75"$$

Allowable percentage ⟵

The resulting numerical value is converted to fraction form

1 is whole number and not relevant in the process

The decimal value is multiplied by the denominator of choice

$$.75 \times 4 = 3$$

The resulting mathematical value is the numerator

$$\frac{3}{4}$$

Add the result with the whole number dimension to achieve the entire dimension

$$1\frac{3"}{4}$$

Procedures

Using Teflon Tape

A Directions for this procedure are based on holding the roll of tape in your right hand and the pipe in your left hand. If reversed, then directions must also be reversed. The first two threads are considered starting threads that must remain exposed. Tape applied over them can enter the piping system as the connection is completed.

B Place the roll in your right hand and remove the tape from the top of the roll and toward the pipe. Place your index finger in the center hole of the roll for ease of rotation. Place your left index finger over the end of the tape, and press the tape against the threads. Begin wrapping the tape in a clockwise motion while pulling the roll away from the pipe. The tape must be applied tightly; it will stretch as you pull the roll toward you.

C Continue rotating the roll around the pipe until three full rotations are completed. Be sure all the threads except the first two threads are covered with Teflon tape. Too little tape might cause the threaded connection to leak, and too much tape is wasteful. When the process is complete, simply pinch the tape with your right index finger and thumb, and pull the roll toward you. Press any loose tape against the pipe, wrapping it in the same direction as applied. The process is then complete and the pipe is ready for assembly.

A

B

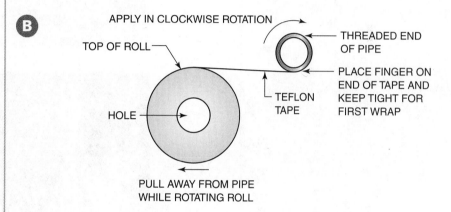

C

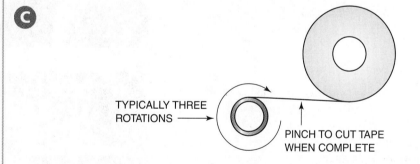

Procedures

Soldering Copper

A Measure the pipe length needed. Mark it with a pencil or scratch the pipe with a sharp object or with the reamer on the back side of the tubing cutter.

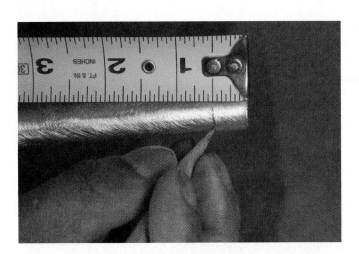

B Cut the copper pipe with the tubing cutter.

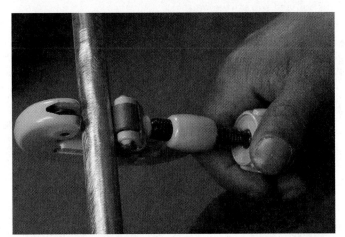

C Insert a reamer into the pipe and twist it to remove any burrs created by the cutting process. Be sure metal shavings do not remain inside the pipe.

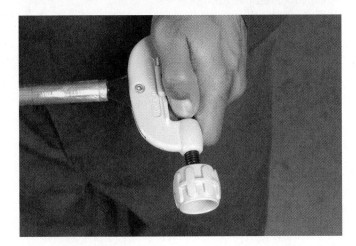

D Clean the pipe end with a sand cloth until the pipe is visibly clean.

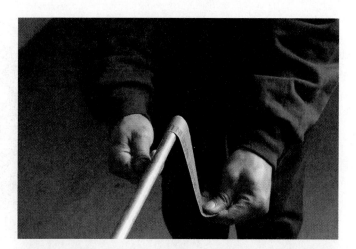

E Insert a wire brush inside the fitting socket, and rotate it clockwise several times until the fitting is visibly clean.

F Apply a thin layer of flux onto the clean, dry, oil-free pipe end.

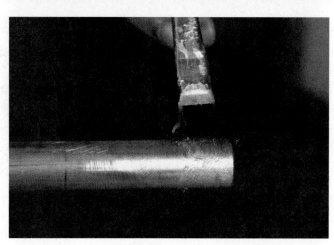

Procedures

Soldering Copper (continued)

G Apply a thin layer of flux into the clean, dry, oil-free fitting socket.

H Light the torch with a striker, not with a match or cigarette lighter.

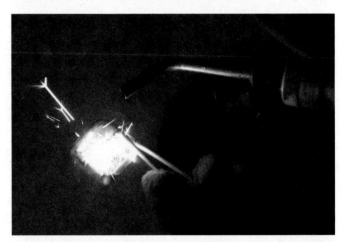

I Place the blue portion of the flame on the rear of the fitting socket.

J Move the torch to evenly distribute the heat. Do not keep the heat in one spot for more than a few seconds at a time to avoid overheating the flux. Remove the heat from the pipe, and wipe the excess flux with a clean, dry rag, as necessary.

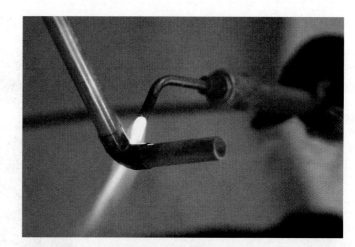

K Apply the heat again, and place the solder onto the edge of the fitting where it connects with the copper tube. The solder is being pulled into the fitting socket by the heat due to capillary action. While maintaining the heat by moving the torch off and on or closer and farther away from the piping, continue to apply the solder. For smaller-diameter piping, such as 1/2" and 3/4", the minimum amount of solder required is equal to the diameter of the pipe. Too much solder will either exit externally from the fitting socket or enter the piping system, which can cause obstructions within the piping system.

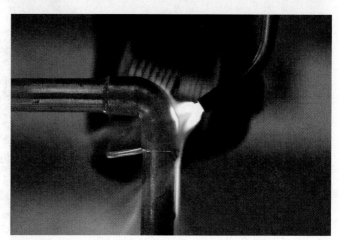

L Pay close attention to the molten solder, gently wiping the excess from the piping with a clean, dry rag. If you accidentally hit the piping or wipe the solder roughly while it is molten, apply more heat to prevent a leak due to the abrupt treatment of the soldered fitting. Once a solder joint is complete, allow it to cool naturally. Cooling the molten solder with water can shock the solder joint and cause it to leak.

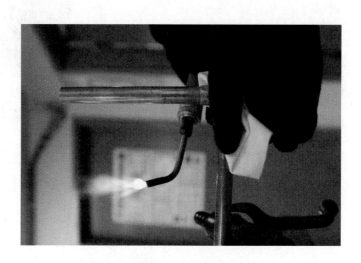

Procedures

Brazing

A The measuring, cutting, and reaming steps for soft-soldering also apply to brazing, but you do not have to clean and flux the pipe ends and fittings before assembling. Once the dirt and oil-free pipe and fittings are assembled, the next step is to ignite the torch with a striker.

A

B Place the flame over the pipe to heat that area first and then heat the fitting socket all the while moving the flame back and forth to each area and around the entire fitting. The goal is to evenly heat the pipe and fitting until they become cherry-red (glowing). Once the pipe and fitting socket are heated, it is important to maintain the high temperature with the torch.

B

Place the silver-solder rod on the fitting socket where it intersects with the pipe.

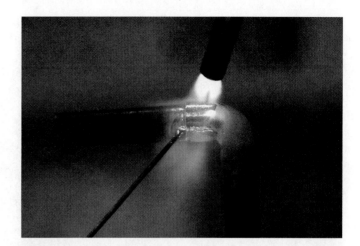

 The solder will be pulled into the socket by the heat, much like it does using flux, but it will not flow if the torch is removed from the joint.

E The goal is to apply enough solder that the ridge of the fitting is capped with silver solder. After the brazing process is complete, allow the joint to cool, and be very careful that the hot copper does not contact your skin. Cooling a silver-solder joint with water does not shock the joint like cooling a soft-solder joint does.

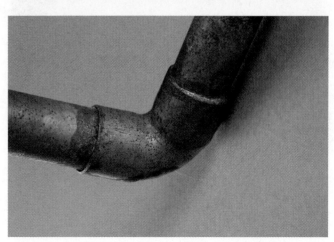

Procedures

Flaring Copper Tubing

A Several styles of flaring tools are available; this procedure uses a popular two-piece type. The procedure for other types may vary slightly. A flaring tool is used to flare various pipe sizes, which are identified as to the proper flaring hole to use. Cut the copper tubing with cutters.

B Ream the inside of the pipe with a reamer. Install the flare nut over the tubing.

C Open (spread) the clamp portion of the flaring tool, and insert tubing into the correct hole of the tool.

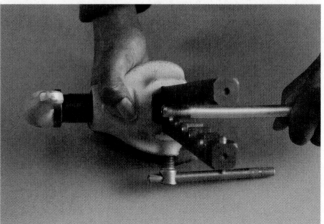

Procedures: Flaring Copper Tubing (continued)

D Install the flaring portion of the tool over the clamp portion. Begin securing the flaring tool, but do not completely tighten the flaring portion to the clamping portion. Align the flaring post directly into the copper tube, making sure that the tubing is placed above the surface of the clamping portion approximately 1/8" (the thickness of a penny). Complete the tightening of the flaring portion to the clamping portion. The flaring process can begin once the tool is tightened together.

E Turn the flaring tool handle clockwise, so the flaring post to the end of the tube is flared. When you have turned the handle completely downward, the flaring process is complete.

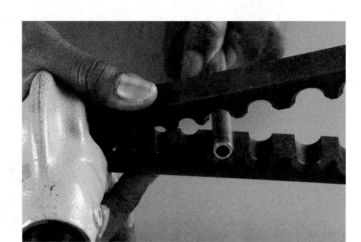

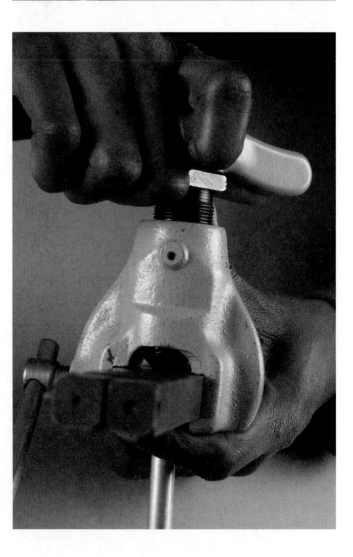

Remove the tool and slide the flare nut over the newly created flare to make sure the flare is the right diameter. If the flared end of the tube does not fit into the nut or if it is too small, the flared end must be cut and the process completed again.

Procedures

Solvent Welding

CAUTION

CAUTION: These instructions do not replace a manufacturer's recommendations; a plumber must read the directions provided with all products.

 Cut the pipe squarely and remove all burrs that result from the cutting process. Manufacturer recommendations may require that the pipe and fitting be cleaned with a solvent cleaner before the primer is applied. With the pipe end sloped downward so the excess primer runs off the end of the pipe instead of down the length of the pipe, apply purple primer to the pipe end. Make sure that it is applied equally to the depth of the fitting socket.

Apply purple primer to the inside of the fitting socket.

 Hold the fitting with the socket facing downward so excess glue does not collect inside the fitting, and apply glue to the inside of the fitting socket.

D Apply glue to the pipe end.

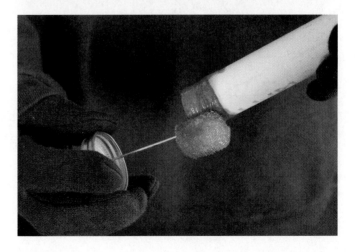

Procedures

Solvent Welding (continued)

E Push the pipe end inside the fitting socket and twist the pipe at least one-quarter of a full turn. Hold the pipe inside the fitting socket for at least one minute. When pressure is no longer applied to the connection, visually inspect it to make sure the pipe is not being forced out of the fitting socket. The pipe might have to be held in place longer than one minute under certain conditions such as warm or cold weather or by manufacturer recommendations. The joint is complete, but can take up to 24 hours to be ready for pressure testing. Do not pressure test a solvent weld before the duration stated by manufacturer recommendations.

E

Review Questions

1. A pipe that extends from a wall during a rough-in phase to serve a fixture is known as a
   a. Pipe extension
   b. Stub out
   c. Rough pipe
   d. Drain

2. The two common methods for testing a water piping system are with an air compressor and a
   a. Hydrostatic pump
   b. Well pump
   c. Circulating pump
   d. Smoke injector

3. Before drilling wall studs, it must be determined whether the stud is
   a. A ceiling joist
   b. A floor joist
   c. Load-bearing or non–load-bearing
   d. Vertical or horizontal

4. A joint compound to seal a threaded connection is known as
   a. Teflon tape
   b. Pipe dope
   c. Primer
   d. Solvent cement

5. The water pipe serving a toilet is
   a. On the left side of the drain
   b. On the right side of the drain
   c. Centered with the drain
   d. None of the above are correct

6. Unless otherwise directed by a manufacturer, hot water is always on the
   a. Right side of a fixture rough-in
   b. Left side of a fixture rough-in
   c. Bottom of the cold water
   d. Top of the cold water

7. A water pipe serving one fixture is a
   a. Fixture branch
   b. Individual fixture supply
   c. Fixture main
   d. Water service

8. The paste used to soft-solder copper tube is known as
   a. Pipe dope
   b. Primer
   c. Cleaner
   d. None of the above are correct

9. Silver soldering is the trade term for a welding process known as
   a. Solvent welding
   b. Brazing
   c. Arc welding
   d. Soft-soldering

10. One method of connecting PEX tubing uses a crimping process, and the other uses a(n)
    a. Expanding process
    b. Soldering process
    c. Solvent-welding process
    d. Threaded process

11. Codes dictate the hanging and supporting of piping based on
    a. Climate
    b. Wall location
    c. Spacing
    d. Job-site location

12 **A piping system that routes individual dedicated piping to each fixture is a(n)**

   a. Individual supply system
   b. Manifold system
   c. Dedicated system
   d. Water service

13 **Most codes dictate that a hole cannot be drilled in a wall stud at the same elevation as a**

   a. Nail
   b. Knot in the board
   c. Notch
   d. Electrical outlet

14 **Some codes dictate that, before applying glue to a solvent-welded connection, a plumber must first use a**

   a. Purple primer
   b. Clear cleaner
   c. PVC saw
   d. Thread sealing compound

15 **A copper stub out is secured with either a piece of wood and a copper 2-hole clip (strap) or a**

   a. Copper clevis hanger
   b. Copper adjustable swivel-ring hanger
   c. Copper stub-out bracket
   d. Nail wrapped around each pipe

16 **A tub and shower faucet is installed during the**

   a. Trim-out phase
   b. Rough-in phase
   c. Underground phase
   d. None of the above are correct

17 **A torch without an integral igniter is ignited with a**

   a. Cigarette lighter
   b. Wooden matches
   c. Barbeque igniter
   d. None of the above are correct

18 **The typical spread of the water piping stub out for a kitchen sink is**

   a. 4"
   b. 6"
   c. 8"
   d. 10"

19 **The typical spread of the water piping stub out for a pedestal lavatory is**

   a. 4" or 6"
   b. 6" or 8"
   c. 10"
   d. 12"

20 **A tool used to fold (spread) the end of copper tubing to create a unique connection is called a(n)**

   a. Crimping tool
   b. Expanding tool
   c. Flaring tool
   d. Copper folding tool

Chapter 13: Drainage, Waste, and Vent Segments and Sizing

Protecting the health of occupants of a building from harmful sewer gas is a primary objective of regulating installations of drainage, waste, and vent (DWV) systems. We have covered DWV fitting types and their uses in previous chapters. This chapter discusses various codes for sizing of specific segments of the DWV system. To properly size any portion of a plumbing system, correct identification of the segment is required. This ensures that the proper chart in a code book is used.

One of the most difficult aspects of sizing a DWV system is that there are some overriding codes based on the location of an installation. Segment identification is vital in locating correct sizing information, but many codes relate to where the segments are installed as opposed to their definition. Some charts in a code book provide numerous options based on the location of an installation; a plumber must choose the correct option. This chapter prepares you to install DWV piping by teaching proper segment identification, definitions, and sizing approaches.

Many new home designs have different approaches to construction, but all follow a similar sequence for installing major elements. In a new residential home, above-ground DWV piping installation typically begins after the roof is installed and the majority of the framing is complete. The plumbing system is typically installed first, followed by heating, ventilation, and air conditioning (HVAC) and electrical systems. This phase of construction is known as the rough-in phase.

OBJECTIVES

Upon completion of this chapter, the student should be able to:

- identify and describe segments of a DWV system.
- recall the basic abbreviations concerning a DWV system.
- size the various segments of a DWV system.
- understand how a basic conventional septic system operates.

Glossary of Terms

air admittance valve one-way valve that allows air to enter a DWV system; used in place of a vent that would normally terminate with another vent or through a roof; abbreviated as *AAV*

branch piping of a DWV system that connects to main portions of a system

branch interval vertical distance along a stack equal to one-story height, but no less than 8'; also the area where horizontal branches connect to a stack

branch vent vent that connects one or more individual vents with a stack vent or vent stack

building drain lowest horizontal main drain of a DWV system; conveys wastewater to a building sewer

building sewer conveys wastewater from building drain to point of disposal

circuit vent special vent serving at least two, but no more than eight, fixture traps; begins at its connection to a horizontal branch and terminates at its connection with the vent stack

cleanout access point to remove obstructions from a DWV system

common vent vent serving two fixture drains located on the same floor and connecting either at the same height or at different heights to the drain

D-box trade name for distribution box

developed length way of measuring the distance a pipe is installed along the centerline of all pipe and fittings

distribution box fabricated box or structure to distribute effluent to drain field or other designated location

drain pipe that conveys wastewater from point of entry within a DWV system

drain field area of installation for perforated piping to drain wastewater (effluent); also called leach field

drainage fixture unit (dfu) abbreviated as dfu, it is based on rate of flow measured in gallons per minute or liters per second into a drainage system from a plumbing fixture used to size pipe

drainage, waste, and vent (DWV) complete system draining soil, waste, and wastewater to a point of disposal; circulates air within the system

effluent wastewater that has been separated from solids, but may contain dissolved sewage solids

fixture branch pipe draining two or more fixture drains to a stack or other drains; in some piping configurations, it can also be known as a horizontal branch

fixture drain drain from a trap serving a fixture that connects to another pipe; also called a waste arm in some design applications

horizontal branch pipe connecting two or more fixture drains or fixture branches to a main portion of a drainage system; in some piping configurations, it can also be known as a fixture branch

hydraulic gradient vertical distance (rise) from the trap weir to the centerline of the connecting vent fitting

individual vent vent serving one fixture trap that terminates to open air or connects with another vent; can serve more than one trap in some design applications, such as common venting

interval equal to one story height, but not less than 8'; relates to a branch connecting to a stack

loop vent special vent similar to a circuit vent except that it terminates connecting to a stack vent instead of a vent stack and is only used on a top floor or at the highest branch interval

open air outside a building or structure; typically known as vent through roof or VTR

P-trap nonrestrictive fitting installed at each fixture that does not have an integral trap; uses a water seal to prevent sewer gases from entering occupied areas; often called a trap; see also trap

perc test trade name for percolation test; a method to evaluate percolation conditions of soil

percolation natural drainage ability of soil; also known as perc

relief vent pipe circulating air between a drainage and a vent system; has several specific areas of installation and is sized based on its use

rough-in phase of construction before finish or trim phase when all piping is installed in floors, walls, and ceilings

septic tank fabricated holding tank or structure to contain sewage and solids

soil stack vertical pipe conveying wastewater that contains fecal matter; the same pipe as a waste stack, can be installed with horizontal offsets

stack vertical pipe that is at least one story in height; can be installed with horizontal offsets

stack vent vertical pipe that connects to a soil or waste stack; extends to open air or connects with another approved vent; can be installed with horizontal offsets

trap nonrestrictive fitting or device installed at each fixture using a water seal to prevent sewer gases from entering occupied areas; see also P-trap

trap adapter fitting used to connect tubular piping to other pipe connections

trap distance distance a trap weir is located from its protective vent

vent stack vertical vent pipe that receives other vents and terminates to open air or with stack vent; can be installed with horizontal offsets

waste stack vertical pipe conveying wastewater only; the same pipe as a soil stack; can be installed with horizontal offsets

wastewater water that does not contain sewage; term often used instead of the word *effluent* by plumbers

weir the portion of a p-trap where the water flow crests from the trap and enters the connecting horizontal drain

Introduction

Designing and installing a DWV system can be a complex and extensive process. Actual pipe and fitting installation results from sizing and code knowledge being applied to achieve the design intent. Pipe routes are determined by fixture locations, fixture requirements, relative codes, construction obstacles, coordination with other trades, and company installation standards. Every project has unique obstacles. A plumber must adjust to changes in construction designs and learn through experience to increase productivity.

Every plumbing fixture connected to a drainage system must be protected by a fitting or device known as a **trap** that uses a water seal to prevent harmful sewer gases from entering a building. To ensure that a drainage system does not become overpressurized and that a vacuum is not created, an adequate vent system is installed to protect a trap seal. If the pressure in a drainage system is too high, sewer gas is forced through the water seal of a trap into an occupied area. If a system's pressure becomes drastically negative, the trap seal can be siphoned. A trap seal can also be jeopardized through wicking, caused by debris in the fixture drain, such as string or hair. Figure 13-1 illustrates a **P-trap.** Additional codes are explained in the supporting text to Figure 13-17 later in the chapter. Most codes dictate that the protective water seal be 2" minimum and 4" maximum measured from the top dip to the **weir.**

Each segment of a DWV system and its role is described in this chapter. Specifics concerning installation and additional code regulations will be discussed in Chapter 14. As you study this chapter, focus on how each segment connects with other segments and how they all work together to create a functioning system. We will begin with major segments.

Major Segments of a DWV System

Imagine placing your thumb over the end of a drinking straw placed in a glass of water (Fig. 13-2). When the straw is removed from the glass, the water remains in the straw because a vacuum is created within the straw. When your thumb is removed, the water drains from the straw because the vacuum is broken. This illustrates the importance of having a vent system connected to the drainage system and how air flow plays a major role in allowing water to flow by gravity.

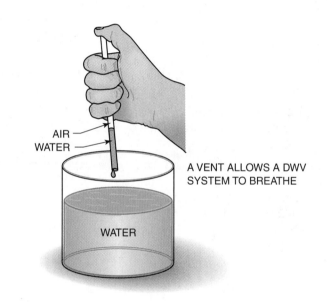

Figure 13-2 Basis of a vent.

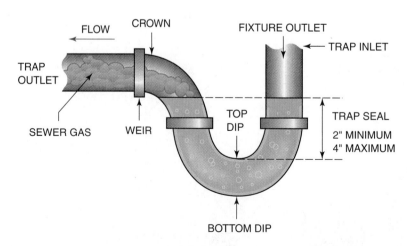

Figure 13-1 P-trap segments.

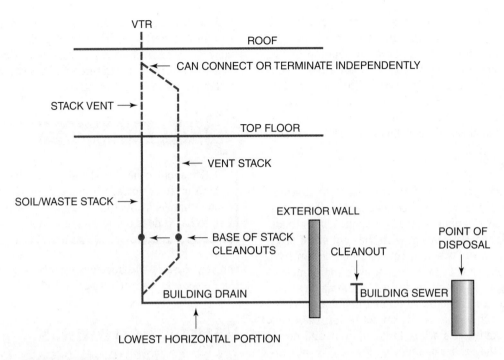

Figure 13-3 Major DWV system segments.

Figure 13-3 illustrates major segments of a DWV system. Following is a list of common abbreviations for the major segments of a DWV system:

BD: Building Drain
BS: Building Sewer
CO: Cleanout
SV: Stack Vent
VS: Vent Stack
VTR: Vent Through Roof
WS: Waste Stack

 FROM EXPERIENCE

In the past, we were required to build separate piping systems that conveyed wastewater and sewage to different points of disposal. Thus, the terms **soil stack** and *waste stack* were used to differentiate the two systems. Today, all wastewater and sewage terminate together, so these terms can now be used interchangeably.

Building Sewer

A **building sewer** is the main pipe conveying sewage and **wastewater** from a DWV system to a point of disposal, or termination. A public sewer system or a private **septic tank** are two common points of disposal that are legal means of termination. Minimum sizes are regulated by code and there must be a cleanout serving the building sewer. A cleanout is usually installed at the beginning of a building sewer with its connection to the building drain. Some codes do not allow a building sewer to be smaller than 4".

Building Drain

The **building drain,** which is the lowest horizontal portion of a drainage system, receives discharge from waste stacks and horizontal branches. It connects with a building sewer where the cleanout is installed at the junction of the two pipes. Many codes state that it must extend at least 2' 6" and no more than 10' from the exterior of the building. The size of building drain is based on the drainage load of the entire system. Minimum sizing is dictated by codes; for example, some codes mandate that the smallest DWV pipe that can be buried below ground is a 2" diameter. Other codes allow smaller DWV pipe to be buried, but never smaller than 1-1/4".

Waste Stack

The **waste stack** is the main vertical pipe, which begins with its connection to the building drain and terminates with its connection to the stack vent. It receives discharge from horizontal branches and must have a cleanout at its base. Its size is based on the total load of all connecting fixtures, and it can be offset to the horizontal position. If a waste stack is transitioned horizontally, strict codes dictate where branches can connect to the **stack** in relation to the horizontal offset portion. Codes also state that a relief vent must be installed to eliminate pressure differences within the stack. A 45° offset in a waste stack is still considered to be in the vertical position. A relief vent is required if a branch connects too close above or below the offset area.

Stack Vent

The vent for the waste stack is known as the **stack vent**. It begins at the highest branch connection to the waste stack and is a dry piping system. It typically extends through the roof, but can connect with the vent stack prior to terminating to **open air.** In a conventional design, the vent stack connects to the stack vent prior to roof penetration. Its size is typically based on the size of the waste stack, but in some codes it is sized based on building drain size. If code allows its size to be reduced, it cannot be less than half the diameter of the waste stack or building drain and never less than 1-1/4". Most codes dictate that the stack vent cannot transition horizontally until it is 6" above the flood level rim of the highest fixture connecting to the waste stack. When a battery-vented horizontal branch is used, most codes allow the stack vent to serve as the required relief vent for certain designs, and to receive the loop vent.

Vent Stack

The **vent stack** is the main vent of a DWV system. It receives all other vents and is a dry piping system. The word *vent* refers to a pipe that circulates air. Unless otherwise indicated, all vents are dry. Several piping configurations allow a drain to be used as a vent. It is known as a wet vent and is discussed later in this chapter. Its size is based on numerous factors including the total discharge load of a system and the length it travels. It can terminate to open air or connect to the stack vent. However, it must connect to the waste stack or building drain in a vertical position, or no greater than 45° from true vertical, so it is not obstructed after a blockage in the drainage system or when moisture settles in the piping system. Most codes require a cleanout to be installed at its base. The vent stack can transition horizontally without requiring a relief vent. However, as with all horizontal vents, it must have adequate slope to prevent moisture from settling and obstructing its airway.

Cleanout

All codes dictate that there must be a **cleanout** installed at the base of every waste stack and at the transition from a building drain and building sewer. In general, a cleanout must be the same size as the pipe when it is serving a stack, building drain, or building sewer, but most codes allow any pipe greater than 4" to be served with a 4" cleanout. Exterior cleanouts in a sidewalk or driveway must be installed flush with the finished grade and must be designed to handle relevant traffic loads. Vertical piping has a DWV fitting known as a test tee or a wall cleanout. The test tee is also used to perform a water test on the system during the installation phase of construction.

FROM EXPERIENCE

The cleanout code that dictates that the drain cleaning cable must be installed in the direction of flow or no more than 90° from the direction of flow is meant to ensure that it does not make a u-turn.

For step-by-step instructions on cleanout procedures, see the Procedures section on pages 361–362.

Minor Segments of a DWV System

Figure 13–4 adds other segments of a DWV system to show their relationship to a system. The minor segments are part of a fully functioning system. Even when many minor segments are not part of a design or installation, the major segments still exist. Major and minor segments are similar to a tree with branches and twigs. A tree's trunk supports its branches and twigs; a DWV system's major segments support its minor segments. Following are the common abbreviations for the minor segments of a DWV system:

BV: Branch Vent
CV: Circuit Vent
FB: Fixture Branch
FD: Fixture Drain
HB: Horizontal Branch
IV: Individual Vent
LV: Loop Vent
RV: Relief Vent

FROM EXPERIENCE

Drain and vent sizes are based on their designated purpose, but no pipe directly connected to a DWV system can be less than 1-1/4".

Fixture Drain

The **fixture drain** serves a single fixture trap and is sized according to the individual fixture load (Fig. 13–5). A fixture **drain** can be horizontal or vertical, but never smaller than

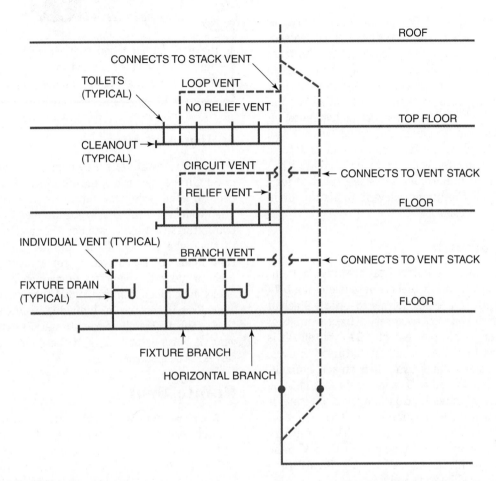

Figure 13-4 Minor DWV system segments.

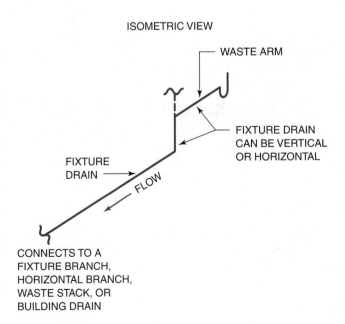

Figure 13-5 Fixture drain example.

the minimum trap size of a fixture. Fixtures producing large loads have different drain-sizing criteria; the drain size can be determined from a sizing chart in a code book. The fixture drain must have adequate slope, which is dictated by codes and by its size. Many codes allow a cleanout for a fixture drain to be smaller than the pipe. Most allow a removable joint in a trap to serve as a cleanout for a fixture drain. Because a toilet can be a point of access to the drain, the toilet can also serve as a cleanout for its fixture drain. In addition, a floor drain or shower drain with a removable strainer can serve as a cleanout for a fixture drain.

 FROM EXPERIENCE

> Most codes allow a 2" fixture drain to be served by a 1-1/2" removable slip joint, and a 1-1/2" fixture drain to be served with a 1-1/4" removable slip joint. Another name for some horizontal fixture drain designs is waste arm, but by definition the fixture drain remains a fixture drain until it connects to another drain pipe.

Fixture Branch

The **fixture branch** is a drainpipe that connects two or more fixture drains to a horizontal **branch** or major segment of a DWV system (Fig. 13-6). It can also be called a horizontal branch because it can connect to a major segment such as a stack or building drain. Its position can be vertical or horizontal. Its size is based on the total load of all the fixtures it serves and whether it is vertical or horizontal. It must have adequate slope, which is dictated by code and size. It must be served with a full-size cleanout, but some codes allow a pipe greater than 4" to be served with a 4" cleanout. A fixture drain cannot serve as a cleanout for a fixture branch. Code exceptions relevant to fixture drain cleanouts are not applicable.

Horizontal Branch

As its name indicates, the **horizontal branch** is a drainpipe that connects horizontally to a major segment of a DWV system (Fig. 13-7). It can connect either to a waste stack or to a building drain. It connects two or more fixture drains or a fixture branch to a main segment of a DWV system. It is sized based on the total load of all the fixtures it serves. Most codes allow a maximum of two toilets on a 3" horizontal branch, but some allow more. Special regulations must be followed for a battery-vented horizontal branch. Typically it can serve only 50 percent of the drainage load of a standard horizontal branch. It must be served with a full-size cleanout, but some codes allow a pipe greater than 4" to be served with a 4" cleanout. Because it is a drain, it must have adequate slope, dictated by code and size.

FROM EXPERIENCE

The horizontal portion that connects to the building drain or waste stack is sized the same way whether it is called a horizontal branch or a horizontal fixture branch.

Individual Vent

An **individual vent** is by definition a vertical extension of a drain that serves one fixture trap (Fig. 13-8). It can connect to other vents or terminate to open air. A piping arrangement known as a **common vent** allows it to serve two fixtures that are located on the same floor that connect to the same vertical drain. Another piping configuration that might not be allowed by some codes is a wet vent; it uses an individual vent for numerous fixtures. Some code books indicate that an individual vent can also be a branch vent. This is because an individual vent serving one drain that connects directly to a waste stack is essentially a branch vent. It must be at least half the diameter of the drain it serves, but no smaller than 1-1/4".

Branch Vent

When two individual vents connect together, they form a **branch vent** that serves as a vent for a horizontal branch and connects to the vent stack or stack vent (Fig. 13-9). The size of a branch vent is based on the size of the horizontal branch, the drainage load of the horizontal branch, and the

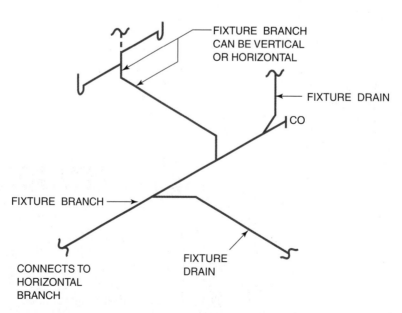

Figure 13-6 Fixture-branch example.

CHAPTER 13 Drainage, Waste, and Vent Segments and Sizing 333

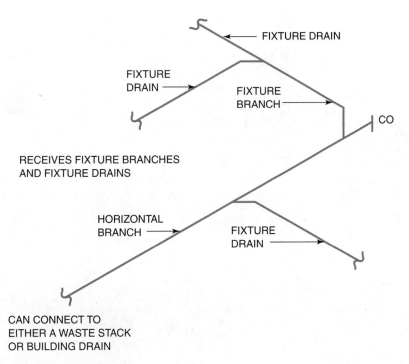

Figure 13-7 Horizontal-branch example.

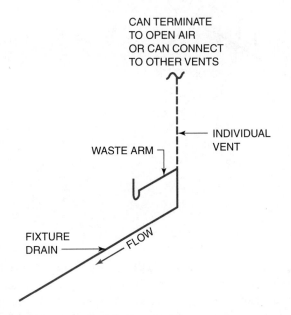

Figure 13-8 Individual vent example.

distance it travels. A vent must have adequate slope to drain condensation or moisture, but is not typically regulated like a drain regarding minimum fall per foot. A circuit vent and loop vent are also branch vents, but their specific sizing requirements differ from those of a standard branch vent.

Circuit Vent

A **circuit vent** is a branch vent, but it is different from other standard branch vents and has specific code requirements (Fig. 13-10). It serves a drainage piping arrangement known as a battery of fixtures. A relief vent must accompany a circuit vent to complete the battery-vented system. Single-family residential constructions usually do not have battery configurations due to the small number of fixtures located within one room. Commercial plumbing installations frequently use batteries of fixtures in large bathroom areas. Circuit vent sizing is based on the horizontal branch size, fixture load, and distance it travels. It typically connects vertically to a horizontal branch before the last fixture and terminates to the vent stack. A circuit vent is also known as a loop vent when it is on the top floor of a building or at the highest **branch interval.** Most plumbing license exams have questions about circuit venting.

Loop Vent

A **loop vent** is a circuit vent that is installed on the top floor of a building or on the highest branch interval (Fig. 13-11). Most codes do not require a relief vent to serve a loop-vented horizontal branch, because it loops back and connects to the stack vent instead of the vent stack. No other fixtures discharge into a stack vent above a loop vent connection. Because the waste stack no longer exists on the top floor, the loop vent can connect to the stack vent instead

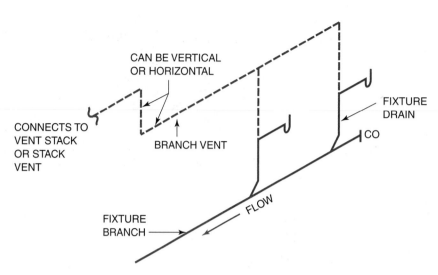

Figure 13-9 Branch-vent example.

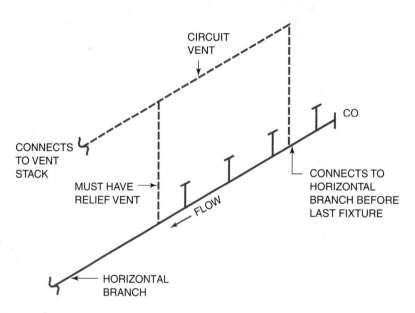

Figure 13-10 Circuit-vent example.

of the vent stack. All sizing and code regulations are the same as for a circuit vent. Some codes still require a relief vent on a loop-vented branch if more than three toilets are connected to the horizontal branch.

Relief Vent

Many types of **relief vents** exist in a DWV system, all with different regulating codes (Fig. 13–12). A relief vent is required when a waste stack transitions from vertical to horizontal. The most common relief vent is one that serves a battery of fixtures. It connects with the circuit vent before the first fixture. Many codes mandate that a relief vent be installed when a venting device known as an **air admittance valve** is installed to serve a horizontal branch. A relief vent used on buildings more than ten stories tall is a yoke vent that connects the waste stack and stack vent to equalize pressures within the waste stack. A relief vent serving a battery of fixtures is half the diameter of the horizontal branch. All relief vent sizes are based on their specific use and are determined by using a code book, but they cannot be less than half the diameter of the drain they serve and no smaller than 1-1/4".

CHAPTER 13 Drainage, Waste, and Vent Segments and Sizing   335

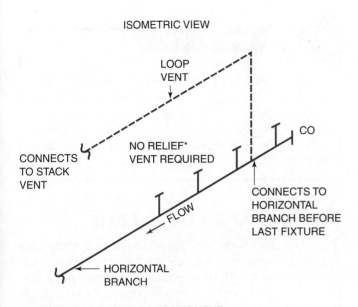

Figure 13–11 Loop-vent example.

 FROM EXPERIENCE

As with all vents, the relief vent cannot transition horizontally until it is 6″ above the flood-level rim of the highest fixture it serves.

Special Venting Arrangements

Special venting arrangements are used when conventional methods do not work. For many special arrangements, the design must be stamped by a registered engineer and submitted to a local code official or department for approval before installation. Even though a battery-vented design is not a standard installation, it typically does not require special permission or submission for approval.

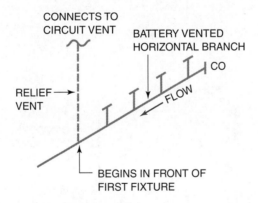

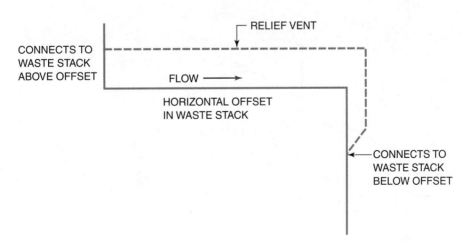

Figure 13–12 Relief-vent example.

Wet Venting

Wet venting is a piping arrangement that uses a single vent for more than one fixture; it is not accepted by all codes. Some designs would be difficult, if not impossible, if all fixtures within the design had to be individually vented. Using a wet-vented system can offer a solution. A single-occupancy bathroom is commonly wet vented. In a typical residential bathroom consisting of a toilet, lavatory sink, and bathtub, the likelihood of all fixtures being used at the same time is minimal. A single vent can serve the bathroom group with the initial size being based on the largest pipe of that group. A chart in a code book is used to size the wet-vented portion of the drain according to the actual fixtures and their dfu values.

Wet-venting arrangements for this type of installation are common where codes allow and do not require approval by a code official. Specific sizing must adhere to code based on the total drainage load and the wet-vent size. One code that typically remains the same for a wet-vent arrangement is that any fixture downstream of a toilet must be individually vented. Figure 13–13 illustrates a wet-venting arrangement that is common for a bathroom group.

FROM EXPERIENCE

The dry vent serving a wet-venting configuration is sized as a branch vent. It typically does not have to be larger than half the size of the largest drain it serves, but it cannot be smaller than 1-1/4". The drain size must be increased for most designs because of its dual use as a drain and vent.

Combination Waste and Vent

The two types of combination waste and vent systems are vertical and horizontal. Both are sized based on their installation position and the drainage load they serve. Only specific fixtures can discharge into this type of system. Code dictates the fixture type and drainage load that can be handled by a certain pipe size. The vertical design is used for typical low-volume fixtures such as an electric water cooler or floor drains that are directly above one another on several floors of a building (Fig. 13–14). The vertical waste stack is

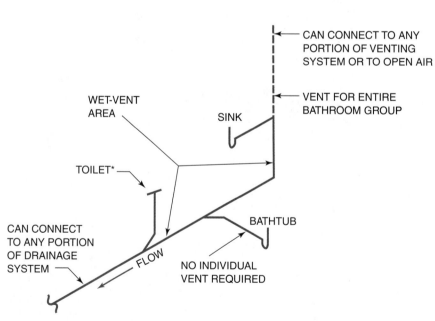

Figure 13–13 Wet-venting example.

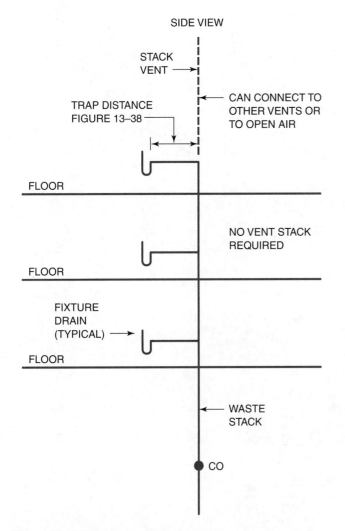

Figure 13-14 Vertical combination waste and vent example.

also the vent and, because a separate vent stack is not required, the stack is classified as a wet vent. It still has a stack vent similar to a conventional waste stack and must have a cleanout at the base of the stack. Code limits the distance a trap can be from the vertical stack based on the size of the fixture drain. (See Figure 13-38 later in the chapter.)

A horizontal combination waste and vent system has stricter regulations than a vertical design (Fig. 13-15). This design is most common in large commercial kitchens and manufacturing plants where there are not enough available walls to install numerous vents. The size and type of fixtures that can be connected to this kind of system are strictly regulated and listed in a code book. Slope is required and dictated by code, because this system is in the horizontal position and often cannot slope more than 1/4" per foot, regardless of the pipe size. The distance a trap can be from the horizontal branch is limited and dictated by a code based on the size of the fixture drain. Most plumbing license exams contain questions about combination waste and vent systems.

Island Venting

A kitchen sink located away from a wall can usually be served with an island vent (Fig. 13-16). A venting device known as an Air Admittance Valve (AAV) can be used instead of an island vent, but it might not be allowed by some codes or areas of installation. Code dictates that a dry vent cannot transition horizontally until it is 6" above the flood level rim of a fixture. Therefore, an island vent arrangement is strictly regulated, because a portion of the vent is placed horizontally below the flood level rim of the fixture. The vent located under a sink can transition horizontally below the flood level of the sink, but the vertical portion away from the sink that rises in a designated location cannot transition horizontally until it is 6" above the flood level rim of the sink. This design might need to be submitted to a code official before it can be installed.

FROM EXPERIENCE

Island vents are rarely used, but questions about them might be included on a plumbing licensing exam to ensure that you know the regulatory aspects of the code.

DWV Sizing

Layout for a DWV system is typically the first process completed on a project; other piping systems are installed around it. Pipe and fitting sizes are larger than for other piping systems in a residential building and code regulations, specific fixture locations, and available space to install piping make it necessary to install or lay it out first. Layout is based on objectives, codes, and sizing requirements. Sizing of piping is initially based on theory, but health issues combined with practical considerations determine how codes actually regulate the sizing of a DWV system. Sizing is a key area that a plumber must understand to become licensed.

Theoretical methods that determine sizing are based on the rate of flow at which a fixture discharges wastewater into a drainage system. The measuring factor is a **drainage fixture unit (dfu).** The average plumber does not perform calculations on a job site, but instead uses the known dfu values of fixtures to locate correct sizing information in a code book. Most dfu values listed in a code book are based on flow rates of certain fixtures, pipe sizes, and the dfu load effects on a DWV system.

A code book provides charts that indicate the dfu value of typical fixtures and particular pipe sizes. Figure 13-17 is an example of a typical fixture-drain and trap-sizing chart based on the dfu value of a common fixture. Specialized piping arrangements such as wet venting, common

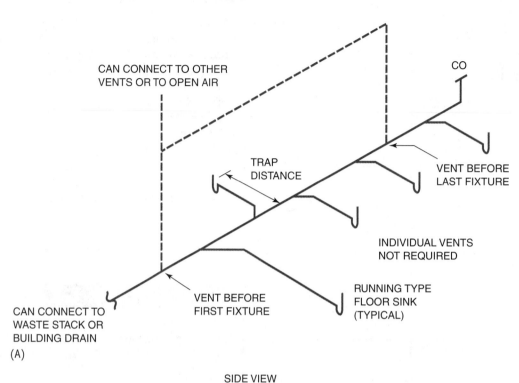

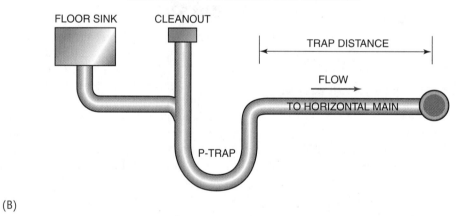

Figure 13-15 Horizontal combination waste and vent system.

venting, and battery venting are not sized with standard charts but require the use of other charts. It is important to remember that a fixture drain serves one fixture trap. Separate sizing charts are needed for other segments of a system. Sizes listed in a chart do not take into account the installation specifics of a job site. In addition, some minimum standards listed in a code book can become obsolete by an overriding code.

FROM EXPERIENCE

The dfu values of fixtures listed in a code book are arbitrarily chosen from the results of theoretical flow calculations of a particular fixture. A single-occupancy bathroom group has a lower dfu value than if all the fixtures were sized separately, because not all fixtures are used at the same time.

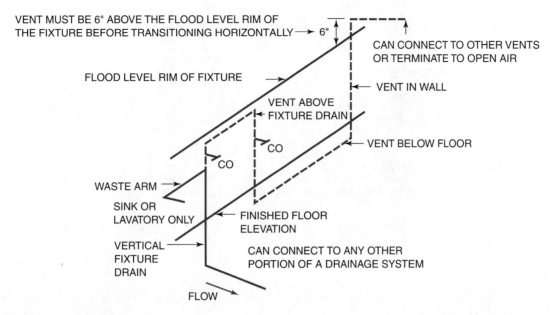

Figure 13-16 Island-vent example.

All horizontal drains must have adequate slope—also called fall or pitch—to evacuate wastewater and solids. Minimum pipe slope is based on the pipe size. Too much slope on a waste arm serving a fixture can place the trap weir above the center of the connecting vent, causing air to be trapped in the waste arm. (See Figure 13-38 later in the chapter.) Trapped air in a waste arm can cause a sluggish drain and a gurgling noise. Charts in a code book list minimum slope for different pipe sizes, and many charts have options for additional dfus to be served if more slope than the minimum standard is provided. Figure 13-18 is an example of a chart that is used to determine minimum slope and the maximum dfus that can be served based on size and slope.

 FROM EXPERIENCE

With some horizontal piping configurations, too much slope can leave the solids behind in the drain. The minimum slope is dictated by code, but the maximum slope is only regulated on a waste arm and on the horizontal branch of a horizontal combination waste and vent system.

A separate chart in a code book shows where an installation occurs based on the segment of a system. A different chart is needed if a drain connects to a building drain than if it connects to a waste stack. A vertical pipe, such as a waste stack, can evacuate wastewater and sewage faster than a horizontal pipe, such as a building drain.

Figures 13-19 and 13-20 are examples of vertical and horizontal sizing charts that have unique features. Horizontal branch or fixture drain sizes are based on whether they connect to a waste stack or building drain. Each chart indicates the maximum dfus that a particular segment can handle without increasing the pipe size. Most charts in a code book have footnotes that indicate exceptions and provide clarifications. Pay close attention to every footnote, because they include important codes and are the only place some codes are listed.

The waste stack and building drain are major segments of a DWV system that receive discharge from the fixture drains and branches they serve. Both are sized based on the total dfu value of all drains and branches, but the building drain must also consider the dfu load of the waste stack. A drain cannot become smaller as the wastewater flows downstream. In many instances the building drain must be larger than the waste stack because a horizontal pipe cannot drain wastewater at the same rate as a vertical pipe. All codes dictate that a drain serving a toilet must be at least 3" in diameter and most codes do not allow more than two toilets to be connected to a 3" horizontal drain. Some codes allow a maximum of six toilets on a 3" waste stack; once again, that is because a waste stack is in the vertical position and can evacuate wastewater faster than a horizontal drain.

Codes have a practical basis for determining sizing, but vary according to the particular segment of a DWV system being sized. Comparing the charts in Figures 13-19 and 13-20 shows the differences in allowable dfus for a particular pipe size based on its vertical or horizontal position. Columns in Figure 13-19 are based on slope of pipe because it refers to horizontal pipe. Columns in Figure 13-20 relate to branches

(Based on 2003 International Plumbing Code)

Fixture Type	DFU Value	Minimum Trap Size
Automatic clothes washing machine/commercial[1]	3	2″
Automatic clothes washing machine/residential	2	2″
Bathroom group/1.6gpf (6.061pf) toilet[6]	5	N/A
Bathroom group/toilet more than 1.6gpf (6.061pf)[6]	6	N/A
Bathtub with or without shower head or whirlpool[2]	2	1½″
Bidet	1	1¼″
Combination sink and tray (Note: This is a single commercial pot sink)	2	1½″
Dental lavatory	1	1¼″
Dental unit or cuspidor	1	1¼″
Dishwasher/domestic (residential)[3]	2	1½″
Drinking fountain (Note: Also for an electric water cooler)	½	1¼″
Emergency floor drain	0	2″
Floor drain	2	2″
Kitchen sink/domestic (residential)	2	1½″
Kitchen sink/domestic (residential) with garbage disposal and/or dishwasher	2	1½″
Laundry tray/1 or 2 compartments	2	1½″
Lavatory	1	1¼″
Shower	2	1½″
Sink	2	1½″
Urinal	4	4
Urinal/1gpf (3.79lpf) or less	2[5]	4
Wash sink/per faucet/circular or multiple use	2	1½″
Water closet (toilet) tank type/public or private use	4[5]	4
Water closet (toilet)/private use/1.6gpf (3.79lpf)	3[5]	4
Water closet (toilet)/private use/more than 1.6gpf (3.79lpf)	4[5]	4
Water closet (toilet)/public use/1.6gpf (3.79lpf)	4[5]	4
Water closet (toilet)/public use/more than 1.6gpf (3.79lpf)	6[5]	4

[1] Use Figure 13–21 when a trap is larger than 3″.
[2] There is not a fixture unit value increase when a bathtub has a shower head and/or a whirlpool.
[3] Use Figure 13–21 for regulations on calculating fixture unit values for fixtures not listed or for fixtures with intermittent flow.
[4] Trap size shall be the same size as fixture outlet.
[5] The fixture unit values must be used when calculating the load of a building drain and building sewers. A lesser value cannot be used unless the lesser value is proven through testing.
[6] For fixtures added to a group value, add that particular fixture unit value to the group value.

Figure 13–17 Fixture drain and trap sizing chart.

(Based on 2003 International Plumbing Code)

Pipe Size	Minimum Slope
2½" or smaller	¼" per foot
3" through 6"	⅛" per foot
8" or larger	1/16" per foot

Figure 13-18 Pipe slope chart.

and different intervals of a waste stack because it is concerned with horizontal branches connecting to a vertical stack. See Figure 13-24 for sizing and code issues for a waste stack that is offset horizontally.

FROM EXPERIENCE

To remember which chart to use when sizing a horizontal pipe connecting to a building drain or waste stack, focus on the position of the connecting pipes. Remember horizontal to horizontal relates to a branch connecting to a building drain, and horizontal to vertical relates to a branch connecting to a stack.

(Based on 2003 International Plumbing Code)

	Slope per Foot			
Pipe Size	1/16" Slope	⅛" Slope	¼" Slope	½" Slope
1¼"	N/A	N/A	1	1
1½"	N/A	N/A	3	3
2"	N/A	N/A	21	26
2½"	N/A	N/A	24	31
3"*	N/A	36	42	50
4"	N/A	180	216	250
5"	N/A	390	480	575
6"	N/A	700	840	1000
8"	1400	1600	1920	2300
10"	2500	2900	3500	4200
12"	3900	4600	5600	6700
15"	7000	8300	10,000	12,000

*Minimum size building drain serving a water closet (toilet) is 3".

Figure 13-19 Building drain, building sewer, and branch sizing chart.

(Based on 2003 International Plumbing Code)

		Stacks[2]		
Pipe Size	Total per Horizontal Branch[1]	Total for One Branch Interval	Total for Stack 3 Stories or Less	Total for Stack More than 3 Stories
1½"	3	2	4	8
2"	6	6	10	24
2½"	12	9	20	42
3"	20	20	48	72
4"	160	90	240	500
5"	360	200	540	1100
6"	620	350	960	1900
8"	1400	600	2200	3600
10"	2500	1000	3800	5600
12"	2900	1500	6000	8400
15"	7000	Footnote 3	Footnote 3	Footnote 3

[1]Does not include building drain branches. Use Figure 13-19 for building drain, building sewer, and branches of building drain sizing
[2]Size stacks based on combined fixture unit totals at each branch interval or story. Stacks can be reduced as the fixture unit load decreases, but can never be reduced less than half the diameter of the largest size of the stack required
[3]Sizing load based on design criteria

Figure 13-20 Waste stack and branch sizing chart.

(Based on 2003 International Plumbing Code)

Drain or Trap Size	Fixture Unit Value
1¼"	1
1½"	2
2"	3
2½"	4
3"	5
4"	6

Figure 13-21 Unlisted fixture sizing chart.

A separate chart is used to size drains that receive indirect discharge from a piece of equipment, such as a pump or ice machine. A floor drain is listed in a chart similar to the one in Figure 13-17. However, once it receives discharge from equipment, it is no longer classified as a floor drain, but as a receptor. Sizing is based on known volumes of wastewater that a specific piece of equipment discharges into the drainage system. If the volume of a piece of equipment is not known, the dfu value of the fixture drain is based on the size of the trap receiving discharge from the equipment. If the discharge is automatically controlled, the dfu value is double its normal value. Figure 13-21 is an unlisted fixture chart used to determine a dfu value before reviewing charts in Figures 13-19 and 13-20.

FROM EXPERIENCE

Notice that the unlisted fixture sizing chart lists the drain sizes in sequence from the smallest to the largest and that the numerical dfu value is in sequence from 1 through 6 correlating with the drain sizes.

For step-by-step instructions on fixture sizing procedures, see the Procedures section on page 363.

Fixture Drain Sizing

A fixture drain installed above ground is sized according to the particular fixture served. A code book indicates the minimum drain and trap size that can serve a particular fixture. Actual piping location can change the minimum size listed in a chart for a drain, especially when DWV piping is installed below ground. Although a drain cannot be smaller than 1-1/4" in diameter, most codes dictate that the smallest drain that can be buried below ground is 2" in diameter. This illustrates that sizing is not always based on dfu theory, but can also be based on practical experience.

Figure 13-17 lists some minimum-size drains and traps for typical fixtures. Figure 13-21 lists values based on a drain and trap size of any fixture that is not listed in other charts in a code book. A separate drain cleanout serving a fixture drain is not required if the drain can be accessed with a removable slip joint connecting the trap to the drain. Because a toilet and a urinal are removable, they can serve as a cleanout for the fixture drain. A fixture drain must have an equal size cleanout with the exception of a 1-1/2" **trap adapter** that can serve a 2" fixture drain and a 1-1/4" trap adapter that can serve a 1-1/2" fixture drain. These sizes are acceptable by most codes. Figure 13-22 illustrates fixture drain regulations.

FROM EXPERIENCE

Some codes allow a maximum of three sinks to be served by a fixture drain with one trap. This is known as a continuous waste. The center of the fixtures cannot be more than 2'-6" apart; an example of this configuration is a double-bowl kitchen sink.

For step-by-step instructions on continuous waste procedures, see the Procedures section on page 364.

Fixture Branch and Horizontal Branch Sizing

A fixture branch and a horizontal branch are often the same pipe, depending on the complexity of a piping configuration. From a sizing standpoint, they use the same approach, which is to calculate the dfu load the branch serves. Some branches connect to a waste stack and some to a building drain. A separate chart in a code book is used for each connection location. For example, Figure 13-19 is used for branches that connect to a building drain, and Figure 13-20 is used for branches that connect to a waste stack. A battery-vented horizontal branch is not sized with methods discussed in this section, but is explained later in the chapter.

Unlike a fixture drain, the branches require a designated cleanout and cannot be served by removing a fixture or trap. The fixture branch illustrated in Figure 13-23 requires a cleanout by code. However, code officials might allow a fixture to serve as a cleanout for similar configurations as long as the fixture branch is not larger than the fixture drain size. If the fixture drain is 1-1/2" and the fixture branch is 2", an official might still allow the fixture as a point of access to the drain. Cleanouts are installed in various locations in a system. The maximum allowable distance between them is dictated in a code book, but they typically cannot be more than 100' apart.

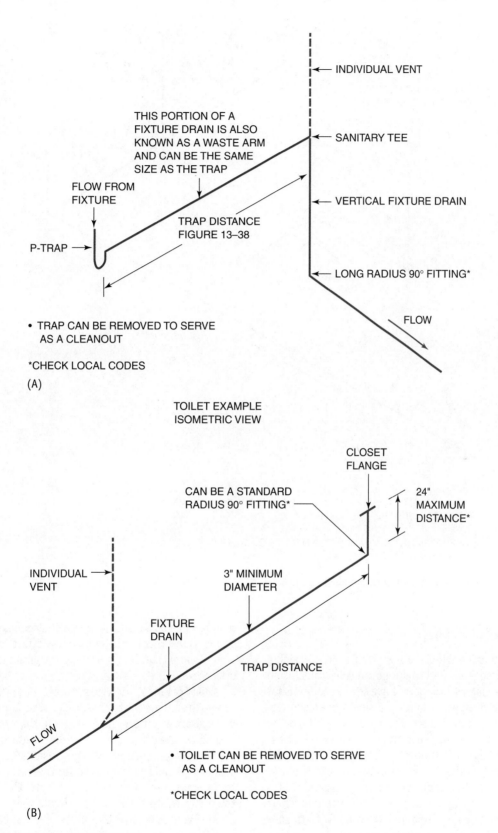

Figure 13–22 Fixture drain regulations.

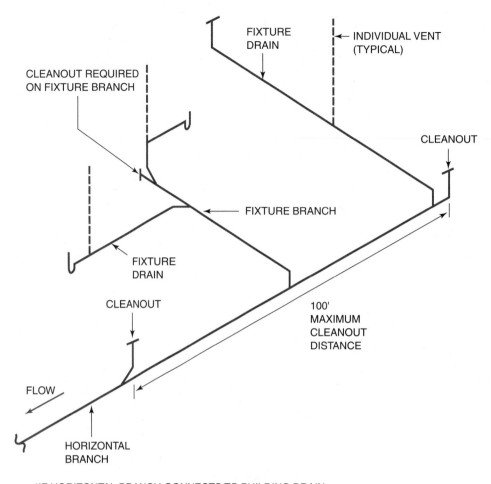

Figure 13-23 Fixture-branch and horizontal-branch regulations.

Waste Stack Sizing

The text supporting Figure 13-20 explains the sizing methods for a waste stack and the importance of calculating all dfus it will evacuate. Most charts have several columns for different building designs, including a single-branch interval, a two- or three-branch interval, and a column for more than three intervals. Some codes allow a waste stack to become smaller the higher it travels up a building, but it can never be less than half the diameter of the base of the stack. Because various configurations exist on construction sites and a waste stack is a major artery of a DWV system, a chart can only provide general sizing information. Offsets in a waste stack are required on many projects, and the offset areas are regulated with additional codes.

If a waste stack transitions 45° or greater from true vertical, special regulations are used to size the waste stack. If any connections occur within the offset area or dedicated protected zone within the offset area, the waste stack can no longer be sized from a chart. Many codes do not allow any connections within the horizontal offset area or protected zone if more than four branch intervals are located above the offset.

A vertical offset in a waste stack is considered to be 45° or less from true vertical. A relief vent is not required on a vertical offset unless there are more than four branch intervals above the offset area. If a relief vent is required, the upper and lower portions of the offset area must be vented separately, and the relief vent for the lower portion connects to the vent stack. The stack vent vents the upper portion of the offset waste stack. One allowable way to avoid installing a relief vent is to size the entire stack, including the offset area, based on building drain regulations, which essentially means oversizing the stack.

A code book has regulations and possibly illustrations pertaining to offsets in a waste stack, many of which require the installation of a relief vent. Figure 13-24 illustrates a waste stack that transitions horizontally. A cleanout must be

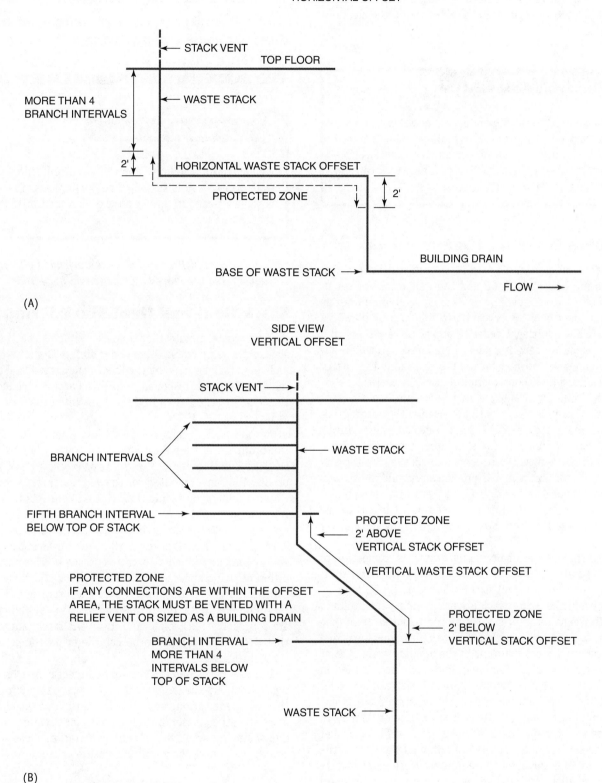

Figure 13–24 Offset in waste stack.

installed at the base of a waste stack. Some codes might dictate that the base of a stack cleanout must be installed above the highest fixture located on the floor where the cleanout is installed.

FROM EXPERIENCE

A code will state that the protected zone above and below a stack offset is either 2' or ten pipe diameters. The reason stack offsets are strictly regulated is because additional turbulence is created within the stack causing excessive positive or negative pressures.

Building Drain and Sewer Sizing

The text supporting Figure 13-19 explains the sizing methods for a building drain and the importance of calculating all dfus that it evacuates. Because a building drain is a major segment and the lowest horizontal portion of a DWV system, it has additional codes to adjust to variations in a project design. A chart lists general sizing scenarios, but additional codes are needed for unique situations that occur during the design or construction phase of a project.

The building drain and building sewer are two different segments of a system, but both are sized with a chart similar to the one in Figure 13-19. Some codes may allow different testing requirements for a building sewer than for a building drain because the sewer is outside the building and can typically be excavated if problems arise in the future. Most codes require a more rigid pipe specification for a building drain than for a building sewer because it is within the footprint of the building, often making it more difficult to replace.

Because a vent stack can originate from its connection to a building drain, certain codes regulate the connection point to make sure the vent stack is not affected by the turbulence created when the waste stack discharge enters the building drain. If a vent stack originates at the building drain, the connection must be no farther than ten pipe diameters from where the waste stack connects to the building drain. A vent stack that connects to a building drain must be installed in the vertical position. Figure 13-25 illustrates various codes for building drains and sewers. If the building drain is underground, most codes dictate that it must be at least 2" in diameter even though the chart in Figure 13-19 lists smaller sizes. The smaller sizes are typically only relevant when the building drain is installed above ground. Most codes dictate that the smallest building sewer be 4"; the building drain can be smaller than the building sewer. According to some codes, the base of a stack cleanout on a 3" minimum-size waste stack can be used as the cleanout for the building sewer if it is located not more than a developed length of 10' from the junction of the building drain and sewer.

CAUTION: Using the incorrect cleanout for traffic loads can cause extensive damage to underground piping.

FROM EXPERIENCE

A cleanout for a building sewer that is installed in a sidewalk or driveway is subjected to traffic loads. Specially designed cleanouts based on the relevant traffic loads should be used.

For step-by-step instructions on vent stack and building drain procedures, see the Procedures section on page 365.

Stack Vent and Vent Stack Sizing

Many codes dictate that a stack vent must be the same size as the waste stack. Other codes dictate that a stack vent and vent stack be sized by the length they travel and the total dfus they are responsible for discharging. In addition, they cannot be less than half the diameter of the waste stack or less than 1-1/4".

A vent stack, which receives vents from a DWV system, can originate in the vertical position from either the waste stack or building drain. The chart for sizing a stack vent is usually the same one used for the vent stack; it is shown in Figure 13-20. A horizontal branch that connects to a building drain does not have to be calculated in the sizing of the waste stack. However, its dfu value does have to be considered in sizing a vent stack if the vent for that same horizontal branch connects to the vent stack. This results in different sizing approaches for a vent stack and a stack vent concerning the dfus connected to each segment. If vents serving a horizontal drain that connects to a building drain are routed to areas other than the vent stack, the dfus of that drain do not have to be considered in the total vent stack dfu value.

If a vent stack and a stack vent connect before terminating to open air, the larger of the two must be used for sizing the roof penetration, and the size must accommodate the stack with the greater dfu value. In colder climates, the minimum size allowable for a roof penetration must be confirmed on a chart showing frost closure codes.

CAUTION: If a vent becomes obstructed with frost, the DWV system will not function as designed.

CHAPTER 13 Drainage, Waste, and Vent Segments and Sizing

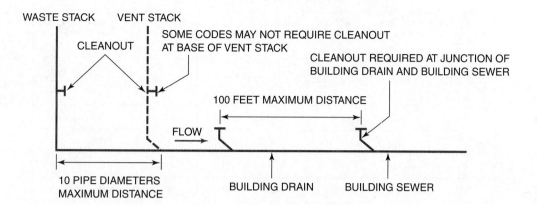

*SIZE BUILDING DRAIN AND BUILDING SEWER USING FIGURE 13–19

*MINIMUM SIZE BUILDING SEWER IS 4" (SEE LOCAL CODE)

*MAXIMUM DISTANCE BETWEEN CLEANOUTS ON STRAIGHT RUN OF PIPING IS 100'

(A)

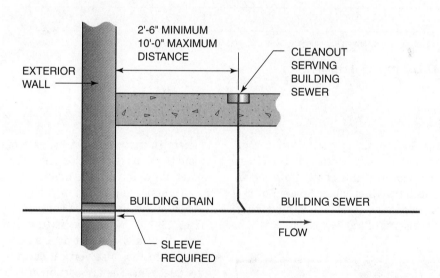

*CLEANOUT MUST BE SAME SIZE AS BUILDING SEWER, EXCEPT A 4" CLEANOUT CAN BE USED TO SERVE PIPES LARGER THAN 4"

*BUILDING SEWERS 8" DIAMETER AND LARGER TYPICALLY REQUIRE A MANHOLE

(B)

Figure 13–25 Building drain and sewer regulations.

(Based on 2003 International Plumbing Code)

Stack Size	DFU Total	Maximum Developed Length of Vent in Feet*					
		1¼" Pipe	1½" Pipe	2" Pipe	2½" Pipe	3" Pipe	4" Pipe
1¼"	2	30	0	0	0	0	0
1½"	8	50	150	0	0	0	0
1½"	10	30	100	0	0	0	0
2"	12	30	75	200	0	0	0
2"	20	26	50	150	0	0	0
2½"	42	0	30	100	300	0	0
3"	10	0	42	150	360	1040	0
3"	21	0	32	110	270	810	0
3"	53	0	27	94	230	680	0
3"	102	0	25	86	210	620	0
4"	43	0	0	35	85	250	980
4"	140	0	0	27	65	200	750
4"	320	0	0	23	55	170	640
4"	540	0	0	21	50	150	580

*Measure developed length from the point vent connects with drain to its termination point

Figure 13–26 Vent stack and stack vent sizing chart.

Figure 13–26 shows a typical sizing chart from a code book for a vent stack and stack vent. Most code books list more pipe sizes than are shown in this chart. However, in this section, we are focusing on your understanding of the essentials and not every pipe size. Figure 13–27 illustrates connections to various segments of a DWV system.

FROM EXPERIENCE

A vent stack that connects to a waste stack must remain in the vertical position. It is completed with a wye and 45° fitting. It must be installed at or below the height of the first branch interval connected to the waste stack.

For step-by-step instructions on vent stack and waste stack connection procedures, see the Procedures section on page 366.

Individual Vent Sizing

An individual vent is generally half the diameter of the drain it serves, but no less than 1-1/4" diameter. Some codes dictate that a vent must increase one pipe size if it travels more than 40 feet and that the increased size begins from its point of origin with the drain.

A piping configuration that can be served by an individual vent is a common-vented piping arrangement (Fig. 13–28). When two fixture drains are located on the same floor and connect at the same height or at different heights to the same vertical drain, it is vented with an individual vent. The dry portion of a common vent is sized the same as an individual vent. If the two fixtures connect at different heights to the same vertical drain, the middle section is a wet vent. The middle section is sized using a different chart in a code book than the chart used to size the individual vent.

Specific regulations exist pertaining to common venting. A toilet cannot be the fixture using the top drain and the dfu value of the upper fixture must comply with a chart in a code book based on the pipe size of the wet vent. Figure 13–29 illustrates common vent configurations for an individual vent. Figure 13–30 is a sizing chart for a common vent indicating the wet-vent size and its maximum dfu allowed. Most code books have a designated chart for common venting even though it is a form of wet venting.

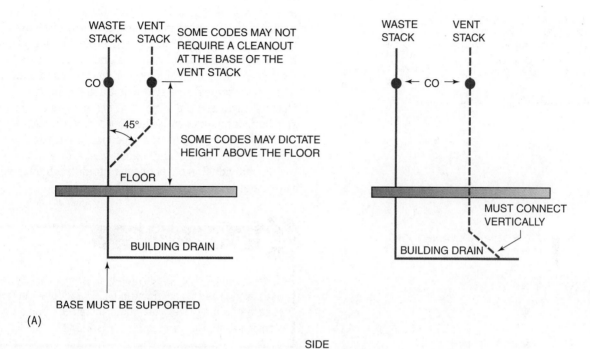
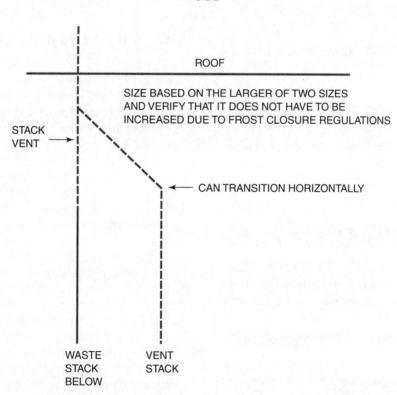

Figure 13–27 Vent stack and stack vent connections.

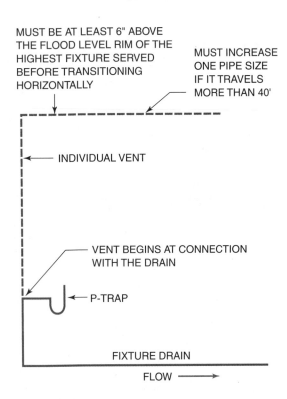

Figure 13–28 Individual vent sizing.

CAUTION: If a toilet is installed on the top fixture drain of a common-vented configuration, the lower fixture trap seal can be jeopardized, and sewer gas can enter an occupied area.

FROM EXPERIENCE

One way to remember that a common vent exists is to recall that two fixtures share a common vertical drain. This configuration is typical if two fixtures are installed side by side or back to back.

Branch Vent Sizing

A branch vent is sized by the diameter of the drainage branch, the dfu value it is responsible for discharging, and the length it travels (Fig. 13–31). A branch vent cannot be less than half the diameter of the drain it serves, and no less than 1-1/4" diameter. Some codes dictate that a branch vent must increase one pipe size if it travels more than 40' and that the increased pipe size must begin at its connection to the horizontal branch. Some codes dictate that a branch vent must be increased one pipe size when two individual vents connect to form the branch vent. For example, if two 1-1/2" individual vents connect to form the branch vent, it must be at least 2" in diameter.

 FROM EXPERIENCE

Even though 2-1/2" pipe is rarely used in a DWV system, it is recognized when sizing pipe. If a 2" pipe must increase one pipe size, you would install 3", but on a licensing exam the correct answer is 2-1/2".

Circuit and Loop-Vent Sizing

A circuit and a loop vent are both classified as branch vents, but they are special piping arrangements that must be sized according to their unique piping allowances. The word *circuit* refers to isolating a portion of a DWV system to create its own little system. It makes its own air-circulating circuit from other segments, even though the drain and vent connect to the same main segments as other branches do. Most codes dictate that any fixture located downstream of a toilet must be individually vented; however, most codes allow for numerous fixtures to be piped in a battery. A toilet can be downstream of another toilet when it is installed in a battery-vented configuration served by a circuit or loop vent, but it must adhere to numerous regulations.

The horizontal battery-vented branch is sized based on the total dfus it serves and must be piped full size throughout its length. Most codes dictate that no more than eight fixtures can be connected to a battery-vented horizontal branch. If more than eight are connected, the excess fixtures must be individually vented. A circuit vent begins in front of the last fixture connected to the horizontal branch and terminates with the vent stack. A code book may have a vent sizing chart for sizing a circuit or loop vent, and code might limit the distance it travels based on a pipe size of the vent.

A relief vent is required on a battery-vented branch served with a circuit vent. A loop vent is the same as a circuit vent except that it does not require a relief vent. However, some codes dictate that a relief vent must be installed if more than three toilets are connected to a battery-vented

CHAPTER 13 *Drainage, Waste, and Vent Segments and Sizing* **351**

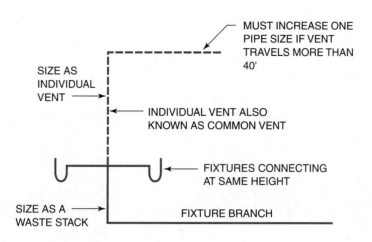

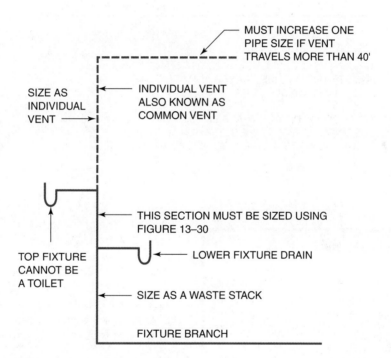

Figure 13–29 Individual and common vent.

(Based on 2003 International Plumbing Code)

Drain Size	DFU Maximum
1½"	1
2"	4
2½"–3"	6

Figure 13–30 Common vent sizing.

branch. The loop vent still connects in front of the last fixture of the horizontal branch, but it connects to the stack vent instead of the vent stack. Some codes dictate that a relief vent for a battery-vented branch must be at least half the diameter of the horizontal branch, but no less than 1-1/2" diameter. This varies from other vents that cannot be smaller than 1-1/4" diameter. Figure 13-32 illustrates two battery-vented horizontal branches and the difference between a circuit vent and a loop vent.

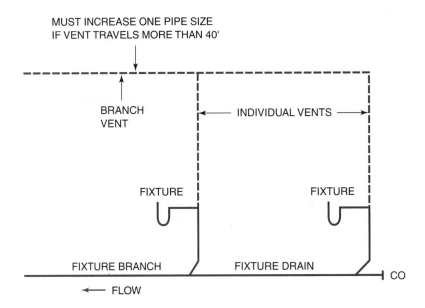

Figure 13–31 Branch vent.

 FROM EXPERIENCE

Most codes dictate that not more than two toilets can be installed on a 3" horizontal branch, but some codes allow more. When a 4" branch is needed to accommodate the number of toilets, the minimum size of the relief vent, circuit vent, and loop vent connected to a 4" battery-vented branch is 2" because a vent cannot be less than half the diameter of the drain it serves.

Wet-Vent Sizing

One regulation that typically remains in effect even when wet venting is that no fixture can be installed downstream of a toilet without being individually vented. If a toilet is one of the fixtures within a wet-vented configuration, the minimum size of the dry vent is 1-1/2" because a vent must be at least half the diameter of the drain it serves. The reasoning behind this vent sizing is that a toilet must have at least a 3" diameter drain. That step in the process only established the dry vent portion of the bathroom group; the size of the wet-vented portion is based on the dfu value of the fixture or fixtures draining into the wet vent. A chart in a code book usually lists the allowable number of dfus that can drain into a particular size wet vent. Figure 13-33 illustrates a wet-vented piping arrangement. Figure 13-34 is a typical sizing chart that might be found in a code book. Any fixtures or branches that are not part of the wet-venting arrangement must connect downstream of the wet-vented configuration.

 FROM EXPERIENCE

The bottom half of the wet-vented portion of a horizontal pipe conveys wastewater and the top half is for airflow. This is why a wet-vent configuration must be larger than a standard size drain that is individually vented.

Air Admittance Valve Sizing

Air admittance valves (AAV) allow air into a DWV system, but do not allow sewer gas to escape into an occupied area. They were initially introduced for use under a kitchen sink or in remote areas where it is difficult to install individual

CHAPTER 13 *Drainage, Waste, and Vent Segments and Sizing* **353**

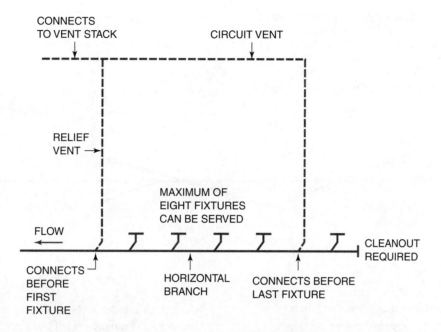

(A)

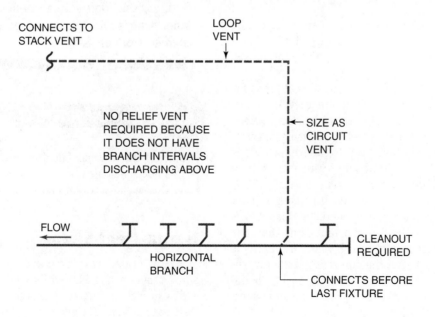

(B)

Figure 13-32 Circuit and loop vents.

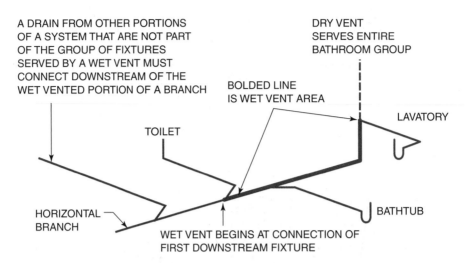

Figure 13–33 Wet venting.

Wet Vent Size	DFU Maximum
1½"	1
2"	4
2½"	6
3"	12

(Based on 2003 International Plumbing Code)

Figure 13–34 Wet vent sizing chart.

vents. A stack vent must terminate to open air. It cannot be served by an AAV to minimize the possibility of a DWV system becoming overpressurized or creating a vacuum by the fixture load.

Many new codes have created more strict regulations for AAV use. For example, a relief vent is now required on a horizontal branch served with an AAV. Remember that a horizontal branch is DWV pipe that serves two or more fixtures, so a relief vent is not required for a fixture drain that is served with an AAV. This type of relief vent differs from relief vents described for offsets in waste stacks and for battery-vented branches. Figure 13–35 illustrates a relief vent installed on a horizontal branch vented with an AAV.

AAV sizing is based on the size of a vent using the same codes as those for individual and branch vents. If the AAV is installed on an individual vent, the pipe size and the AAV selection would be determined as shown in Figure 13–28 and would need to comply with all codes for an individual vent. Some codes dictate that only a certain number of dfus can be served with a particular size of AAV. A chart in a code book is used for sizing based on the dfu the AAV serves. An AAV cannot be installed in supply or return HVAC air plenums or be used in certain specialty DWV systems such as acid neutralization systems. Access must be provided to replace an AAV; wall boxes with louvered access covers are sold for this purpose. The AAV must be in an area where there is adequate air to enter a DWV system through it. The height above the trap and above the insulation in an attic is regulated as illustrated in Figure 13–36.

CAUTION: If an AAV serves a washing machine and the drainage configuration connects close to other fixtures, the high-volume demands may jeopardize other trap seals because pressure cannot be discharged from an AAV.

 FROM EXPERIENCE

Suspect the AAV has failed if sewer gas odor is present in an occupied area.

Island-Vent Sizing

A sink located away from walls is a candidate for an AAV, but if they are not allowed by code, the sink can be piped with a special venting arrangement known as an island vent. Most special DWV designs must receive approval from code officials before installation can begin. Most codes limit island venting to sinks and lavatories, which includes dishwashers and garbage disposals when used for a kitchen sink installation.

CHAPTER 13 *Drainage, Waste, and Vent Segments and Sizing* **355**

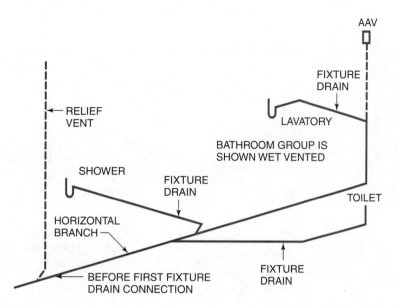

Figure 13-35 Air admittance valve relief vent.

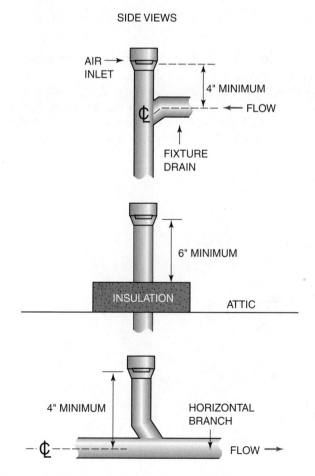

Figure 13-36 Air admittance valve installation.

If an island vent is installed on a bottom floor and the pipe is installed underground, most codes dictate that a below-ground DWV pipe be at least 2″ diameter. If the island-vent pipe is not installed below ground, drain and vent sizing is based on codes for standard fixtures. A full-size cleanout must be installed to serve the lowest portion of an island vent under the sink. The island vent piping must be installed vertically as high as possible under the sink before transitioning horizontally.

All fittings used in the vent piping below the flood level rim of the fixture must comply with drainage codes for their flow patterns. The island vent under the sink must be a vertical continuation of the fixture drain, and the fixture drain must connect on the top half of a horizontal drain. Vertical means that a pipe is in the true vertical position or no more than 45° from true vertical. Figure 13-37 illustrates an island-vent configuration.

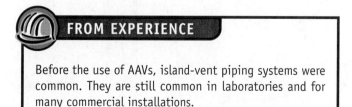

Before the use of AAVs, island-vent piping systems were common. They are still common in laboratories and for many commercial installations.

For step-by-step instructions on island-vent installation procedures, see the Procedures section on pages 367–368.

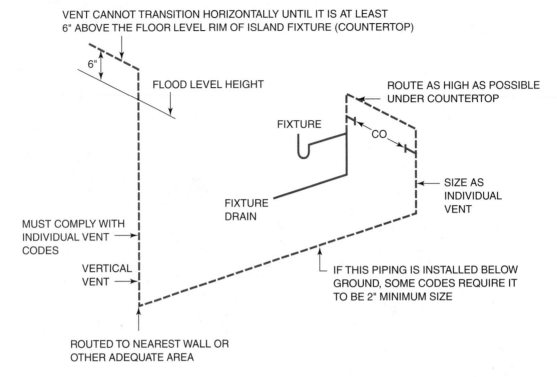

Figure 13-37 Island-vent sizing.

Trap Distance

An important layout consideration based on DWV codes is the location of a fixture trap in relation to its protective vent. Codes that regulate the maximum distance a trap can be from a vent—**trap distance**—are based on the size of the drain and trap. The trap distance is determined along the pipe route as opposed to simply measuring the distance from the vent to the center of a fixture. This is known as **developed length.** A code book has a chart indicating the maximum distance. Figure 13-38 illustrates trap distance, and Figure 13-39 shows a typical chart for determining allowable trap distance.

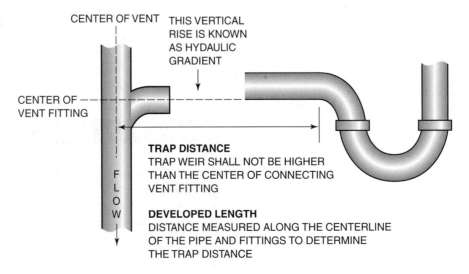

Figure 13-38 Trap distance and developed length.

(Based on 2003 International Plumbing Code)

Trap Size	Drain Size	Slope/Foot	Distance
1¼"	1¼"	¼"	3'-6'
1¼"	1½"	¼"	5'
1½"	1½"	¼"	5'
1½"	2"	¼"	6'
2"	2"	¼"	6'
3"	3"	⅛"	10'
4"	4"	⅛"	12'

Figure 13-39 Trap distance chart.

The horizontal fixture drain extending from the vent to the trap is also known as a waste arm. Some codes regulate the maximum vertical distance a trap weir can be from the centerline of the connecting vent fitting. This depends on the slope of the waste arm. This distance is known as **hydraulic gradient.** Some codes dictate that the total vertical rise of the waste arm cannot exceed the diameter of one pipe the size of the waste arm. For example, a 1-1/2" waste arm cannot rise more than 1-1/2" from the center of the vent fitting to the trap weir.

The closest a trap can be to a vent is also regulated; that code is known as crown venting. The weir of a trap cannot be less than two pipe diameters from the connecting vent; that is, a 2" trap weir cannot be less than 4" from the center of the vent.

CAUTION

CAUTION: Too much slope can cause air to become trapped in the waste arm resulting in gurgling or in sluggish draining of the fixture.

FROM EXPERIENCE

Hydraulic relates to fluid and gradient relates to slope; hydraulic gradient is simply a term to describe the total slope of the horizontal fixture drain.

For step-by-step instructions on crown venting procedures, see the procedures section on page 369.

Septic Systems

Areas that are not incorporated into a city or other formal municipality use private sewage-disposal systems. Many years ago, outhouses were used for toilet facilities and are still used in some rural areas of the country. However, pollution and health concerns within communities led to the prohibition of outhouses and dictated the use of septic tank systems.

Further advancement in septic system design, combined with trial and error, has changed the installation requirements of private disposal systems in the past few decades. Many private disposal systems use pumps to discharge **effluent** to a drain field and do not require a distribution box. In conventional septic tank systems, the ground absorbs wastewater (effluent) that flows from the tank by gravity. Figure 13-40 is an illustration of a conventional gravity septic tank system. Further breakdowns of each area are shown throughout this section.

FROM EXPERIENCE

Garbage disposals are typically not installed on a septic tank system. Advancement in biodegradable agents has allowed them to be used, but they are not common. Any food particles that escape from the septic tank could obstruct the distribution box or drain field piping.

The negative impact on groundwater and streams from septic systems continues to be debated and many environmental organizations warn that excessive development in certain areas can contaminate the earth. Numerous installation options are available for many specific obstacles on a desired building location. As with most plumbing-related codes, strict regulations that dictate installation practices vary among states and, in many instances, among counties of a state.

The operation of a conventional septic system is simple. It requires proper soil conditions that must be adequate for **percolation** of effluent into the surrounding soil. The trade name for that process is perc. If the ground does not perc due to a high water table or conditions that do not allow proper water absorption, such as rock, alternative private disposal systems must be considered. Soil conditions must be inspected by local authorities through a **perc test** or a site evaluation to ensure that the treatment qualities of a particular property are adequate. Figure 13-41 is an illustration of a typical septic tank design. Actual choice of a design is based on the plumbing system served and local codes.

358 SECTION THREE *Layout and Installation*

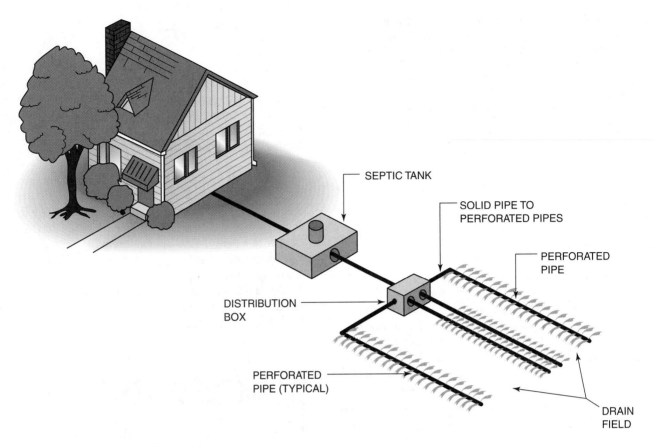

Figure 13-40 Conventional septic system.

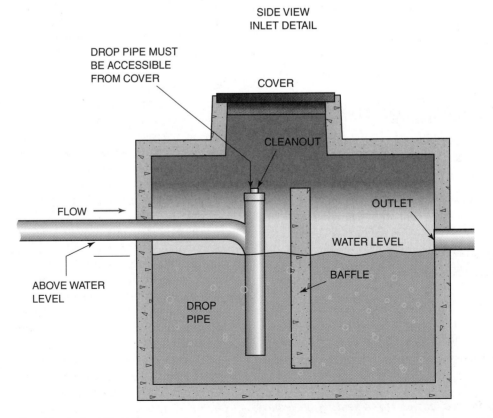

Figure 13-41 Conventional typical septic tank.

FROM EXPERIENCE

A perc test is simply a way to establish how fast soil absorbs water. One basic approach is to establish the minimum depth and diameter that a hole must be dug from a local regulating department, then dig the hole and fill it with water. Track the time it takes for the water to absorb into the ground to determine the percolation of the soil.

Many states adopt guidelines for sizing septic systems based on occupancy of a dwelling. Residential application guidelines based on federal standards estimate that a single occupant generates 60 gallons per day (gpd) of wastewater. This allows for all uses of a plumbing system, such as preparing food, washing clothes, using a toilet, and bathing. Other methods are also used, but the 60 gpd per occupant is the most common. Sizing is also based on the number of bedrooms per residential dwelling; it assumes that two people occupy one bedroom using 120 gpd per bedroom.

The maximum rate of flow from a building must be able to adequately percolate into the soil to install a septic tank system. In a conventional gravity system, sewage-laden wastewater flows from a building to a holding area known as a septic tank. The solids are retained in the tank, and the effluent flows to a distribution box that distributes it equally to several perforated pipes for drainage into the ground. Figure 13–42 shows a typical **distribution box,** or **D-box.** Actual design is based on the number of pipes it must serve and local regulations.

FROM EXPERIENCE

The distribution box is usually made of concrete and ordered according to the number of outlets needed based on a particular drain field design.

Effluent evaporates in many arid regions of the country, but local regulations usually do not take evaporation into account in determining the size of a **drain field.** With an evapotranspiration system, 100 percent of the effluent evaporates. This type of system needs a larger drain field, because it has to hold more water than a conventional drain field. A conventional drain field drains regardless of the moisture

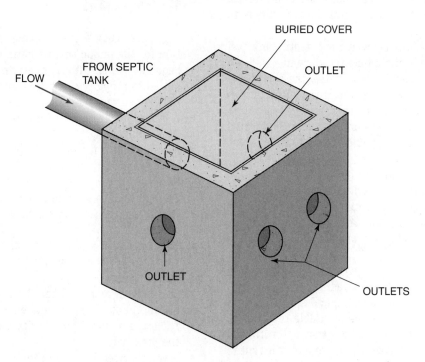

Figure 13–42 Conventional distribution box.

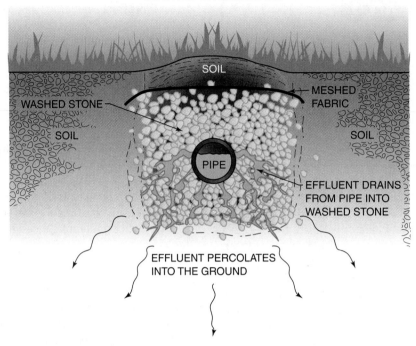

Figure 13-43 Conventional gravity drain field.

present in the atmosphere, but an evapotranspiration system must rely on the evaporating qualities of the atmosphere to function properly.

The distance between a water well and a septic tank system is strictly regulated. A drain field too close to a well can contaminate drinking water. In most states, the septic system must be at least 50 feet away from a well. Regulations also exist regarding location of a septic system in relation to adjacent properties. A well on one property must be located a minimum distance from a septic system on an adjacent property.

Perforated pipes are surrounded by gravel or stone, and the actual design is regulated by codes for specific locations. Many areas with rocky soil mandate that sand be used in conjunction with gravel or stone. Areas that are prone to flooding have their own methods or allowable designs for septic systems.

Figure 13-43 is an illustration of a typical drain field, also known as a leach field or by the misleading term filter bed. A drain field does not filter wastewater, and any solids that enter the piping system can eventually disable the system. This is one area of a septic tank design that is scrutinized extensively. Improper installation can render the entire system useless and can result in extensive repair or replacement. Periods of heavy rainfall can also hamper the percolation ability of soil that becomes saturated and cannot properly absorb effluent.

Summary

- A plumber must focus on being the front line of protection for the public against harmful sewer gas.
- Segment identification of certain drains and vents is required to properly size a DWV system.
- DWV codes vary by region, state, and local area.
- A drainage fixture unit (dfu) is the basis for sizing a DWV system.
- A code book is required to locate the allowable dfu load on a particular pipe size.
- Minimum trap size is dictated by code and located in the appropriate table in a code book.
- Fitting use is based on the flow position of a drainage system.
- The three flow positions of a drainage system are horizontal to horizontal, vertical to horizontal, and horizontal to vertical.
- Cleanouts are required based on changes in direction and distance, and they are required at the base of a stack.
- The slope required on a horizontal drain varies based on the pipe size.
- A vent is sized based on the drain size, the distance it travels, and its classification.
- Most codes dictate that one 3" minimum vent be extended to open air.
- A wet vent is a drain for allowable fixtures being used as a vent for other fixtures.

Procedures

Positioning Cleanouts at Base of Stacks

A Cleanout in direction of flow: This prevents a drain cleaning cable from having too much tension when entering the drain.

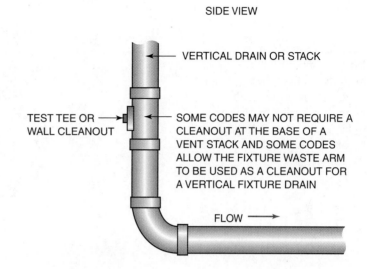

B Cleanout 90° from direction of flow: The cleanout can be installed 90° from the direction of flow. This illustration shows the cleanout facing you, but it can also be installed 180° from this view.

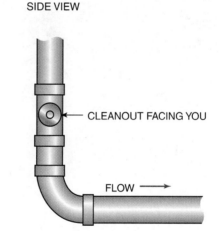

Procedures

Positioning Cleanouts at Base of Stacks (continued)

C Cleanout installed incorrectly: This incorrect installation would force the drain-cleaning cable to make a U-turn. This would result in greater tension on the cable, which could cause it to break or become lodged in the drain.

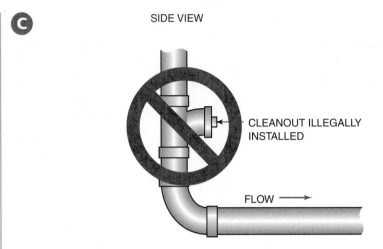

D Correct installations

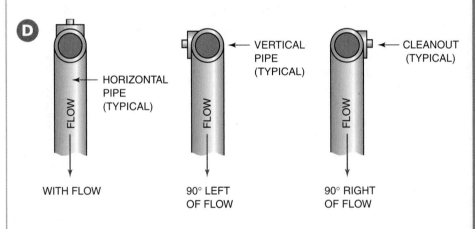

CHAPTER 13 *Drainage, Waste, and Vent Segments and Sizing*   **363**

Procedures

Unlisted Fixture Sizing

- Use Figure 13-21 to locate the trap and drain size for five dfus.

- Because the floor drain is not serving as it is listed in Figure 13-17, it must be sized as a waste receptor using this method.

A Some codes dictate that if the trap is located underground and the chart indicates a pipe size less than 2" in diameter, you must install 2" pipe, because that is the smallest size that can be buried below ground.

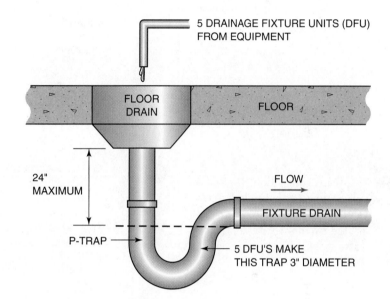

Procedures

Continuous Waste

A Two sink connection: Two sinks connect to a single trap with one fixture drain. The centers of the sink outlets cannot be more than 30" apart. You can purchase a continuous waste outlet assembly or create your own.

B Three-sink connection: Three sinks connect to a single trap. A sanitary cross can be used to create the two horizontal continuous waste extensions at the same level.

C Three-sink connection: This configuration allows you to install a sanitary tee on its back, even though it is illegal to install it in that position anywhere else in a DWV system. Fittings installed on the fixture side of a p-trap do not have to comply with the fitting codes that regulate DWV fittings on the downstream side of the trap.

Note: Any of these configurations is allowable with a maximum of three sinks. Remember that sink outlets cannot be more than 30" from center to center.

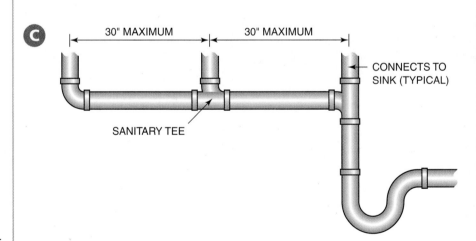

Procedures

Vent Stack and Building Drain Connection

A Vent stack connection to building drain: When establishing a vent stack from the building drain, the center of the vent stack cannot be more than ten pipe diameters away from the waste stack. This illustration shows how a wye fitting and a 45° fitting are used to make sure the vent stack connection is in the vertical position. A combination wye and 1/8 bend fitting (combo) is also acceptable.

A
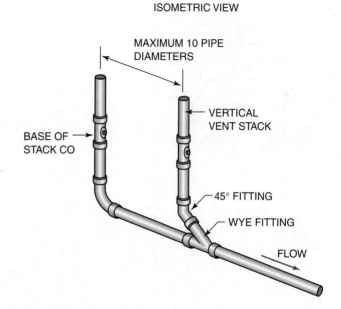
ISOMETRIC VIEW

B Incorrect vent stack connection to building drain: This illustration shows an installation that is illegal because the vent stack is connected to the building drain in a horizontal position. If the vent stack must offset away from the building drain, the connection must remain vertical or no more than 45° from true vertical.

B
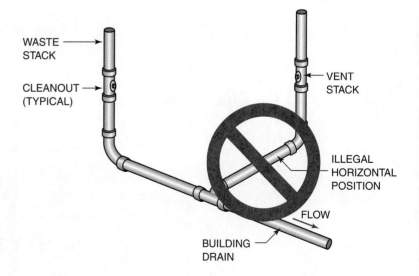
ISOMETRIC VIEW

366 **SECTION THREE** *Layout and Installation*

Procedures

Vent-stack and Waste-stack Connection

A Vent-stack connection to waste stack: A vent stack must originate in a vertical position. Vertical is considered no more than 45° from true vertical. A wye fitting is installed with a 45° fitting to create the vent stack. The vent stack must be established at the same height as or below the first branch interval on the waste stack.

B Incorrect vent-stack connection to waste stack: This installation violates code. It places the vent stack connection in the horizontal position. Horizontal is considered to be more than 45° from true vertical.

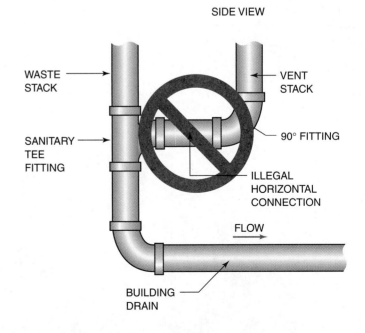

Procedures

Island Vent

A Island-vent drain connection to branch: An island-vented drain must connect to the top half of a drainage branch in the vertical position or no more than 45° from the true vertical. Because the code does not specifically state that the connection must be vertical, your local inspector may allow a variation of this connection.

B Island-vent under-sink connection: The sink connection and waste arm are installed like other sink installations. The dry vent is routed as high as possible under the sink. Code normally dictates that a dry vent must be at least 6″ above the flood level rim of the fixture it serves before transitioning horizontal. This routing of a dry vent below the flood level rim of a fixture is the only exception to that code. A cleanout must be installed to serve both the drain and vent.

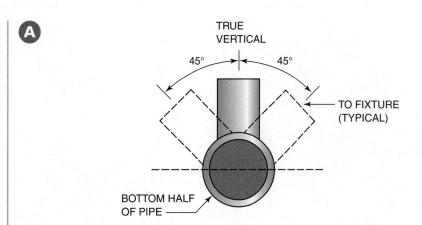

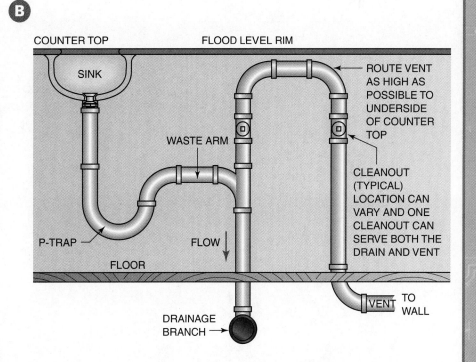

Procedures

Island Vent (continued)

C Island-vent termination: After routing the island vent from the sink location to a wall or other designated area, the vent must remain vertical until it is at least 6" above the flood level rim of the fixture. All vent fittings for the entire island-vent system must comply with drainage codes, which means the 90° fittings below the flood level rim may have to be a long radius-type based on most codes.

C
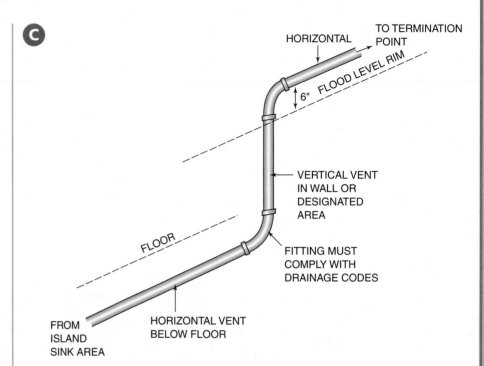

Procedures

Crown Venting

A Vertical example: Trap-distance codes state the maximum distance a trap weir can be from the connecting vent. Crown venting code regulates how close the weir can be. The weir cannot be closer than two pipe diameters from the vent. If the pipe in this illustration is 2" diameter, the minimum trap distance is 4".

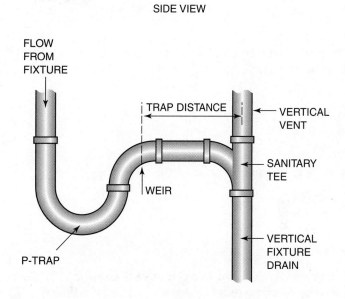

B Horizontal example: Because codes allow certain fixtures to be installed using a wet vent, the fixture in this illustration uses the horizontal branch as a drain and vent. This configuration is often overlooked when considering trap distance and crown venting codes. If the diameter of the p-trap in this illustration is 3", the minimum trap distance is 6".

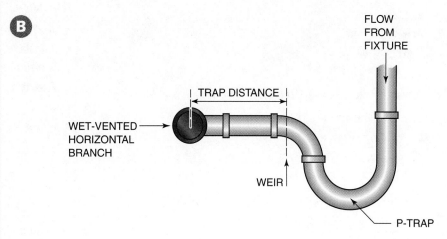

Review Questions

Each code-related question has a selection for none of the above. If none of the answer selections are correct according to your state code, select none of the above and write in your answer.

1. Drainage from a building drain to a public sewer or private septic tank flows through a

a. Horizontal branch
b. Building sewer
c. Waste stack
d. Fixture drain

2. A stack vent is an extension of the

a. Vent stack
b. Building drain
c. Circuit vent
d. Waste stack

3. A loop vent is the same as a circuit vent except it does not require a relief vent and it connects to the

a. Stack vent
b. Branch vent
c. Vent stack
d. Individual vent

4. A waste stack is a vertical extension of the

a. Building drain
b. Building sewer
c. Stack vent
d. Vent stack

5. A drain serving a single-fixture trap is known as a

a. Fixture branch
b. Fixture drain
c. Horizontal branch
d. Individual drain

6. The connection of a building drain and building sewer requires a

a. Relief vent
b. Cleanout
c. Wet vent
d. Horizontal branch

7. A drain that is also used as a vent for other fixtures within a group is called a

a. Drain vent
b. Vent drain
c. Waste vent
d. Wet vent

8. A kitchen sink located away from a wall is a fixture that can be vented with

a. A 2" vent
b. An island vent
c. A return vent
d. A vent drain

9. Every plumbing fixture directly connected to a drainage waste and vent system must have

a. A trap
b. A handicap faucet
c. An aerator
d. A wall anchor

10. Minimum size pipe that can be directly connected to a DWV system is

a. 1"
b. 1-1/4"
c. 1-1/2"
d. None of the above—provide answer

11 **A dry vent cannot transition horizontal until it is above the flood level rim of a fixture it serves a distance of**

a. 2"
b. 4"
c. 6"
d. None of the above—provide answer

12 **The minimum size a vent can be, without knowing any specifics, is less than half the diameter of the**

a. Drain it serves
b. Largest vent
c. Vent stack
d. None of the above—provide answer,

13 **The minimum size DWV pipe that can be buried below ground is**

a. 2"
b. 3"
c. 4"
d. None of the above—provide answer

14 **The minimum size of a building sewer is**

a. 2"
b. 3"
c. 4"
d. None of the above—provide answer

15 **The minimum and maximum trap seal depth is**

a. 1" and 2"
b. 2" and 3"
c. 2" and 4"
d. None of the above—provide answer

16 **The code that regulates the minimum distance a trap weir can be from the connecting vent fitting is called**

a. Trap distance
b. Crown venting
c. Developed length
d. Hydraulic gradient

17 **The code that regulates the maximum distance a trap weir can be from the connecting vent fitting is called**

a. Trap distance
b. Crown venting
c. Developed length
d. Hydraulic gradient

18 **A drainage pipe must slope to drain adequately. The minimum per foot slope is determined based on the**

a. Specific job conditions
b. Preference of a company
c. Type of piping
d. Size of piping

19 **A horizontal branch connecting to a building drain is considered to be a horizontal to**

a. Vertical connection
b. Horizontal connection
c. Upward connection
d. Downward connection

20 **Sizing for a horizontal branch that connects to waste stack is based on**

a. Drainage fixture units
b. Size of building drain
c. Size of stack vent
d. Size of vent stack

21 **The minimum size pipe that can serve a toilet is**

a. 2"
b. 3"
c. 4"
d. 6"

22 The maximum distance between cleanouts on a straight run of piping is

a. 100 feet
b. 200 feet
c. 150 feet
d. None of the above—provide answer

23 A cleanout must usually be full size, but some codes allow a pipe greater than 4" in size to be served by a

a. Contractor
b. 2" cleanout
c. 4" cleanout
d. None of the above—provide answer

24 Most codes dictate that if an individual vent exceeds 40' in length, it must increase

a. One pipe size
b. Two pipe sizes
c. At the connection to the vent stack
d. None of the above—provide answer

25 An air admittance valve must be located above the center of the fixture drain at least

a. 2"
b. 3"
c. 4"
d. None of the above—provide answer

26 A circuit and a loop vent are specialized vents but are classified as

a. An individual vent
b. A branch vent
c. A wet vent
d. A common vent

27 The maximum distance a cleanout can be from the exterior of a building at the junction of the building drain and building sewer is

a. 3'
b. 8'
c. 10'
d. None of the above—provide answer

28 A septic system that flows without the use of any mechanical equipment is a

a. Gravity system
b. Low-pressure pump system
c. Vaulted system
d. Residential system

29 Effluent flows from a conventional gravity septic tank to a

a. Distribution box
b. Drain field
c. Filter bed
d. Leach field

30 Soil-evaluation site inspections are done to determine the treatment quality of soil and in many instances require a

a. Perc test
b. Property deed
c. Gravel availability report
d. Chemical treatment report

31 A septic tank system must be located a safe distance, dictated by local regulations, from a

a. Drain field
b. Septic tank
c. Distribution box
d. Water well

Chapter 14: Drainage, Waste, and Vent Installation

We have discussed basic drainage, waste, and vent (DWV) fittings and their sizing and code information. The design intent of a residential home determines the actual layout and installation of a drainage and vent system. The drilling and notching codes for water distribution systems that we discussed in an earlier chapter apply to the installation of DWV piping as well. DWV piping is larger than water piping, so drainage and vent piping routes are usually established first. Residential blueprints do not typically indicate a pipe route. The fixture layout and pipe route must adhere to code regulations and design intent while remaining productive. Limited wall and ceiling space combined with strict drilling regulations poses challenges for routing drainage and vent systems in a house. This chapter discusses those difficulties and provides information to help you face common job-site challenges.

OBJECTIVES

Upon completion of this chapter, the student should be able to:

- know basic layout considerations based on common fixture types and codes.
- be able to recognize pipe route installations based on structural obstacles.
- apply fitting information from other chapters to the correct installation practices.
- know the correct testing methods for passing a plumbing inspection.
- recognize that company preference can dictate installation practices.

Glossary of Terms

fixture group a bathroom consisting of a tub or shower, toilet, and lavatory

half bathroom a bathroom consisting of only a toilet and a lavatory

rough word used to describe the rough-in dimension from the back wall to the center of the toilet drain

test ball a testing accessory that is injected with air after being inserted into a drainage pipe to allow the system to be filled with water

test cap used to seal a pipe for testing a DWV system; several types are available

test plug a testing accessory that inserts into a pipe end for testing purposes; several types are available

trim-out finish phase of construction when fixtures are installed; also known as trim

waste arm a horizontal fixture drain that begins with the vent connection and terminates at the fixture connection

Layout Considerations

Chapter 10 introduced you to layout considerations for various bathroom and kitchen designs. A plumber must consider all aspects of a plumbing-system route before beginning the layout process. Every job site has unique layout challenges. A plumber begins by identifying and laying out the fixture locations. Knowing where the sewer enters the house establishes the flow direction of the building drain. The location of the major **fixture groups** determines where waste stacks or fixture drains are routed and how they will connect to other piping segments of the DWV system. Knowing where the vents terminate through the roof and what fixtures will use air admittance valves (AAV) is essential for establishing the vent routes. If your local code allows wet venting, actual pipe routes are established for particular groups of fixtures. The fittings allowed by codes can also determine a pipe route and the physical size of fittings can limit the chosen pipe route. Holes should not be drilled until the entire route is mapped out. This will prevent increased labor costs and eliminate unnecessary holes in studs and joists. Table 14–1 lists some common DWV layout considerations that are not specific to any particular project.

FROM EXPERIENCE

Each job site will have unique layout considerations. A plumber must approach every piping route with a thorough understanding of codes and the ability to visualize an entire pipe route.

In this section, we will look at a basic DWV layout of the above-ground rough-in for a house that provides some challenges. A plumber typically arrives on a job site when the framing is complete. In addition, any one-piece tub and shower units have typically been ordered and delivered to the job site. A blueprint indicates the fixture locations, but usually not the area of the building that connects with the building sewer. The building sewer and building drain connect at the low point of the building drain. When the plumber locates that, the main piping artery of the drainage system can be established. A plumber determines the building drain route based on job-site conditions. Some job sites require the piping system to begin at the building drain exit location; other conditions dictate that the piping be installed so the last connection of the building drain is with the building sewer. Many plumbers install the piping on the floors above and stub all vertical drains into a basement or crawl space. Then they install the building drain. Personal preference plays a large part in the actual layout process. Most companies focus more on the productive installation of a particular job site or task than on whether the building drain is installed before the vertical piping.

Figure 14–1 is a blueprint of a home constructed with a crawl space. We will use it as an example for discussing layout. If this home had an unfinished basement, the layout might be similar to the crawl-space design. If a concrete slab were used, the pipe route would be below ground. An actual blueprint provides more information than the illustration in Figure 14–1, such as dimensions of rooms, but this illustration is focused on the location of plumbing fixtures, wall locations, and joist directions. Each fixture and bathroom group is explained in this chapter, and the piping configurations and sizes are based on possible code allowances. The illustrations are only examples; they do not represent a particular code and are not intended for use on a job site.

FROM EXPERIENCE

A plumber must be most concerned about the location of the toilet in each bathroom group. Because the toilet requires a minimum 3″ pipe, it poses the greatest challenge in routing the drain to its required location.

Table 14–1 Common DWV Layout Considerations

Consideration	Notes
Sewer entry location	What side of the house
Number of bathrooms	Total fixtures and location of each bathroom in house
Bathroom fixture layout	Fixture relation to each other and types
Kitchen sink layout	Location in house and whether there is garbage disposer and DW
Washing machine layout	Location in house
Wall relation to bathroom groups and other fixtures	Wall types; sizes and location of wall studs
Ceiling joist direction and relation to bathroom groups	In relation to bathtub and toilet fixture drain requirements
Fixture types	Specific fixture rough-in requirements
Venting code allowances	If AAV and wet venting allowed
Vent terminations	Penetration locations through roof

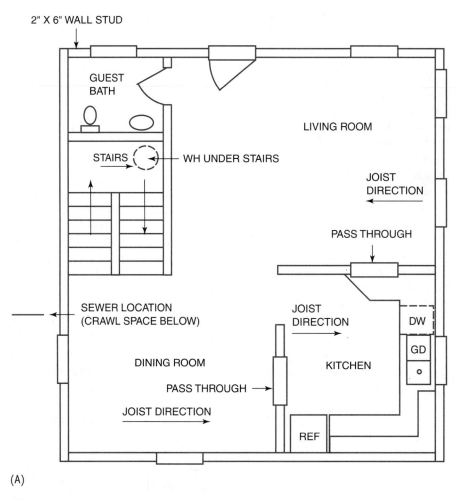

Figure 14-1 Plumbing floor plan.

Scope of Work

The scope of work is the total knowledge of the entire project, including the type of materials and fixtures, and the design. You must locate all the fixtures throughout the house to understand the scope of work required to complete the DWV system. Then scan the entire bathroom or other related areas to determine a pipe route, focusing on the conflicts with structural designs such as beams, load-bearing walls, or joists. The unfinished crawl-space area below the first floor would be a desirable pipe route due to its ease of installation. If the area below the first floor were a finished basement, the pipe route would have to coordinate with the design intent of that space. The location of underground vertical riser piping penetrating the concrete floor must take into account the pipe routes established to serve upper floors.

Guest Bathroom Layout

The bathroom in Figure 14-1 with only a toilet and a sink is known as a **half bathroom** (half bath) because it only has a toilet and a lavatory. A full bath has a toilet, lavatory, and bathtub or shower. The toilet and pedestal lavatory are located along an interior wall with a stairwell directly behind the fixture. The water heater sits below the stairwell landing. Access must be provided, so piping can be installed in that area that is not in the center of the wall. Notice the first-floor guest bathroom fixture layout, and the direction of the floor joists in the living room area. The stairwell wall continues through the floor above, which means that it might be possible to route the vent pipe vertically up to the attic. If the vent cannot be installed in a continuous manner in a wall, it can be routed within the second floor joists.

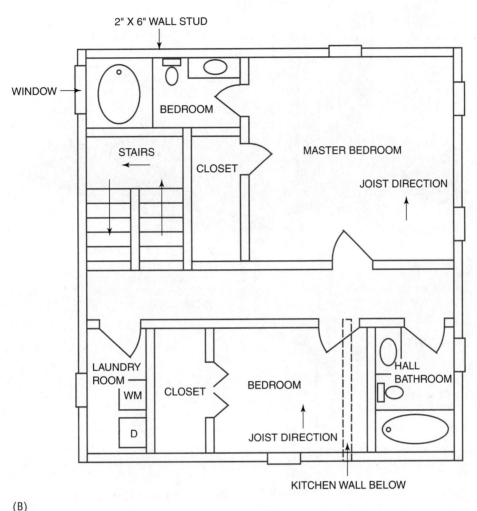

(B)

Figure 14-1 Continued

Figure 14-2 illustrates the possible piping installation for the guest bathroom.

 FROM EXPERIENCE

The rough-in height of a pedestal lavatory drain must be confirmed by actually installing the sink to avoid conflict with the pedestal design. Some pedestals require that the drain be a specific height above the floor.

Kitchen Sink Layout

The kitchen sink is centered directly below a window, and the exterior wall studs are load bearing. If your local code allows an AAV, it can be installed below the kitchen sink to minimize labor costs. Eliminating the need to drill holes through the wall studs saves labor, but, more importantly, the structural integrity of the wall is protected. If code does not allow an AAV, the vent pipe can be routed in the exterior wall studs up through the second floor where it connects with the venting system above or terminates independently to open air through the roof. The vertical drain pipe penetrating the floor can be located in the wall, or it can penetrate the floor where the cabinet will be installed. Where the

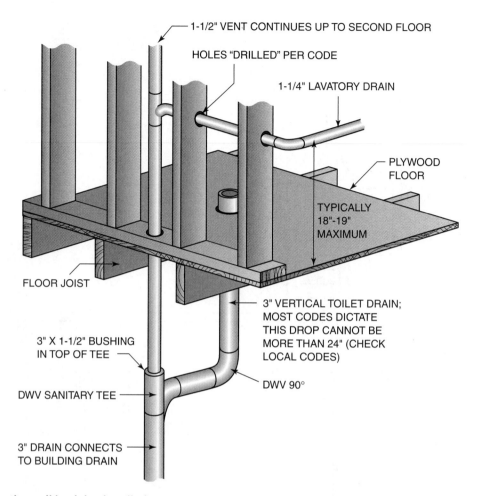

Figure 14–2 Half bath possible piping installation.

pipe penetrates through the floor is typically coordinated with where the garbage disposer is located in the sink. The desired drain location is on the opposite side of the disposer to allow for more creative piping possibilities during the **trim-out** phase. The rough-in using an AAV that has piping penetrating through the cabinet base only requires the pipe to stub up from the crawl space. The remaining piping is completed when the sink is installed during the trim-out phase. A typical sink cabinet base is 36″ wide, but a plumber must confirm those dimensions on a blueprint. Figure 14–3 illustrates a possible piping arrangement for the kitchen sink using an AAV installed below the sink.

 FROM EXPERIENCE

A code book may allow 1-1/2″ minimum pipe to serve a kitchen sink, but most plumbers install 2″ pipe if a sink has more than one bowl or if a garbage disposer is installed.

Master Bathroom

It is not uncommon for an architect to locate the plumbing fixtures on the exterior wall of a house, which poses a challenge for the plumber. When this is the case, it is important to know drilling and notching codes. For example, a 3″ pipe that serves a toilet needs a 3-5/8″ hole to be drilled through a joist, but some drilling and notching codes may not allow it where the layout requires. Fortunately, the joists in the example are traveling in a direction that allows the toilet piping to be installed between two joists. A plumber must approach a piping route to a bathroom group based on the piping requirement for the toilet. If wet venting is legal in your area, the lavatory drain and the vent for all the fixtures can be routed as a single pipe. If wet venting is not legal, a more difficult bathroom layout than the one illustrated in this lesson may be required. Individual vents would need to be installed. Exterior walls that have a 3″ waste stack might have to be constructed with 2″ × 6″ wall studs instead of 2″ × 4″ studs to accommodate the 3″ pipe and fittings. If a 2″ × 4″ wall is constructed, a plumber might have to find an alternative pipe route be-

CHAPTER 14 *Drainage, Waste, and Vent Installation* **379**

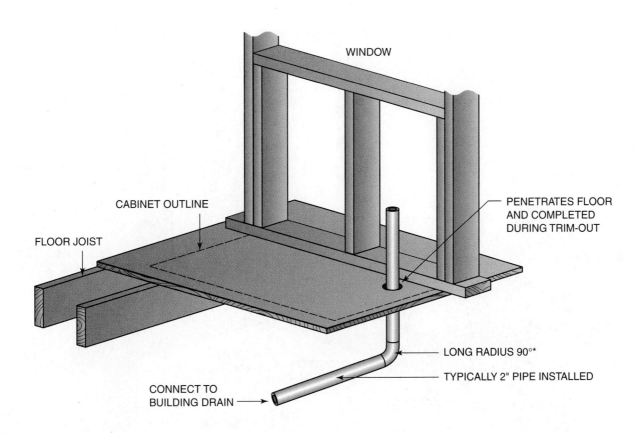

* CHECK LOCAL CODES

Figure 14-3 A kitchen sink rough-in varies based on whether an air admittance valve can be installed or if the vent must be routed to open air.

cause he or she might not be allowed to cut the outside band joist enough for the 3″ sanitary tee to be installed, as illustrated in Figure 14-4. An alternative would be to install the vertical piping in the interior wall of the stairwell, because the piping could still be routed in the direction of the second floor joists. Figure 14-4 shows the drilling layout and a piping riser diagram, which is one way to handle the piping situation for Figure 14-1.

 FROM EXPERIENCE

A cleanout must be installed at the base of every waste stack. It must be accessible and must have clearance to remove the cleanout plug and operate a drain cleaning machine.

Hall Bathroom

When reviewing the hall bathroom design in Figure 14-1, notice the indication of the kitchen wall below. The direction of the joists often dictates the pipe route. It might not be possible to drill through a joist to install a 3″ pipe, because large-diameter holes pose a threat to the structural integrity of a joist. Figure 14-5 illustrates that a 3″ pipe will not fit through a floor joist in some instances without violating codes pertaining to the hole size that can be drilled. Most codes dictate that a joist cannot be drilled in the top and bottom 2″, and many codes do not allow large-diameter holes in the middle one-third of the joist span. A plumber often has to route the piping systems serving a bathroom separately to avoid violating drilling and notching codes or to increase productivity. Figure 14-5 illustrates that the toilet in the hall bathroom can be served by routing the 3″ pipe in the same direction as the floor joists. This type of piping configuration would require wet venting if allowed. The bathtub located in the hall bathroom could be piped two different ways. The bathtub piping had to be routed using a separate waste stack in Figure 14-5. Most codes dictate that 2″ pipe must slope 1/4″ per foot. Connecting it to the same piping as the toilet would place the drilled holes for the lavatory through the floor joist in violation of the top 2″ of the floor joist. Figure 14-5 shows that a plumber must have a

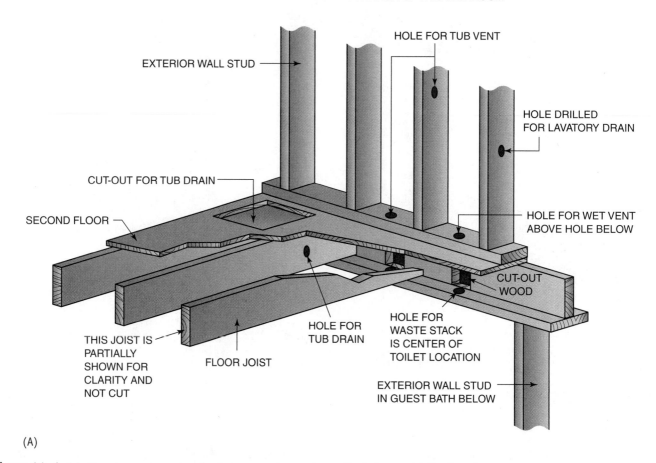

(A)

Figure 14-4 A bathroom group can be piped as one system and coordinated around toilet location.

complete knowledge of codes to lay out a piping route and remain creative in the process. Not all bathrooms can be constructed as designed. A plumber might have to provide an architect or general contractor with possible design changes.

 FROM EXPERIENCE

Many bathroom designs are similar. Attention to codes and drawing on past experience initiates a productive thought process.

Laundry Room

Figure 14-1 shows a washing machine located on the second floor of a house. When a piping system must be installed on a level of a building other than the first floor, it must be routed to accommodate the design, which increases labor costs. As with all piping systems routed in a residential house, the direction of the floor joists is a primary factor in deciding where to install the piping. A plumber has two choices for this design. One is to drill through each joist; the other is to route the drain serving the washing machine in the same direction that the joists are installed. A dry vent must originate in the vertical position—true vertical to a 45-degree angle. Figure 14-6 illustrates the best approach based on saving labor and protecting the structural integrity of the joists. Your code may dictate the minimum and maximum vertical distance of the standpipe from a washing machine box to the inlet of a p-trap. Most codes state the minimum distance as 18" and the maximum as 42". The actual installation of the p-trap is determined by code, company preference, and specific job conditions.

 FROM EXPERIENCE

A washing machine drain connected to a drain serving a toilet should be as far downstream of the toilet as possible. The discharge of a washing machine can cause a disturbance in a piping system and can siphon the trap seal from a toilet bowl.

CHAPTER 14 Drainage, Waste, and Vent Installation

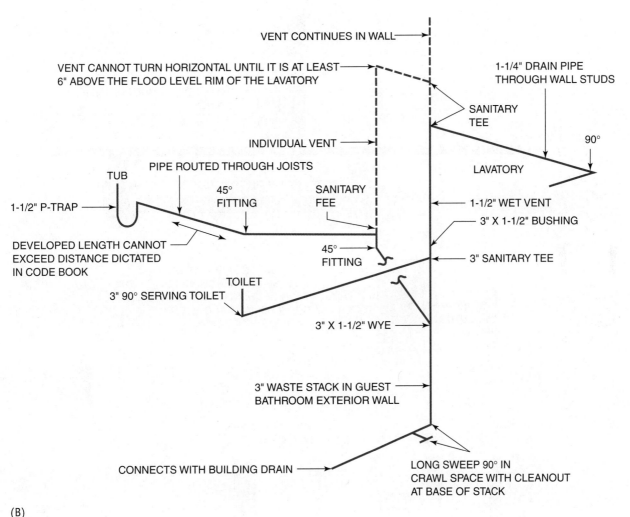

(B)

Figure 14-4 Continued

Building Drain

Now that the piping layout of each fixture has been addressed, the building drain route can be established. Because the house in Figure 14-1 is built with a crawlspace, the route of the building drain is not as crucial as if it were routed in a finished basement. A building drain is at the lowest horizontal portion of a DWV system and routes drainage discharge from all other drains to the building sewer. The building drain installation in Figure 14-7 is based on Figures 14-1 through 14-6. This isometric view of the piping system does not include any variables that would be required on a job site, such as the offsets around structural supports, heating and air conditioning units, or other obstacles a plumber might encounter.

 FROM EXPERIENCE

The total number of drainage fixture units and toilets that are connected to the building drain determines its size. The minimum size of the building drain is dictated by code, and a toilet must be at least 3" in diameter.

Venting System

The entire venting system includes every vent. Many connect together to form a single vent; others remain as individual vents. The use of an AAV, if allowed by code, terminates

382 SECTION THREE *Layout and Installation*

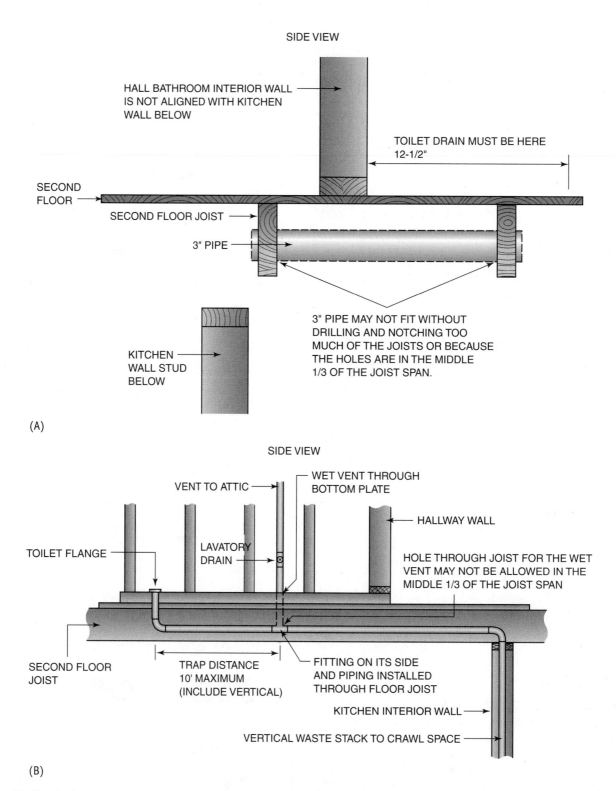

Figure 14-5 A bathroom group may require separate waste stacks because of structural conflicts.

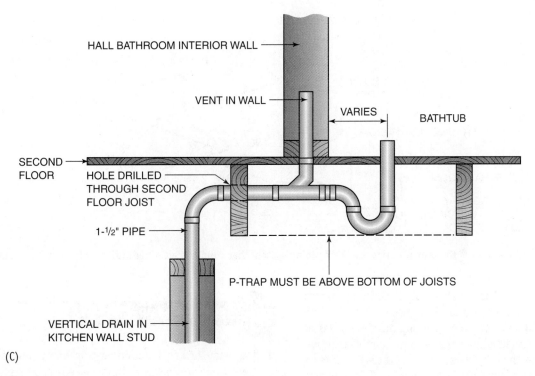

(C)

Figure 14–5 Continued

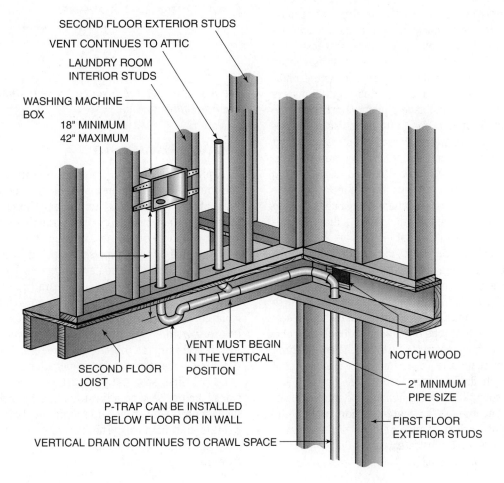

Figure 14–6 Installing piping in the same direction as the floor joists saves labor and does not weaken the structural integrity of the joist.

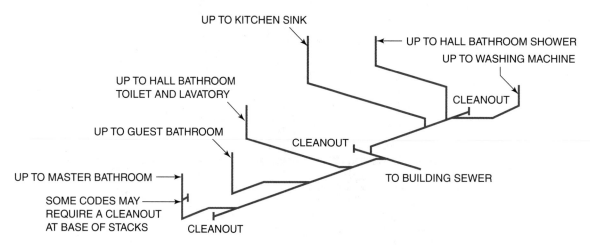

Figure 14-7 A building drain is the lowest horizontal DWV piping system that connects all other drains to the building sewer.

the vent where the AAV is installed. Other vents terminate through the roof; that penetration is known as vent through roof (VTR). It is common for vents serving different areas of a building to be connected, but the most productive merging of vents is often in the attic. Most codes dictate that at least one vent in a system must be a minimum of 3" in diameter for the entire length of the vent. The 3" vent can be a continuation of a 3" or larger drain. As with any dry vent, it must remain connected to the drainage system in the vertical position. The height of the vent above the roof varies according to climate conditions such as snowfall. A vent is routed throughout a house in walls and ceilings as a continuation of the drain it protects. Drilling and notching regulations are the same for venting as for drainage piping. Vent fittings are responsible for circulating air and draining moisture, as opposed to discharging wastewater. Most codes dictate that all fittings used in a vent system must adhere to drainage regulations. However, many codes allow variances in the fitting selection between the two systems; they allow standard-radius fitting in a vent system where long-radius fitting would normally be dictated if the piping were a drain. Terminating the vent pipe through the roof requires an oblong hole to be cut with a reciprocating saw from inside the attic. A plumber uses a level to align the pipe with the angled slope of the roof and then draws an oval line to indicate the cut area. A hole can be drilled and a reciprocating saw used to complete the cut, or a plunge cut can be made using the saw blade at an angle. The vent pipe is installed through the roof, and a plumber or roofing contractor installs the vent flashing on the roof. A roof sealant is placed under the flashing, and galvanized roofing nails are driven through the flashing and into the roof plywood. The most common type of roof flashing is made with neoprene. The center portion, where, the pipe passes through is flexible and conforms tightly around the vent pipe. Figure 14-8 illustrates a riser diagram of the vent system used in this example.

 FROM EXPERIENCE

Too many AAV installations within one piping arrangement can cause pressure to increase in a DWV system. A vent system must be able to breathe, and an AAV does not discharge positive pressures.

See page 402 for the layout of a vent through the roof in the Procedures section.

Fixture Rough-In

A plumber installs a piping system in a house based on the rough-in requirements of each fixture or group of fixtures. Each fixture has unique requirements. A plumber must use either typical company rough-in heights or manufacturer rough-in sheets to install drainage and vent piping. The size of each pipe serving a fixture is based either on the minimum requirements dictated by code or a company's use of a larger size based on preference. A 1-1/4" pipe is the smallest size allowable in a DWV system, but most contractors do not install piping smaller than 1-1/2". In other words, pipe sizing is based on codes and then adjusted according to job-site preferences. Many drilling and notching codes do not allow an adequate hole size to install a 1-1/2" pipe, so a 1-1/4" must be used in certain situations, such as serving a lavatory. The following information focuses on specific rough-in requirements based on common heights and distances. The information may vary on a job site based on preference and certainly on a particular fixture installed.

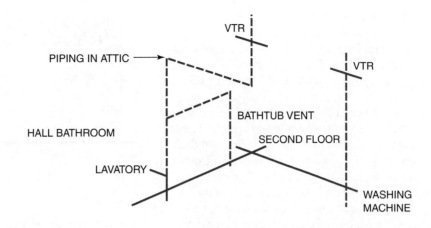

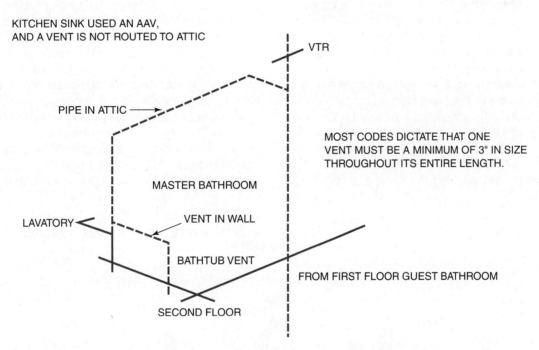

Figure 14–8 A venting system can connect together throughout a house or in the attic, and a vent pipe can terminate through the roof independently.

 FROM EXPERIENCE

The rough-in process is one of the most crucial aspects of construction. Errors made during the rough-in phase are expensive to correct if they are discovered during the trim-out phase.

Toilets

The most common kind of toilet installed in a residential home has a two-piece design, with a separate tank and bowl. Some custom homes install one-piece toilets in the guest bathroom and the master bathroom. The two toilets have different water supply requirements, but the drain connection remains the same for both. The two most common rough-in dimensions for drainage piping are 10" and 12" from the back wall, but 14" might be used for some toilet designs. The rough-in dimension is indicated when ordering by using the word **rough.** If a plumber does not specify a rough-in dimension, it is assumed that the drain is 12" from the finished back wall. Because rough-in occurs before the finished walls are installed, a plumber must know the thickness of the wall finish. Code dictates that the center of the drain serving a toilet must be a minimum of 15" from a side wall or the side of another fixture for non-handicap installations. One side of a handicap installation must be exactly 18″ from the finished sidewall that has the grab bar and a minimum of 18″ from the other side wall or any partition or fixture. Handicap codes must be reviewed before installing a rough-in to make sure it adheres to code.

Before the final rough-in of a toilet drain, the plumber also needs to know the floor thickness. The toilet connects to a drainage system with a closet (toilet) flange. The flange must be installed so it is resting on top of the finished floor. Many new homes have either a vinyl or tile floor, and the toilet is installed after the flooring is complete. If vinyl is used, a plumber does not have to be concerned with additional floor thickness. If tile is installed, a plumber must know the exact thickness if the flange is installed during the rough-in phase. Many plumbers install the flange during the trim-out phase to ensure an exact installation. Others install it during rough-in but leave the flange above the plywood; a tile contractor then installs the finished floor under the flange. A closet flange has four designated holes to receive non-corrosive wood screws. It must be fastened securely to the wood flooring to prevent the toilet from moving after it is installed. The hole cut into the floor must be properly sized to allow the piping and flange installation. A hole that is cut too large does not allow the securing wood screws to be installed. Figure 14–9 illustrates the rough-in dimensions for a toilet and Figure 14–10 shows a closet flange installation.

FROM EXPERIENCE

A toilet rough-in is one of the most crucial installations for a plumbing system. An error can result in the most difficult corrective action needed during the trim-out phase.

Lavatory

The two most common residential lavatory installations are pedestal and countertop designs. A pedestal sink has exposed piping; the center of the drain rough-in is installed so it is in the center of the sink. A plumber must route the horizontal drain piping (**waste arm**) through the wall studs if the vertical piping is not located in the same area of the sink center. The piping to a countertop lavatory is not visible from the room, so a plumber can install the rough-in piping away from the center of the sink and complete the piping in the cabinet. A two-piece p-trap has a 90° portion and a J-bend portion that allows the J-bend to swivel away from the center of the stub-out piping. The size of the trap dictates the dimension of the trap swing. A 1-1/4″ trap is

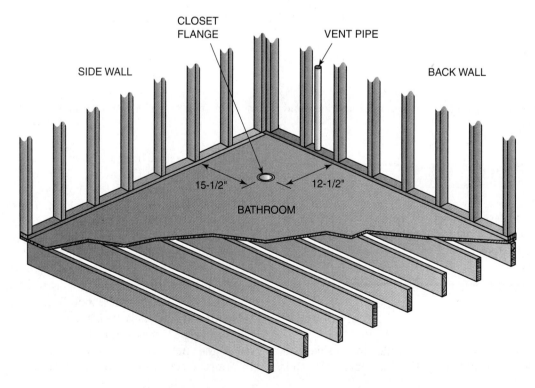

Figure 14–9 A toilet rough-in has specific code and manufacturer requirements and must be installed based on wall-finish thickness.

CHAPTER 14 *Drainage, Waste, and Vent Installation* **387**

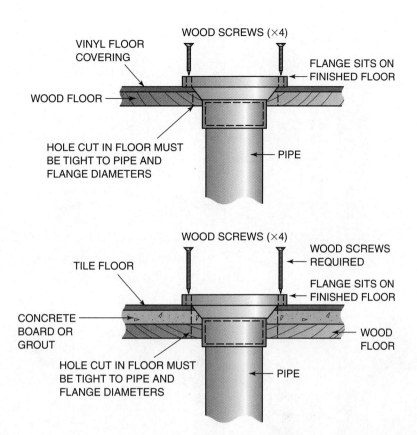

Figure 14–10 A closet flange secures to the wooden floor to eliminate movement when the toilet is installed.

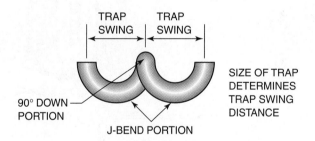

the most common size installed for a lavatory. The average trap swing of a 1-1/4" p-trap is 3". A plumber can install the center of a drain rough-in 3" on either side of the center of a countertop sink. This can save labor when the vertical piping is routed in a different wall-stud cavity than the center of the sink. By not drilling an additional hole in a wall stud, less labor is used and the structural integrity of the wall is protected. Many pedestals have a small designated area in which to install the drain connection that does not accommodate the swing p-trap. Figure 14–11 illustrates two-piece, p-trap swing possibilities. Figure 14–12 compares a countertop and pedestal lavatory rough-in and shows that for a pedestal lavatory rough-in, a plumber must be more exact.

FROM EXPERIENCE

A plumber should review the manufacturer rough-in data to be sure of the recommended height of the drain stub-out serving a pedestal sink.

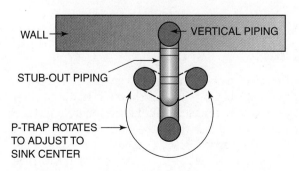

Figure 14–11 A two-piece p-trap allows a plumber to swing the J-bend portion away from the 90° portion to align the piping with the fixture center.

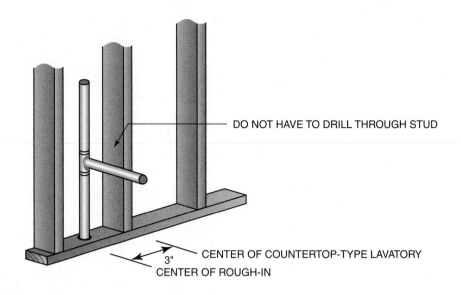

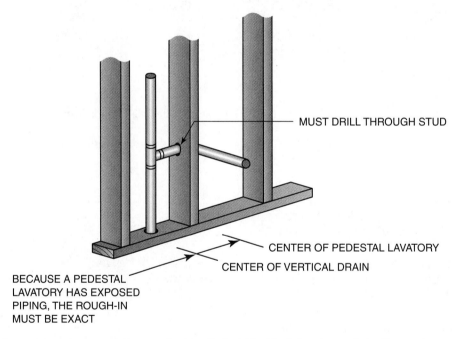

Figure 14-12 A countertop lavatory drain must be installed within 3" of the center of the fixture, but a pedestal lavatory must be installed in the center of the fixture.

Bathtub

If a two-piece p-trap is used, the center of the rough-in piping can vary, similar to what is shown in Figure 14-11. A bathtub is typically connected during the rough-in phase of construction. The direction of the horizontal piping that connects to the vertical stack depends on job-site conditions. The location of the bath waste and overflow (BWO) depends on the tub design. The drain is usually connected to the p-trap within 2" of the wall. This close proximity to the wall as well as the physical size of a p-trap and the connecting sanitary tee often dictates that the rough-in of the vertical drain must be installed away from the center of the tub. If a horizontal waste arm is routed from a location that is not near the tub, the maximum allowable distance from a trap to the center of the vent is regulated by code. These distances are listed in a chart in your local code book, most of which dictate that the 1-1/2" drain with a 1-1/2" p-trap cannot travel more than 5'. Some BWO designs have a dual connection option. A plumber can connect a dual option

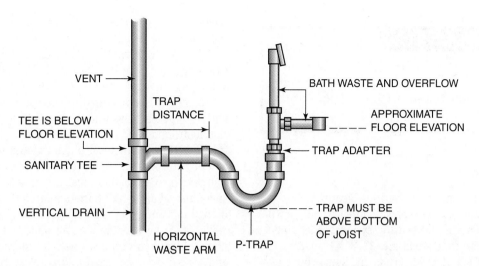

Figure 14–13 A bathtub is typically installed during the rough-in phase of construction. A plumber connects a bath waste and overflow to the drainage piping.

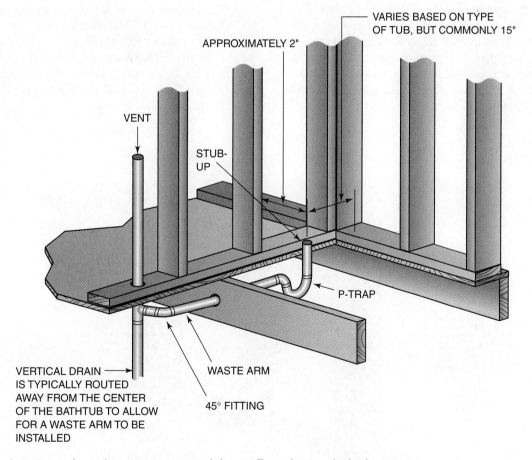

Figure 14–14 A waste arm is used to create an exact piping configuration to a bathtub.

BWO to the p-trap using a trap adapter or a female adapter. Plastic female adapters should not receive metal male threads, because the female adapter could split. Figure 14–13 illustrates the basic connection to a bathtub and p-trap. Figure 14–14 shows a waste arm and p-trap serving a bathtub rough-in.

FROM EXPERIENCE

A waste arm should not be installed with too much slope (fall), because air can become trapped in it and cause a gurgling noise.

Kitchen Sink

Kitchen sinks are installed in various locations, but the most common one is on an exterior wall directly below a window. Installing the drain and venting system poses a challenge if the exterior wall studs are 2″ × 4″ boards. The window has several load-bearing studs and some codes may not allow the required hole sizes to be drilled. In addition, the vent cannot transition horizontally until it is 6″ above the flood-level rim of the countertop, which is the rim of the kitchen sink. It is typically 36″ above the finished floor. The wall space between the countertop and the bottom of the window typically does not accommodate the horizontal transition of the vent pipe. Most drilling codes do not allow more than 40 percent of a load-bearing stud to be drilled out. A 2″ × 4″ wall is 3-1/2″ wide. Therefore, a 1.4″ (1-3/8″) hole is the largest diameter that can be drilled, which does not accommodate the outside diameter of a 1-1/4″ pipe. Some codes allow 60 percent of a load-bearing wall to be drilled if it is doubled with another wall stud. However, no more than two doubled load-bearing studs can be drilled consecutively. Sixty percent of 3-1/2″ is 2.1″ (2-1/16″), which allows a 1-1/2″ pipe to be installed. A typical window frame is supported by at least two wall studs and some inspectors require that additional studs be installed before drilling the window framing studs. Many local plumbing and framing inspectors do not allow 2″ holes to be drilled through 2″ × 4″ exterior wall studs, so a plumber must use an AAV. The introduction and approval of an AAV has eliminated the need to install an individual vent to serve a kitchen sink. If a vent must be installed and the window framing wall studs can be drilled, a plumber has two options. One is to place the vertical drain and vent in the wall cavity adjacent to the window and to install a waste arm to the sink location. The other is to install the vertical drain below the window and offset a 1-1/4″ vent at a 45° angle into the wall cavity adjacent to the window. The 45° installation of the vent means the holes must be drilled at an angle into the wall studs, which is more labor intensive than drilling straight through a wall stud. Figure 14–15 shows an AAV on a drain serving a kitchen sink. Figure 14–16 illustrates a piping configuration installed with a waste arm. Figure 14–17 shows a piping configuration with a vent installed at a 45° angle.

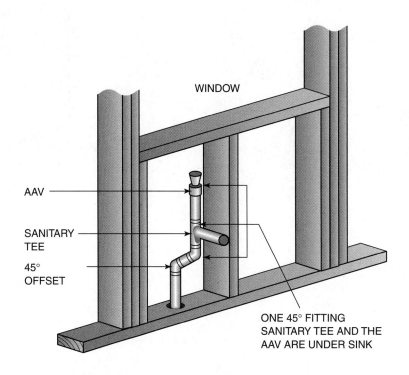

A PLUMBER OFFSETS OUT OF WALL DURING ROUGH-IN PHASE AND INSTALLS THE REMAINING ASPECTS AFTER CABINET AND SINK ARE INSTALLED, AND DURING THE TRIM-OUT PHASE

Figure 14–15 An air admittance valve may be allowed by local codes to avoid installing an individual vent for a kitchen sink.

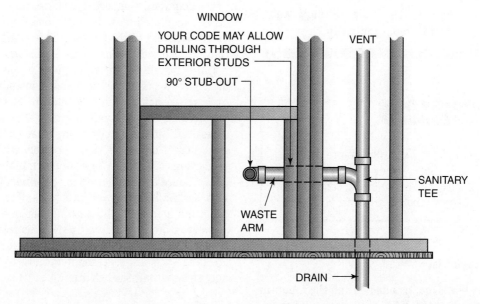

Figure 14-16 If the exterior load-bearing walls can be drilled based on your local codes, a waste arm can be used to serve a kitchen sink.

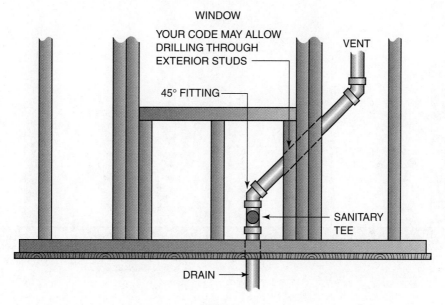

Figure 14-17 If the exterior load-bearing walls can be drilled based on your local codes, a vent can be routed at a 45° angle into a wall stud cavity adjacent to window.

CAUTION: Drilling holes at an angle through wall studs can cause the drill to react violently.

 FROM EXPERIENCE

A plumber must know the type of sink before installing the rough-in piping. If it is not known, locate the drain rough-in far to one side of the cabinet and complete the under-sink drain piping during the trim-out phase.

Washing Machine

Washing machine boxes are installed during the rough-in phase between two wall studs. Most codes dictate that the minimum size drain, whether plastic or metal, is 2″. Plastic types are solvent-welded. Most metal ones have female threads that receive a male adapter to connect to plastic DWV pipe. Most codes do not allow plastic boxes to be installed in a fire-rated wall. The p-trap is either installed within the wall stud cavity below the washing machine box or below the floor. The vertical pipe that connects from the washing machine box to the p-trap inlet pipe is the standpipe. Most codes state that the standpipe cannot be shorter than 18″ or longer than 42″ and that distance is measured to the trap weir. The height of the washing machine box may also be regulated by code, with some codes stating that the overflow height of the box must be at least 36″ and not more than 42″ from the floor. Some codes dictate that a washing machine installed on any floor with an occupied floor below must have a safety pan. A plumber installs the drain piping for a safety pan during the rough-in phase, and the pan is installed during the trim-out phase. The safety pan drain is typically 1″ copper or plastic pipe made of water distribution materials; it does not have to comply with drainage fitting regulations. The drain typically terminates on the exterior of the house and cannot directly connect to the drainage or vent system. Check your local codes for termination locations and heights of the washing machine safety pan drains above the ground. Figure 14–18 illustrates a washing machine drain rough-in using a p-trap below the floor. Figure 14–19 is a washing machine rough-in with a p-trap installed between the wall-stud cavities.

 FROM EXPERIENCE

Most drilling and notching regulations do not allow a hole to be drilled through a 2″ × 4″ wall stud to install a 2″ pipe; the piping must remain within a single wall cavity between two studs.

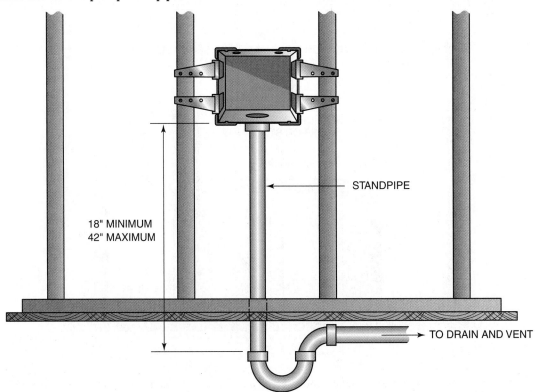

Figure 14–18 Some washing machine box installations may have the p-trap below the floor.

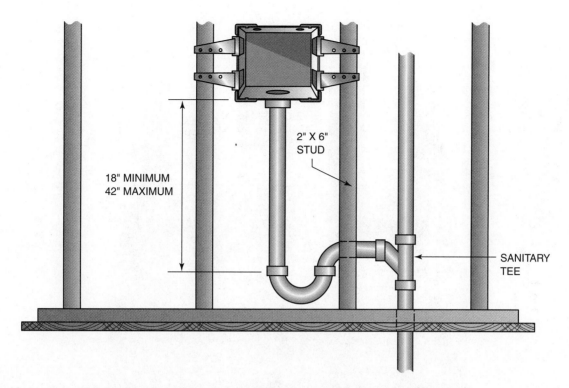

Figure 14-19 Some washing machine box installations may have the p-trap above the floor.

Shower Pan

A shower has a p-trap similar to a washing machine. Some codes allow a 1-1/2" trap to connect a shower; others dictate a 2" minimum trap size. The drain assembly for a one-piece shower unit or a pre-molded shower base is typically provided with the unit. The piping exiting the shower floor connects with the p-trap located below the floor. A tiled shower located on a wood floor or above a finished area is required by most codes to have a safety pan installed, so a plumber must install a three-piece shower drain. The most common type of safety pan is a PVC liner, which is installed by a plumber during the rough-in phase. The actual installation techniques may vary by preference, and the location of the securing nails to the wall studs may vary based on local codes. Some tiled showers have a seat and a plumber must protect that wood structure from water damage with the same PVC liner material. Galvanized roofing nails are frequently used to secure the pan to the wood stud and seat. A threshold is constructed in the entryway of a shower and a plumber terminates the safety pan on or over the threshold. Figure 14-20 illustrates a shower safety pan liner.

See pages 403–405 for a step-by-step procedure for installing a safety pan liner.

Hangers and Supports

Like water distribution piping, drainage and vent systems must be supported. Codes determine the spacing between hangers based on horizontal or vertical installation positions. Additional support is installed where necessary to ensure that a pipe does not sag. Band iron is the most popular means of supporting residential DWV piping, because plastic piping is very lightweight and band iron increases productivity. Galvanized nails or drywall screws are often used to fasten the band iron to the wood structure. Band iron is available in rolls from 10' to 100' in length. Some plumbers use 1/4" bolts and nuts to provide a higher quality installation, but that method increases material and labor costs. The bolts are typically 1" long and are galvanized or have another type of corrosion-resistant coating. Other types of hangers were discussed in Chapter 12. The attachment methods for various hanger installations are the same for DWV piping as for water distribution piping. When piping is hung in an attic, additional wood must often be added to the ceiling joist. Small pieces of wood in varying thicknesses can be added to the top of the joists to allow the pipe to slope. Another method is to cut longer boards and secure them to the sides of the ceiling joists to control the slope and then use

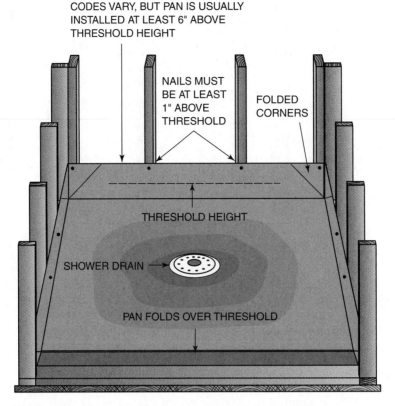

Figure 14-20 A plumber installs a safety pan for a tiled shower located on a wood floor.

Table 14-2 Maximum Hanger Spacing for Residential DWV Piping

CAUTION: These values are only examples and not intended to reflect your local codes.

Pipe Type	Vertical Installation	Horizontal Installation
All-plastic DWV pipe	10'[1]	4'
Cast iron	15'	5'[2]
Galvanized steel	15'	12'

[1] 2" and smaller pipe must have additional support half the distance indicated when installed vertically through a floor (mid-story support)

[2] If 10' length of cast-iron pipe is installed, this distance can be increased to 10'

band iron to secure the pipe to the wood. A more labor-intensive method of supporting horizontal pipe from a ceiling or floor joist is to drill the appropriate-size hole through a board and then secure the board to the joists. Table 14-2 lists common hanger spacing requirements. Figure 14-21 illustrates two common horizontal hanging methods using band iron. Figure 14-22 illustrates two common methods for supporting vertical piping with band iron.

 FROM EXPERIENCE

Hanger spacing distances are based on straight runs of piping, and a hanger is typically required after a piping offset even if the distance is less than indicated in a code book.

See page 406 for a step-by-step procedure for using wood as a hanger and support accessory.

Pipe Protection

Like water distribution piping, DWV piping must be protected against nails and screws used to install wall board and wood trim during the final phases of construction. A metal stud guard (nail plate) that protects the piping is available in several sizes; most have a nail-in feature to drive into the wood boards with a hammer. A 3" pipe that penetrates the top plate of a 2" × 4" wall requires the top plate to be cut completely in half. Removing this top plate can weaken the wall, and an angled iron plate may

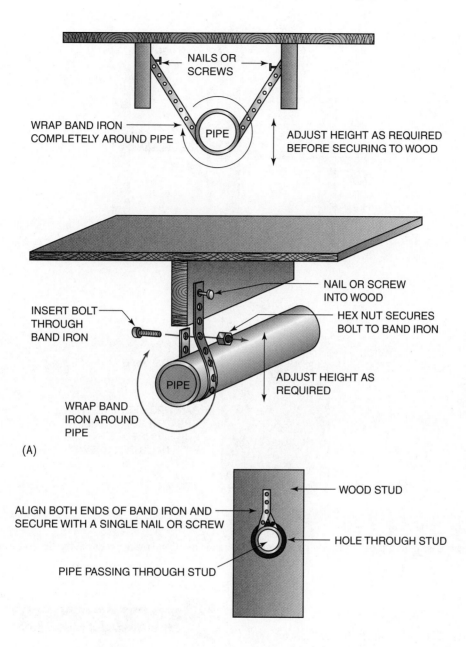
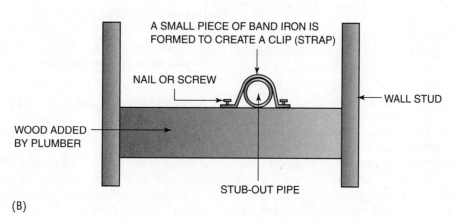

Figure 14-21 Horizontal plastic DWV pipe is usually supported with band iron to increase productivity. Several common methods are used.

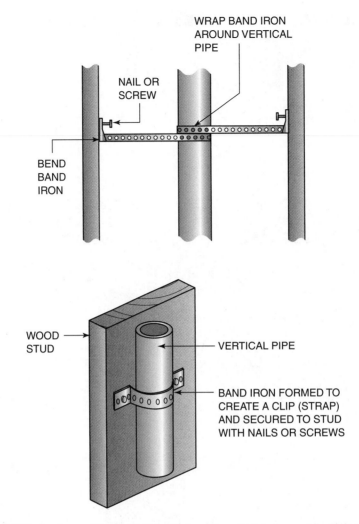

Figure 14–22 Vertical plastic DWV pipe is usually supported with band iron to increase productivity. Several common methods are used.

be required by code. The plate must fasten to the wall studs on both sides of the cut area and to the floor or ceiling joists.

Most codes dictate the burial depth of drainage piping. In colder climates, drainage piping must be buried deeper than in warmer climates. A drainage pipe can freeze if not adequately protected from cold temperatures. A vent through a roof can become closed with frost, so most colder regions have a frost enclosure code that dictates the size vent pipe that can penetrate a roof. Because special protective measures may be required based on certain local conditions, a plumber must know local codes. Earthquake regions are regulated with seismic codes and may require unique pipe protection codes. Figure 14–23 illustrates some common pipe protection concerns that may be included in your local code.

 FROM EXPERIENCE

A plumber must recognize that wall insulation must be installed in an exterior wall and attempt to locate DWV piping close to the inside portion of a wall stud.

Connection to Dissimilar Pipe

A connection between dissimilar pipes is not as common on a new construction site as on a site that has been repaired. Connecting the plastic DWV pipe inside the house to a municipal sewer may require a transition between two different materials. Many custom homes use a cast-iron pipe installed

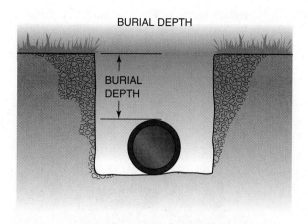

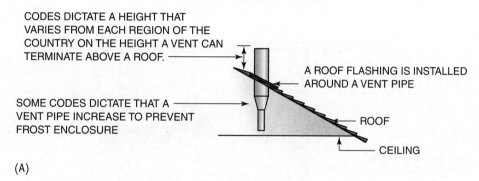

(A)

Figure 14–23 Protection of DWV piping is crucial so sewage spills and sewage gas does not exit the system and so the system functions properly.

vertically in a wall to provide a more silent water flow through the pipe. The transition between cast iron and plastic can be achieved two ways. A transition coupling is a rubber sleeve with stainless steel tightening clamps that creates a watertight connection. A transition coupling may not be legal in a house; many codes only allow their use below ground. A cast-iron pipe installed in the vertical position places all the weight of the pipe downward onto the connecting pipe. All codes approve of a specialty fitting designed to connect plastic DWV pipe to no-hub cast-iron pipe. The adapter is solvent-welded (glued) onto the plastic pipe and a no-hub clamp (coupling) connects the adapter to the cast-iron pipe. Threaded male and female adapters used to connect various dissimilar pipes are approved by all codes. Copper DWV pipe is not often used for residential drainage and vent installations, but, when it is, it can be connected to other materials with a brass DWV male or female adapter. A unique clamp (coupling), similar to a no-hub clamp, is available and approved by most codes to connect DWV copper to cast iron. A no-hub clamp and transition couplings are offered in various sizes, including reducing sizes to connect two different-sized pipes together. Figure 14-24 illustrates common dissimilar DWV connections between plastic and cast-iron piping in a residential plumbing system.

Connection between Schedule 40 plastic pipe and SDR plastic pipe typically requires a solvent-welded transition bushing or reducer.

Testing DWV Systems

DWV pipe and fitting must be tested using a test method determined by codes. The most common is a water test in which a plumber fills the piping system with water and then visually inspects it. Most codes dictate that a 10′ head of water must be used to apply about 5 pounds per square inch (psi) of pressure above the highest connection. Some codes allow a plumber to install a 3′ head of water above the

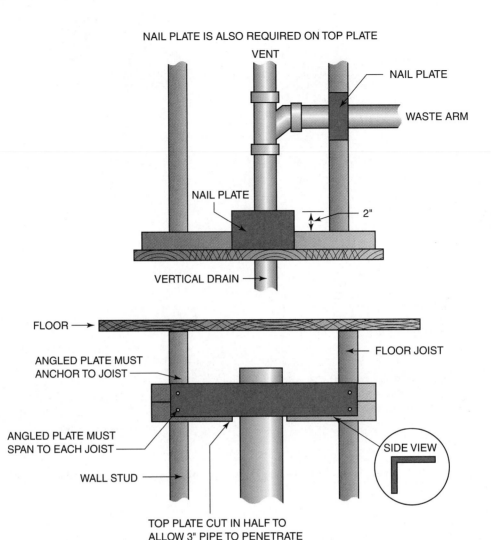

(B)

Figure 14-23 Continued

highest drainage connection and do not require that the vent system be tested. Still other codes state that the entire drainage and vent system must be filled with water up to the highest vent through the roof. Regions experiencing extremely cold temperatures must fill the system the same day as the inspection, so the DWV system does not freeze. Some codes allow an air test to be applied with 5 psi when water is not available, but most plastic piping manufacturers caution against using air for testing their products. Caps, test balls, and **test plugs** are used to seal pipe ends and portions of piping systems to allow the pipe to retain the water or air. Solvent-welded **test caps** are used to seal plastic piping ends. With a test tee, a blow-up rubber **test ball** is inserted into the pipe and air is injected into the ball to expand inside the pipe. Test balls are available in the same sizes as DWV pipes and can also be used to seal a pipe end. Test balls are available in two different lengths and most cannot be subjected to more than 40 psi or whatever is stated by the manufacturer. Rubber test caps with a stainless steel clamp and worm drive screw, similar to a transition coupling, can also be used to seal a pipe end. Figure 14-25 shows several test ball and cap uses. Many other kinds are available; contractors choose based on preference. Figure 14-26 illustrates the 10′ head test that is used to inspect a system for leaks and prepare it for a plumbing inspection.

A plumber must test all elements of a DWV rough-in. If alterations occur after a plumbing inspection, that portion of the system must be tested and possibly inspected again.

CHAPTER 14 *Drainage, Waste, and Vent Installation* 399

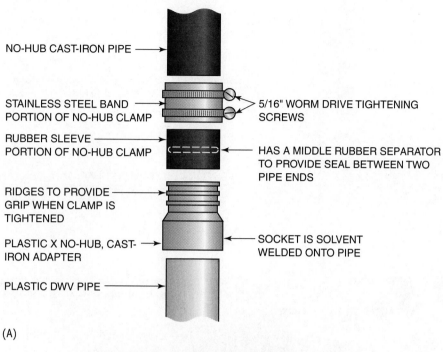

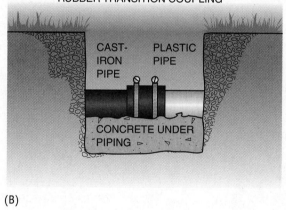

Figure 14–24 Dissimilar pipe connections are performed using approved connectors.

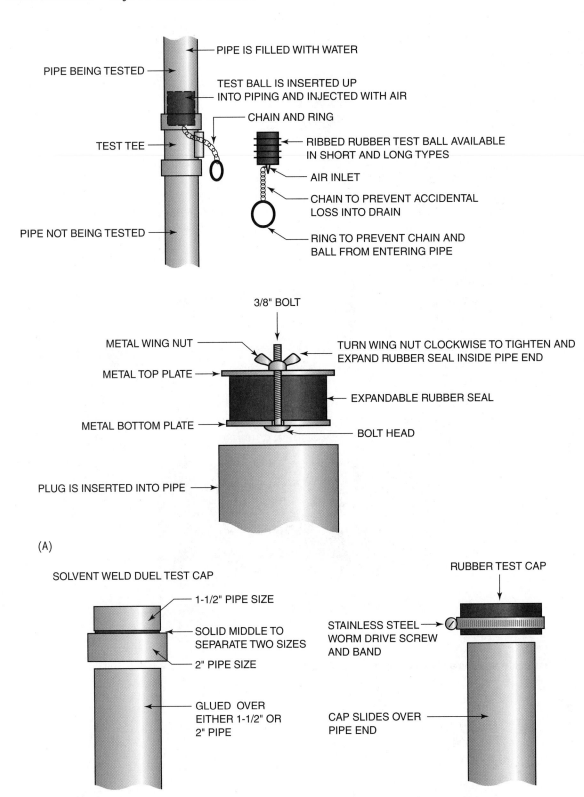

Figure 14–25 Various methods and products are used to seal pipe ends and portions of a DWV system to complete a test.

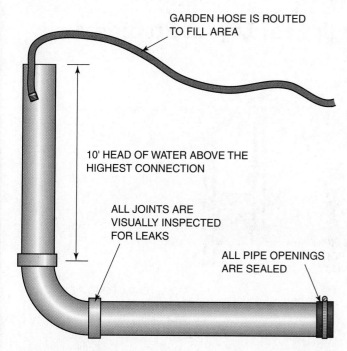

Figure 14-26 Most codes require a 10' head of water to test a DWV piping system.

Summary

- Installation of a DWV system is based on codes, fixture types, fixture locations, floor joist direction, and wall-stud layout.
- A plumber must view the entire pipe route before committing to drilling holes.
- The scope of work is the specific installation requirements of a job.
- Many codes allow an air admittance valve (AAV) to eliminate some of the vents that terminate to open air.
- The height of a vent pipe that terminates above a roof varies based on local code.
- The distance a trap can be installed from its vent is known as trap distance.
- A trap distance is measured based on its developed length.
- The slope of a waste arm (horizontal fixture drain) is based on hydraulic gradient.
- A dry vent must remain vertical until it rises at least 6" above the flood level rim of the highest fixture served by that vent.
- Vertical is considered true vertical to 45° from true vertical.
- A shower safety pan is required when a tiled shower base is installed on a wood floor.
- Nail guards (plates) are required when a pipe is installed close to the edge of a wall stud.
- A DWV system is usually tested with water, but some codes allow air to be used.

Procedures

Vent Through Roof Layout Procedure

A Use a level to vertically align the pipe route with the roof. This can be performed with a 2" × 4" board and a torpedo level if the distance is too great to use an actual level. Continue until you have created an oblong mark on the plywood.

B The result of the layout process is an outline of where to cut the plywood so the vent can be extended through a roof.

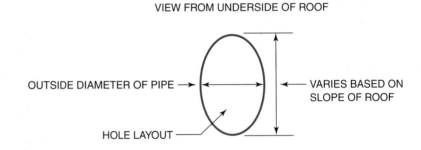

A — MARK PLYWOOD
ALIGN PIPING WITH ROOF BY USING A LEVEL
MOVE TO ALL FOUR SIDES OF PIPE IN 90° INCREMENTS

B — VIEW FROM UNDERSIDE OF ROOF
OUTSIDE DIAMETER OF PIPE
VARIES BASED ON SLOPE OF ROOF
HOLE LAYOUT

CHAPTER 14 *Drainage, Waste, and Vent Installation* **403**

Procedures

Shower Pan Liner Installation Procedure

A Sweep the floor clean and remove all sharp edges of wood and nails from the work area. With the bottom portion of the floor drain installed flush with the plywood floor, remove the clamping collar and strainer portion of the drain assembly, but leave the four bolts threaded into the lower portion. Apply a bead of plumber's putty around the drain assembly and flatten it to form a seal. This putty will act as a seal both when testing the drainage system and for the four clamping collar bolts.

A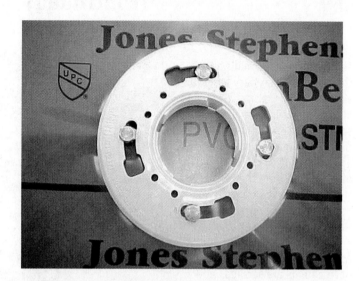

B Spread the pan liner onto the wood floor and situate it to make sure it covers the entire shower area and about 6" of the wall above the threshold height. Some codes dictate that the liner must extend up the wall at least 2" above the threshold height. Any excess can be trimmed with a utility knife. Begin working in one of the rear corners of the shower, folding the liner in a triangular shape. Fold the triangular-shaped piece of liner to one side of the wall to form a 90° corner.

B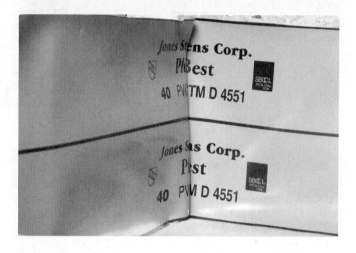

Procedures

Shower Pan Liner Installation Procedure (continued)

C The location of the nails above the threshold may vary based on your local code. Using a hammer, drive a galvanized roofing nail through the folded liner to secure it in place, making sure the nails are at least 1" above the threshold. Working from that corner of the shower to the opposite rear corner, install the liner evenly. Nail the liner to the wall studs and fold the liner in that corner as previously described. Once both corners are secured, work your way toward the front of the shower on both sides until you have completed the installation up to the threshold. Care should be taken not to tear the liner.

C

D Cut the liner to form around the threshold. Some plumbers fold the liner completely around the threshold and stop at the floor outside the shower. Most tile contractors place a concrete-type board over the liner and drive nails into the top of the threshold. A plumber can also install nails there, if local codes allow, without jeopardizing the integrity of the safety pan.

D

E Cut small slots where the bolts are installed in the shower drain body under the liner and carefully slide the liner over the bolt heads. Install the clamping ring over the bolts and tighten in place. This will compress the putty and form a seal, so the drainage system can be filled with water to test for leaks. Fill the shower pan with water up to the threshold height to test for leaks. A plumbing inspector will visually inspect the water level to ensure the wood floor below the liner is dry. Most plumbers leave the water in the pan and let the tile contractor cut the liner directly above the drain opening before performing their work. Leaving the water in the liner keeps other contractors from walking on the liner and possibly damaging the safety pan. Keeping the water in the pan until the tile contractor arrives also proves that the installation was correct and not damaged.

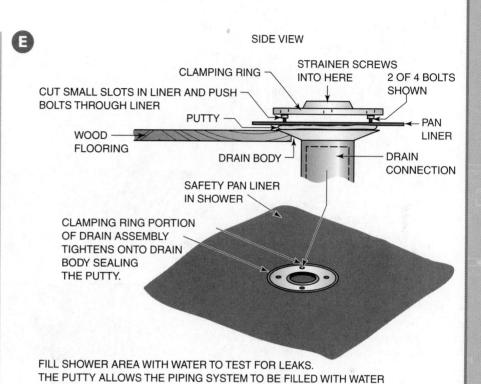

E

SIDE VIEW

FILL SHOWER AREA WITH WATER TO TEST FOR LEAKS. THE PUTTY ALLOWS THE PIPING SYSTEM TO BE FILLED WITH WATER AND WILL NOT LEAK FROM THE UNDERSIDE OF SAFETY PAN LINER

Procedures: Using Wood for Pipe Support

A There are four common methods for supporting sloped piping from the top of a joist. Productivity and specific job conditions are considered when selecting any of the methods. Band iron can be used to secure the piping to the wood support, but if a hole is drilled through the wood support, band iron is usually not used. Screws or nails are used to fasten the support to the side of the joist with the actual positioning of the support based on the desired slope of the pipe being installed.

B The same four methods are used to install sloped piping on the under side of a joist. The installation procedure is the same, and where the support is fastened to the side of the joist is dictated by where the pipe is installed. The number of supports required is based on code, job conditions, and preference. Job conditions often require more supports than what is required to satisfy code.

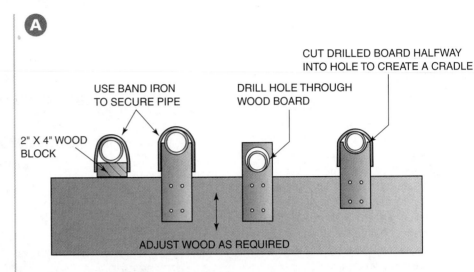

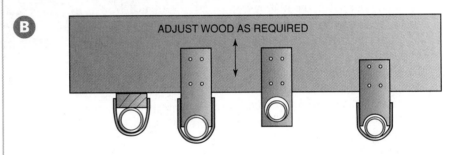

Review Questions

1. The drainage fixture rough-in typically begins before the
   a. Wall studs are installed
   b. Water piping rough-in
   c. Permit is obtained
   d. Plywood floors are installed

2. A DWV rough-in layout procedure includes the total vision of the system installation and begins with
   a. Specific fixture requirements
   b. Obtaining a permit
   c. Present climate conditions
   d. The fixture installation phase

3. Cast-iron pipe is used for vertical installations in many homes to
   a. Satisfy code
   b. Create a structurally sound piping system
   c. Provide a more silent water flow
   d. Allow for a smaller wall-size construction

4. The highest typical rough-in height of a lavatory sink drain is
   a. 12" to 13"
   b. 14" to 15"
   c. 18" to 19"
   d. 24" to 25"

5. If a toilet is ordered without specifying a rough-in dimension, you will receive a
   a. 10" rough toilet
   b. 12" rough toilet
   c. 14" rough toilet
   d. 16" rough toilet

6. The closest a toilet can be installed to a finished side wall is
   a. 10"
   b. 12"
   c. 14"
   d. 15"

7. The minimum-size drain that can be routed to a toilet is
   a. 2-1/2"
   b. 3"
   c. 2"
   d. 1-1/2"

8. A tiled shower installed on a wood floor requires a
   a. Shower pan
   b. Overflow drain
   c. One-piece shower drain
   d. Seat installed in the shower

9. Most codes require a DWV test to have a(n)
   a. 5' head of water
   b. 8' head of water
   c. 10' head of water
   d. 15' head of water

10. The most common hanger material used for supporting a DWV system to increase productivity is
    a. Band iron (strapping)
    b. Clevis hangers
    c. Split-ring hangers
    d. Adjustable swivel-ring hangers

11. Most states dictate that a washing machine standpipe cannot be less than 18" or more than

a. 30"
b. 36"
c. 42"
d. None of the above are correct

12. Sixty percent of 3-1/2" is

a. 2.1"
b. 2.6"
c. 1.7"
d. 5.8"

13. The smallest size pipe that can be used for a drain or vent rough-in is

a. 1"
b. 1-1/4"
c. 1-1/2"
d. 2"

14. Most codes that allow air to be used for testing a DWV system dictate a minimum of

a. 3 psi
b. 5 psi
c. 10 psi
d. 15 psi

15. Forty percent of a 2" × 4" wall stud is

a. 1.4"
b. 2.1"
c. 2.6"
d. 1.6"

16. Most codes dictate that at least one vent must be full size throughout its length and a minimum of

a. 2"
b. 3"
c. 4"
d. 1-1/2"

17. To increase productivity, a plumber should attempt to install a drain pipe

a. In the same direction as a joist
b. Using cast-iron pipe
c. Using clevis hangers instead of band iron
d. Perpendicular to the joist direction

18. Some codes dictate that a washing machine located above an occupied floor below must have a

a. Safety pan and drain
b. Safety pan only
c. Automatic water disconnect
d. 1-1/2" standpipe

19. A closet flange must be

a. plastic
b. secured to the floor
c. metal
d. 2" minimum size

20. The underside of a closet flange should be installed

a. on top of the finished floor
b. completely below the finished floor
c. 1/2" below the finished floor
d. 1/2" above the finished floor

Chapter 15 | Fixture and Equipment Installation

Most plumbing fixtures and equipment are installed during the trim-out phase of construction. The installation sequence depends on the project and the preference of the contractor. Most bathtubs are installed during the rough-in phase, but some large-capacity tubs are installed into a platform during the trim-out phase. A water heater is typically installed during the trim-out phase or after the water distribution testing. The final drain and water connections to fixtures and equipment are not subjected to the same testing requirements as the rough-in piping. A performance test is used to test for leaks. The operating water pressure of a water distribution system is applied to each fixture. The fixtures are then filled with water, which is allowed to flow through the drainage system while it is visually inspected for leaks. The methods for connecting water and drain piping to a fixture depend on the specific fixture, but many fixtures have similarities. All residential toilets have a wax seal where the fixture connects to the drainage system, and they only have a cold water connection. Sinks all have a hot and cold water connection and a p-trap connecting the drain outlet of the fixture to the drainage system. The water supply and drains for most bathtubs and showers are installed during the rough-in phase and only need to have the trim items installed. All sinks and toilets have an isolation valve called a stop. A wall or floor escutcheon is installed before the stop to conceal the hole opening of the pipe through the wall or floor. An icemaker and washing machine box only require the trim plate to be installed. A plumber installs a dishwasher while installing the kitchen sink.

OBJECTIVES

Upon completion of this chapter, the student should be able to:
- know basic fixture and equipment installations.
- know the tools and material required before beginning an installation.
- understand the sequence of installing materials to complete a task in a productive manner.
- recognize that company preference can dictate installation practices.

Glossary of Terms

closet bolts non-corrosive bolts inserted into a closet flange to anchor the toilet to the drainage system

closet flange also called a toilet flange; installed to connect a toilet to the drainage system

compression a type of connection used for water distribution tubing

continuous waste a drain assembly used to connect a double-bowl kitchen sink to a single p-trap

escutcheon floor or wall plate installed around a stub-out pipe to conceal the penetration

slip joint a type of drainage connection between the rough-in stub-out and tubular p-trap

trap adapter a slip-joint style fitting installed to connect a tubular p-trap to the drainage system; desanco is another name for a trap adapter

tubular a pipe that is smaller in diameter and has a thinner wall thickness than pipe used for drainage piping

wax seal a seal that ensures water does not leak and sewer gases do not escape when connecting the toilet to the drainage system

Escutcheons and Stops

Every ceiling, wall, or floor stub-out must have an **escutcheon** installed to conceal the pipe penetration. The penetration through a fire-rated ceiling, wall, or floor creates a small opening that must be sealed with an approved fire caulk before the escutcheon is installed. Some codes also dictate that a pipe penetrating a non–fire-rated ceiling, floor, or wall must be sealed to prevent insects from entering a home through the pipe penetration. Escutcheons are available in plastic or metal, and exposed escutcheons are color-coordinated with the faucet or other trim items. A split escutcheon can be installed around a pipe after the stop is installed, but it is not visually attractive. Escutcheons are ordered for the pipe being served in either **tubular** or iron pipe size.

Every fixture except a bathtub and a shower faucet must have an isolation valve or a stop. The boiler drains that are installed in the washing machine box during the rough-in phase serve to isolate a washing machine; the wall-trim plate conceals the wall opening. An icemaker box is similar to a washing machine box, and the isolation valve is also part of the rough-in installation. Sink stops are installed below the fixture and connect to the stub-out pipe. They do not have to be color coordinated with the faucet finish if they are located in a cabinet. A toilet stop is installed onto the stub-out pipe and is coordinated with the trim finish of the faucets in the bathroom. A dishwasher might be isolated with a separate stop under the kitchen sink, or a dual stop might isolate the dishwasher and hot water supply to the kitchen sink faucet. The type of pipe used as a stub-out determines the connection of a stop to the stub-out pipe. A PEX stub-out pipe requires a stop compatible with that pipe. The connection method is the same as the one used to install the PEX piping during the rough-in. A copper stub-out pipe connects to a stop with either a compression connection or a solder joint. A stub-out (up) pipe through the floor typically requires a straight stop; a wall stub-out pipe usually uses an angle stop. The outlet side of a stop is a **compression** connection regardless of the stub-out pipe installed. Table 15–1 lists some common tools and items required to connect a stop to a stub-out pipe. Some plumbers use specialty tools to perform this task. Figure 15–1 shows a copper stub-out pipe with an escutcheon installed. Figure 15–2 illustrates an angle stop and a straight stop installed on a copper pipe with

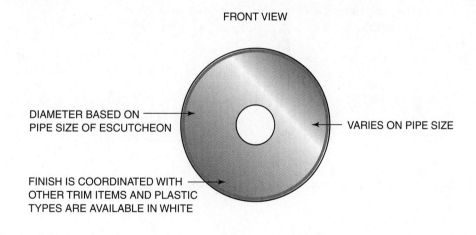

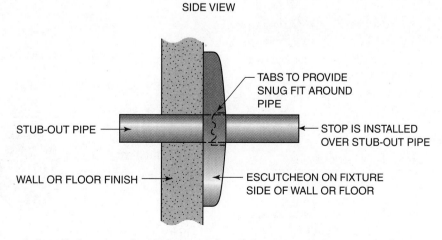

Figure 15–1 An escutcheon is installed to conceal a pipe penetration of the water supply to certain fixtures.

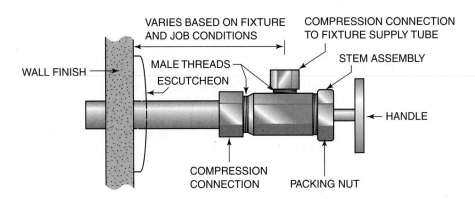

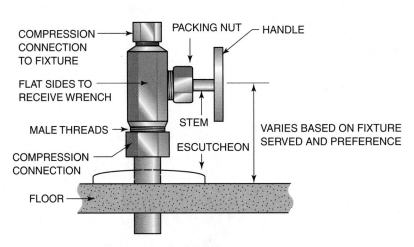

Figure 15-2 An angle stop and straight stop are installed based on stub-out location.

Table 15-1 Common Tools and Items Required for Installing a Compression Stop

Quantity	Tool or Item
1	Tubing cutter based on type of stub-out pipe (copper or flexible tubing)
2	Adjustable wrenches or compatible sized open-end wrenches
1	Bucket or container to capture water exiting cut pipe
1	Rag to clean spilled water or debris

a compression connection. Figure 15-3 illustrates a dual-type stop that can connect a single stub-out pipe to two different fixtures.

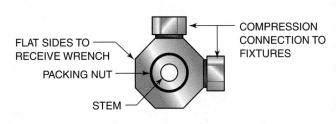

Figure 15-3 A dual stop is used to connect two adjacent fixtures to a single stub-out pipe.

FROM EXPERIENCE

Turning a metal escutcheon in the opposite direction and installing it temporarily over a pipe will bend the tabs inward and allow an easier fit around the stub-out pipe. Remove the temporary installation and install in the correct direction.

FROM EXPERIENCE

When installing a stop that is soldered onto copper, the handle and stem assembly is typically removed so the rubber washer is not damaged. The items are reinstalled after the stop has cooled.

See pages 426–428 for a step-by-step procedure for installing an escutcheon and compression stop.

Toilets

The water supply is located on the left side of all toilets, with the stop and escutcheon installed the same as for other fixtures. The stop and escutcheon can be installed before the toilet or, to provide a more exact installation, the toilet can be installed first. Because the water supply connection to the tank is typically closer to the floor, the stop must be positioned more exactly for a one-piece toilet than for a two-piece toilet. A two-piece toilet is sold with a compatible tank and bowl, and a plumber must combine the two items. Some plumbers install the bowl to the drainage system before installing the tank; others assemble the set before installing the toilet. The tank secures to the bowl with non-corrosive bolts designed specifically for this purpose and provided with the tank. Most tanks use a two-bolt pattern; some have three bolts. A rubber gasket is provided with each tank bolt to seal the bolt penetration through the bottom of the tank. Most toilet tank designs require a plumber to install a foam-rubber gasket, provided with the tank, that prevents the water that is leaving the tank and entering the bowl from leaking. The gasket, called a close-couple gasket or tank-to-bowl gasket, is installed over the threaded portion of the flush valve on the bottom side of the tank. Some manufacturers install rubber gasket that serves as the seal between the tank and bowl. In this case, the plumber does not have to install a separate close-couple gasket. The manufacturer information provided with each toilet should be reviewed for specific instructions about the fixture being installed.

The water connection to a toilet requires a supply tube, known as a tank supply, that is designed to mate with the fill valve. A tank supply is an exposed chrome-plated, soft copper tubing with a flat end that connects the stop to the fill valve with a compatible securing nut that is provided with the tank. Flexible plastic and braided stainless steel versions of a tank supply are available, each with a distinct connecting end that is beveled instead of flat. The soft copper tank supply is bent by hand or with a bending tool to create the required offset to connect to the fill valve and stop. The trade name for the fill valve is ball cock. It regulates the water level in the toilet tank.

The toilet is installed onto the **closet (toilet) flange** and sealed with a manufactured wax ring often called a **wax seal.** The wax seal is either installed to the underside of the toilet or installed directly on the closet flange. Most manufacturer instructions recommend installing the wax seal onto the toilet, but many plumbers place the wax ring onto the flange and set the toilet over it. A plastic accessory called a horn is molded into some wax seals, but wax seals without the horn are more common and do not restrict the flow from the toilet. A closet flange allows two different anchoring selections to secure the toilet to the closet flange. One option has a large slot that allows a plumber to adjust the closet bolts slightly in case the flange is not installed to exact dimensions from the back wall. In the second option, small slots require the plumber to install the closet flange to an exact dimension from the back wall. Most plumbers use the larger slot option. Once the flange is installed to the drain pipe, the next step is to install the **closet bolts.** A pair of 1/4" closet bolts are sold as a set along with the required flat washers and hex nuts to anchor the toilet to the closet flange. Plastic decorative bolt caps and retainer washers that conceal the nut and washer of the closet bolt set are sold with the toilet. Each closet bolt is cut with a miniature hacksaw after the nuts are securely fastened to allow the bolt caps to be installed. A toilet seat is purchased by color and bowl design—either round or elongated. Residential toilet seats have a hinged lid and installation is self-explanatory. Some codes require the base of a toilet to be caulked where it sits on the finished floor; if so, the caulk should be the same color as the toilet. Table 15–2 lists the common tools and items needed to install a toilet. Figure 15–4 shows a closet flange and a method of securing the closet bolts to the flange. Figure 15–5 illustrates the connection of the stop and fill valve using a soft copper tank supply tube.

FROM EXPERIENCE

A toilet installation is a one-person activity, and a plumber must straddle the toilet with both legs to lift and install. The toilet should be assembled in the immediate work area to avoid carrying the assembled toilet a long distance.

Table 15-2 Common Tools and Items Required for Installing a Toilet

List assumes stop and escutcheon are already installed

Quantity	Tool or Item
1	Tubing cutter based on type of tank supply used (copper or flexible tubing)
2	Adjustable wrenches or compatible-sized open-end wrenches
1	Adjustable pliers (channel lock or water pump type)
1	Flat-head screwdriver
1	Miniature hacksaw to trim bolts (blade only from a standard hacksaw will suffice)
1	Tube of caulking that matches the toilet color (if required)
1	Rag to clean work area

FROM EXPERIENCE

A wax seal may have to be warmed on a cold day. This can be done by placing the seal near the heater vent of a vehicle. This allows the wax to thoroughly compress during the toilet installation. Apply your body weight onto the toilet bowl to compress the wax seal. Do not overtighten the bolts or you may crack the toilet bowl.

See pages 429–431 for a step-by-step procedure for installing a two-piece toilet.

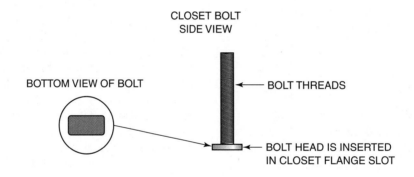

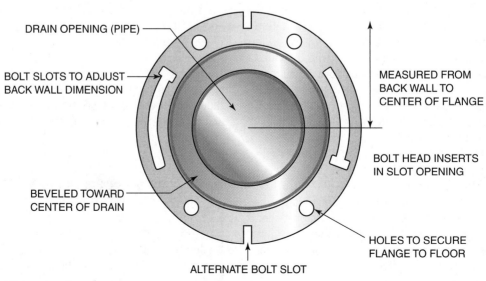

(A)

Figure 15-4 A closet flange and set of bolts are installed to secure a toilet to the drainage system.

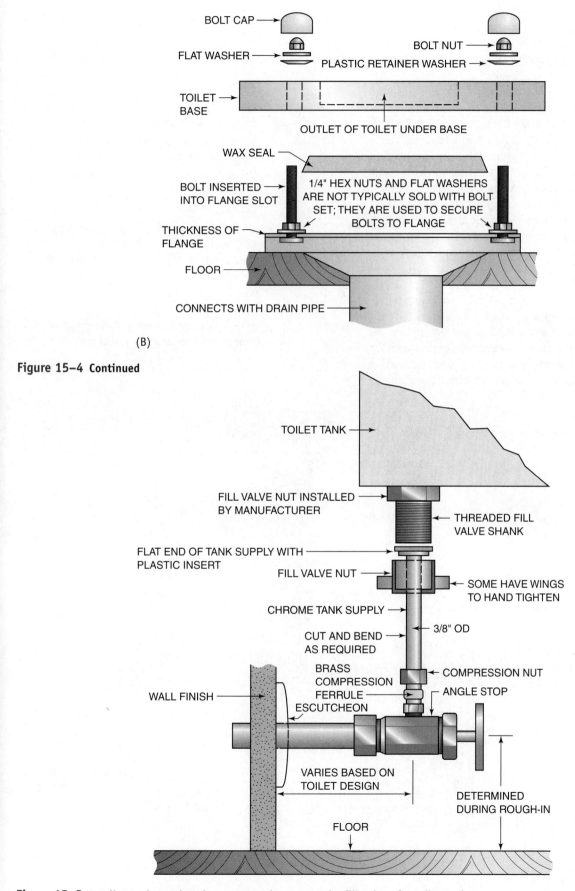

(B)

Figure 15-4 Continued

Figure 15-5 A toilet tank supply tube connects the stop to the fill valve of a toilet tank.

Lavatories

Most residential lavatory sinks are pre-molded with the countertop from cultured marble. A general contractor typically provides and installs the combined countertop and sink, and a plumber installs the faucet and drain to create a functioning plumbing system. There are numerous faucet designs, each of which is installed according to manufacturer instructions. A plumber must review the manufacturer information when installing a specific faucet for the first time to ensure that the installation is correct. Some faucets use a specific supply tube known as a lavatory supply; other faucets use a compression coupling (union) connection. Every fixture is protected by a p-trap to prevent sewer gas from entering an occupied space. The stub-out piping serving a lavatory is either 1-1/4" or 1-1/2". A plumber typically installs a 1-1/4" p-trap; therefore, a 1-1/2" × 1-1/4" **trap adapter** is sold with many p-trap assemblies. If a 1-1/2" p-trap is installed, it will have a 1-1/2" trap adapter. A reducing **slip joint** washer is required to connect to the 1-1/4" drain from the sink. Most residential lavatories use a pop-up drain assembly, which is sold with the faucet. The manufacturer installation instructions include specific information about the drain as-

Table 15-3 Common Tools and Items Required to Install a Cultured-Marble Lavatory

List assumes stop and escutcheon are already installed

Quantity	Tool or Item
1	Tubing cutter based on type of tank supply used (copper or flexible tubing)
2	Adjustable wrenches or compatible-sized open-end wrenches
1	Adjustable pliers (channel lock or water pump type)
1	PVC saw to cut the rough-in pipe
1	Can of purple primer and glue to connect trap adapter to rough-in pipe
1	Copper tubing cutter to cut 1-1/4" threaded tailpiece
1	Basin wrench to install faucet and lavatory supply tubing
1	Can of pipe dope or roll of Teflon tape for threads on pop-up assembly items
1	Rag to clean work area

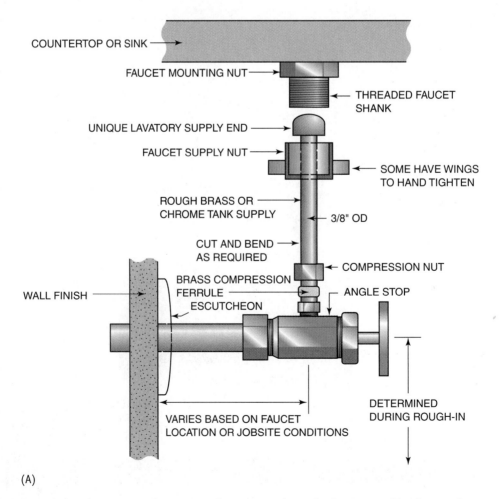

(A)

Figure 15-6 A lavatory supply tube connects the stop to a faucet, or a piece of soft copper tubing is routed to some faucet designs.

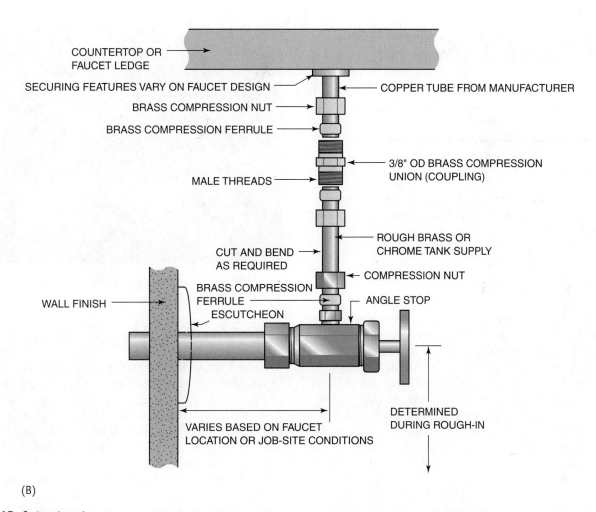

(B)

Figure 15–6 Continued

sembly. The overflow drain on a residential lavatory is provided by the sink manufacturer. If the tailpiece provided with the pop-up assembly is too short to connect with the trap inlet, a longer threaded 1-1/4" tailpiece or extension tailpiece is installed. If it is metal, it is cut to the desired length with a copper tubing cutter. Plastic tailpieces are cut with the appropriate PVC saw. Table 15–3 lists some common tools and items needed to install the faucet and drain assembly to a residential countertop-style lavatory. Figure 15–6 illustrates two common methods used to connect a residential lavatory water supply to a faucet. Figure 15–7 illustrates the drain connection to a lavatory.

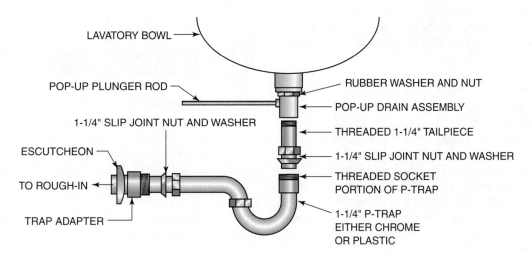

Figure 15–7 A p-trap, installed to prevent sewer gas from entering an occupied area, is connected to the lavatory outlet.

FROM EXPERIENCE

Installing the lavatory supply tubes before the faucet is more productive and eliminates the need to tighten connections with a basin wrench.

See pages 432–437 for a step-by-step cultured-marble countertop style lavatory faucet and drain installation procedure.

Kitchen Sinks

The same tools are required to install a kitchen sink as a lavatory sink with the possible additions of a larger pair of pliers and a Phillips screwdriver. There are numerous styles of kitchen sinks and many have unique installation requirements. Most install into a cut opening of a countertop and rest on the countertop. Some plumbers cut the hole into the countertop, but usually the general contractor provides the opening based on dimensions provided by a plumber. Kitchen sinks in residential construction are most often stainless steel or cast iron. The weight of a cast-iron sink provides the stability needed to maintain its permanent position on the countertop. A plumber simply applies caulking to the edge of the cutout area of the countertop and places the cast-iron sink into the hole. The excess caulking is cleaned with a wet rag. A stainless steel sink must be fastened with clips provided with the sink. Putty or caulking is applied under the rim of the sink or around the cutout area of the sink. The sink is then inserted into the cutout and attached from under the countertop with the fastening clips. The faucet is typically installed onto the sink before the sink is installed onto the countertop to increase productivity. Basket strainers can be installed in the drain openings of a stainless steel sink before the sink is installed. Basket strainers are installed after a cast-iron sink because drain openings are required to safely install a heavy sink.

The drain connection arrangement below the sink varies based on the rough-in location of the stub out, the number of bowls, whether a garbage disposer or a dishwasher is installed, and company preference. Code allows a single p-trap to connect a double-bowl sink, and a plumber typically uses an arrangement known as a **continuous waste**, which is purchased as a kit. If a garbage disposer is installed in one side of a double-bowl sink, a plumber typically installs a separate p-trap for each drain opening. If a dishwasher is installed, the discharge hose is connected to a designated connection on the side of a garbage disposer. If there is no

Figure 15–8 Various drain assemblies are used to connect a sink to the p-trap.

garbage disposer, the dishwasher connects to a tailpiece designed for connecting the dishwasher hose to the sink drain. The electrical connection to a garbage disposer is installed by an electrician, not a plumber. Additional garbage disposer information is included in Chapter 8. If an air admittance valve (AAV) is the vent for the sink, it is installed during this phase of construction.

The type of water supply to the faucet varies depending on the type of faucet installed. Some connections have a lavatory supply; others require a compression coupling (union) similar to the one shown in Figure 15–6. A plumber can install a separate stop to serve a dishwasher or use a dual stop on the hot water rough-in stub out. Figure 15–8 includes photographs of a continuous waste and dishwasher tailpiece. Figure 15–9 shows a countertop with a cutout for receiving a kitchen sink. Figure 15–10 illustrates how a stainless steel sink is fastened to the underside of a countertop.

FROM EXPERIENCE

Use an adhesive latex caulk to seal a kitchen sink to a countertop. It is easily cleaned from the edges of the sink with water and a rag. Adhesive caulk provides a more secure fastening of the sink to the countertop than does regular caulk.

See pages 438–441 for a step-by-step procedure for installing a stainless steel double-bowl sink.

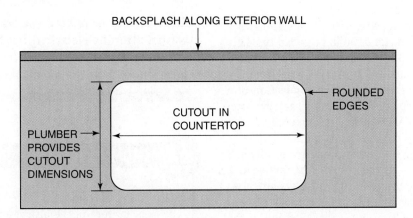

Figure 15-9 A countertop is cut to exact dimension to receive certain types of kitchen sinks.

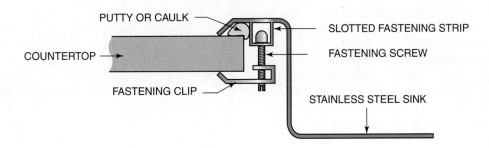

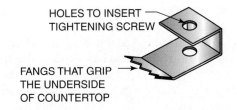

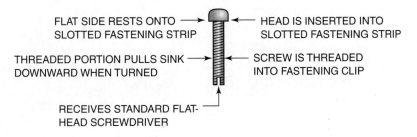

Figure 15-10 A stainless steel sink fastens to the underside of a countertop.

Dishwashers

A plumber provides the water and drain connections to a dishwasher. A general contractor or building owner typically provides the dishwasher. The hot water supply is routed from under the kitchen sink and the drain hose from the dishwasher connects to either a dedicated connection of a garbage disposer or a tailpiece designed for that purpose. A manufacturer provides installation information, and a plumber must review that data before installing a specific unit. Most dishwashers have a rubber connector on the end of the hose that connects to the disposer or tailpiece. A plumber secures that connection with hose clamps. Two short wood screws are provided with the dishwasher to anchor it to the underside of the countertop. Damage to the countertop can result if longer screws are used. The dishwasher must be installed so the sides are flush with the face of the cabinets and the top of the door is level with the countertop. A dishwasher has leveling legs that are adjusted to accomplish that. A hole is drilled in the side of the sink base cabinet as high as possible to route the drain hose to the sink area. The drain hose is connected at the factory to a pump located under the dishwasher. Some codes require an air gap device to be installed in a hole dedicated for that purpose on a kitchen sink. If so, the drain hose is routed through that device and then to the sink. A smaller hole is drilled at the bottom of the cabinet to route the water supply from under the sink to below the dishwasher. The water supply connection for a dishwasher is 3/8" female iron pipe size (FIP) and a brass compression 90° fitting is manufactured to connect the 3/8" soft copper tubing to the dishwasher. This fitting is used in various places within a piping system, but it is often referred to as a dishwasher elbow (ell). The water supply connects to a factory-installed solenoid valve that is electrically activated and located under the dishwasher. The electrical connection to a dishwasher is installed by an electrician, not a plumber. Figure 15-11 illustrates a compression fitting used to connect the water supply to a dishwasher. The drain connecting methods and additional dishwasher information were presented in Chapter 8.

CAUTION: Never remove the access panel to a dishwasher after the electricity has been energized without disconnecting the electrical supply at the proper location, such as the circuit breaker panel.

Place the scrap cardboard from the dishwasher box on the floor in front of the dishwasher installation area to protect the floor from damage. Adjust the height of the dishwasher after inserting it into its designated space.

Laundry Sink

The same tools are required for installing a laundry sink as a lavatory, with the possible exception of a Phillips screwdriver and a level. The two most common types of laundry sinks are wall mounted or have four legs that are secured to the floor. A drill might be needed to install anchors into the floor. If the sink is installed on a concrete floor, a hammer drill may be required to drill holes into the floor to install anchors. The water supply connections are similar to a lavatory installation. There are knock-out spots on laundry sinks where the faucets are installed that must be removed by a plumber. The faucet ledge can be drilled to install a faucet that is different from those that are typically designed for a laundry sink or to install accessories such as a soap dispenser. The drain connection is the same as for any sink and also has a p-trap. The trap size is 1-1/2" and it has a flanged tailpiece to connect the sink strainer to the p-trap. A wall-mounted version requires a plumber to install wood backing in the wall during the rough-in phase. A hanger bracket that is installed during the trim-out phase is provided with the

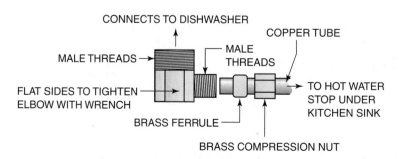

Figure 15-11 A brass compression fitting known as a dishwasher elbow is used to connect the hot water supply to a dishwasher.

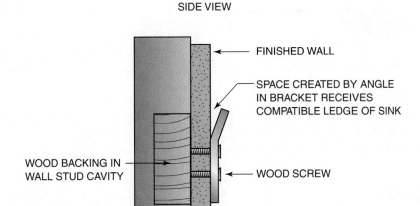

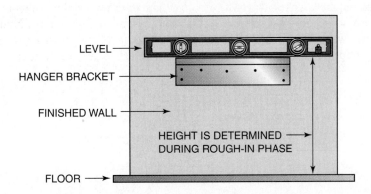

Figure 15–12 A hanger bracket is installed on the finished wall to install a wall mounted laundry sink.

sink. The hanger bracket is installed with wood screws at a height determined during the rough-in phase. The sink is typically not more than 36″ from the rim to the floor. The sink is suspended from the hanger bracket and two side panels are installed. These provide additional stability when the sink is filled and conceal the piping behind the sink. Additional information is provided in Chapter 8. Figure 15–12 illustrates a hanger bracket installed on a finished wall for a wall-mounted laundry sink.

FROM EXPERIENCE

Never use wood screws that are more than 2″ long when installing the hanger bracket for a wall-hung sink to avoid penetrating pipe or wiring on the opposite side of the wood backing in the wall.

Bidet

The same tools are required to install a bidet as a lavatory, but a basin wrench may not be required. A bidet connects to the drainage system much like a lavatory using a 1-1/4″ trap adapter. If the bidet drain rough-in is through a concrete floor, then the p-trap is installed below the floor. If the drainage rough-in has a stub-out through the wall, then a plumber must proceed according to manufacturer recommendations and install the p-trap and associated drain piping. The faucet and pop-up assembly are also installed according to manufacturer instructions and are similar to a lavatory faucet installation. The water supply serving a bidet is typically a 3/8″ supply tube similar to that of a lavatory. Some bidets have a seat similar to a toilet seat. The base of a bidet typically has two mounting holes, which are often concealed under the fixture, to anchor it to the floor according to jobsite conditions. A wood floor uses non-corrosive wood screws.

FROM EXPERIENCE

Set the bidet where it will be installed and mark the floor where it will be anchored. Remove the bidet. Install the necessary anchors or pre-drill the floor to install wood screws and complete the installation. Connect the drain and water supply after the bidet is secured to the floor.

Water Heaters

Local codes and manufacturer instructions must be followed when installing a water heater. Many codes dictate, and all manufacturers recommend, the use of an expansion tank on the incoming water supply. An isolation valve must be installed on the cold water piping near the inlet of the water heater so the expansion tank is on the downstream side of the isolation valve. If local codes require a check valve, it should be installed between the isolation valve and the expansion tank. A typical residential water heater has 3/4" male or female threaded water supply connections. Most codes dictate that copper or other metal piping be installed to connect the rough-in piping to the water heater. Some codes allow copper flexible connectors designed for that purpose to be used. Dielectric unions are often used and may be dictated by code to prevent electrolysis when copper is connected to the steel connection points of a water heater.

The physical location of the water heater often determines the extent of the actual installation. Most codes dictate that any water heater located above a finished area must be installed in a safety pan. The pan drain piping would typically be installed during the rough-in phase. The discharge piping from the relief valve is installed by a plumber, with the termination point depending on local codes and the location of the water heater. Most relief valve connections on a residential water heater are 3/4" female and require a plumber to install a 3/4" male adapter. Most codes dictate that the relief valve discharge piping be copper or some other metal, so the extreme hot water discharge does not damage it. If your local codes have unique installation requirements, such as earthquake protection, you would need to adhere to those installation requirements. Figure 15-13 illustrates some general installation features of a water heater. Figure 15-14 shows a copper water pipe connection with a dielectric union. The tools required to install a water

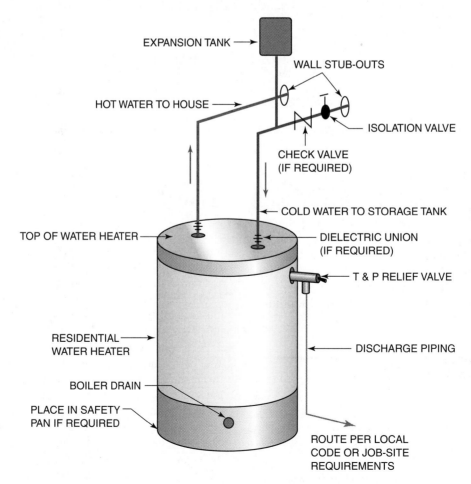

Figure 15-13 A water heater installation has similarities regardless of the type of heater.

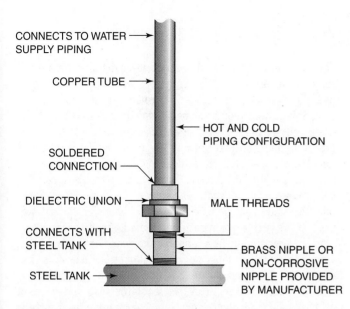

Figure 15–14 Water supply connection to a water heater often requires a dielectric union.

heater depend on the piping methods used. All pipe connections can be tightened with a pair of pliers, an adjustable wrench, or a small pipe wrench. If copper tube is used, you will need all the tools required to solder copper fittings. Flux, solder, and sand cloth are required for copper connections, and if PEX or CPVC is connecting to the copper, you will need the tools required to install those systems.

CAUTION

CAUTION: Never solder a fitting close to a water heater connection. Excessive heat will damage the dip tube. Some water heaters have lined inlet and outlet pipe nipples that are installed by the manufacturer; heat from soldering will damage the lining.

 FROM EXPERIENCE

Fabricating the copper water heater piping beforehand that connects directly to the heater increases productivity.

Electric Water Heaters

On an electric water heater, a plumber connects the hot and cold water piping from the rough-in stub outs to the designated inlet and outlet connection on the top of the heater. The type of piping used dictates the actual installation steps. Converting dissimilar materials such as PEX to copper must be performed using approved methods. An electric water heater can be installed with minimal clearance to combustible material. A manufacturer indicates any dangerous conditions pertaining to the specific heater. Electric water heaters are not as heavy as gas water heaters, but they are fragile and two people should install them in place. A hand truck or other equipment-moving aids should be used to protect the water heater from damage and to avoid injury to the lifters. Most codes allow the safety pan for an electric water heater to be plastic due to the lack of heat generated externally from the water heater. An electrician connects the electrical wiring to the water heater, but all internal operating devices and wiring are installed by the manufacturer. After the water supply connections are complete, the tank is filled with water and all air is removed by opening a hot water faucet. The electrical current is then energized to make sure hot water is being produced. Figure 15–15 shows how to position an electric water heater so access is available for future repair.

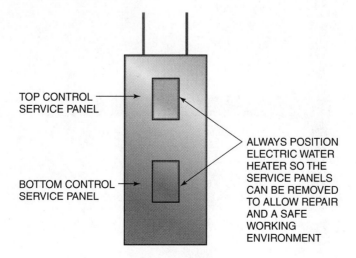

Figure 15–15 The service panels on electric water heaters must be positioned for easy access.

CAUTION: Connecting the electrical portion of a water heater without being licensed might violate your local code.

CAUTION: Never attempt to remove the electrical service panels until the electric current has been disconnected, such as from the circuit breaker panel.

Always situate an electric water heater so the service panels can be removed for safe access for future repair.

Gas Water Heaters

The tools required for a gas water heater installation are the same as for an electric water heater with the exception of needing two pipe wrenches to connect the gas piping system. Some codes require a special gas license to connect the gas supply to the heater. The water supply connection methods and code information may be the same for a gas water heater and an electric water heater. Codes vary pertaining to the gas supply connections and venting regulations. The specific water heater dictates many of the installation methods. One code that is different for gas than for electric water heaters is that the safety pan must be metal rather than plastic. A manufacturer's label on the water heater indicates what the clearance must be from combustible materials. The gas supply pipe arrangement is fairly typical for most residential water heaters. The gas regulator (gas valve) has a 1/2" female threaded connection. The manufacturer installs all operating devices and piping in the burner assembly area, and a plumber simply provides a single gas connection. Most codes dictate that a drip leg be installed to collect any small particles or moisture present in the gas supply system. A drip leg is assembled using Schedule 40 black-steel pipe nip-

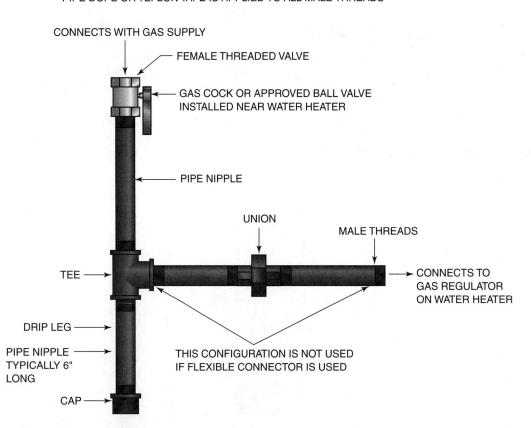

Figure 15-16 A gas piping connection to a water heater often uses a drip-leg style configuration.

ples and black malleable pipe fitting. Flexible metallic connectors designed for this purpose may be allowed by local codes. A gas cock or another approved isolation valve is installed near the water heater. The venting requirements of a gas water heater are dictated by code, type of heater, and job-site conditions. A direct-vent gas water heater has exact installation requirements. The heater is placed according to manufacturer instructions and code regulations. After a direct-vent water heater is situated, the venting portion is complete, but many codes mandate that an HVAC contractor install the venting (flue) system. After the flue portion is installed, the plumber connects the gas and water supply systems. When the tank is filled with water and all air is removed from the tank and system, the gas supply is activated and tested for leaks. A specially formulated soapy solution is applied to the gas pipe connections to inspect for visible signs of a leak. Once the gas system is tested and inspected, the water heater can be operated and tested. Figure 15–16 shows a typical gas piping configuration serving a residential gas water heater.

CAUTION: Never attempt to place a gas water heater in service without inspecting the final gas connections for leaks. A gas leak can cause a deadly explosion.

 FROM EXPERIENCE

A solution of dishwashing soap and water can substitute for gas leak detection solution. A gas leak will cause the soapy solution to bubble.

Summary

- Most fixture installations are done during the trim-out phase of construction.
- A toilet is installed onto a closet flange with a wax seal and a set of closet bolts.
- A two-piece toilet is assembled by a plumber on a job site.
- A p-trap installed on all sinks connects the fixture outlet to the rough-in stub-out pipe.
- A trap adapter is installed to connect the stub-out pipe to the p-trap outlet.
- A kitchen faucet is typically installed before installing a kitchen sink.
- A plumber usually installs a dishwasher.
- The hot water supply to a dishwasher is usually the same supply as the kitchen sink.
- The drain from a dishwasher usually connects to the drain from a kitchen sink.
- A garbage disposer has a designated port to connect the dishwasher drain hose.
- Some codes require a dishwasher drain hose to be routed through an air-gap device.
- A water heater installation is based on the specific type of heater and relevant codes.
- Fixture installations are tested with a performance test, not a pressure test.

Procedures

Escutcheon and Stop Installation onto a Copper Stub-Out Pipe

A Clean the exterior of the stub-out pipe to remove any paint or other construction debris that has adhered to the pipe. Apply fire-caulk or other sealant around the stub-out pipe, if required. Measure where the stub-out pipe will be cut and mark the pipe. Angle stop outlets can often be installed directly beneath a faucet or toilet connection to avoid bending the supply tube that connects the stop with the fixture. Sometimes the available space beneath a fixture means that the supply tube must be bent to connect to the fixture.

A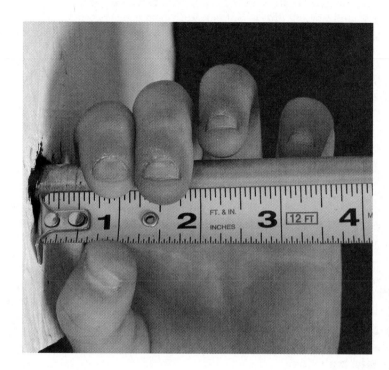

B Turn off the water supply to the house or the portion of the water supply system that serves the stub-out pipe. Open the boiler drain on a washing machine box, or tub and shower faucet, to relieve the pressure and to drain water from the piping system. Place the bucket or container below the stub-out pipe. If the penetration is through the floor, place a rag around the base of the stub-out pipe. If the stub-out piping is located below an opened faucet, you could be draining the piping system into your work area when you cut the pipe. Be prepared to capture all water that will exit the pipe. Do not attempt to cut completely through the stub-out pipe without knowing if the water pressure is alleviated from the system and without being prepared for the amount of water that will drain out of the pipe. Make a small cut on the bottom of the pipe or on the side that faces the bucket or container to allow all water to drain out safely. Once the water stops draining into the container, finish cutting the stub-out piping.

C When reaming the copper tube, be sure that the metal shavings do not enter the stub-out pipe. If a metal escutcheon is used, place it over the pipe in the opposite direction than it will be installed to slightly bend the tabs inward. Remove it and install it over the stub-out pipe in the intended direction. Depending on the type of stop installed, complete installation onto stub-out pipe.

B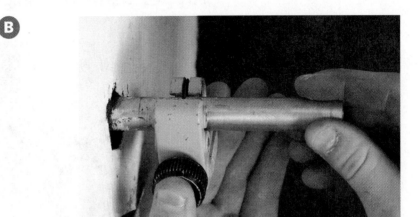

C

Procedures

Escutcheon and Stop Installation onto a Copper Stub-Out Pipe (continued)

 Cut, bend, and align supply tube to faucet or toilet. Install compression nut and ferrule onto supply tube.

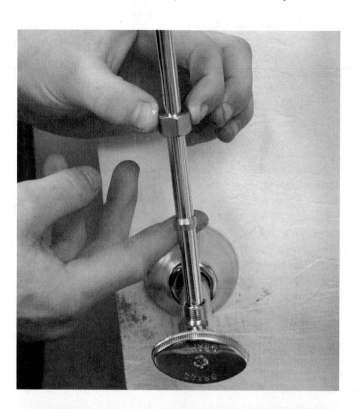

 Using two wrenches that will not scar the finish of the stop, such as adjustable wrenches or open-ended box wrenches in opposing directions, tighten the compression nut to the stop. Do not overtighten the nut, but be sure it is securely tightened to the stop. The intent is to compress the ferrule around the copper pipe. The stop handle is turned clockwise to close and the connection is tested for leaks by applying water pressure to the piping system.

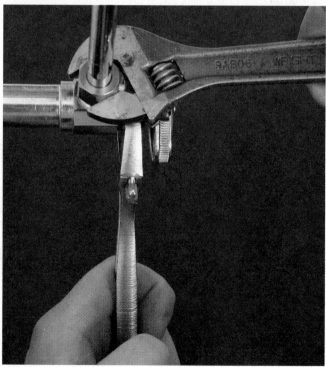

Procedures

Installing a Two-Piece Toilet

A Make sure the closet flange is installed and the closet bolts are securely fastened to the flange, similar to what is shown in Figure 15–4. The stop and escutcheon can be installed before or after installing the toilet. Company preference dictates whether to place the wax seal onto the closet flange or directly onto the toilet. Your personal preference dictates whether you assemble a two-piece toilet before setting it onto the flange. In this procedure, the wax seal is placed onto the flange rather than the underside of the bowl.

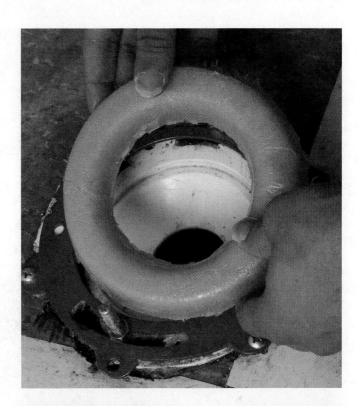

B Place the close-couple gasket over the threaded portion and nut of the flush valve located on the underside of the tank. Insert the tank bolts and washers through the designated holes of the toilet tank. Some tank designs require the tank bolts to be secured to the tank before securing the tank to the bowl. Place the tank onto the bowl, aligning the bolts with the designated holes in the top of the bowl. Place the flat washers over the bolts and tighten the hex nuts to secure the tank to the bowl. The tightening process must secure the tank evenly to the bowl. To avoid breaking the fragile tank or bowl do not overtighten the bolts, but be sure to tighten them enough so the tank-to-bowl gasket is firmly sealed.

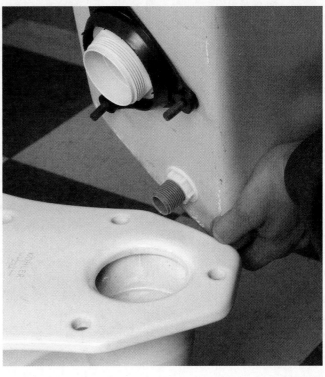

Procedures

Installing a Two-Piece Toilet (continued)

C Place the wax seal onto the closet flange so it is centered over the drain opening and between the closet bolts. Stand in front of the toilet and grasp it with both hands. Set the toilet while aligning the designated holes in the base of the bowl over the closet bolts. Apply even downward pressure until the wax seal is compressed. The toilet should be level and resting firmly against the floor. If the floor is uneven, plastic shims sold for this purpose can be added under the toilet base where necessary to level the bowl.

C

D Insert the plastic retainer washer that is provided with the toilet. Identifying text on one side indicates which side should face up (or down). If the retainer washer is installed incorrectly, the plastic bolt caps will not remain in place. Place the flat metal washers over the closet bolts to rest on top of the retainer washers.

D

Tighten the closet bolt nuts onto the bolts, alternating between the bolts to evenly tighten the toilet to the closet flange. With a hacksaw, cut the excess portion of each closet bolt flush with the top of the nut. Place the bolt caps over the bolts until they snap over the retainer washers. Install the tank supply as shown in Figure 15-5. Turn on the water supply by rotating the stop handle counterclockwise to fill the toilet tank. Flush the toilet to inspect for leaks. Install the toilet seat and, if code or your company requires the base of the toilet bowl to be caulked to the floor, neatly apply the caulking.

 The process is complete and the toilet should be firmly set onto the floor. Water level adjustments may be required; a plumber must read the manufacturer's instructions for the proper adjustment procedure.

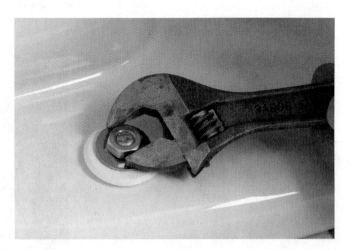

Procedures

Cultured-Marble Lavatory Faucet and Drain Installations

A The general contractor typically has the countertop securely fastened to the cabinet and a plumber must install the faucet, pop-up drain assembly, and necessary piping. The plumber must review manufacturer installation information. This procedure uses a faucet design with a lavatory supply tube similar to the one in Figure 15–6 (see page 416). The actual steps or sequence of installation on a job site are based on preference and job-site conditions and might vary from this procedure. Install the lavatory supply tubes onto the threaded faucet shanks with the nuts provided with the faucet. This step can also be performed after the faucet is installed onto the countertop. Be sure the plastic seal provided with the faucet is installed, or apply putty or caulking around the faucet holes in the countertop before inserting the faucet into the designated holes. This will prevent water from leaking from the countertop into the cabinet. Insert faucet through countertop holes.

B Tighten the faucet from below the sink with a basin wrench or other required tools.

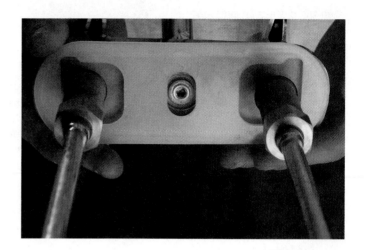

C Bend lavatory supply tubes to align with the stops below the sink.

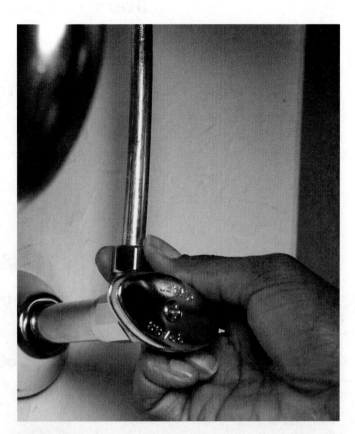

D Mark the supply tube to the required length and cut with the appropriate cutting tool. Install the 3/8" nut and ferrule over the supply tube and insert the supply tube into the stop.

Procedures

Cultured-Marble Lavatory Faucet and Drain Installations (continued)

 Tighten the compression nut and ferrule. This completes the water supply connection to the faucet. Water can be supplied to test for leaks after the faucet is in the closed position.

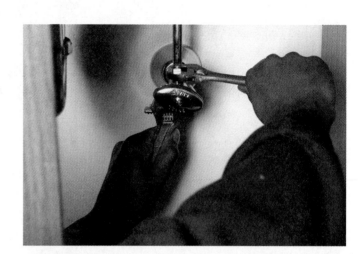

 There are numerous pop-up designs, each with unique installation features. Oil-based putty cannot be used with some cultured-marble products, so a plumber may need to use caulking. Apply putty or caulking to the underside of the drain-assembly flange.

 Install the securing nut onto the threaded portion of the drain assembly followed by the rubber gasket, with the beveled side of the gasket in the up position. Apply pipe dope to the male threads of the pop-up drain assembly and insert it up through the drain opening of the sink from below. Hand tighten the drain flange onto the threads of the drain assembly. Tighten the securing nut from below to compress the putty, or the caulking and gasket. Make sure the side connection designated for the pop-up rod is facing toward the faucet.

H Insert the pop-up lift rod into the designated hole through the center of the faucet.

I Insert the pop-up plunger into the drain opening of the drain assembly.

Procedures

Cultured-Marble Lavatory Faucet and Drain Installations (continued)

J Insert and tighten the plunger rod into the designated opening on the side of the drain assembly, making sure that it activates the plunger.

K Connect the link extension to the lift rod and plunger rod and adjust as required to ensure that the plunger fully operates when the lift rod is activated.

L Cut rough-in stub out to desired length, based on job site conditions, so that the p-trap can be installed.

WALL

OUTLINE OF CABINET

CUT STUB-OUT TO DESIRED LENGTH FROM THE BACK WALL

DRAIN STUB-OUT

M Install escutcheon over the stub-out pipe if job-site conditions dictate the installation at this stage. The escutcheon might have to be installed after the trap adapter for some installations. Solvent-weld the trap adapter to the stub-out pipe. Measure and cut the p-trap outlet to its required length to align the p-trap inlet with the pop-up drain outlet. Install the p-trap into the trap adapter. The vertical distance from the trap inlet to the pop-up drain outlet determines whether the tailpiece provided with the drain assembly needs to be cut or extended. Install the tailpiece into the drain assembly and p-trap.

N Turn on the water supply and fill the sink. Drain the sink to test for leaks in the drainage system.

Procedures

Installing a Stainless Steel Kitchen Sink into a Countertop

A Install the faucet on the sink based on manufacturer's instructions. Install the basket strainer into each drain opening of the sink using putty or caulking on the underside of the basket strainer flange.

A

B Install the garbage disposer mounting flange into the dedicated sink drain hole using putty or caulk per manufacturer instructions, but do not mount the motor portion of the disposer at this time.

B

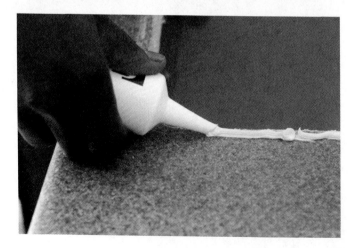

C Clean the countertop of dust and debris with a clean moist rag and dry the surface. Apply putty to the underside of the rim of the sink, or apply a latex non-siliconized adhesive caulk to the edge of the cutout area of the countertop.

C

D Install sink into cutout area.

D

E Install fastening clips similar to those shown in Figure 15–10, tightening them from the middle of the sink outward to the edges. Do not overtighten. The intent is to securely fasten the sink until all gaps are removed between the sink edges and the countertop. Clean excess putty or caulking from the edges of the sink. If caulking was used, clean with a wet rag.

E

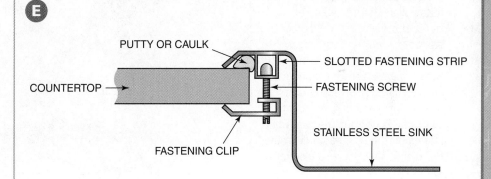

Procedures

Installing a Stainless Steel Kitchen Sink into a Countertop (continued)

F This step assumes that the stops and escutcheon are installed according to the step-by-step procedures explained previously in this chapter. Connect the water supply tubing from the stops to the faucet as required, similar to Figure 15–6. This procedure is based on a double-bowl sink with a garbage disposer. Cut the drain rough-in stub-out piping as needed and install the escutcheon and necessary piping and trap adapter. Align the p-trap with the flanged tailpiece. Cut the tailpiece if it is too long or install an extension tailpiece if it is too short. Tighten all slip-joint connections and continue with the garbage disposer drain connection.

G Many kinds of garbage disposers are available, and a plumber must follow manufacturer's instructions. If a dishwasher is being installed, it typically discharges into the drainage system through the garbage disposer connection. Lay the garbage disposer on its side and use a chisel and hammer to lightly knock out the plug installed by the manufacturer inside the dishwasher drain connection port.

F

G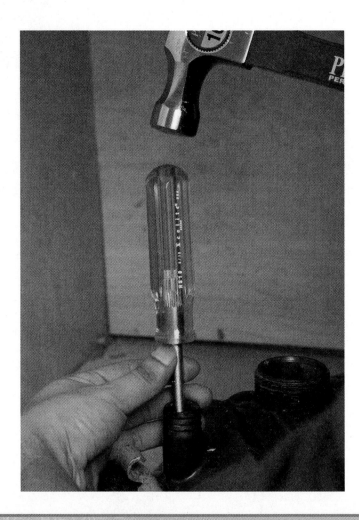

H Remove the plug from inside the disposer by tilting it upside down. Install the flanged elbow facing downward into the disposer outlet. Mount the garbage disposer motor portion onto the mounting flange from below the sink, aligning the drain outlet of the flanged elbow in the direction it will discharge into the p-trap.

I Install the required piping below the sink to place the p-trap in alignment with the flanged elbow. Install the extension tailpiece to connect the p-trap and flanged elbow. Once the dishwasher is installed and the drain hose connected, the drain system can be tested by filling both sinks with water and draining them to visually inspect for leaks.

Review Questions

1. The wall plate used to conceal a pipe penetration is called
   a. An escutcheon
   b. A cover plate
   c. A pipe conceal
   d. A concealing plate

2. A valve used to isolate the water supply under a fixture is called a
   a. Gate valve
   b. Stop
   c. Ball valve
   d. Fixture isolator

3. The name for the water supply tube that connects to a toilet is
   a. Fill valve connector
   b. Tank supply
   c. Toilet supply connector
   d. Lavatory supply

4. A toilet is sealed to the drainage system to prevent sewer gas from entering a building with a
   a. Sewer gas preventer
   b. Closet flange and wax seal
   c. Air admittance valve
   d. P-trap

5. A lavatory tailpiece that connects a pop-up drain assembly to the p-trap is
   a. 1-1/4"
   b. Always plastic
   c. Always metal
   d. 2"

6. A lavatory p-trap connects to the rough-in stub-out pipe with a
   a. Trap adapter
   b. Rough-in adapter
   c. Trap connector
   d. Basket strainer

7. A stainless steel kitchen sink installed into a counter is sealed with either caulk or
   a. Wax
   b. Putty
   c. Solvent cement
   d. Primer

8. The drain opening in a kitchen sink that does not have a garbage disposer requires a
   a. Basket strainer
   b. Pop-up assembly
   c. Wax seal
   d. Closet flange

9. A piping arrangement that connects two bowls of a sink to a single trap is called a
   a. Two-sink arrangement
   b. Double-bowl connector
   c. Continuous waste
   d. Pop-up assembly

10. A dishwasher drain hose discharges into a tailpiece designed for that purpose or into the
    a. Stub-out pipe on the downstream side of the p-trap
    b. Designated connection to a garbage disposer
    c. Vent serving the kitchen sink
    d. Sink bowl

11. The most common types of laundry sinks either have four legs and are mounted to the floor or
    a. Have two legs and are mounted to a wall
    b. Are wall mounted
    c. Are installed into a countertop
    d. Have a single leg and are wall mounted

12 **To prevent electrolysis when copper is connected directly to a water heater requires a**

a. Copper male adapter
b. Copper female adapter
c. Dielectric union
d. Copper union

13 **An electric water heater**

a. Does not require a vent (flue) pipe
b. Requires a vent (flue) pipe
c. Is available in a direct-vent type
d. Is always 30-gallon capacity

14 **A residential gas water heater**

a. Requires a 120-volt electrical circuit
b. Requires a 240-volt electrical circuit
c. Does not require electricity
d. Has to be 50-gallon capacity minimum

15 **If a dishwasher is installed with a dedicated connection to a garbage disposer, a plumber must**

a. Knock out the plug inside the connector
b. Route the dishwasher hose lower than normal
c. Use a tailpiece designed for that purpose
d. Install a wax seal

16 **A flanged tailpiece used on a kitchen sink installation attaches to the**

a. Garbage disposer mounting assembly
b. Rough-in pipe to install a p-trap
c. Pop-up assembly
d. Basket strainer

17 **The minimum size p-trap used for a kitchen sink is**

a. 1-1/4"
b. 1-1/2"
c. 2"
d. 2-1/2"

18 **The minimum size p-trap used for a laundry sink is**

a. 1-1/4"
b. 1-1/2"
c. 2"
d. 2-1/2"

19 **A stop that serves both the dishwasher and the hot water supply to a kitchen sink faucet is known as a**

a. Multi stop
b. Dishwasher stop
c. Kitchen sink stop
d. Dual stop

20 **The connection method used to install the 3/8" outlet of an angle stop is known as**

a. A compression joint
b. A flare joint
c. A solder joint
d. Threaded joint

SECTION FOUR

Trouble-shooting

SECTION FOUR
TROUBLESHOOTING

- **Chapter 16:** *Plumbing Repairs and Troubleshooting*
- **Chapter 17:** *Hydronic Heat*

Chapter 16: Plumbing Repairs and Troubleshooting

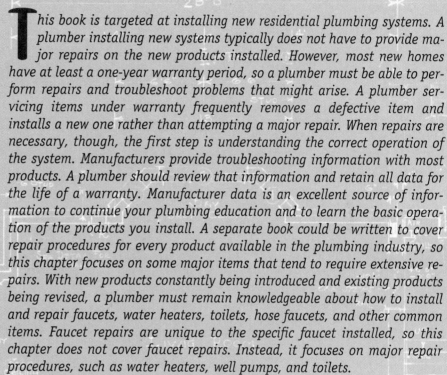

This book is targeted at installing new residential plumbing systems. A plumber installing new systems typically does not have to provide major repairs on the new products installed. However, most new homes have at least a one-year warranty period, so a plumber must be able to perform repairs and troubleshoot problems that might arise. A plumber servicing items under warranty frequently removes a defective item and installs a new one rather than attempting a major repair. When repairs are necessary, though, the first step is understanding the correct operation of the system. Manufacturers provide troubleshooting information with most products. A plumber should review that information and retain all data for the life of a warranty. Manufacturer data is an excellent source of information to continue your plumbing education and to learn the basic operation of the products you install. A separate book could be written to cover repair procedures for every product available in the plumbing industry, so this chapter focuses on some major items that tend to require extensive repairs. With new products constantly being introduced and existing products being revised, a plumber must remain knowledgeable about how to install and repair faucets, water heaters, toilets, hose faucets, and other common items. Faucet repairs are unique to the specific faucet installed, so this chapter does not cover faucet repairs. Instead, it focuses on major repair procedures, such as water heaters, well pumps, and toilets.

Troubleshooting a system for high or low pressure or noisy operation is done by inspecting and then eliminating possible causes to determine the reason for the problem. This chapter includes some troubleshooting and reasoning methods for specific problems that may arise. The information provided does not relate to any specific manufacturer or product. Safety must always be the first concern when troubleshooting or approaching any repair. In addition, knowing how a device will react within a system is vital to a correct, safe repair.

OBJECTIVES

Upon completion of this chapter, the student should be able to:

- know basic and safe troubleshooting approaches.
- understand that a troubleshooting approach is based on a specific product or system.
- recognize that a manufacturer warranty applies to the replacement of a defective product.
- know that a manufacturer of a specific product provides repair information based on that product.
- understand basic water heater, well pump, and toilet repair approaches.

Glossary of Terms

circuit electrical wiring system from a power source to equipment or a device

circuit breaker a device that disconnects (isolates) an electrical circuit

foot valve a check valve device installed at the bottom of a vertical-drop pipe in a well serving an above-ground pump; ensures that water remains in the pipe after a pumping cycle

heating element an immersed heating device that heats water in a storage tank when energized with electricity

high limit a safety device serving a water heater that disconnects an energy source when water temperatures approach unsafe temperatures

meter electrical testing tool used to test voltage, amperage, and continuity of an electrical system

ohm (Ω) a unit of measure describing the resistance of flow of an electrical current

submersible pump a pump motor and impeller assembly that is submersed in a well below ground to provide water to a piping system

volt (V) a measure of electromotive force (EMF) often described as electrical pressure, but a plumber relates it to the amount measured on a voltage meter, such as 120 volts or 240 volts

watts a measure of electrical power or the true power in a circuit, but a plumber relates it to a heating element of an electric water heater element, such as 4500 watts

Electric Water Heaters

Electric water heaters were discussed in Chapter 8, but this chapter is designed to dissect each component and test its function to properly diagnose a problem. A plumber must understand the basic operation of an electric water heater to troubleshoot a malfunction. Most residential electric water heaters are 240-volt, non-simultaneous with two **heating elements.** In a non-simultaneous heating cycle, only one of the elements operates at a time. The elements are identical and positioned so they are referred to as an upper and a lower element. The temperature-regulating components are the thermostat and the **high-limit** device. The two thermostats are different and are also classified as upper and lower. The high-limit device disrupts the electrical current to the thermostats and elements if the water temperature reaches an unsafe level. The high limit and upper thermostat are typically manufactured as a single unit, but their operating features remain independent. The manufacturer installs the internal wiring to devices, but a plumber must disconnect it when replacing items during a repair. A plumber must know the operating sequence to be able to correctly diagnose an internal problem with the devices and elements. Figure 16-1 illustrates the positioning of the heating elements and temperature-regulating devices in relation to one another. A plumber must drain a gas or electric water heater before removing any device that is immersed in the water.

FROM EXPERIENCE

Numerous water heater designs exist, but the configuration of most elements and thermostats is the same regardless of the manufacturer.

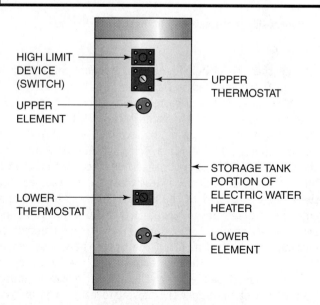

Figure 16-1 A residential electric water heater has heating elements and temperature-regulating devices.

See pages 469-470 for a step-by-step procedure for draining a water heater.

High-Limit Device

The high-limit device is commonly referred to as a high-limit switch or high limit. A plumber should never route wiring differently than the manufacturer because the sequence of operation will not be as designed. The wire providing electricity to a water heater is called a leg. The two wires that connect to the high-limit device are identified as line voltage one (L1) and line voltage two (L2). The screws that secure the wire connection to the electrical devices are known as terminals or posts. One wire provides 120 **volts** of electricity to one side of a heating element and the second wire completes the circuit by providing an additional 120 volts to the same element that connects to another dedicated post. The sequence of operation depends on the temperature of the water in the tank. The hot water rises to the top of the storage tank, and the cold water is routed to the bottom of the tank through a dip tube.

The routing of electricity after each wire connects to the high-limit device is different between L1 and L2. Both legs are connected to their dedicated post in the high-limit switch and their continuation is only allowed if the high-limit switch is not tripped due to unsafe conditions. The high-limit device has a bimetal disk housed internally that reacts to high temperatures within the tank. If the tank temperatures become too hot, the bimetal disk disconnects the electrical power through the device. The reset button on a high-limit switch can return the bimetal disk to its original form and allow the electricity to flow through the device. A plumber should always assume that the reset button was tripped due to unsafe conditions and should never simply reset the device without ensuring that a problem does not exist. Residential electric water heaters have surface-mounted thermostats and high-limit devices. The temperature is actually sensed on the exterior of the steel tank. A plumber must be sure that the devices are in contact with the clean exterior shell of the tank and that there is not an air gap between the devices and the tank.

The high-limit switch's four terminals (posts) are identified by the manufacturer as 1 through 4 (Fig. 16-2). One 120-volt wire (L1) connects to post 1, and the other 120-volt wire (L2) connects to post 3. If the high-limit device allows electricity to flow through, L1 is routed from post 1 to post 2. The manufacturer provides a metal jumper that connects post 2 of the high-limit device to post 1 of the upper thermostat. Post 4 of the high-limit device used in this chapter has two wires connected because there are two heating elements. The electrical current to post 4 is provided from post 3. One of the post 4 wires is routed directly to the upper element and the other is routed directly to the lower element. This provides constant electrical current to each element, but they cannot begin a heating cycle until the

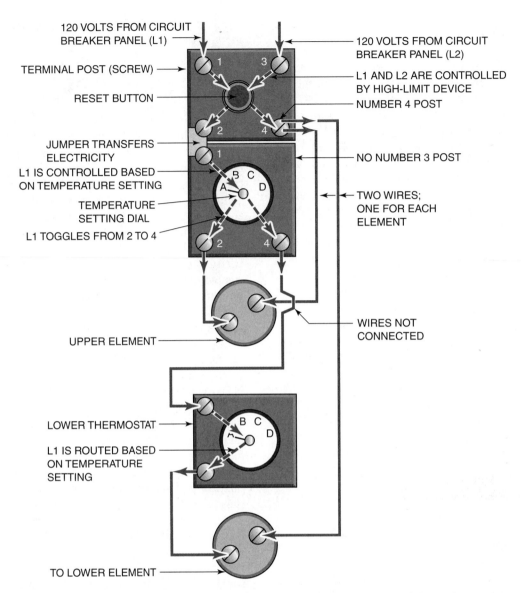

Figure 16–2 A 7-pole, non-simultaneous configuration is commonly used for residential electric water heaters.

additional 120 volts are provided from L1. The number 4 post of the high-limit switch is often described as the element post because its only purpose is to provide 120 volts of electricity directly to each element.

Upper Thermostat

The temperature on each thermostat can be adjusted with a standard screwdriver. Most safety standards do not allow a plumber to set the temperature above 120°F. The upper thermostat of a seven-pole design has three posts. These three posts combined with the four posts of the high-limit switch make the seven-pole design (Fig. 16–2). The space where the number 3 post would normally be installed is left empty by the manufacturer. The upper thermostat has a temperature setting that is either identified alphabetically from A through D or identified as warm, hot, and very hot. The alphabetical identification correlates to a specific water temperature as defined by the manufacturer in the user guide. The voltage to the upper thermostat is provided by the jumper from the high-limit switch and only distributes the electricity of L1. If the upper thermostat senses that the water in the top of the tank is less than the set temperature, the voltage is routed to the upper element. At this point, both L1 and L2 are energized in the same location, and 240 volts are applied to the upper element, which allows the heating cycle to begin. When the upper thermostat senses that the water is at the set temperature, the L1 voltage is routed to post 4 of the upper thermostat. A wire is routed from post 4 of the upper thermostat to post 1 of the lower thermostat.

Lower Thermostat

The lower thermostat only has two posts, numbered 1 and 2. It also has a temperature setting feature like the one on the upper thermostat. L1 (120 volts) provides electricity to post 1 of the lower thermostat from post 4 of the upper thermostat (Fig. 16-3). If the lower thermostat senses that the water in the bottom portion of the tank is less than the desired temperature, L1 is transferred to post 2. A wire is connected from post 2 of the lower thermostat to the lower element. When post 2 is energized, the lower element is fully energized with 240 volts, and the heating cycle occurs. When the lower thermostat senses that the water is at the set temperature, the voltage is removed from post 2 and remains energized to post 1. As hot water exits the top of the water heater, cold water is routed to the bottom of the tank through the dip tube. The lower thermostat senses the cold water entering and again routes the electricity to post 2, which energizes the lower element to begin heating the cold water. If more water is used than the lower element can heat (recover), the upper thermostat senses that the water temperature in the top portion of the tank is lower than its set temperature. The upper thermostat then removes L1 voltage from post 4 of the upper thermostat and routes L1 to post 3, which energizes the upper element and begins a heating cycle. Figure 16-2 illustrates a typical seven-pole non-simultaneous configuration like the one described here. Figure 16-3 is a side view of the devices in a residential electric water heater. The thermostats are secured in place and held against the surface of the tank with a retainer clip. Shown in the illustration is a plastic safety guard that must be placed back in the correct hole/slot after testing or replacing a thermostat or element. The safety guard minimizes the possibility of electric shock. The lower thermostat and element have a smaller version of the plastic safety guard, but it is not illustrated in Figure 16-3.

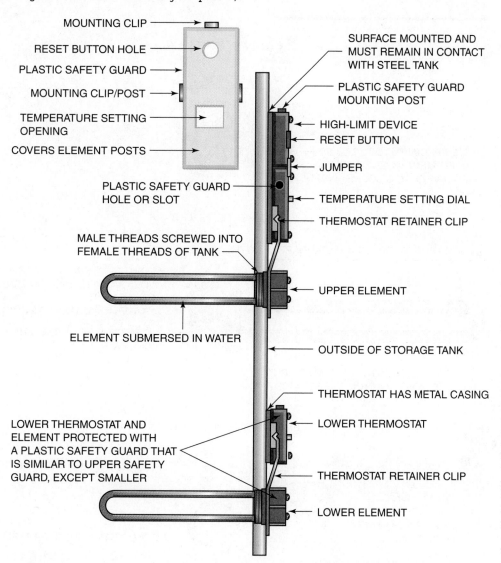

Figure 16-3 Heating elements are submersed in the water of a storage tank, and thermostats and high-limit devices are surface mounted.

CAUTION: Your local code may not allow a plumber to perform electrical work without proper certification. Never route wires differently than a manufacturer does. Incorrect wiring can create a very dangerous situation.

FROM EXPERIENCE

A simultaneous configuration has more than one element heating at the same time and requires greater amperage protection. The non-simultaneous electric water heater has a slower recovery rate than other electric water heaters.

Heating Elements

Residential electric water heating elements typically screw in, but bolt-in types are used for some water heaters. They vary in length, but 12" is the most common. A rubber washer, similar to a 1-1/4" slip-joint washer used for a p-trap, is installed over the male threads of the heating element before threading the element into the steel tank. The electrical and wattage rating of the element is indicated by the manufacturer on the element so it can be identified from the exterior. Most screw-in water heater elements are removed with a 1-3/8" socket. A specialty tool is available for that purpose. Because the element is located within the outer shell of the water heater, it is difficult to remove it without using some type of socket tool. Figure 16–4 illustrates a typical element and a special socket tool.

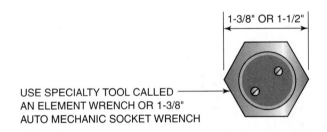

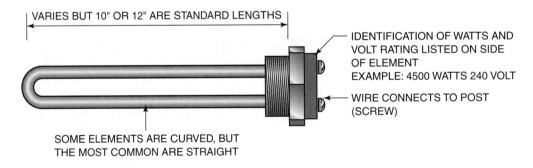

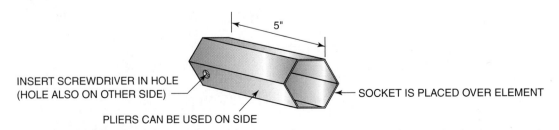

Figure 16–4 Heating elements are identified by their rating and capabilities and are removed with a socket tool.

FROM EXPERIENCE

Various element types and lengths are available. A plumber must be sure the voltage and **wattage** rating of the element is correct before installing it. Refer to the manufacturer's operation and maintenance manual provided with the water heater to confirm that a different length and design of a replacement element is allowed by the warranty.

See pages 471–472 for a step-by-step screw-in element replacement procedure.

Electrical Source

Basic electrical knowledge is required to understand an electrical source. A plumber should never attempt a diagnosis or repair without extensive training, and certification or licensing. As previously explained, a 240-volt water heater has two wires, each supplying 120 volts of electric current. A third wire—the ground wire—which is green or bare copper, is secured to the water heater with a dedicated screw provided by the manufacturer. Most codes require an electrician to connect the electric wiring to the water heater manufacturer's wires in a dedicated area, typically on top of the heater. The electrical wiring is routed by an electrician from a **circuit breaker** panel directly to the water heater. The electrical wiring is typically dedicated to serve only a water heater and is known as a dedicated **circuit.** A circuit breaker panel is where a dedicated circuit is manually disconnected (shut off) by toggling a circuit breaker. The circuit breaker is either off or on and is self-explanatory. Warning tags or other specially designed lock-out accessories are required by OSHA, and their use is based on specific job sites or company standards. Always follow safety procedures. Figure 16–5 illustrates the basic wiring schematic from a circuit breaker panel to a residential 240-volt water heater. The amperage (amp) rating of a circuit breaker indicates the capabilities of that electrical circuit.

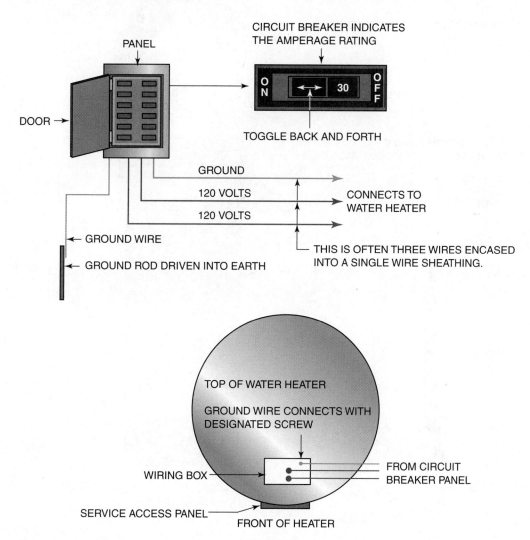

Figure 16–5 A dedicated circuit provides electricity from a circuit breaker panel to an electric water heater.

454 SECTION FOUR *Troubleshooting*

FROM EXPERIENCE

The wire routed by an electrician is connected to the manufacturer wiring on top of the water heater with specially designed wire nuts.

Electrical Tests

A plumber must perform voltage and amperage tests while the electricity is energized (live), so extreme caution must be taken to avoid electrocution. A residential 240-volt, non-simultaneous electric water heater draws 18.75 amps when it is functioning properly. The formula that determines the amperage load of an element is the wattage rating divided by the voltage applied (W ÷ V = A). The dedicated circuit wiring only provides electricity to the water heater and is served with an individual circuit breaker. Most circuit breakers serving a residential non-simultaneous water heater have a minimum 30 amp ratings. The circuit breaker interrupts the electricity to the heater if a circuit overloads; that interruption is known as a tripped breaker.

An electrician learns the theory of voltage and amperage, but for basic water heater troubleshooting, a plumber needs to be more educated about safety and practical determination of a repair than theory. Amperage can be viewed as the flow of current, similar to the flow of water through a pipe, and the resistance caused by the flow is what increases the amperage load. When the resistance decreases (fewer amps), the current increases (higher voltage). A plumber can check the number of amps present with an amperage **meter.** An amperage meter that snaps around the wire (Fig. 16-6) can

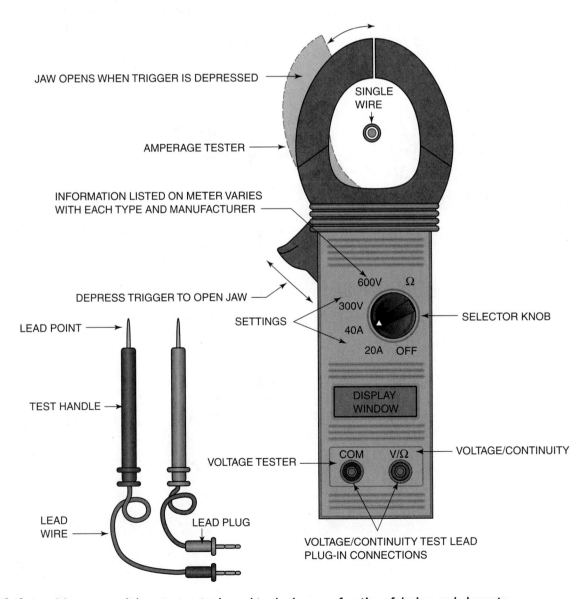

Figure 16-6 A multi-purpose volt/amp test meter is used to check proper function of devices and elements.

be used for water heater repair. The service panel on the front of a water heater must be removed, but the circuit breaker must be in the off position before removing the access panel. Once the access panel is removed, the circuit breaker can be toggled to the on position. The amperage meter must be set to the correct position for checking the amperage load. Voltage tests are performed with a different function of a testing meter than amperage tests. The sequence of testing for 120 volts or 240 volts is based on knowing the correct operation of each device. Knowing where 240 volts must be energized for a correct heating cycle based on tank temperatures allows a plumber to correctly diagnose a problem.

A continuity test can also be used to determine whether an element is still intact. It is performed with the electricity shut off (disconnected) and all wires removed from the elements. The symbol for ohm (Ω) on a testing meter indicates that the meter is being used for a continuity test. The plumber then touches each testing lead to each post of an element. If the display window is digital, it should register to infinity or maximum numerical display (100, 1000, etc.) depending on the meter used. If a needle-type meter is used, the needle will move to the maximum reading possible. The element does not have to be removed from the water heater to perform the continuity test. A plumber must never use the testing meter to test voltage when the selector knob is in the continuity test mode. Most test meters have an internal fuse that protects it from voltage applied while set in the continuity test mode, but permanent damage to the meter can result with improper use. A multi-purpose electrical testing meter is useful for testing voltage, amperage, and continuity. Figure 16-6 shows a multi-purpose meter that snaps around a wire that is energized with electricity. It has test leads to check for voltage and continuity. Figure 16-7 illustrates the placement of testing leads to test voltage in specific areas of an electric water heater. Table 16-1 lists the test reading based on a residential seven-pole non-simultaneous water heater. Table 16-2 lists some common residential electric water heater problems and solutions. Refer to a manufacturer troubleshooting guide for specific information about a particular water heater.

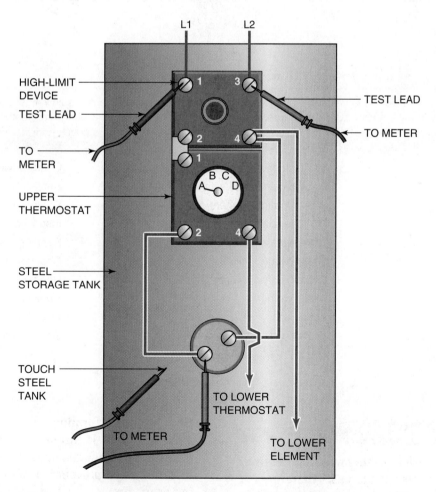

Figure 16-7 Testing a high-limit, thermostat, or element for voltage is performed with test leads touching specific terminal posts and knowing correct settings to use on a voltage meter.

Table 16-1 240-volt, Non-Simultaneous Water Heater Readings at Each Terminal Post of Each Device and Element

CAUTION: Be sure meter is set to a minimum 300-volt setting. Never attempt an electrical test without being qualified to perform such a test. Misuse of electricity can result in personal injury or death.

- Upper T = Upper thermostat
- Lower T = Lower thermostat
- HL = High-limit device

Device	Black Lead	Red Lead	Reading	Notes
All	Any post	Steel tank	120 volts	Touch red lead to steel tank
All	Steel tank	Any post	120 volts	Touch black lead to steel tank
High limit	# 1 Post	# 3 Post	240 volts	Power is correct from breaker
High limit	# 2 post	# 4 post	240 volts	Power is through reset button
Upper T/HL	# 1	# 4	240 volts	At least one element is fully energized
Upper T/HL	# 2	# 4	240 volts	Upper element is fully energized
Upper T/HL	# 2	# 4	120 volts	Upper element is not fully energized
Upper T/HL	# 4	# 4	240 volts	Lower element is fully energized
Lower T/HL	# 1	# 4	240 volts	Lower element is fully energized
Lower T/HL	# 1	# 4	120 volts	Lower element is not fully energized
Lower T/HL	# 2	# 4	240 volts	Lower element is fully energized
Lower T/HL	# 2	# 4	120 volts	Lower element is not fully energized

Table 16-2 Electric Water Heater Troubleshooting Chart

CAUTION: This list does not include every possible problem or solution that could be present. It is only a general listing of common occurrences and solutions. Always refer to the manufacturer's troubleshooting guide for a particular water heater.

Problem	Possible Cause	Possible Solution
No hot water	No electricity from source	Check power source
No hot water	Electrical problems with thermostats	Replace thermostats
No hot water	Failed heating element(s)	Replace element(s)
Little hot water	Dip tube failure	Inspect dip tube
Little hot water	Lower element failure	Replace lower element
Little hot water	Thermostat(s) failure	Replace thermostat(s)
Water too hot	Thermostat(s) failure	Replace thermostat(s)
Reset button tripped	Water too hot	Replace thermostat(s)
Rotten egg smell of water	Anode rod	Inspect and contact manufacturer
Popping noise when heating	Scale build-up on element(s)	Replace element(s)

CAUTION: Because wiring is energized for testing, all electrical tests are dangerous. Therefore, a plumber must be thoroughly trained and licensed as an electrician or possess other required certification before performing any tests. Any information presented here is intended to provide an overview of testing procedures. Consult manufacturer's information before proceeding with any electrical testing or repairs.

FROM EXPERIENCE

A continuity test is safer than other tests for making sure an element is intact, because voltage is not applied. A continuity test is not as reliable as an amperage test for determining if an element is operating correctly.

Gas Water Heaters

Various types of gas water heaters are used in the residential industry, several of which are explained in Chapter 8. A plumber must review manufacturer information concerning diagnosis and repair. A gas regulator that controls the gas flow to a burner assembly has several important operating features. Most codes do not allow a plumber to disassemble a gas regulator; most have security screws that require a special tool to remove the cover. A residential plumber involved with new installations typically replaces a malfunctioning gas water heater rather than perform extensive repairs if it is covered by a manufacturer warranty.

This chapter focuses on the basic operation of a gas regulator and not on a specific type. Thermocouple and gas regulator replacements are common repairs that could be performed instead of replacing a water heater. Gas water heaters with a power vent or electronic ignition require electricity, but conventional draft and direct-vent types do not. A heater having electronic ignition does not have a pilot flame; it is ignited much like a gas barbeque grill, except the process is automated. Water heater technology has advanced in the past decade. To prevent an explosion, burner assemblies must be protected against fumes entering from the atmosphere. For example, gasoline emits flammable vapors and should not be stored near the water heater. A plumber must replace the burner assembly access cover gasket or other sealing components properly after servicing a burner assembly. Figure 16-8 illustrates the basic gas flow through a gas

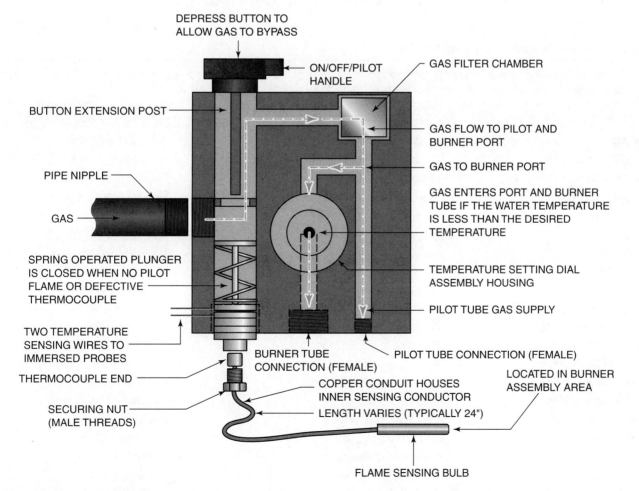

Figure 16-8 Many residential gas water heaters are manually ignited per manufacturer instructions. A pilot button is essentially a safety by-pass for the gas regulator.

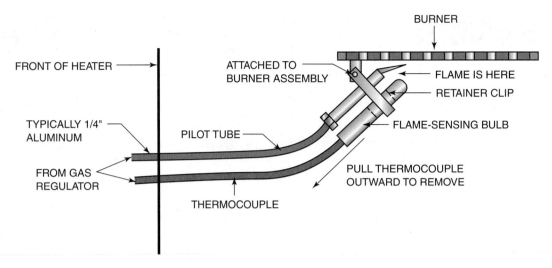

Figure 16-9 A thermocouple is replaceable and typically secured in place with a retainer clip for easy removal.

regulator using a pilot, and Figure 16-9 shows a thermocouple detail view in relation to a pilot tube and burner assembly. The spring-operated plunger of a gas regulator is held in the open position, allowing gas to flow through the regulator when the thermocouple senses that a pilot flame is present. If the pilot flame does not stay lit, the thermocouple or regulator could be defective. Venting problems could also cause the pilot to be extinguished. Table 16-3 lists some common troubleshooting items for a conventionally vented gas water heater.

Table 16-3 Gas Water Heater Troubleshooting Chart

CAUTION: This is only a general listing and does not represent every possible problem or solution that could be present in a basic water heater. Always refer to the manufacturer troubleshooting guide for a particular water heater.

Problem	Possible Cause	Possible Solution
No gas flow to pilot	No gas from meter	Turned off at meter
No gas flow to pilot	No gas from regulator	Debris in regulator
No gas flow to pilot	Faulty regulator	Replace regulator
No gas flow to pilot	Crimped pilot tube	Repair or replace tube
No gas flow to pilot	Leak in pilot tube	Replace tube
No pilot flame	No gas flow from regulator	See above
No pilot flame	Defective thermocouple	Replace thermocouple
No pilot flame	Air in gas piping	Purge air from piping
No pilot flame	Defective regulator	Replace regulator
No gas flow to burner	Defective regulator	Replace regulator
No gas flow to burner	Crimped burner tube	Repair or replace tube
No gas flow to burner	Blockage in burner tube	Remove and clean
No gas flow to burner	Water is too hot	Run water to cool tank
No gas flow to burner	Defective high-limit device	Replace device
Relief valve leaking	Water is too hot	Run water to cool tank
Relief valve leaking	Defective relief valve	Replace relief valve
Relief valve leaking	Excessive pressure	Install expansion tank
Relief valve leaking	Excessive pressure	Check PRV to system
Low water temperature	Thermostat set too low	Adjust temperature
Low water temperature	Dip tube failure	Inspect and replace
Low water temperature	Defective thermostat	Replace thermostat

Table 16-3 Continued

CAUTION: This is only a general listing and does not represent every possible problem or solution that could be present in a basic water heater. Always refer to the manufacturer troubleshooting guide for a particular water heater.

Problem	Possible Cause	Possible Solution
Slow recovery	Sediment in tank	Drain and flush tank
Slow recovery	Dirty burner assembly	Clean burner
Slow recovery	Poor flame	Adjust burner air
Slow recovery	Poor flame	Need combustion air
Not enough hot water	Improper size heater	Calculate demand
Not enough hot water	See low temperature/slow recovery	See low temperature/slow recovery
Popping/banging noise	Calcium build up	Drain and flush tank
Popping/banging noise	Sediment build up	Drain and flush tank
Banging noise	Check valve slamming	Install shock absorber
Banging noise	Solenoid to equipment	Install shock absorber
Fume odor	Poor draft on flue system	Flue pipe installation
Fume odor	Poor draft on flue system	Termination location
Gas odor	Leak on piping system	Soap all piping joints
Soot build up	Poor draft on flue system	See fume odor information
Soot build up	Poor draft on flue system	Install draft hood fan
Soot build up	Poor combustion air	Install air-supply ducts
Soot build up	Poor burner flame	Clean and adjust burner
Flame back flash	Negative air pressure	Isolate heater air

FROM EXPERIENCE

Natural gas and propane require different regulator orifices. A plumber must ensure that the correct regulator is installed based on the gas type.

See page 473 for a step-by-step thermocouple replacement procedure.

See pages 474–475 for a step-by-step gas regulator procedure.

Well Pumps

Many pump installations are performed by a licensed well contractor instead of a plumber. The two basic types of well pumps are above ground and submersible, and variations of each are available. A plumber must refer to the manufacturer's troubleshooting guide for a particular pump. Because a pump is operated with electricity, a plumber must use extreme caution when working around a pump that is energized with electricity. The multi-purpose volt/amp meter illustrated in Figure 16–6 can also be used to check the voltage and amperage of an electrical system that operates a pump. Information about well pump connections and water service piping is found in Chapter 11. The well pump system design varies depending on the region of the country, the job-site conditions, and the preference of the contractor. A **submersible pump** must be manually removed from a well for replacement. The wiring is routed along the side of the drop pipe in the well. An above-ground pump is usually located in the house or in a dedicated pump house. The operating control used for both pump designs is called a pressure switch. A pressure switch provides electricity to a pump motor when the pressure in a piping system drops below the low set pressure and disconnects the electrical source when it reaches the high set pressure. The most common pressure switches in residential applications have a 20 psi differential. A 20/40 psi is the most common, but 30/50 and 40/60 psi types are also used. The adjustment features and electrical requirements are based on the pressure switch and pump installed. Some above-ground pumps have a built-in tank design, and on some, the pressure switch is located on the pump and installed at the factory. An expansion/storage tank is used if one is not built in. The most common design is similar to a water heater expansion tank; however, this type stores water for use as opposed to simply accommodating the increase in system pressure (expansion). If a well pump system does not have a tank, or if the tank has lost its air storage, the pressure switch will function rapidly,

460 SECTION FOUR *Troubleshooting*

(A) SHALLOW-WELL JET PUMP

(B)

(C) DEEP-WELL JET PUMP

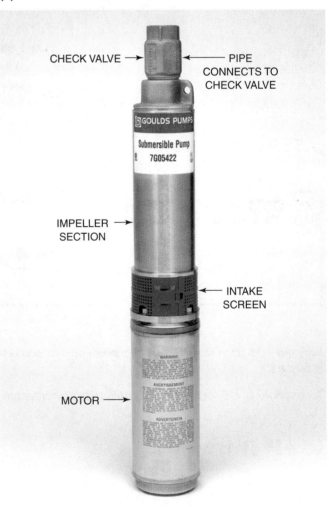

(D) SUBMERSIBLE PUMP

Figure 16–10 A submersible and an above-ground pump are two types often used for residential installations. *Courtesy of Goulds Pumps.*

indicating a problem with the tank. An optional specialty fitting known as a tank tee can be installed to situate the pressure switch, pressure gauge, and pressure relief valve in a central location near the tank. An above-ground pump has a **foot valve** located in the well and connected to the end of the drop pipe. A foot valve is a check-valve fitting that ensures that the water in the piping system does not drain back into the well through the drop pipe. If an above-ground pump does not produce any water, the foot valve might have failed or might be lodged in the open position from debris, rock, or sand. A well casing must be sealed with either a plug or a cap. Some designs in which piping enters the top of the well casing usually have plugs; designs with a pitless adapter typically use a cap to seal the well casing. Both the caps and plugs must have designated areas for the wiring to enter the well so it is protected against damage. Figure 16–10 illustrates several pump designs used for residential applications. Figure 16–11 is a photograph of a typical pressure switch and Figure 16–12 is a basic illustration of the electrical transfer feature of a pressure switch. Figure 16–13 shows an expansion/storage tank that can be used for any well pump that does not have a self-contained tank. Figure 16–14 contains photographs of various tank tees. Figure 16–15 has photographs of numerous well casing plugs and caps. Figure 16–16 has two different foot valve designs exposing the internal components. Table 16–4 lists common troubleshooting causes and solutions.

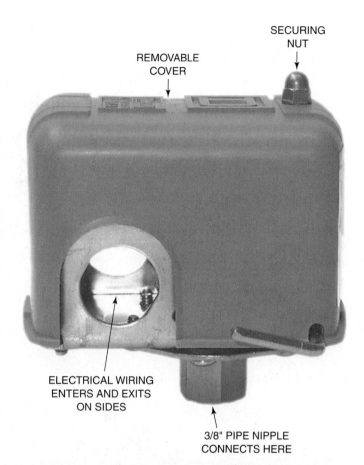

Figure 16–11 A pressure switch controls the on and off cycle of a pump by sensing the pressure within a piping system. *Courtesy of Simmons Manufacturing Company.*

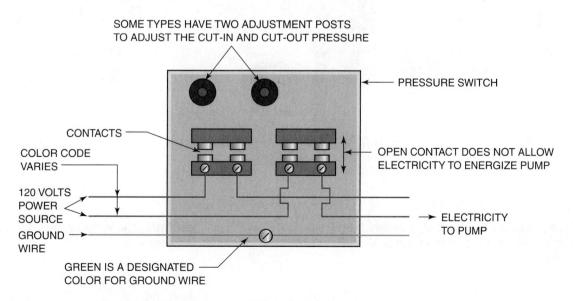

Figure 16–12 A pressure switch provides electricity to and disconnects it from a pump.

462 SECTION FOUR *Troubleshooting*

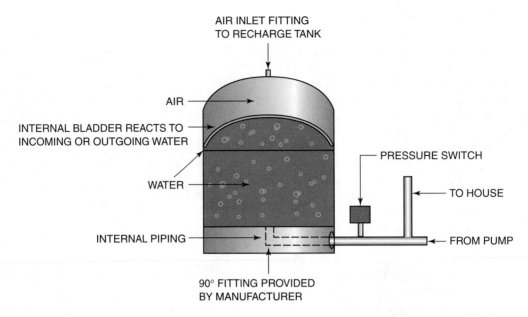

Figure 16–13 An expansion tank, also called a storage tank, receives incoming water from the pump and discharges water as required between pumping cycles.

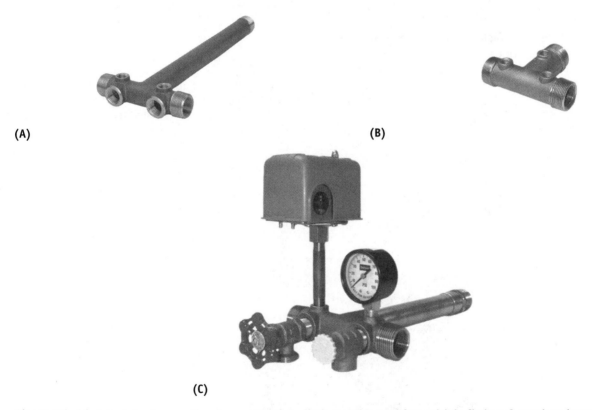

Figure 16–14 A tank tee is an optional accessory that eliminates the need for multiple fittings for various items such as a pressure switch, pressure-relief valve, and pressure gauge. *Courtesy of Simmons Manufacturing Company.*

CHAPTER 16 *Plumbing Repairs and Troubleshooting* **463**

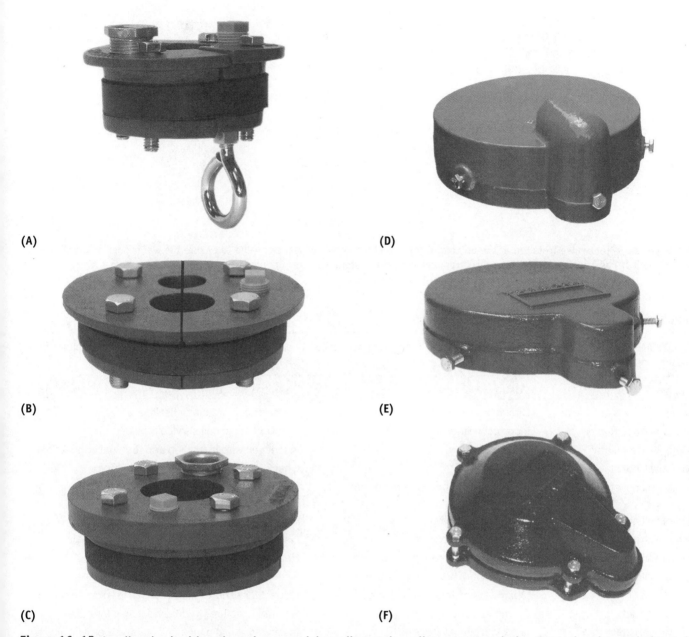

Figure 16–15 A well casing is either plugged or capped depending on the well pump system design. Some plugs have a designated eyelet to affix the rope to the pump, and all have an area for wiring to enter the well casing. *Courtesy of Simmons Manufacturing Company.*

FROM EXPERIENCE

A plumber involved with pump installations and repairs sees a variety of designs on a job site. He or she should focus on the components that are similar regardless of the system design and their basic relation to each installation.

See pages 476–478 for a pressure-switch replacement procedure.

See pages 479–480 for an expansion/storage tank replacement procedure.

(A) (B)

Figure 16-16 A foot valve installed at the bottom of the drop pipe serves as a check valve to ensure the water does not drain back into the well. It has a screen to prevent debris from entering the system. *Courtesy of Simmons Manufacturing Company.*

Table 16-4 Well Pump Troubleshooting Chart

CAUTION: This general listing does not include every problem or solution that could be present. Always refer to a manufacturer troubleshooting guide for a particular well pump.

Problem	Possible Cause	Possible Solution
Motor will not start	Circuit breaker is tripped or faulty fuse	Check circuit breaker; check fuse
Motor will not start	Incorrect voltage	Check voltage; call service provider
Motor will not start	Damaged wiring	Inspect all wire insulation and terminal connections
Motor will not start	Faulty motor	Check for high amperage; replace motor
Motor will not start	Poor alignment between motor and impeller	Check for high amperage; align motor with impeller
Motor will not start	Faulty pressure switch	Check for corrosion; clean contacts
Motor will not start	Faulty pressure switch	Check for voltage transfer between terminals
Motor will not start	Plugged pressure switch nipple	Remove, inspect, clean, or replace nipple
Motor will not start	Faulty control box (if applicable)	Check for voltage transfer between terminals
Motor will not start	Obstruction in impeller housing	Inspect for sand or debris; check for high amperage
Motor will not start	Overload is tripped	See manufacturer troubleshooting information
Motor starts often	Leak in piping system	Inspect visible piping first; test buried or well piping
Motor starts often	Foot valve or check valve not closing properly	Inspect for obstruction, or worn seat or disk
Motor starts often	Pressure switch setting is incorrect	Check pressures at the pump and on/off differential
Motor starts often	Faulty pressure switch	Replace pressure switch
Motor starts often (Bladder type tank)	Tank is water logged	Depress air inlet valve, inspect for water in tank
Motor starts often (Bladderless tank)	Tank is water logged	Drain tank
Overload is tripped	Control box is defective (if applicable)	Check voltage transfer between terminals; replace box
Overload is tripped	Voltage is incorrect	Check voltage supply; contact electrical supply provider

Table 16-4 Continued

CAUTION: This general listing does not include every problem or solution that could be present. Always refer to a manufacturer troubleshooting guide for a particular well pump.

Problem	Possible Cause	Possible Solution
Overload is tripped	Control box has overheated (is applicable)	Install in a shaded or partially sunny area
Overload is tripped	Control box has overheated (if applicable)	Install in an area away from heated area
Overload is tripped	Motor is defective	Check amperage; replace motor if required
Overload is tripped	Defective wiring	Inspect all wiring and terminal connections
Overload is tripped	Pump is wired incorrectly	Review wiring diagram
Overload is tripped	Motor is in an area not properly ventilated	Provide adequate ventilation to motor
Motor always runs	No water flow due to sand or obstruction	Inspect check valve to verify that it is opening
Motor always runs	No or low water flow due to loose shaft connection	Check impeller connection to shaft
Motor always runs	No water flow due to broken shaft	Inspect shaft
Motor always runs	Leak in piping system	Inspect visible piping and test all hidden piping
Motor always runs	Malfunctioning control box (if applicable)	Test voltage transfer between terminals
Motor always runs	Pressure switch is defective	Inspect points to see if they are "welded" together
Motor always runs (Submersible pump)	Intake screen is plugged	Pull pump and inspect screen
Motor always runs (Above-ground pump)	Foot valve screen is plugged	Pull drop pipe and inspect screen
Motor always runs	Low well water level	Pull pump and/or piping. Check water level and depth
Motor always runs	Low well water level	Pump is oversized for well capacity or yield
Motor always runs	Well has collapsed	Attempt to free pump or piping from collapsed soil
Motor always runs	Well has collapsed	Have well cleaned and lined or drill new well
Circuit trips or fuse blows on start up	Obstructed impeller from sand or debris	Inspect impeller for obstruction
Circuit trips or fuse blows on start up	Defective wiring	Inspect wiring
Circuit trips or fuse blows on start up	Wiring is grounded to casing	Perform **ohm** test on well casing and wiring
Circuit trips or fuse blows on start up	Motor winding is defective	Perform ohm test on winding
Circuit trips or fuse blows on start up	Defective capacitor (three-wire type pump)	Locate in control box, perform ohm test
Circuit trips or fuse blows on start up	Defective relay coil (three-wire type pump)	Locate in control box, perform ohm test
Circuit trips or fuse blows on start up	Control box is overheated (if applicable)	Locate in shaded area or away from heat
Circuit trips or fuse blows on start up	Voltage supply is incorrect	Perform voltage test Call electrical service provider

Toilets

Two-piece toilets are usually installed in residential homes, but custom installations might have a one-piece toilet. Both types have toilet tanks that store water used for flushing the waste and water from the fixture. Siphon jet is the most common kind of flushing action. A typical water-saver toilet tank holds 1.6 gallons per flushing cycle. The fill valve, also called a ballcock, controls the water level in the tank. There are two basic types of fill valves—a ball float and a float that rides vertically along a shaft. An approved fill valve has an anti-siphon feature that most codes dictate must be installed at least 1" above the top of the overflow/fill tube. An overflow/fill tube is part of a device known as a flush valve, which prevents water in the tank from overflowing if the fill valve malfunctions. The overflow/fill tube also provides a route through which water from the fill valve can flow to replenish the trap seal in the toilet bowl after each flushing cycle (see Fig. 16-19).

When activated, a toilet tank handle, typically installed on the left side of the tank, raises an accessory known as a tank flapper. A chain connects the tank flapper to the tank handle. The flapper rests on the compatible seat of the flush valve, creating a seal so water does not leak into the bowl during periods of non-use. There are various kinds of fill valves, tank flappers, and tank handles. Many are compatible with most toilet designs, but some are specific to a particular toilet design. The water enters the toilet bowl through the rim and is distributed into the bowl area through rim holes. A siphon-jet toilet has angled rim holes to create a swirl (vortex) and an additional hole below the water level that acts as a jet to initiate the discharge from the toilet bowl. If a toilet does not flush completely, a plumber should suspect that an obstruction is present in the rim or holes that is jeopardizing the flushing cycle. The holes can often be cleaned with a thin metal rod or the end of a clothes hanger.

A defective tank flapper can cause water to slowly enter the toilet bowl during a non-use period. One way to check the flapper is to isolate the water supply to the tank and return after a short period of time to visually check whether the water level has lowered (dropped). Another method is to add some food coloring to the tank water and see if the water in the bowl becomes colored. A slight crack in the overflow/fill tube can cause the same symptoms as a defective tank flapper. The seat of the flush valve that mates with the underside of the tank flapper can also have a defect, such as a gouge or nick, that does not allow the flapper to seal properly. If a leak occurs during a flushing cycle where the tank and toilet connect, the close-couple gasket is defective. A leak between the tank and bowl connection when a flushing cycle is not occurring can indicate that the tank bolts are not tight, the tank bolt washers are deteriorated or defective, the flush valve rubber gasket is defective, the flush valve is not properly tightened, or there is a crack in the tank. A leak around the base of the toilet where it rests on the floor is caused by either a wax seal leak or a crack in the bowl. Replacing a toilet handle is a simple process. The threaded portion has left-hand threads, which means that a clockwise rotation is needed to loosen the securing nut. Figure 16-17 illustrates the relationship among the components housed in a toilet tank. Figure 16-18 illustrates a popular ballcock (fill valve) design with unique features for adjusting the water level in a tank. Figure 16-19 illustrates a typical flush valve assembly installed in a toilet tank.

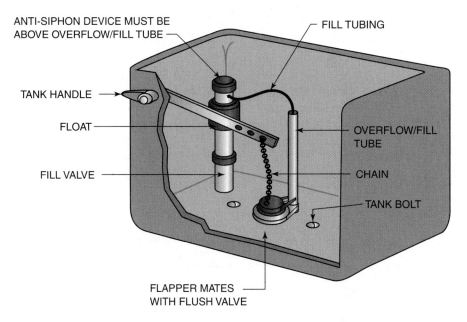

Figure 16-17 A toilet tank holds water for flushing the toilet and houses a fill valve, flush valve, flapper, and toilet handle.

CHAPTER 16 *Plumbing Repairs and Troubleshooting*   **467**

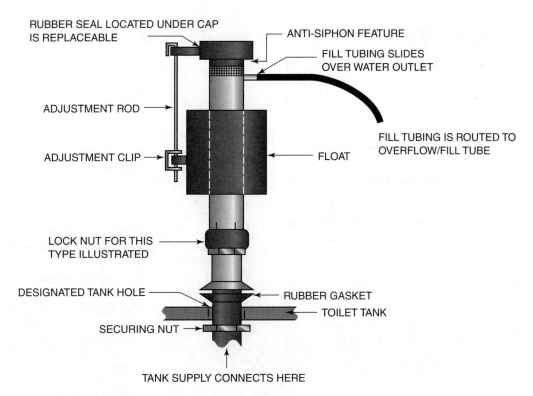

Figure 16–18 Various ballcock designs are available to regulate the water level in a toilet tank.

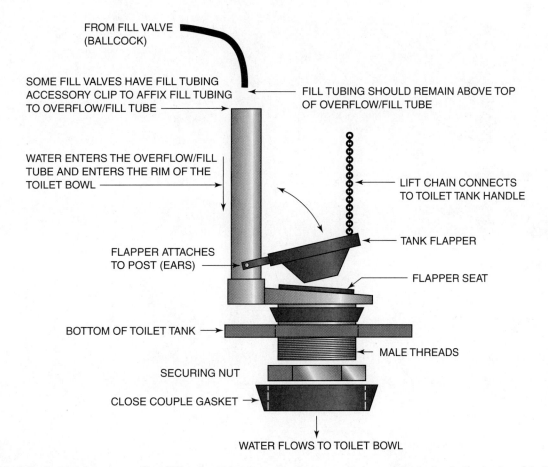

Figure 16–19 A flush valve has an overflow/fill tube. It is where the flapper rests to seal the water in the tank and is the passageway for the water exiting the toilet tank.

FROM EXPERIENCE

The toilet tank must be removed from the bowl to replace the flush valve and close-couple gasket. The fill valve, flapper, and tank handle can all be replaced with the tank on the bowl.

See pages 481–482 for a step-by-step procedure for ballcock replacement.

See pages 483–485 for a step-by-step procedure for flush valve replacement.

Summary

- Fixtures and devices have manufacturer warranties that vary depending on the item.
- A plumber typically responds to repair problems even though the warranty is by a manufacturer.
- Some states or general contractors dictate the minimum warranty period of an installation.
- Troubleshooting gas and electric water heaters must be performed by qualified individuals.
- Basic electrical knowledge is necessary to safely troubleshoot an electric water heater.
- An electric water heater storage tank must be filled with water before applying electricity.
- A gas water heater must be filled with water before igniting the gas supply.
- A plumber must have an electrical voltage/amperage meter to troubleshoot an electric water heater.
- Troubleshooting well pumps must be performed by a qualified individual.
- A well pump pressure switch controls the on/off function of a well pump.
- Refer to manufacturer information to assist in repair diagnosis.

CHAPTER 16 *Plumbing Repairs and Troubleshooting*

Procedures

Water Heater Draining

A For electric water heaters, disconnect (shut off) the electricity via a circuit breaker or other means relevant to the particular job site. For gas water heaters, shut off the gas cock or other isolation valve serving the water heater. Turn off the cold water isolation valve serving the water heater. Open the hot water side of several faucets in the house to relieve the pressure from the piping system.

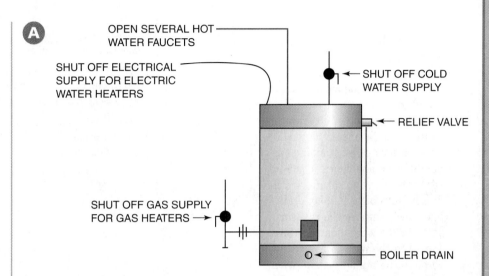

B Attach a garden hose to the boiler drain on the bottom of the water heater. Be sure that the hose is routed to a safe location and that the water draining from the hose will not cause a hazard or damage. Open the boiler drain to allow all or part of the water to drain from the tank. If you are replacing a temperature and pressure relief valve, only the top 6 inches need to be drained. If you are replacing the top element of an electric water heater, only the top one-third of the tank must be drained. If you are replacing the lower element of an electric water heater or the gas regulator of a gas water heater, the entire tank must be drained. Specialty pumps can be purchased to expedite the draining process, but a typical residential water heater should drain in 10 to 15 minutes.

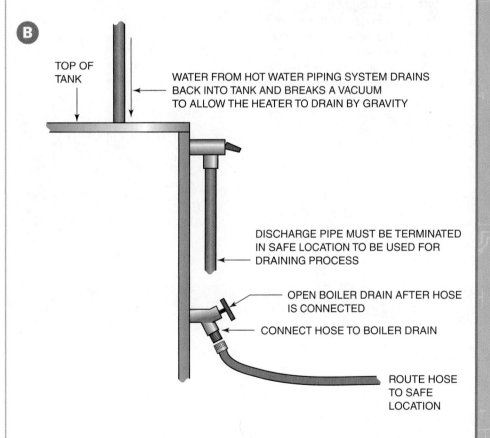

Procedures

Water Heater Draining (continued)

C After completing a repair, close all opened hot water faucets and turn on the hot water faucet of the bathtub so the air will be allowed to escape when you turn on the water supply to the water heater. Avoid using a faucet with an aerator or other straining capabilities so any debris in the water heater does not obstruct the faucet. Turn off the boiler drain and slowly turn on the isolation valve to the water heater. Allow the air in the tank to escape through the piping system. When water is flowing from the open hot water faucet, shut it off and turn on the electricity or gas supply to the water heater. If it is a gas water heater, ignite the pilot and turn on the gas to the burner assembly.

- Close faucets
- Open tub faucet or other fixture that does not have an aerator
- Turn off boiler drain
- Slowly open cold water valve
- Allow air to escape from faucet
- Close faucet
- Turn on electricity or gas
- If gas type, ignite pilot
- Ensure repairs were successful

Procedures

Screw-In Element Replacement

A Your local code might require that the electrical portion of this procedure be performed by a licensed electrician. This procedure refers to an upper element. The same procedure is followed for the lower element except that the entire water heater must be drained to replace the lower element. Disconnect (shut off) the electricity by shutting off the circuit breaker or by using other means. Remove the access panel on the side of the water heater using an appropriate screwdriver with a rubber-coated handle. Approach the process as if the electricity is still energized in case it was improperly disconnected. Using a pair of needle-nosed pliers with a protective handle coating, remove any insulation and expose the plastic safety guard covering the thermostat and element (do not grasp insulation or plastic safety guard with your hands).

B Using a voltage meter similar to the one in Figure 16–6, test to make sure the voltage has been correctly disconnected by touching the meter leads to posts 1 and 3 of the high-limit switch. Drain the water heater as described in the previous procedure. Loosen both terminal screws from the element and remove the wires from the element. Using the element socket tool or another tool of choice, loosen the element by turning it counterclockwise. Remove the element from the water heater slowly, making sure the water has been successfully drained from the heater.

A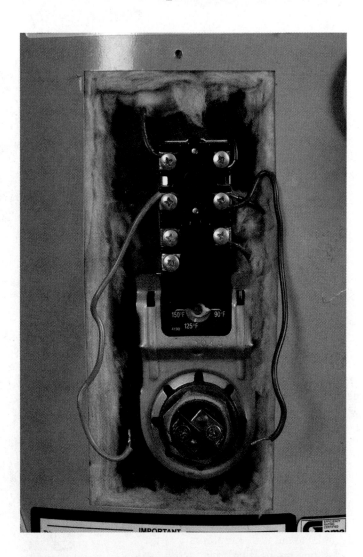

B

Procedures

Screw-In Element Replacement (continued)

C Install a rubber washer over the screw-in element threads. Insert the element into the tank and tighten it with an element tool or other socket-type tool. Do not overtighten or the rubber washer may be damaged.

C

D Turn off boiler drain, turn water supply on, and fill tank. Check for leaks and ensure that the rubber gasket is sealing the connection with the tank. If no leaks are present, connect both wires to the element. Install the plastic safety guard and place the insulation over it. Replace the access panel on the side of the water heater. Turn the electricity back on. You should have warm water within 20 minutes if no water is used during the heating cycle.

D

Procedures

Thermocouple Replacement

A Your local code might require that this procedure be performed by a licensed gas contractor. The actual procedure depends on the specific gas water heater burner assembly that is used. Refer to the manufacturer instructions for your heater. The gas supply does not have to be isolated for this procedure, because no gas can pass through a regulator if the thermocouple is not sensing a pilot flame. If your company standards require the gas cock or another isolation valve to be in the off position, do so at this time. Remove the burner assembly access panel. If the panel has a gasket sealing the combustion under the access panel, pay close attention not to damage it.

B Loosen the thermocouple securing nut from the gas regulator and remove that end of the thermocouple from the regulator.

C This step often requires a plumber to lie down on the floor. Removing thermocouples from some water heater designs is more difficult than from others. Using a pair of needle-nosed pliers or other compact pliers, grasp the flame-sensing bulb, not the copper extension from the regulator. Pull the thermocouple flame-sensing bulb downward (outward), which will remove it from the retainer clip. Now reverse the previous steps. If the access panel gasket is required to seal the burner assembly area, make sure that it is intact and in excellent condition. If the gas supply was turned off, turn it back on. Ignite the pilot and burner assembly according to manufacturer instructions. The process is now complete.

A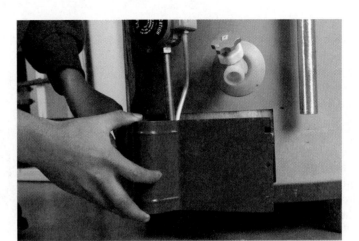

B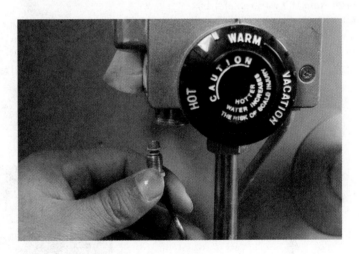

C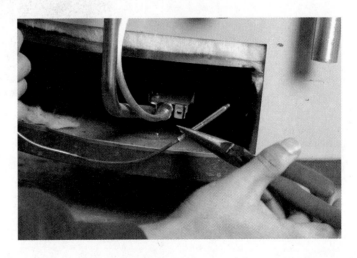

Procedures

Gas Regulator Replacement

A Turn off the gas supply from the gas cock or other isolation valve and make sure the pilot light and gas burner are not ignited. Drain the water heater completely as described previously. Loosen the pipe connection slightly that is supplying gas to the regulator to make sure the gas supply is properly isolated. Disconnect the pipe from the regulator. Remove the thermocouple, burner tube, and pilot tube by loosening their securing nuts from the regulator. After the water heater is completely drained, remove the regulator from the water heater by rotating it counterclockwise with a pipe wrench or other grasping tool. Pull the regulator out of the tank.

B Apply pipe dope or Teflon tape to the male threads of the regulator.

C Insert the regulator into the water heater and hand tighten. Use a pipe wrench or other grasping tool to tighten the regulator so it is level and in the correct position to connect all devices and piping. Turn off the boiler drain, turn on the cold water supply, and remove all air from the tank, as described previously. Check for water leaks where the regulator was installed in the tank.

C

D Connect the gas supply pipe, thermocouple, burner tube, and pilot tube. Turn the gas supply on and, using a soapy solution, test the pipe connection to the regulator for leaks. Follow manufacturer instructions for igniting the pilot. Test the connection between the pilot tube securing nut and the regulator for gas leaks using a soapy solution. Follow manufacturer instructions to ignite the burner. Test the securing connection of the burner tube to the regulator for leaks using a soapy solution. The process is complete and you should have warm water in 10 minutes if no water is used during the heating process.

D

Procedures

Pressure Switch Replacement

A Your local code might require that this procedure be performed by a licensed electrician and licensed pump repair technician. The pressure switch in this example is connected to a pipe nipple. Some are mounted on an above-ground pump; others are configured a different way. Refer to manufacturer information for correct procedures for your specific installation. Disconnect (shut off) the electricity from the circuit breaker panel or by some other means. Always approach an electrical component as if it still has electricity applied, even though you have disconnected the power source. Loosen the securing nut on the pressure switch cover and remove it. Check the voltage using a meter similar to the one in Figure 16–6. Touch one meter lead to a wire terminal and the other to the exposed ground wire to make sure electricity is disconnected. Loosen and remove all wires from the securing posts (screws) inside the pressure switch. Loosen and remove the electrical connectors that secure the wiring entering and exiting both sides of the pressure switch.

A

Connect a garden hose to the boiler drain to drain the piping system, and open a hose bibb or faucet to allow air to enter the piping system. The piping system could have been drained before the electrical work to expedite the replacement, but never allow water to puddle or be present in your workspace when working around electricity. Using an adjustable wrench or tool of your choice, loosen and remove the pressure switch from the connecting pipe nipple. You may have to use two wrenches or pliers. Hold the pipe nipple securely while removing the pressure switch with the other wrench to avoid damaging other portions of the system or loosening the pipe nipple.

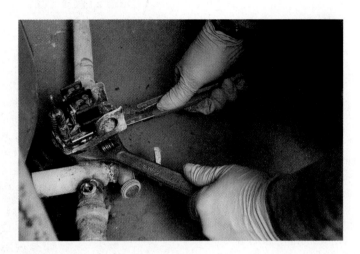

 Apply Teflon tape or pipe dope to the male threads of the pipe nipple. If pipe dope is used, be careful that excess pipe dope does not get in the pipe nipple, because it will then enter the internal portion of the pressure switch.

Procedures

Pressure Switch Replacement (continued)

D Hand tighten the pressure switch onto the pipe nipple. Using two wrenches or pliers, securely fasten one wrench onto the pipe nipple while rotating the pressure switch clockwise with the other wrench. Tighten the pressure switch so it is aligned in the same position as the one removed to allow for proper electrical wiring connections. Install the wiring using electrical connectors to both sides of the pressure switch.

D

E Connect the wiring to the posts (screws) according to manufacturer information. Install the pressure switch cover. Shut off the boiler drain and any open faucets. Turn on the electricity. Open a hose bibb or any faucet that does not have an aerator. Bleed all air from the piping system and turn off all open faucets. Make sure the pressure switch is operating properly and refer to manufacturer information for any needed adjustments in pressures. The process is complete.

E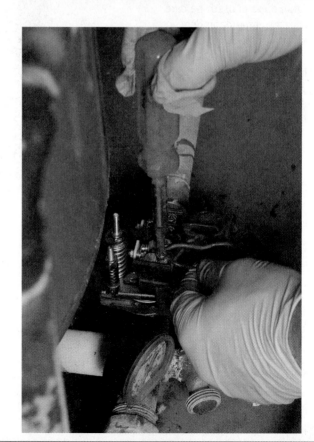

Procedures

Expansion/Storage Tank Replacement

A Your local code might require that this procedure be performed by a licensed well pump technician. If the electrical wiring serving the pressure switch has to be removed, your local codes may require that this be done by a licensed electrician. This procedure is based on only one of several piping arrangements that might be used. The procedure begins by assuming that the electrical wiring is disconnected (shut off) and removed from the pressure switch, and that the piping system is drained. Cut the piping on both sides of the tee or as required to remove the tank from the piping system.

B Turn the tank onto its side to allow access to the underside of the tank. Using a pipe wrench or other grasping tool, remove the pipe or tank tee from the 90-degree fitting.

C Apply Teflon tape or pipe dope to the male threads of the tank tee fitting. If pipe dope is used, be careful that excess pipe dope does not enter the inside of the piping system. Tighten the male threads into the 90° fitting of the tank using a wrench to avoid damaging the tank by exerting too much pressure.

A

B

C

Procedures

Expansion/Storage Tank Replacement (continued)

 D Place tank back in its original location. Connect the piping system that was cut. If PVC or other solvent-welded connections are used, follow manufacturer instructions concerning the curing time required before applying water pressure to the system.

 E Check the air pressure in the tank with an automobile tire gauge or other preferred means at the air inlet fitting located at the top of the tank. Most manufacturers recommend that the air pressure in the tank be 2 psi below the low pressure (cut-on) of the pressure switch. If the air pressure must be increased, add air with a bicycle pump through the air inlet fitting located at the top of the tank; if air pressure must be decreased, release it from the top of the tank. Connect the wiring to the pressure switch as required based on your job site. Turn the system back on. Check all connections for leaks and make sure the system is functioning properly. The process is complete.

Procedures

Ballcock Replacement

A Turn off the water supply by isolating the stop below the toilet. Flush the toilet to remove the water from the tank. Use a sponge to remove residual water in the bottom of the tank. Place a rag on the floor to absorb any small amounts of water that may drain from the tank. Remove the tank supply nut from the ballcock shank where the supply tube connects to the ballcock on the underside of the tank. It may be necessary to remove the supply tube from the stop, which is a compression type connection.

B Remove the ballcock securing nut located on the underside of the tank. Remove the ballcock from the inside of the tank. Clean any debris from the inside of the tank near the hole designated for the ballcock.

A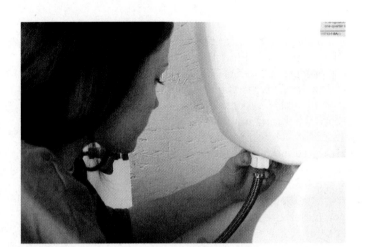

B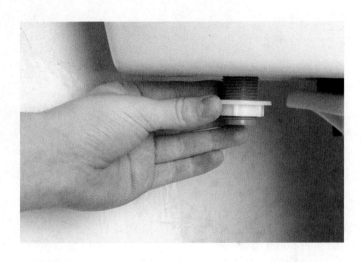

Procedures

Ballcock Replacement (continued)

C Follow manufacturer instructions for the ballcock being installed. This procedure is based on just one of the many types available. Install the rubber gasket over the ballcock shank, and insert the ballcock through the designated hole located inside the tank. Install the ballcock securing nut over the shank on the underside of the tank. Note where the fill tubing outlet is located inside the tank before completely tightening the securing nut.

D The fill tubing outlet should point in the direction of the overflow/fill tube portion of the flush valve. If the tank supply tube was removed from the stop, reinstall it. Tighten the tank supply nut onto the shank of the ball cock. Install the fill tubing from the ball cock to the overflow/fill tube. Turn on the stop to supply water to the tank. Adjust the float to the desired water level for the toilet tank being filled. Check for leaks. The process is complete.

Procedures

Flush Valve Replacement

A This procedure begins with the water isolated, the tank sponged dry, and the ballcock disconnected from the water supply tube. Loosen the tank bolts by removing the tank bolt nuts from the underside of the toilet bowl. This process typically requires a flat-head screwdriver and a small pair of pliers, socket tool, or any other tool of choice that can accommodate the confined space under the toilet tank. Remove the tank from the toilet bowl. Remove the close-couple gasket from the threaded portion of the flush valve located on the underside of the tank. Notice the position of the overflow/fill tube inside the tank in relation to the ballcock fill tube and the tank bolts.

A

B Loosen and remove the flush valve securing nut from the underside of the tank using a large pair of pliers or other tools. Remove the flush valve from the inside of the tank. Clean any debris from around the designated hole in the bottom of the tank.

B

Procedures

Flush Valve Replacement (continued)

 Install a rubber gasket over the threaded portion of the flush valve. Insert the flush valve through the hole inside the tank.

 Tighten the securing nut over the threads of the flush valve on the underside of the tank, placing the overflow/fill tube in the same location as the old flush valve.

 Be sure to tighten the securing nut adequately. A leak will not be noticeable until the very last step of the process, and then the tank will have to be disassembled from the bowl to fix the leak. Place the close couple gasket over the threads of the flush valve (if that is the design of the toilet you are repairing). Some types will also cover the securing nut (recessed type).

Place the tank back onto the toilet bowl and tighten the tank bolts with the securing nuts. Reconnect the water supply tubing, turn on the water supply, and check for leaks. After the tank fills, flush the toilet and check for leaks between the tank and bowl.

Review Questions

1. To disconnect an electrical circuit means to
   a. Disassemble the wiring
   b. Shut off the electric source
   c. Disconnect the wire nuts
   d. Leave the power source activated

2. A dedicated electrical circuit to a water heater provides electricity to
   a. Only the water heater
   b. The water heater area
   c. The water heater and wall receptacles
   d. The circuit breaker

3. To avoid electrocution, the thermostats and elements are covered by
   a. An access panel
   b. An insulation or Styrofoam
   c. A plastic safety guard
   d. A piece of cardboard

4. A multi-purpose electrical testing meter is used for testing
   a. Voltage
   b. Amperage
   c. Both a and b are correct
   d. Only water heaters

5. A typical residential electric water heater is a non-simultaneous
   a. 20-gallon type
   b. 240-volt type
   c. 120-volt type
   d. 40-gallon type

6. The most common residential electric water heater element is
   a. 4500 watts
   b. 2500 watts
   c. 1500 watts
   d. 6000 watts

7. A safety device on a residential electric water heater that has a reset button is known as a
   a. Temperature and pressure relief valve
   b. Backflow preventer
   c. Thermostat
   d. High-limit device

8. Residential gas water heaters use a thermocouple to sense
   a. A pilot flame is present
   b. The water temperature
   c. The gas pressure
   d. A flue temperature

9. The operating control device that provides and disconnects the electricity to a well pump is a(n)
   a. Foot valve
   b. Operating switch
   c. Pressure switch
   d. Check valve

10. A specialty well tank fitting designed to locate accessories in a central area near the tank is known as a
    a. Tank tee
    b. 90° elbow
    c. Specialty tank fitting
    d. Pressure switch

11. The water in the drop pipe of a well serves an above-ground pump using a
    a. Foot valve installed at the bottom of the pipe
    b. Check valve installed near the pump
    c. Gate valve installed near the pump
    d. Foot valve installed above ground

12 A fill valve in a toilet tank is also known as a

a. Tank valve
b. Ballcock
c. Toilet float valve
d. Ball valve

13 A toilet handle secured to a toilet tank with a securing nut has

a. Right-hand threads
b. Left-hand threads
c. Plastic threads only
d. Metal threads only

14 The gasket used to connect a toilet tank to the toilet bowl is known as a

a. Close-couple gasket
b. Tank gasket
c. Bowl gasket
d. Full-face gasket

15 A green-colored wire routed to a water heater is typically the

a. 120-volt wire
b. 240-volt wire
c. Ground wire
d. Main power wire

16 To replace a gas regulator on a water heater, the tank must be

a. Drained completely
b. Drained halfway
c. Full of water
d. Drained at least 6" from the top

17 An upper element of a residential electric water heater is connected to a high-limit device with a

a. Wire
b. Terminal
c. Jumper
d. Wire nut

18 The 4 post of the high-limit device of a residential electric water heater is often referred to as the

a. Element post
b. Lower thermostat post
c. Upper thermostat post
d. Safety post

19 The typical psi differential of a well pump pressure switch is

a. 20
b. 30
c. 40
d. 50

20 Before energizing an electric water heater, the

a. Tank must be filled with water
b. Tank must be drained
c. Faucets must be open
d. Wires must be disconnected

Chapter 17 | Hydronic Heat

Although this book is primarily concerned with residential plumbing systems, this chapter is an introduction into residential heating systems—specifically hydronic heating systems. Other heating systems are found in various parts of the country. A different license is often required to install heating systems.

In this chapter, we will concentrate not on the source of heat, but on the medium being heated. In the case of the hydronic system, this medium is water instead of air, which is used in furnaces. Water can be heated to generate hot water or, if heated above 212°F at atmospheric conditions, steam. Either hot water or **steam boilers** can be used to provide heat, but this text will only address the issue of hot water, because steam is not often found in new residential installations.

OBJECTIVES

Upon completion of this chapter, the student should be able to:
- explain the concept of hydronic heating.
- list the three most commonly used heat sources in boilers.
- describe basic boiler construction.
- identify component parts of a boiler.
- explain how a boiler operates.
- describe various components that maintain the desired water temperature in a boiler.
- explain the difference between a one-pipe and a two-pipe hot water system.
- discuss the difference between direct-return systems and reverse-return systems.
- describe the operation and function of centrifugal pumps.
- explain the function of boiler controls and safety devices.
- explain the function of an expansion tank.
- explain the point of no pressure change.
- check the pressure in an expansion tank.
- explain primary-secondary pumping.
- explain the concept of zoning.
- explain how a radiant heating system creates comfort.
- explain how a radiant heating system operates.
- install a boiler.
- service boilers.

Glossary of Terms

air cushion the air above the semipermeable membrane in an expansion tank

air separator device that removes air from the system

air vent device that removes air from a hydronic system; air vents can be manual or automatic devices

air vent fitting used to remove air, either manually or automatically, from a hydronic system

aquastat electrical component that opens and closes its contacts to energize and de-energize electric circuits in response to the temperature sensed by the device

automatic air vent a fitting that automatically removes air from a hydronic heating system

balancing valve a manually controlled valve used to increase resistance and reduce water flow through a given branch circuit

baseboard heating see radiator or terminal unit

boiler equipment designed to heat water using electricity, gas, or oil as a heat source for the purpose of providing heat to an occupied space or potable water

boiler feed valve valve that reduces the pressure entering the structure to the pressure required by the hydronic system; automatically feeds water into the system to maintain the desired water pressure

centrifugal pump pump that moves water through a piping circuit by means of centrifugal force

circulator see centrifugal pump

closed loop a system that is closed or isolated from the atmosphere

compression tank see expansion tank

diaphragm-type expansion tank see expansion tank

direct-return configuration of a water heating system in which the first terminal unit supplied with hot water is the first one to return to the boiler and vice versa

expansion tank system piping component that provides additional space for expanding water to occupy

feet of head the pumping capacity of a pump; 1 foot of head is the equivalent of a 0.433 psig difference between the inlet and outlet of the pump (1 psig = 2.31 feet of head)

flow check valve see flow control valve

flow-control valve prevents backward and gravity circulation through loops not requiring flow

hydronics heating systems that circulate hot water or steam through piping arrangements located in the areas being heated

manifold station location of the manifold for radiant heating loops

manual air vent fitting that, when opened manually, will remove air from a hydronic heating system

monoflo tee see diverter tee

one-pipe hot water hydronic piping configuration that uses a main hot water loop and diverter tees to connect the terminal units to the system

outdoor reset control used on hydronic systems that decreases the temperature of the water as the outdoor temperature increases

PEX tubing cross linked polyethylene; see polyethylene tubing

point of no pressure change point in a hydronic hot water system where the expansion tank is connected to the piping system; the pressure at this location cannot be affected by circulator operation

polyethylene tubing tubing material used for buried water loops in radiant heating systems as well as geothermal heat pump systems

pressure-reducing valve valve that reduces the pressure of the water entering the structure to the desired pressure in the hydronic system

pressure relief valve spring-loaded valve that opens when the pressure in a hydronic system exceeds the rating of the valve

radiant heating system heating system that attempts to regulate the heat loss of the individual as opposed to the rate of heat loss of the structure

radiator heat emitters or terminal units that transfer the majority of their heat to the occupied space by radiation

relief valve see pressure relief valve

reverse-return configuration of a water heating system in which the first terminal unit supplied with hot water is the last one to return to the boiler and vice versa

standard expansion tank see expansion tank

steam boiler equipment that heats water to the point of vaporization for the purpose of providing heat to an occupied space

terminal unit a radiator or section of baseboard

two-pipe system hydronic piping configuration that uses one pipe as the supply and one pipe as the return; can be configured as a direct return or a reverse return

volume factor provides the number of gallons of water per linear foot of a piping material

volute portion of the circulator housing that carries water from the pump

zone valve thermostatically controlled valve that opens and closes to regulate the flow of hot water to the terminal units in the occupied space

zoning process of dividing a structure into separate areas, each of which has its own means to regulate the temperature in the space

Theory of Hydronic Heating Systems

Unlike furnaces that use blowers to move heated air through duct systems, hydronic systems rely on circulating water or steam to deliver heat to the remote locations where heating is desired. Instead of using large ducts, as is the case in air conditioning or forced-air systems, heated water is circulated by heating it at one location and then pumping the heat-laden water to the remote locations via a piping arrangement. The heat is then transferred to the occupied space by heat exchangers located in the space. These heat exchangers, typically radiators or sections of baseboard, are referred to as terminal units. Once the water has passed through the terminal units, it is returned to the heat source, where it is reheated and once again pumped through the piping arrangement. This cycle (Fig. 17–1) continues until the occupied space reaches the desired temperature. This system is referred to as a **closed loop** because the water circuit is closed to, or separate from, the atmosphere.

CAUTION: Please note that Figure 17–1 represents only the water flow through a simple hot water hydronic system and does not contain other system components that are necessary for safe and proper system operation.

In hot water hydronic systems, many different piping configurations and controls can be used to satisfy the system requirements. A number of these components are common to all or most system configurations; others are specific to certain configurations. As we make our way through the remainder of this chapter, we will discuss a variety of components, controls, and piping configurations that are designed to enhance system performance as well as meet the heating requirements for the structure.

The Heat Source

The heat used in hydronic systems can be generated by a number of different methods including burning fossil fuels, collecting solar energy from the sun, converting electrical energy directly into heat energy, and using vapor-compression, reverse-cycle refrigeration, or heat pumps. The most common method is burning fuels such as natural gas, manufactured gas, or oil.

Heat energy is transferred from the heat source to the water in the **boiler.** The boiler is the piece of equipment that facilitates the generation of the heat energy as well as the transfer of this heat energy to the water flowing through it. Figure 17–2, Figure 17–3, and Figure 17–4 show gas, oil, and electric boilers. Boilers are typically constructed from either cast iron or steel. The most common boiler for residential applications is the cast-iron boiler.

Cast-iron boilers are often made up of individual sections (Fig. 17–5) that are bolted together to form the heat exchange surface between the water and the burning fuel. The more sections a boiler has, the higher the capacity of the boiler, with all other factors remaining the same. Residential cast-iron boilers usually hold between 10 and 15 gallons of water.

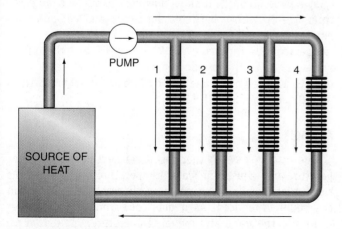

Figure 17–1 Hot water, two pipe direct return hydronic system.

Figure 17–2 Gas-fired boiler. *Courtesy Weil-McLain.*

492 SECTION FOUR *Troubleshooting*

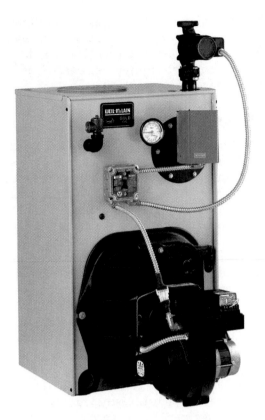

Figure 17-3 Oil-fired boiler. *Courtesy Weil-McLain.*

Figure 17-4 Electric boiler. *Courtesy Weil-McLain.*

CAUTION: Cast-iron sectional boilers must be used in closed-loop systems, because the cast iron can corrode if exposed to dissolved air for long periods of time.

Figure 17-5 Cast-iron boiler sections. *Courtesy Weil-McLain.*

Aquastat

The **aquastat** (Fig. 17-6) is a temperature-sensing switch that is responsible for cycling the boiler on and off to keep the water in the boiler close to the desired temperature. The temperature of the water in the boiler varies as the boiler cycles on and off. The temperature at which the boiler turns on is called the cut-in temperature, and the temperature at which it shuts down is the cut-out temperature. The difference between the cut-in and cut-out temperatures is called the differential. If the desired water temperature is, for example, 170 degrees F and the differential is 10°F, the boiler will cycle off when the water reaches 170 degrees F and will cycle back on when it drops to 160 degrees F (170°F − 10°F).

Reset

The **outdoor reset** thermostat senses the outdoor temperature and adjusts the water temperature in the boiler. As the outdoor temperature increases, the water temperature in the boiler is reduced. As it gets colder outside, the boiler maintains the water at a higher temperature. The reason for this is simple. When it is warmer outside, we require less heat inside, and vice versa.

CHAPTER 17 Hydronic Heat 493

Figure 17-6 Aquastat. *Photo by Bill Johnson.*

Figure 17-8 Low-water cutoff. *Photo by Eugene Silberstein.*

Low-Water Cutoff

The low-water cutoff (Fig. 17-8) is responsible for de-energizing the boiler in the event the water level in the system falls below the desired level. In some states, a low-water cutoff is required by law, but even if it is not, it is good field practice to equip all systems with one. This is especially true for hot water boilers that serve radiant systems (covered later in this chapter) because the tubing is usually below the boiler and often buried in concrete. A leak in a buried tube can cause the boiler to drain which, in turn, can cause it to dry fire.

Expansion Tank

Closed-loop hot water hydronic systems are, ideally, air free. As water is heated, it expands and, if the system is truly air free, the relief valve on the system will open to release the excess pressure. To prevent the relief valve from constantly opening and flooding the floor, an additional volume or space must be provided to accept the extra volume generated when the water is heated. The system component that accomplishes this is the expansion tank (Fig. 17-9), which ideally, is located on the supply side of the boiler near the inlet of the circulator.

There are two types of expansion tanks. One is the **standard expansion tank,** which is simply a large tank located above the boiler (Fig. 17-10). Initially, the air in the tank is at atmospheric pressure. As water is added to the boiler, water is pushed into the tank, which compresses the air in the tank, creating an **air cushion.** As the water is heated and its volume increases, more water is pushed into the tank, compressing the air even further (Fig. 17-11). The main problem with this type of expansion tank is that eventually it becomes water-logged, or completely filled with water, and the air cushion is removed. This eliminates the extra space that was previously available to the expanding volume

The reset control measures two temperatures: the outside air temperature and, typically, the water supplied by the boiler. At the time of installation, the control must be manually set. Once set, the water temperature will change in response to the outside temperature, depending on the selected reset curve. A representational reset curve is shown in Figure 17-7. On this curve, the boiler will maintain a water temperature of approximately 155 degrees F when the outside temperature is 20 degrees F.

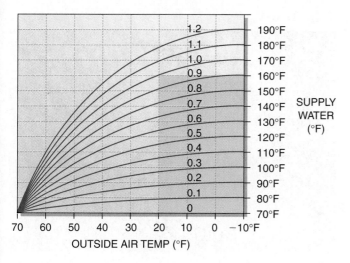

Figure 17-7 Representational reset control curve.

Figure 17-9 Expansion tank. *Photo by Eugene Silberstein.*

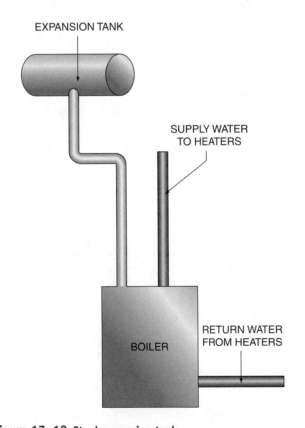

Figure 17-10 Steel expansion tank.

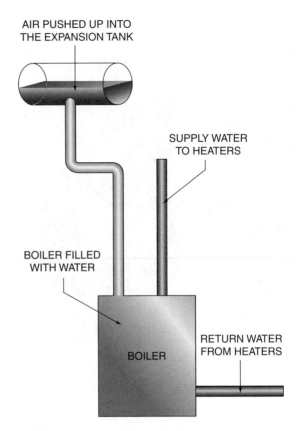

Figure 17-11 Pressure in the tank increases as water is added to the system.

of water. When this occurs, the pressure relief valve on the system opens, releasing the excess pressure. The other type of expansion tank, which is used in nearly all new residential construction installations, is the diaphragm-type expansion tank.

The **diaphragm-type expansion tank** is divided into two sections separated by a rubber, semi-permeable membrane (Fig. 17-12). One side of the tank contains air and the

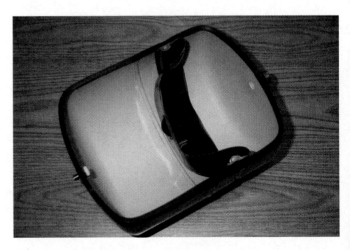

Figure 17-12 Cutaway of a diaphragm-type expansion tank. *Photo by Eugene Silberstein.*

CHAPTER 17 Hydronic Heat 495

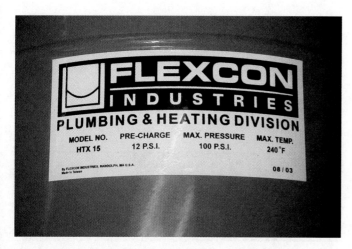

Figure 17-13 Expansion tank data tag. *Photo by Eugene Silberstein.*

other side is open to the water circuit. The air portion of the tank is initially pressurized by the manufacturer. Typically, for residential applications, a pressure of 12 psig is acceptable. The tank pressure at the time of manufacture is noted on the tank itself (Fig. 17-13). Note that the point at which the expansion tank is connected to the piping circuit will *always* remain at the same pressure. This is referred to as the point of no pressure change.

The **point of no pressure change** is the one place in the piping system where the pump cannot affect the pressure in the system. Consider a piping circuit that contains only the **expansion tank** and a pump (Fig. 17-14). Assume that the pressure in the expansion tank prior to installation is 10 psig and the pump generates a pressure difference of 15 psig between its inlet and outlet when operating. When the loop is filled, the water pressure will be 10 psig. When the pump cycles on, no water can be added or removed from the expansion tank. Water cannot enter the loop because it is filled, and water cannot be added to the loop since water is not compressible. Water cannot leave the loop and enter the expansion tank because this would create a vacuum in the loop. The water would then be pulled back out of the expansion tank and into the loop. Since water cannot be added to or removed from the expansion tank under the conditions just described, the pressure at the point where the tank is connected to the piping circuit cannot change.

 FROM EXPERIENCE

The pressure of the air in the expansion tank will increase either when the water is heated or when additional water is introduced to the boiler. In either case, water will be forced into the expansion tank, increasing the pressure of the air in the tank. This will increase the pressure at the point in the water circuit where the expansion tank is connected to the piping circuit.

Refer back to the example and the piping in Figure 17-14. If the pump is started and it generates a pressure difference of 15 psig, it will create a pressure of −5 psig at the inlet of the pump (Fig. 17-15). Since the pressure at the point of no pressure change will be 10 psig, the pump must make up the pressure difference at the inlet of the pump. The vacuum created at the inlet of the pump will permit air to enter the piping circuit.

Placing the pump on the other side of the expansion tank (Fig. 17-16) will eliminate this problem. The pressure at the inlet of the pump will be 10 psig and the pressure at the pump's outlet will be 25 psig. As the water circulates through the piping arrangement, friction will cause the pressure of the water to drop, and the pressure will once again be equal to 10 psig at the point of no pressure drop. Notice that, in this example, the pressure in the system has been

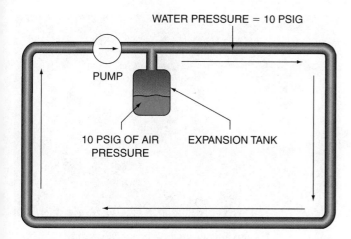

Figure 17-14 Simple loop with expansion tank and circulator.

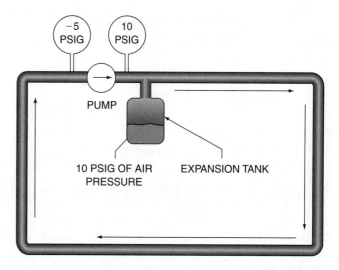

Figure 17-15 Pumping toward the expansion tank can result in a vacuum in the piping circuit.

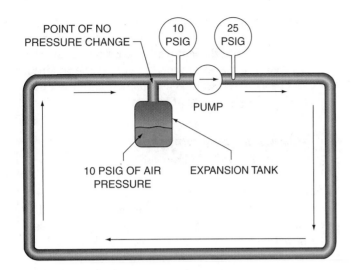

Figure 17-16 Point of no pressure change.

increased to 25 psig, which prevents the piping circuit pressure from falling below atmospheric pressure. In addition, air will remain in solution at higher pressures, reducing the negative effects of air in the water circuit.

 FROM EXPERIENCE

It is always good field practice to check the pressure in the expansion tank prior to installation. A bicycle tire gauge can be used to check the pressure and, if the pressure is too low, it can be raised with a bicycle tire pump.

The pressure in the expansion tank will match the static fill pressure in the system, which is the pressure in the filled system before the boiler is fired. The actual required air side pressure can be calculated using the following formula:

$$Pa = (H_1 - H_2)(Dc/144) + 5$$

where:

Pa = Air side pressure in the expansion tank (psig)

H_1 = Height of the highest system pipe above the base of the boiler (ft)

H_2 = Height of the opening in the expansion tank above the base of the boiler (ft)

Dc = Density of the water at its initial, cold temperature (lb/ft^3)

Consider the following example in which cold water is introduced to the boiler at 60°F. The density of water at 60 degrees F is 62.4 lb/ft^3. The density of water between the temperatures of 50 degrees F and 250 degrees F can be found by substituting the temperature of the water for "T" in the following formula:

$$\text{Density} = 62.56 + 0.0003413(T) - 0.00006255(T^2)$$

The expansion tank opening is 6 feet above the base of the boiler, and the highest pipe is 20 feet above the base of the boiler. The required pressure in the expansion tank can be found:

$$Pa = (22 - 6)(62.4/144) + 5 = 16(0.433) + 5 = 6.93 + 5 = 11.93 \text{ psig}$$

 FROM EXPERIENCE

Average-sized homes will often require an expansion tank with an air side pressure of 12 psig.

Another factor in determining the expansion tank to be used on a particular system is the volume of the expansion tank. The minimum required volume can be calculated as well. It requires an estimate of the volume of water contained in the system. An average-sized home with a cast-iron sectional boiler will often require an expansion tank with a volume of 4 to 5 gallons.

Refer to pages 521–523 for step-by-step procedures to estimate the volume of water in the system and to calculate the minimum required expansion tank volume.

Centrifugal Pumps

The pumps used to move water through hydronic systems are often called **circulators.** Because these circulators operate on centrifugal force, they are more accurately referred to as **centrifugal pumps.** The centrifugal pump (Fig. 17–17) is made up of a motor, a linkage, and an impeller (Fig. 17–18). The impeller slaps against the water and throws it toward the outside of the chamber or housing, called the **volute,** where it is forced from the pump assembly at a higher pressure. The pressure increase at the outlet of the pump pushes water around the piping circuit until it eventually is

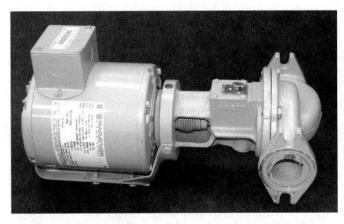

Figure 17-17 Centrifugal pump. *Courtesy Ferris State University. Photo by John Tomczyk.*

CHAPTER 17 Hydronic Heat 497

Figure 17-18 Impeller and volute on a circulator.

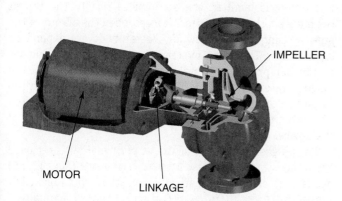

Figure 17-19 Cutaway of a centrifugal pump. *Courtesy of Bell & Gossett.*

pushed back into the inlet of the pump by the very pressure differential created by the pump. A cutaway of a centrifugal pump is shown in Figure 17-19.

Pumping Capability of the Centrifugal Pump

The centrifugal pump is responsible for creating a pressure difference between the water at its inlet and the water at its outlet. It is this pressure differential that facilitates water flow in the piping circuit. However, to ensure proper flow through the piping circuit, the pump must be able to overcome the resistance that exists in the piping itself. Consider a pump operating with no resistance to flow. If it was provided with an unlimited water supply at its inlet and was able to discharge the pumped water immediately to the atmosphere with no resistance at the pump's outlet, the volume of water moved by the pump would be the pump's maximum capacity (Fig. 17-20). Adding a section of vertical pipe to the pump's outlet will reduce the pumping capacity of the pump (Fig. 17-21). As the amount of vertical pipe at the pump's outlet increases, the capacity of the pump will continue to decrease until a point is reached at which the volume of water moved by the pump will be zero (Fig. 17-22). When we

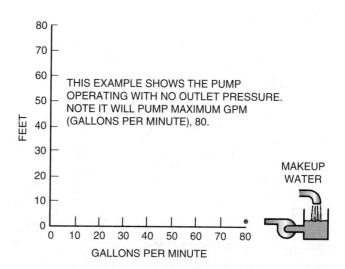

Figure 17-20 With no resistance at its outlet, the pump can move a large volume of water.

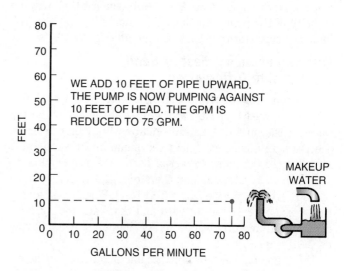

Figure 17-21 As the resistance at the pump's outlet is increased, the volume of water pumped will decrease.

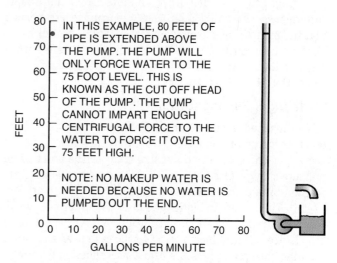

Figure 17-22 Point at which the pump can no longer overcome the resistance of the piping at its outlet.

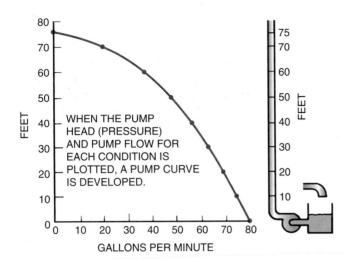

Figure 17-23 Sample pump curve.

plot the feet of pipe, or **feet of head**, against the pumping capacity of the pump in gallons per minute, we get the performance curve for that particular pump (Fig. 17-23).

Pumping Capacity, Feet of Head, and Pressure Differential

As a centrifugal pump operates, a pressure differential is created between the water at the inlet of the pump and the water at the outlet. A definite relationship exists between the pressure difference across the pump and the pumping capability of the pump in feet of head. One psig is equal to 27.7 inches of water column; therefore, 27.7 inches is equal to 2.31 feet (27.7 inches/12 = 2.31 feet). Finally, we can see that 1 psig = 2.31 feet of head and that each foot of head results in a pressure change across the pump of 0.433 psig (1/2.31 = 0.433).

The pumping head of a particular pump can be determined by using the performance curve for that pump if the flow rate through the pump is known. Assume that a pump has a capacity of 40 gallons per minute. From the sample performance chart in Figure 17-23, it can be established that the pump is overcoming approximately 57 feet of head. We can then conclude that this pump will operate with a pressure differential of 24.7 psig between the pump's inlet and outlet (57/2.31 = 24.7 psig).

Centrifugal Pump Location

As discussed in the expansion tank section of this chapter, the circulator should be located so that it is *pumping away* from the point of no pressure change. The point of no pressure change, discussed earlier, is the point where the expansion tank is connected to the system piping circuit (Fig. 17-16). This results in higher pressures in the system, which will not only help keep air in solution, but will also reduce the chances of introducing water to the system as a result of lowering the system pressure below atmospheric pressure.

Continuing with the pumping-away theory, it is possible to install the circulator and the expansion tank on the return side of the boiler. The main problem with this, though, is that once the pump is energized, the pressure at the outlet of the pump will be added to the pressure in the boiler. This might cause the relief valve on the boiler to open. This is more likely to happen in larger structures, but, to be on the safe side, it is good practice to install both the circulator and the expansion tank on the supply side of the boiler.

Air Vents and Air Separators

One of the worst enemies of a hot water hydronic heating system is air. For a hot water hydronic system to operate properly, any air in the system must be removed. The **air separator** (Fig. 17-24) which ideally is located at the point of no pressure change, should be able to remove smaller air bubbles, called microbubbles, from the system. In older air separators, the water flowed through a straight, horizontal section of pipe, resulting in laminar or linear flow. When the water flow is laminar, the air bubbles rise to the top of the pipe. These older air separators scooped the air out of the pipe by using a baffle. New air separators do not depend on laminar flow. They work by a process called collision and adhesion. Simply put, a metal mesh is placed in front of the flowing water (Fig. 17-25). The air bubbles collide with the metal and cling to it by surface tension. The air bubbles then rise to the top of the air separator and leave from the top of the device.

The air then passes to the **air vents,** where it is removed from the system. The **automatic air vent** (Fig. 17-26) opens and closes automatically, but the "coin operated" **manual air vent** (Fig. 17-27) must be opened manually.

Pressure-Reducing Valve (Water-Regulating Valve)

The **pressure-reducing valve** (Fig. 17-28) automatically drops the pressure of the water entering the structure to the pressure at which the boiler is designed to operate. If

Figure 17-24 Air separator. *Courtesy of Bell & Gossett.*

CHAPTER 17 *Hydronic Heat* **499**

Figure 17-25 Wire screen in the air separator. *Photo by Eugene Silberstein.*

Figure 17-27 "Coin-operated" air vent. *Courtesy of Bell & Gossett.*

Figure 17-28 Pressure-reducing valve. *Courtesy of Bell & Gossett.*

the pressure in the boiler drops below the desired pressure, the valve will open to feed more water to the boiler. The feed valve should be piped so water is introduced to the system at the point of no pressure change (Fig. 17-29). Installing the feed valve on the return side of the boiler can cause problems if the circulator is installed on the return side as well. Assume the boiler piping configuration is as shown in Figure 17-30. The air side of the expansion tank is pressurized to 12 psig, the **boiler feed valve** has been factory set to feed water when the system pressure drops below 12 psig, the pump operates with a pressure drop of 6 psig, and there is a 1 psig pressure drop through the boiler.

When the pump starts, the pressure at the outlet of the pump will be 13 psig, and the inlet pressure of the pump will

Figure 17-26 Automatic air vent. *Courtesy of Bell & Gossett.*

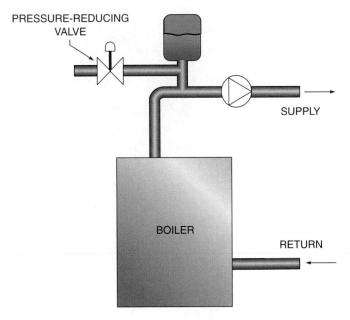

Figure 17-29 Proper location for the pressure-reducing valve.

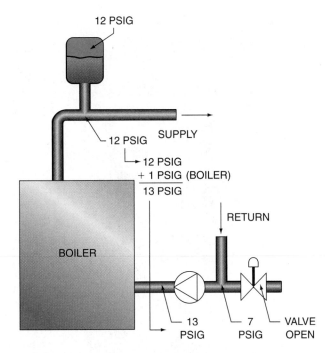

Figure 17-31 Feed valve will open as a result of a false reading.

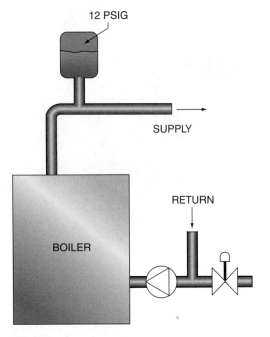

Figure 17-30 Potential problems can arise with this configuration.

be 7 psig (Fig. 17-31). Since the feed valve is sensing a pressure below 12 psig, water will be fed to the boiler until the pressure is increased to 12 psig, the set point on the feed valve. This results in a 5 psig increase to the boiler. This added water will be pushed into the expansion tank because it has no place else to go. There it will push against the air cushion, increasing the air pressure in the tank.

Once the circulator cycles off, the static pressure in the system will be 17 psig, which is 5 psig higher than the original fill pressure. This will cause the air pressure in the expansion tank, which was originally 12 psig, to increase to 17 psig (Fig. 17-32). In addition, the tank contains more water than originally intended and is now too small for the system. The increased pressure in the expansion tank puts more stress on the diaphragm and can result in premature failure of the tank.

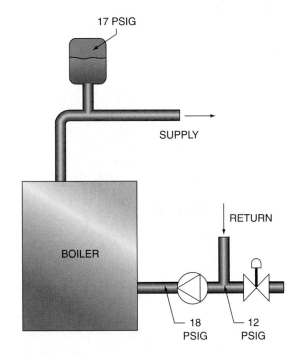

Figure 17-32 Static air pressure in the expansion tank is higher than desired.

CHAPTER 17 Hydronic Heat 501

Pressure Relief Valve

The **pressure relief valve** (Fig. 17–33) is designed to open if the pressure in the system reaches the set point on the valve. Once the valve opens, the pressure in the system will be relieved and, when the pressure drops to an acceptable level, the valve will close again. The American Society of Mechanical Engineers (ASME) Boiler and Pressure Vessel Codes require that a pressure relief valve be installed on *every* hot water boiler. Since all hot water boilers have a pressure relief valve and most have a pressure-reducing valve, many manufacturers combine these two components into a single unit. A complete sample piping arrangement at the supply side of the boiler can be seen in Figure 17–34.

Zone Valves

Circulators can be used to circulate water through various loops in the structure. If water flows through all of these loops or none of them, depending on whether the system is on or not, it is said that there is one zone in the structure.

Figure 17–33 Safety relief valve. *Courtesy of Bell & Gossett.*

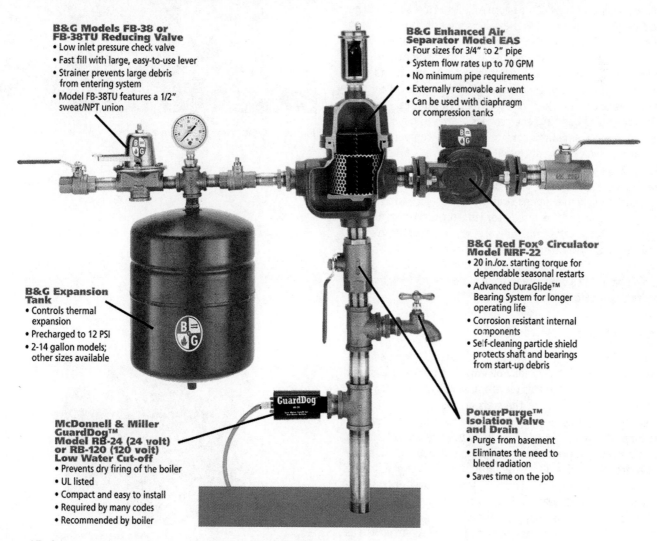

Figure 17–34 Hot water supply manifold. *Courtesy of Bell & Gossett.*

Figure 17–35 Zone valve. *Courtesy of Bell & Gossett.*

Figure 17–36 Zone valves on a multiple zone loop. *Photo by Eugene Silberstein.*

If, however, water flow can be controlled so that some of the loops are getting heat while the others are not, it is said that the structure has multiple zones. In the case of a one-zone structure, it is likely that some parts of the structure will be too hot, some too cool, and others the desired temperature. Zoned structures allow the occupants of each area to control the temperature in that area with a thermostat located within the area. **Zoning,** the process of dividing a structure into areas with separate control over the temperature in each space, is often controlled with **zone valves** (Fig. 17–35). Zone valves open and close in response to the temperature in the space. In operation, when one of the zones calls, the zone valve opens and trips an end switch that starts the circulator. The circulator runs until the thermostat is satisfied. Once satisfied, the valve closes, turning off the circulator. Refer to later sections in this chapter for limitations on and applications for zone valve usage. Zone valves can also be used to control multiple zones being fed by a single circulator (Fig. 17–36).

Flow-Control Valve

From the name, it might be concluded that the flow-control valve (Fig. 17–37) controls flow in a hydronic system. The fact is, the **flow-control valve** is a **flow check valve** that prevents hot water in the system from flowing through a heating loop when no flow is desired through that particular loop. If the individual loops have positive shutoffs, such as zone valves, flow-control valves are not needed. If, however, water flow through the individual loops is controlled by a circulator pump—also discussed later—flow controls are needed to prevent gravity flow through the loops. Since gravity circulation can happen on either end of the secondary circuit, it is a good idea to use two flow-control valves for each circuit; one on the supply and the other on the re-

Figure 17–37 Flow control valve. *Courtesy of Bell & Gossett.*

turn. Always make sure the arrows on the flow-control valves point in the right direction.

FROM EXPERIENCE

It is good field practice to install a valve after each flow-control valve. This can be either a full-port ball valve or a gate valve. This will make future servicing of any of the components between the boiler and the flow-control valve much easier.

Balancing Valves

In hot water **hydronic** systems, the water flow through each branch in the piping circuit must be within the required range to help ensure that the occupied space is kept at the desired temperature. One method that will keep the water flow even through each branch of the circuit is the use of **balancing valves** (Fig. 17-38). Balancing valves, installed in each branch of the circuit, are manually adjusted. Refer to later sections in this chapter for balancing valve application. Since they vary from manufacturer to manufacturer, refer to the literature supplied with the particular valve for information regarding setting procedures.

Series Loop System

In a series loop system (Fig. 17-39), all of the heaters are piped in series with one another. The main advantage of the series loop system is that it is very economical from a first-cost standpoint. The main drawback is that the terminal

Figure 17-38 Balancing valve. *Photo by Bill Johnson.*

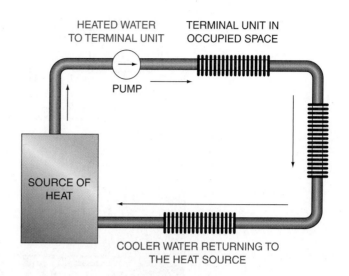

Figure 17-39 Series loop piping circuit.

units that are fed last are often cooler than those that are fed first. This is because, as heat-laden water flows through the loop, british thermal units (btus) are transferred from the water to the air, thereby lowering the temperature of the water in the piping circuit.

FROM EXPERIENCE

Zone valves cannot be used on a series loop system. If a zone valve were installed on this type of system, hot water would flow through the loop *only if all* of the zone valves were in the open position.

One-Pipe System

In a one-pipe system (Fig. 17-40), one main piping loop extends around, or under, the occupied space and connects the boiler's supply to the boiler's return. Each individual terminal unit is connected to the main supply loop with two tees, which are discussed next. The one-pipe system is so named because the same pipe not only supplies water to the terminal units but also carries the return water back to the boiler.

Resistance to Water Flow

Consider a portion of the piping circuit that includes one terminal unit as well as the portion of main loop piping between the tees (Fig. 17-41). The water entering the tee at point A flows through the tee. It can leave the tee through point B, which leads to the **terminal unit,** or through point C, which allows the water to continue through the main water path connected between the boiler's supply and the return pipes. The amount of water flow through the terminal unit is determined by the relationship

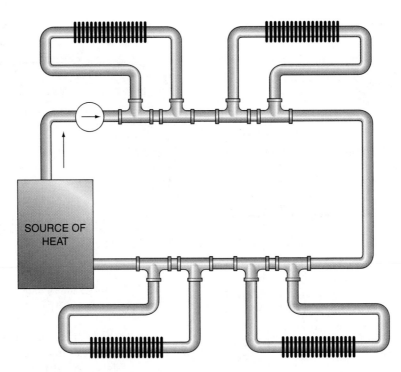

Figure 17-40 One-pipe system.

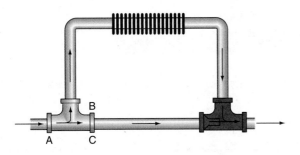

Figure 17-41 Common piping between the tees. Diverter tee located at the return side of the terminal unit.

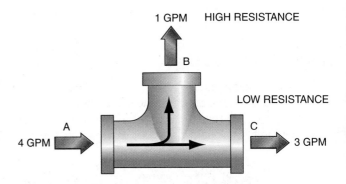

Figure 17-42 The higher resistance at point "B" results in lower flow through that branch, while the lower resistance at point "C" results in greater flow through that branch.

between the resistance in the terminal unit piping circuit and the resistance in the section of main loop piping between the tees.

For example, if the resistance in the terminal unit branch is three times that of the main line loop section, the amount of water flow through the terminal unit branch will be one-third of that through the bypass, or main loop. If the water flow through the main circuit is 4 gallons per minute, 3 gallons per minute will flow through the main loop, but only 1 gallon per minute will be permitted to flow through the terminal unit (Fig. 17-42). It is, therefore, possible for all or most of the water to bypass the terminal unit branch if the resistance of that circuit is much higher than the resistance of the piping section between the two tees.

A number of factors must be considered when laying out or installing a **one-pipe hot water** system. These factors include:

- The length of the terminal unit branch circuit
- The length of the piping between the tees (the distance between the two tees)
- The size of the piping material in the branch circuit
- The size of the piping between the tees

When dealing with one-pipe hot water systems, always be aware of the following:

✓ The longer the pipe, the more resistance to flow there will be
✓ The shorter the pipe, the lower the resistance will be
✓ The smaller the diameter of the pipe, the more resistance to flow there will be
✓ The larger the diameter of the pipe, the less resistance to flow there will be

✓ The farther apart the tees are from each other, the more resistance to flow there will be between them
✓ The closer the tees are to each other, the less resistance to flow there will be between them

The Diverter Tee

The diverter tee (Fig. 17-43) also referred to as a **monoflo tee,** is designed to increase the resistance to water flow in the main water loop. This increase in resistance will direct more water through the terminal unit branch circuit. The diverter tee is constructed with a cone inside (Fig. 17-44) that, in essence, reduces the diameter of the pipe in the main loop, thereby increasing its resistance. This pushes more water through the terminal unit loop. A typical loop using a diverter tee was shown in Figure 17-41. Notice that the tee at the outlet of the terminal unit is a diverter tee.

When working with diverter tees, remember the following:

✓ If the terminal unit is located above the main hot water loop and the length of the piping in the terminal unit branch is not too long with respect to the main hot water line, one diverter tee should be used on the return side of the terminal unit (Fig. 17-41).

✓ If the terminal unit is located above the main hot water loop and the length of the piping in the terminal unit branch is too long with respect to the main hot water line, two diverter tees should be used: one on the supply side of the terminal unit and one on the return side of the terminal unit (Fig. 17-45).

✓ If the terminal unit is located below the main hot water loop, two diverter tees should be used: one on the supply side of the terminal unit and one on the return side of the terminal unit (Fig. 17-46).

✓ If the terminal unit is above the main hot water loop, the tees should be no closer than 6 or 12 inches, depending on the manufacturer. (Bell and Gossett allows them to be as close as 6 inches; Taco as close as 12 inches.)

✓ If the terminal unit is below the main hot water loop, the tees should be spaced as wide as the ends of the terminal unit or **radiator.**

✓ If there are multiple terminal units and some are above and some are below the main hot water loop, it is best to alternate the tees. For example, from left to right: supply upper, supply lower, return upper, return lower (Fig. 17-47).

✓ Manufacturers place markings on the outside of their diverter tees so that the location of the inner cone can be identified. Be sure to install the diverter tees as recommended by the installation literature.

Sizing Diverter Tee Systems

When sizing the diverter tees, it is important to know not only the total heating load, but also the requirements of each radiator in the system. In addition, knowing the temperature drop from supply to return is needed. For residential

Figure 17-43 Diverter tee.

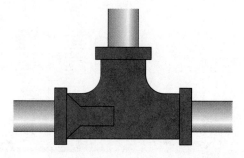

Figure 17-44 Cross-sectional view of a diverter tee.

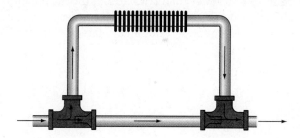

Figure 17-45 Two diverter tees are needed if the radiator is located above the main loop and there is significant resistance in that branch.

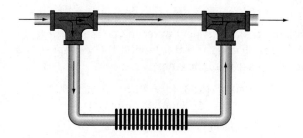

Figure 17-46 Two diverter tees are needed if the radiator is located below the main loop.

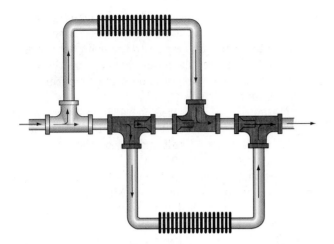

Figure 17-47 Alternate tees if there are terminal units both above and below the main loop.

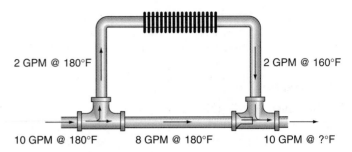

Figure 17-48 A flow of 2 gpm is needed through this terminal unit.

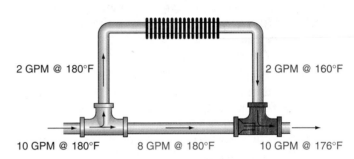

Figure 17-49 The temperature of the water leaving the tee is 176 degrees F.

applications, a 20 degrees F temperature drop is typical. On a hot water system with a 20 degree F temperature drop, each gallon per minute of water flow will carry approximately 10,000 btu/hr of heat.

Assume that there is a system with a total heating load of 100,000 btu/hr with a 20 degree F temperature drop between supply and return water temperatures. Since 1 gpm will carry about 10,000 btu/hr, this system will have a water flow of 10 gpm. If the first radiator requires 20,000 btu/hr, there should be 2 gpm of water flow through that branch, (Fig. 17-48). With this information, the diverter tee that will facilitate the desired flow can be selected.

Once the water flows through the first radiator, the temperature of the water in the main loop after the second tee will be somewhat lower than the water in the main loop before the first tee. The temperature of the water in the main loop after the second tee can be calculated using the following formula:

$$(Flow\ 1) \times (Temp\ 1) + (Flow\ 2) \times (Temp\ 2) = (Flow\ 3) \times (Temp\ 3)$$

where:

Flow 1 = the flow of liquid 1 entering the tee
Temp 1 = the temperature of liquid 1 entering the tee
Flow 2 = the flow of liquid 2 entering the tee
Temp 2 = the temperature of liquid 2 entering the tee
Flow 3 = the flow of liquid 3 leaving the tee
Temp 3 = the temperature of liquid 3 leaving the tee

Assuming that the supply water in the previous example is 180 degrees F and the water leaving the radiator is 160 degrees F, we get the following:

$$(8\ gpm) \times (180°F) + (2\ gpm) \times (160°F) = (10\ gpm) \times (Temp\ 3)$$
$$Temp\ 3 = (1440 + 320)/10$$
$$Temp\ 3 = 176°F$$

These results are summarized in Figure 17-49.

Two-Pipe Direct Return

Two-pipe systems use two pipes (see following figure). The supply pipe carries water to the terminal units, and the return is responsible for bringing it back to the boiler. In a two-pipe, **direct-return** system, the terminal unit that is fed with hot water first is also the first to return its water to the boiler. Simply stated, the terminal unit closest to the boiler offers the least amount of resistance to water flow because that circuit or branch is the shortest. This statement assumes that all terminal units, baseboard sections, or radiators in the piping arrangement are identical to each other.

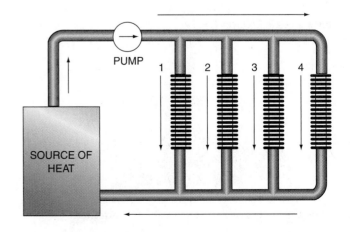

Consider the two-pipe, direct-return system in Figure 17–50. Water flowing through terminal unit 1 will have to travel a distance of 24 feet after leaving the boiler before returning to the boiler. Water traveling through terminal units 2, 3, and 4 will have to travel 26, 28, and 30 feet, respectively. Terminal unit 1 will therefore have the most water flow through it, given the lower resistance of that piping branch circuit.

To generate even and equal water flow through each terminal unit, balancing valves (Fig. 17–33) can be piped in series with each terminal unit (Fig. 17–51). These valves are manually adjustable and, after they have been properly adjusted, they should be left alone. The valve in series with heater 1 will be closed a little more than the valve in series with heater 2, which, in turn, is closed a little more than the valve in series with heater 3, and so on. The closing of the valves closer to the boiler will direct more water to the heaters located further from the boiler.

Two-Pipe Reverse Return

In a two-pipe, **reverse-return** system (Fig. 17–52), the first heater supplied with hot water is the last heater to return its water to the boiler. In other words, the heater with the shortest supply line from the boiler will have the longest return line. Configured in this manner, the distance traveled by the water through the individual heaters will be exactly the same.

Consider the two-pipe reverse-return system in Figure 17–53. Note that, by adding the lengths of pipe between the boiler and terminal unit 1, and from terminal unit 1 back to the boiler, the water must travel a distance of 31 feet (3′ + 9′ + 5′ + 1′ + 1′ + 1′ + 1′ + 10′ = 31′). The distances through terminal units 2, 3, and 4 are all 31′ as well. Assuming that all terminal units in the piping arrangement are the same, the flow through each heater will be the same, and no balancing valves will be needed.

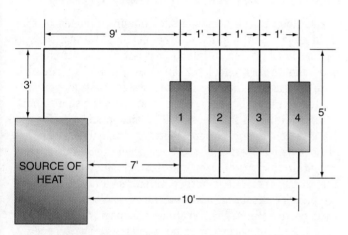

Figure 17–50 Water flowing through radiator #1 must travel 24 feet; water flowing through radiator #4 must travel 30 feet.

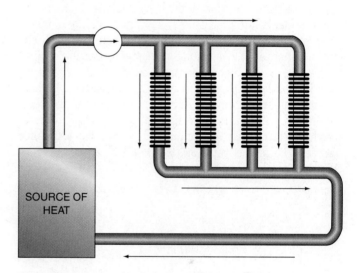

Figure 17–52 Two-pipe, reverse-return system.

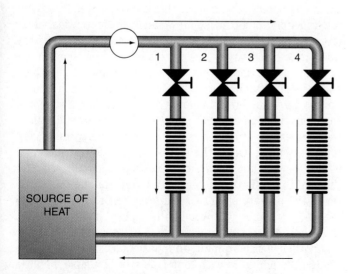

Figure 17–51 Balancing valves used to balance flow through the radiators.

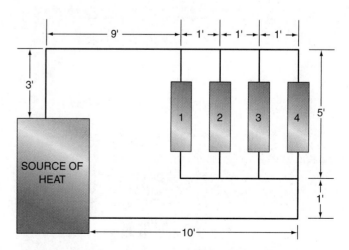

Figure 17–53 The length of the piping through any one branch is exactly the same as the length through any other branch.

Primary-Secondary Pumping

In primary-secondary piping, there are two distinct piping circuits, just as there were two piping circuits in the one-pipe arrangement that used diverter or monoflo tees (Fig. 17-40). One main circuit flows only through the boiler, the other flows through the boiler and one or more of the terminal units. The two circuits meet at the tees. The piping between the tees is common to both circuits. In a one-pipe system there is just one circulator, which is engaged in primary-secondary flow.

The farther apart the tees are placed, the more likely it is that water will flow from the boiler's circuit to the convector's circuit. The more resistance there is in the main loop between the tees, the greater the water flow will be through the convector branch. It is the resistance in the common piping that affects the flow through the parallel branch. The key to understanding primary-secondary pumping is understanding the resistance in the piping common to both circuits. Higher resistance results in a greater pressure drop in the circuit, just as higher resistance in an electric circuit results in a higher voltage drop across the resistance. If there is a large pressure drop along the run between the two tees, more water will flow into the branch, which is why diverter tees have the built-in restriction. Similarly, if the resistance between the tees is very low, there will be almost no flow through the convector loop (Fig. 17-54).

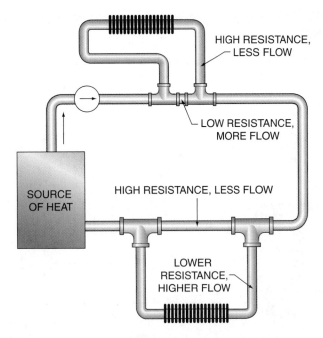

Figure 17-54 More resistance between the tees results in more flow through that radiator, and vice versa.

Primary-Secondary Common Piping

The transition from primary-secondary flow, as in the case of the one-pipe system, to primary-secondary pumping is made by eliminating the pressure drop between the primary and the secondary circuits. To accomplish this, standard tees are used instead of diverter tees, and the tees are piped close together. The section of piping located between the tees is, as mentioned previously, the common piping, which is a part of both loops.

As a rule, the shorter the common piping, the better the primary-secondary system will work. Ideally, the common piping should not be more than two feet long, but it can be as short as is physically possible. If the common piping is longer than two feet, the resistance of the common piping will increase, so the common piping should be as short as possible.

Primary-Secondary Circuit Piping

Each loop in the circuit piping contains a circulator that could serve something that is either putting heat into the system, such as a boiler, or taking heat out of the system, such as a radiator. Each circulator takes care of *only* the circuit it is in. This means that even if your system is large, your circulators are probably going to be small. It is the short length of common piping that makes all of this possible. As long as the common piping is kept short, each circulator will run without being aware that there are other circulators operating within that system. You can mix and match circulators of different sizes and none of them will affect any of the others. The primary circulator, however, will run only if the secondary circulator starts. The two always run together, and the primary's job is to make hot water available to the secondary. The secondary circulator usually removes only a portion of what the primary supplies. The rest of the primary flow continues on and mixes with the water that is returning from the secondary circuit. The result is that the water that flows back into the boiler is hotter than it would be if the secondary flow returned by itself. This layout allows systems designers the freedom to be creative. Consider the following example.

Figure 17-55 shows a primary circulator in the main loop and a secondary circulator that feeds water to the terminal units. The secondary circulator draws hot water from the common piping, just as it would from a boiler, and pumps that hot water through the three radiators that it serves. The secondary pump treats the main loop as the heat source or boiler. It takes hot water from the primary circuit, sends it through the radiators as needed, and returns the cooler water to the primary circuit. In addition, the secondary pump sees only the pressure drop of the piping that goes to and from those three radiators.

Another possible piping arrangement is shown in Figure 17-56. The piping in each loop is relatively simple, and the circulators in each loop are small because they only have to

CHAPTER 17 *Hydronic Heat* 509

ondary loop circulators are off, water will only circulate through the primary loop. As water is pumped toward a tee, the water will bypass the loop, since the resistance between the tees is much less than the resistance of the loop. The pressure drop through the common piping is practically nonexistent compared with the pressure drop through the secondary loop, so the primary flow stays in the primary circuit until the secondary circulator starts.

Once the circulator in the secondary loop starts, water will flow through that loop. How much water will flow depends on the pumping capacity of the circulators. Consider a primary circuit with a 20 gpm circulator and a secondary circuit with a 10 gpm circulator (Fig. 17–57). A short length of common piping connects them. The primary circulator is on and the secondary circulator is off. There is a water flow

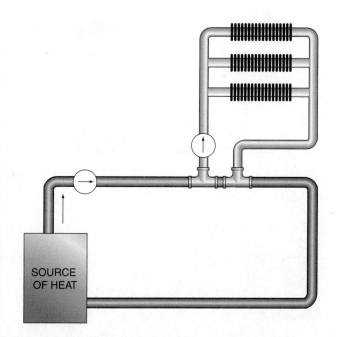

Figure 17–55 Primary-secondary arrangement where one circulator feeds three radiators in the secondary.

take care of that particular loop. This results in lower pressure drops throughout the system and reduces the velocity of the water flowing through the loops. Even if the system were very large, the circulators would stay small.

The Circulator Pumps

It has been established that the primary circulator is responsible for moving water through the main loop, and the pumps in the loops are responsible for moving water through the secondary loops. The pressure drop in the primary loop will be relatively low, because the resistance between the tees is kept low. If the main loop circulator is on and the sec-

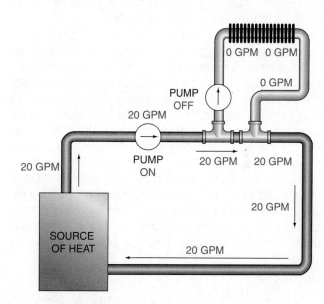

Figure 17–57 When the secondary pump is off, there is no flow in the secondary loop.

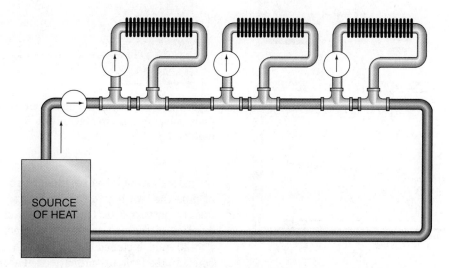

Figure 17–56 Primary-secondary arrangement where there are three separate secondary circuits, each with its own circulator.

of 20 gpm into the first tee in the common piping. If 20 gpm of water enters the tee, then 20 gpm must come out. Since the secondary circulator is off, the flow will all take place in the primary loop. No water is going to flow through the secondary loop because the secondary circulator is off and the resistance of that loop is much higher than the resistance of the common piping.

If the secondary circulator is turned on, both pumps are on and there is still 20 gpm flowing into the first tee on the common piping, but the flow splits with the secondary circulator running (Fig. 17-58). Since the secondary circulator moves 10 gpm of water, we have 10 gpm flowing up to the secondary circuit, and the other 10 gpm going straight through the common piping. At the second tee, the 10 gpm from the secondary loop will combine with the 10 gpm from the common piping to give us 20 gpm at the outlet of the tee, which returns the water to the boiler.

Now let's assume that both circulators have the same flow rate, each moving 20 gpm. When the primary circulator runs and the secondary circulator is off, all of the water will flow across the common piping and continue on through the primary circuit (Fig. 17-59). Once again, the resistance of the piping circuit is responsible for this. When the secondary circulator starts, it draws all the water out of the first tee and sends it around the secondary circuit. No water flows across the common piping. The entire 20 gpm enters the second tee in the common piping and continues on its way (Fig. 17-60).

Finally, let's use the 10 gpm circulator as the primary circulator and the 20 gpm circulator as the secondary circulator. When the primary circulator is on and the secondary circulator is off, all the water flows along the primary. No wa-

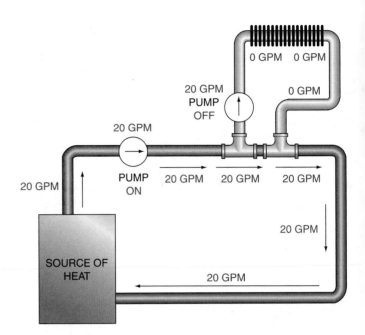

Figure 17-59 When the secondary pump is off, there is no flow in the secondary loop.

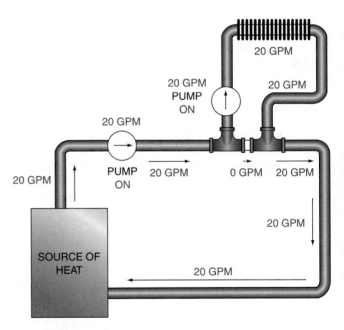

Figure 17-60 When both pumps are on and they have the same pumping capacity, there is no flow through the common piping.

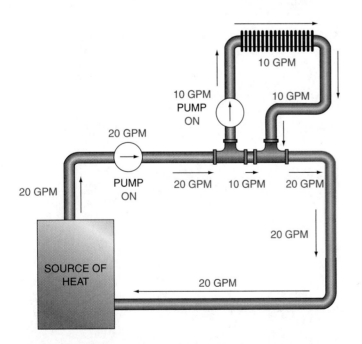

Figure 17-58 Water flows through the secondary loop when both pumps are on.

ter moves through the secondary circuit because the length of pipe the two have in common is so short. When the secondary pump starts, the secondary circulator will draw 20 gpm out of the first tee in the common piping (Fig. 17-61). The 20 gpm includes the 10 gpm that is entering the first common tee from the primary circulator and another 10 gpm from its own circuit. This portion of the 20 gpm is moving *backwards* across the common piping.

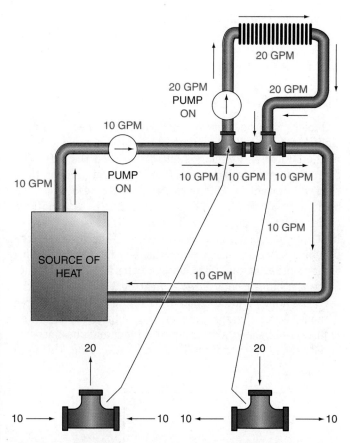

Figure 17-61 Water can flow through the common piping in either direction, depending on the pumping capacity of the pumps.

 FROM EXPERIENCE

Refer back to the section on one-pipe systems for the formula used to calculate the temperature of the water that leaves a tee.

Mixing Valves in Primary-Secondary Pumping

One of the many benefits of primary-secondary piping systems is that each loop can be used to supply water at different temperatures to different zones, depending on the system requirements. For example, one part of the structure might be heated with baseboard and need 180 degrees F water. Another portion of the structure might have radiant heat (covered later in this text) and require water to be 110 degrees F. The tees in the primary-secondary piping circuit act as mixing valves. The temperature of the water leaving a tee can be calculated much like the temperature used in the one-pipe system using diverter tees is calculated. Just as hot water and cold water are mixed in the kitchen faucet to sup-

ply warm water, the water leaving the tee will be cooler than the hot water and warmer than the cold water. Carefully selecting the size of the pumps in both the main loop and the secondary loops will help achieve the desired water temperatures in each loop.

Expansion Tanks in Primary-Secondary Systems

It has been determined that the best configuration for the expansion tank and circulator is with the circulator on the supply side of the boiler, pumping away from the point of no pressure change at the expansion tank. In primary-secondary systems the same holds true, so it may seem logical to have an expansion tank in each circuit that has a circulator. Such is not the case. The primary pump is pumping away from the **compression tank,** but the secondary circulator does not have a separate compression tank in its circuit.

When the water in the secondary circuit gets hot and expands, the extra water moves into the primary circuit, because the compression tank is there. Because of this, the common piping will always be the secondary circulator's point of no pressure change or, to put it another way, its compression tank. For this reason, whenever possible, a system should be piped so that the secondary circulator pumps *into* the secondary circuit and away from the common piping.

Radiant Heating Systems

The concept of radiant heat differs a great deal from that of forced air, which is provided by furnaces. Forced air, or convective, heating systems have blowers that create convection currents to carry heat-laden air to the desired location through duct systems. **Radiant heating systems,** on the other hand, concentrate on heating the shell of the structure or occupied space as opposed to the air within that space. In addition, the calculations for radiant heating are entirely different from those for convective heating. The purpose of convective heating is to determine the *rate of heat loss from the room* by conduction, convection, and radiation when maintained in the desired condition; radiant heating involves the *regulation of the heat loss per square foot from the human body*.

The Human Body Is a Radiator

Under normal conditions, the human body produces approximately 500 Btus per hour (Btuh). Each Btu is the amount of heat it takes to raise the temperature of one pound of water one degree Fahrenheit. The body only needs 100 Btuh to stay alive. The excess heat must be shed, which makes the body act as a radiator. By controlling the rate at which we shed heat, we can control our comfort level. Giving up heat too fast will make us feel cold, and giving up heat slowly or not at all will make us feel hot and uncomfortable.

Cold 70

Cold 70 is the feeling one gets when there is a big difference between the temperature of the air in a room and the temperature of the surfaces in that room. You experience this common phenomenon every time you do your grocery shopping. In your neighborhood supermarket, there is a cereal isle, a frozen-food aisle, and a deli counter. Which of these aisles do you think is the warmest? Which is the coolest? The next time you go shopping, take a thermometer with you. You will find that the temperature of each aisle is exactly the same. So why do you feel cold in the frozen food aisle? That is the phenomenon of Cold 70. The air temperature in the frozen food aisle and the temperature of the surrounding surfaces in the same aisle are wildly different from one another, and your body is very quick to give up heat to the cooler surfaces in that aisle. This is what makes you cooler in the frozen food aisle. You feel warmer when you get close to the chicken rotisserie at the deli counter, because the rotisserie is hotter than you are. It is radiating heat at you and also preventing your body from losing its own heat, making you uncomfortably warm. The concept behind radiant heat is to provide ideal comfort.

What is Ideal Comfort?

Ideally, if our bodies can shed and absorb heat at a controlled rate, we will be comfortable. Typically, a room that is in the 68 degree F range permits us to shed the extra 400 btuh that our body generates. Temperatures higher than that will prevent us from releasing the heat, resulting in an uncomfortable feeling. On the other hand, temperatures colder than 68 degrees F will result in the rapid release of heat from our body, making us feel cold. Ideally, then, the occupied space should be approximately 68 degrees F, with the exception of the areas closer to the floors and ceiling. Since the human body loses a great deal of heat from the feet, the floor should be somewhat warmer, in the 85 degree F range, while the space up by the ceiling can be cooler. Figure 17–62 shows a cutaway of a room and the temperature changes as we go from the floor to the ceiling. This tem-

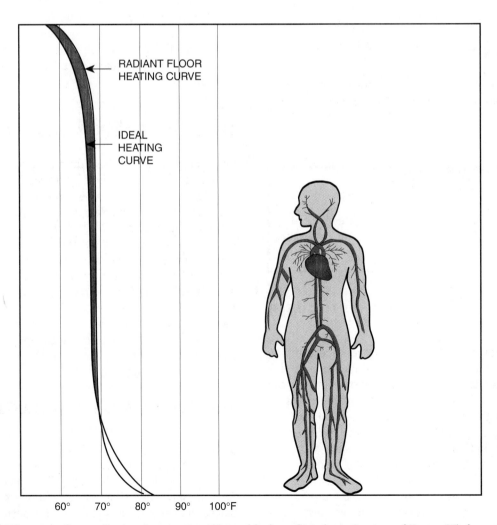

Figure 17–62 Heating curve for a radiant system compared to an ideal comfort chart. *Courtesy of Uponor Wirsbo.*

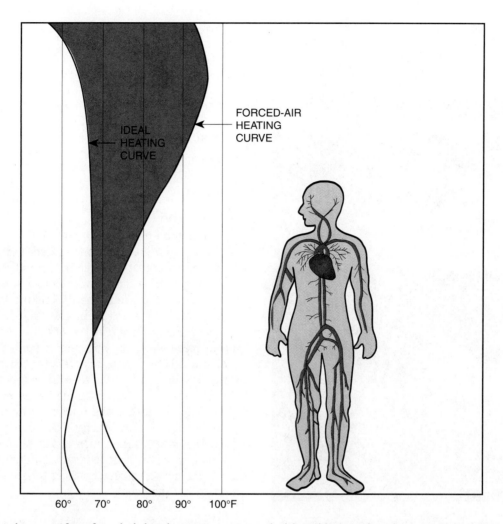

Figure 17–63 Heating curve for a forced-air heating system compared with an ideal comfort chart. *Courtesy of Uponor Wirsbo.*

perature pattern represents an ideal temperature pattern, which is very closely simulated by radiant heat. The heating patterns generated by **baseboard heating** and forced-air heating systems are similar to those shown in Figure 17–63.

The Radiant System

Like hot water hydronic systems, radiant heating systems rely on the circulation of hot water through a series of piping arrangements located in the occupied space. There are three main differences between the two. The first is visibility. In radiant heating systems, the piping that carries the heat-laden water is hidden in the floors, walls, and/or ceilings of the structure; therefore, it is not visible to the occupants. The second major difference is the piping material. Instead of using copper pipe to carry the hot water to the occupied space, radiant heating systems often use flexible **polyethylene tubing,** more commonly known in the industry as **PEX tubing.** PEX tubing (Fig. 17–64) is flexible and can be easily installed using a minimum of fittings (Fig. 17–65). The third difference is the temperature of the water required by the system. Typical hot water hydronic sys-

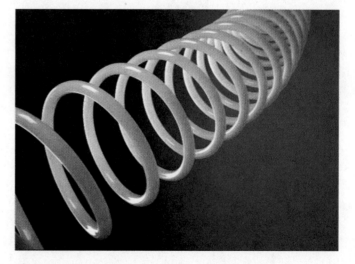

Figure 17–64 PEX tubing. *Courtesy of Uponor Wirsbo.*

tems circulate water in the 180 degree F range. Radiant heating systems circulate water with an average temperature of 110 degrees F.

Radiant Heating Piping

The majority of radiant heating piping systems being installed in new residences are located in the floor, although radiant heat can be installed in walls and ceilings as well. This text will concentrate on floor installation. There are three options for installing the tubing for radiant systems. They are

- Slab on grade
- Thin slab
- Dry

Slab on Grade

The slab-on-grade configuration is the most popular choice for new construction projects that will have concrete floors. Since the concrete has not yet been poured, the tubing for the radiant heat loops will be located within the concrete slab itself. The PEX tubing is secured to the steel reinforcement mesh in the floor using either wire straps or plastic clips (Fig. 17–66). This will hold the tubing in place when the concrete is poured. Generally, the tubing will be on 12-inch centers for these jobs. Near cold surfaces such as windows, the distance between the tubes should be less. Insulation should be placed under the concrete slab to increase the effectiveness of the system. If possible, at least one-inch-thick polystyrene should be used. The edges of the concrete slab should have more insulation, since the heat tends to migrate toward the cooler walls. The goal is to keep the heat toward the center of the room where the people are.

(A)

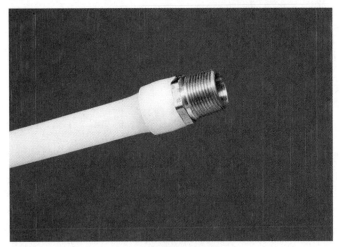
(B)

(C)

Figure 17–65 Fittings used to connect PEX tubing. *Courtesy of Uponor Wirsbo.*

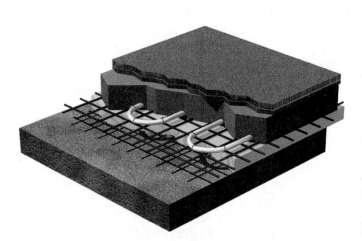

Figure 17–66 Cutaway of a slab-on-grade radiant piping layout. *Courtesy of Uponor Wirsbo.*

FROM EXPERIENCE

When securing PEX tubing to the steel reinforcement mesh, be sure to use clips or straps that are approved for use with the particular tubing being used. Using incorrect strapping materials can result in chemical reactions that will damage the tubing.

In addition, there should always be a vapor barrier under the slab. It can be polyethylene and should be at least 6-ml thick. The main purpose of the vapor barrier is to keep the ground water from robbing the slab of its heat. The distance between the tubing and the top of the slab is not critical, but remember that the lower the tubing is from the top of the slab, the higher the water temperature will have to be.

Figure 17-67 Cutaway of a thin-slab radiant piping layout. *Courtesy of Uponor Wirsbo.*

FROM EXPERIENCE

It is good field practice to pressurize the tubing to at least 50 psig while pouring the concrete. This will prevent the tubing from collapsing and will also make finding leaks much easier. It is better to pressurize with air than with water if at all possible, since leaking air will not affect the composition of the concrete.

FROM EXPERIENCE

Insulate the space below the tubing to R-11 if the room below the floor is heated. Use R-19 if it is over an unheated basement, and R-30 over a crawl space. The colder the space below, the greater the insulation should be.

If weight is a definite issue, there are alternatives to standard concrete. Gypsum concrete is a mixture of very fine sand, gypsum, cement, and bonding agents that is lighter than regular concrete. For a thin slab installation of 1-1/2" concrete, the gypsum concrete will weigh about 13 pounds per square foot compared with the 18 pounds per square foot for regular concrete. The one good thing about this alternative is that, because it is so thin, it levels itself. The main drawbacks are that it is more expansive than concrete and will leak through any holes in the floor. One other alternative to concrete is Portland-base cement mixed with a material called a super-plasticizer and other agents. This material costs about one-third of what gypsum concrete costs.

Thin Slab

When the radiant system is to be installed in an existing structure, a thin slab of concrete on top of the existing floor is an option (Fig. 17-67). The tubing can be stapled to the top of a frame floor, onto which concrete is poured to a thickness of approximately 1-1/2" to 2" with a minimum of 3/4" of concrete over the tubing. If this option is being considered, it is important to make sure that the framing can handle the extra weight.

CAUTION

CAUTION: Consult the architect or engineer on the job when laying out a poured concrete radiant heat installation in an existing structure. An inch and a half of concrete will add about 18 pounds per square foot to the load of the floor. If the structure is not capable of handling this weight, other options should be explored.

Concrete-Free Installations

Another option for installing radiant systems is to staple the tubing under the floor. This is a less expensive way to get the job done. Usually on a staple-up job, the tubing is attached right to the bottom of the floor (Fig. 17-68) by placing a staple every six inches or so. The tubing should be in contact with the floor, because the heat transfer starts as conduction and becomes radiant only when it enters the room where the people are. The heat has to conduct through the walls of the tubing and enter the floor. The

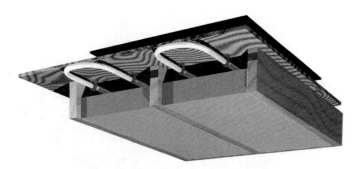

Figure 17-68 Cutaway of a staple-up radiant piping layout. *Courtesy of Uponor Wirsbo.*

main drawback to this configuration is that only a thin edge of the tubing actually touches the floor, in contrast to the entire surface of the tubing being in contact with the concrete slab. As a result, it will be necessary to circulate hotter water through the tubing arrangement to compensate for poor thermal contact between the tubing and the floor.

 FROM EXPERIENCE

With this type of installation, foil-faced insulation is used to bounce the radiant energy off the tubing and up onto the bottom of the sub-floor so the top of the finished floor heats evenly. It is also crucial that several inches of air space be left between the tubing and the foil, so the radiant waves of energy can diffuse down within the joist bay and bounce back up onto the underside of the floor.

 **CAUTION**

CAUTION: Be aware of the electrical wiring in the joist bays when doing a staple-up job. The temperature tolerance of older wiring is not as high as newer wiring.

When considering a staple-up radiant heating system under a wood floor, keep the following in mind:

- ✓ The best type of wood for hydronic radiant floor heating is laminated softwood with a layer of hardwood on the top
- ✓ Nonlaminated or solid wood flooring is a poor choice, given that the floor will expand and contract, likely causing the wood to crack
- ✓ Installing "floating floors" will reduce the chance of the floor cracking as a result of the expanding and contracting of the wood
- ✓ Narrower floor boards are less likely to warp
- ✓ Avoid tarpaper in the floors; it smells very bad when heated
- ✓ Keep the wood floor no warmer than 85 degrees F

Tubing

There are options as to the tubing you bury or staple up. PEX, cross-linked polyethylene (Fig. 17-64) has a great track record for successful installations both here and in Europe. One of the benefits of PEX tubing is that it has memory, so if it becomes kinked, it will return to its original shape when heated. The linking of the molecules in PEX tubing varies from manufacturer to manufacturer and from PEX type to PEX type. The "cross-linking" happens in the manufacturing process, and differing processes can affect the properties of the final product. Various types of PEX tubing, including PEX-A, PEX-B, and PEX-C, all have manufacturing differences, but they have all been approved for use by the American Society for Testing and Materials (ASTM). Always weigh system requirements when selecting the type of PEX tubing to use on a particular job. Other materials that can be used are PEX/Aluminum PEX, polybutylene, and rubber.

Manifold Station

The **manifold station** (Fig. 17-69) is where all the supply and return tubes come together at the manifold. When they are in a poured concrete slab, it is important that all of

Figure 17-69 Radiant manifold. *Courtesy of Danfoss.*

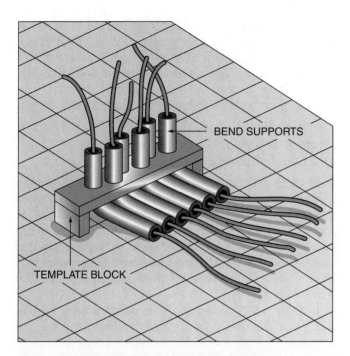

Figure 17-70 Template block and bend supports.

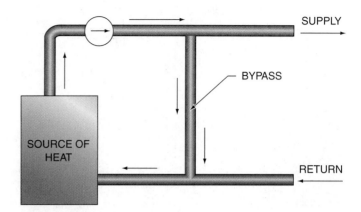

Figure 17-71 Bypass loop.

the tubes penetrate the slab within the confines of future wall locations. Miscalculations at this stage of the project will result in some or all of the tubes penetrating the floor outside the finished wall. The tubes that extend from the slab must be properly supported while the concrete is actually being poured. In addition, they should be protected from kinking at the point where they bend up into the wall. Template blocks and bend supports are used for these purposes. The template block (Fig. 17-70) holds the tubes; the bend supports are sections of tubing into which the tubing is fed.

Water Temperature and Direct Piping

The desired floor temperature in a radiant system is in the range of 85 degrees F, so the temperature of the water flowing through the piping circuit does not need to be higher than 120 degrees F and, in many cases, can be as low as 100 degrees F. If the boiler is only being used to supply water to the radiant loop, a direct piping system can be used. The direct piping arrangement is similar to that discussed at the beginning of this chapter, but the temperature of the supply water is much lower. In this case, the water is heated to 105 degrees F, circulated through the radiant system, and returned to the boiler at 90°F. One major problem with this configuration is that, if the temperature of the return water is lower than the temperature at which the flue gases condense, there will be continuous condensation. This will result in damage to the chimney, flue pipe, and heat exchanger in the boiler. Since electric boilers do not have flue gases, they can be used as the heat source for radiant systems without the possibility of damage from condensing flue gases. Some other common piping configurations are discussed next.

Bypass Lines

One way around this condensation problem is to use a bypass line. The bypass line takes some of the hot water that leaves the boiler and blends it with the water that is coming back from the radiant loop. This increases the temperature of the water returning to the boiler. A diagram of this arrangement is shown in Figure 17-71.

 FROM EXPERIENCE

Most boiler manufacturers include simple drawings of bypass piping arrangements with their installation and operating instruction booklets.

Thermostatic Bypass Valves

The one drawback of the bypass line just discussed is that it is not automatic. The thermostatic bypass valve, an automatic device, protects the boiler against low-temperature water. It has three ports and contains a thermostat similar to the one in a car's radiator. The thermostat keeps the boiler water circulating around the boiler until it reaches a certain minimum temperature. Then the thermostat lets the water out into the system (Fig. 17-72).

Thermostatic Mixing Valve

Some radiant systems use a thermostatic mixing valve to ensure that the temperature of the water feeding the radiant loop is at the desired level. The mixing valve has two inlets, one hot and one cold, as well as one outlet. As the hot and cold water enter the valve, they mix, and the resulting temperature of the water at the outlet is somewhere between the temperature of the cold water and the temperature of the hot water. If there is almost no cold water entering the

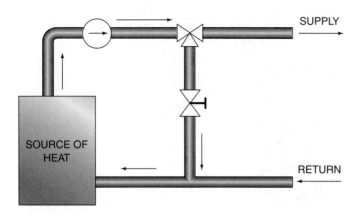

Figure 17-72 Thermostatic bypass valve.

valve, the temperature of the water leaving the valve will be close to the hot water temperature. Similarly, if there is almost no hot water entering the valve, the temperature of the water at the outlet of the valve will be very close to the temperature of the cold water. The thermostatic mixing valve automatically opens and closes the hot and cold inlets to obtain the desired temperature at the outlet of the valve. When used to control the temperature of a radiant system, the piping arrangement can look something like Figure 17-73.

In this configuration, the boiler is supplying water at 180 degrees F to the thermostatic mixing valve. Only a small amount of this water is actually flowing through the mixing valve, while the majority of the water is being returned to the boiler. The water temperature at the outlet of the radiant loop is 90 degrees F and the desired temperature of the water at the inlet of the radiant loop is 110 degrees F. This will cause the hot water port on the mixing valve to open slightly so some of the 180 degree F water will mix with the 90 degree F water to reach the desired temperature of 110 degrees F at the outlet of the mixing valve. If the temperature at the outlet of the radiant loop dropped to 80 degrees F, the hot water port on the mixing valve would open more to keep the temperature at the inlet of the radiant loop at 110 degrees F. The amount of water that leaves the radiant loop and flows back to the boiler is exactly the same as the amount of hot water that enters the loop through the mixing valve.

Primary-Secondary Pumping with High- and Low-Temperature Circuits

In the event that the structure has both high- and low-temperature circuits, the same boiler can service both requirements by using a primary-secondary configuration. Assume that one zone in the house is heated by a radiant loop and two other areas are heated by high-temperature baseboard. By taking the secondary high-temperature loops off the primary, the operation of the radiant zone is not affected (Fig. 17-74).

In Figure 17-74, water at 180 degrees F is being supplied to the first high-temperature baseboard loop and the temperature being supplied to the second baseboard loop is very close to 180 degrees F as well. Each of these baseboard circuits has its own circulator, just like the primary-secondary pumping system described earlier in the chapter. The water in the main loop after the two baseboard loops is approximately 165 degrees F, because of the heat given up by the baseboard. At the thermostatic mixing valve, the 165 degrees F water will mix with the water at the outlet of the radiant loop to give us water at 110 degrees F at the inlet of the radiant loop. In addition, if the radiant portion of this system feeds multiple radiant zones, automatic valves can be used to close off individual branches of the radiant portion of the system.

Installing and Starting the Hydronic System

Always follow manufacturer instructions and relevant codes. All information included in this book pertaining to installing a system is general and not for any specific product. Installing and putting a hot water hydronic system into operation involves the following steps:

- Setting and installing the boiler
- Installing the piping
- Wiring the unit
- Filling the system
- Starting up the system

Installing the Boiler

When setting and installing the boiler, keep the following in mind:

✓ Make certain the boiler is level
✓ Locate boiler as close as possible to the chimney

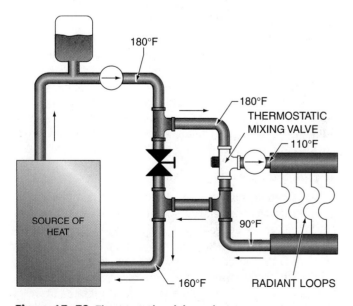

Figure 17-73 Thermostatic mixing valve.

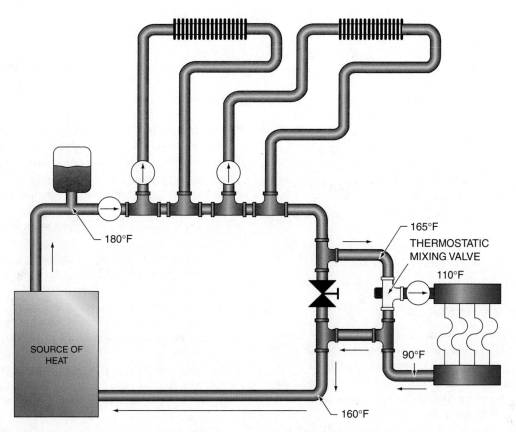

Figure 17–74 Primary-secondary pumping system used for both high-temperature and low-temperature water loops.

- ✓ Make certain the flue piping is installed according to the manufacturer's literature
- ✓ Make certain that the amount of air introduced to the area around the boiler is sufficient for combustion and dilution
- ✓ Make certain all packing material has been removed from the boiler
- ✓ Install the pressure relief in the tap specified by the manufacturer
- ✓ Provide enough clearance around the boiler for future service
- ✓ Make certain that a disconnect switch is located on or very close (within 2 feet) to the boiler
- ✓ Check the combustion of the boiler on initial start up
- ✓ Properly set the pressure-reducing valve prior to filling
- ✓ Properly check and adjust the air pressure in the expansion tank prior to filling the system

Installing the Piping

When installing the piping circuit, keep the following in mind:

- ✓ Keep vertical piping vertical and keep horizontal piping horizontal. In addition to ensuring the proper alignment of fittings and components, customers will appreciate a neat-looking job, and so will you and your boss.
- ✓ Always try to keep the number of fittings to a minimum. Excess fittings increase the cost of the job and the chance of water leakage.
- ✓ For the sake of neatness and ease in wiring, group similar components, such as circulators or zone valves, together. Also keep them at the same height and have them all pointing in the same direction. Neatness counts!
- ✓ Use lots of valves. Although this increases the cost of the materials for the job, time and money will be saved when performing service in the future. Valves placed before and after a circulator, for example, will save the technician the trouble of draining the entire system to change a defective impeller.
- ✓ Properly support the piping and components to avoid sagging.
- ✓ Dry-fit pipe fittings prior to soldering. This avoids excessive measuring, recutting, and soldering.
- ✓ Test all tubing for leaks (radiant heat systems) prior to pouring concrete!

Wiring the System

When wiring the system, keep the following in mind:

- ✓ Do not perform any electrical wiring unless you are qualified and licensed

- ✓ Color code the wires whenever possible to avoid confusion and incorrect wiring
- ✓ Read and follow the wiring diagrams supplied with the individual components to ensure their safe and proper operation
- ✓ Make certain that individual components will operate properly when used in conjunction with other components
- ✓ Ensure that the controls being used are designed to perform the desired tasks prior to installation; many electronic controls cannot be returned after they have been either removed from the package or installed
- ✓ Follow all local electric and fire codes regarding wiring methods used
- ✓ All wiring should be located above piping in the event of a water leak; this will prevent water from leaking onto the wires

Filling the System

Before filling the system, you must determine how much water pressure will be needed to lift water to the highest point in the system. To do this, measure the distance between the circulator and the highest pipe in the system. Remember that 1 psig of water pressure will lift water 2.3 feet straight up, no matter what size the pipe is.

Once you have determined what you need to fill the system, add 3 psi pressure to that so there will be some pressure at the highest point. This additional pressure will help compensate for errors in calculation as well as help remove air from the system.

Check the air pressure inside the compression tank (diaphragm-type expansion tanks). The air pressure should equal the pressure you plan to use on the water side of the diaphragm. Check the air pressure before installing the tank and filling the system with water. When there is water inside the tank, you will not be able to get an accurate pressure gauge reading on the air side of the diaphragm.

Refer to page 524 for step-by-step procedures for filling and purging the system.

Firing the System

When starting up the system, keep the following in mind:

- ✓ Check the operation of all safety components and accessories on initial start up
- ✓ Test the operation of all circulators, zone valves, thermostats, etc.
- ✓ Make certain the thermostats control the operation of the correct zone valves and circulators
- ✓ Make certain that the boiler cycles on and off on the aquastat
- ✓ Test the operation of all safety controls
- ✓ Test the combustion efficiency of the boiler
- ✓ Determine carbon dioxide and carbon monoxide levels and compare them to acceptable levels

Summary

- Hydronic systems circulate hot water or steam to the occupied space for heating purposes.
- Hot water is typically generated in the boiler, which can be gas fired, oil fired, or electric.
- Cast-iron boilers are made up of bolted sections that form the heat exchange surface.
- Water temperature in the boiler can be maintained by the aquastat or reset control.
- The low-water cutoff cycles the boiler off in the event that the water level falls below a safe level.
- The expansion tank allows for the expansion of heated water.
- The point where the expansion tank is connected to the system is referred to as the *point of no pressure change*.
- The centrifugal pump is responsible for circulating the water through the system. It is located so that it pumps away from the point of no pressure change.
- Air in the system is removed by using air separators and air vents.
- The pressure-reducing valve opens and closes to maintain the desired water pressure in the system. It is also piped into the system at the point of no pressure change.
- The pressure relief valve is a safety device that opens to relieve system pressure if it rises above the preset limit.
- Flow-control valves prevent gravity flow through hot water loops when the circulators are off.
- Zone valves are used to control water flow to different zones in the structure in order to maintain different temperatures in those areas.
- Common piping configurations include the series loop system, one-pipe system, two-pipe direct return, and two-pipe reverse return.
- One-pipe systems use monoflo or diverter tees to balance water flow.
- Two-pipe direct return systems often use balancing valves to distribute water flow evenly through each branch.
- Primary-secondary pumping systems use circulators in both the primary and secondary loops.
- The secondary loop uses the primary loop as the expansion tank.
- Radiant heat systems require low water temperatures in the range of 105 degrees F to 120 degrees F.
- Radiant heating systems often use mixing valves to maintain the desired water temperature.
- Primary-secondary systems can be used to provide both high-temperature water for baseboard terminal units and low-temperature water for use in radiant loops.

Procedures

Estimating the Volume of Water in the System, Assuming That Type M Copper Piping is Used

- Determine the number of linear feet of each size type M copper pipe used in the piping system.

- **A** Multiply the number of linear feet used by the appropriate **volume factor** in the chart shown here. The volume factor provides the number of gallons of water per linear foot of the corresponding piping material.

- Add all of the products obtained in the previous step. This will be the number of gallons of water contained in the piping circuit, not including the radiators.

- Referring to the radiator manufacturer's literature, determine the volume of water contained in the radiators. If the manufacturer's literature is not available, estimate 1 gallon of water for each radiator. Standing cast-iron radiators will require a higher estimate. Add this to the total from the previous step.

- Obtain the volume of the boiler from the manufacturer's literature. If this information is not available, estimate 1 to 2 gallons for a small copper tube boiler, or 10 to 15 gallons for a cast-iron sectional boiler. Add this figure to the total obtained in the previous step. The result is an estimate of the volume of water contained in the system.

A

Tubing Size/Type	Volume Factor (Gallons/Foot)
½" Type M Copper	0.01319
¾" Type M Copper	0.02685
1" Type M Copper	0.0454
1¼" Type M Copper	0.06804
1½" Type M Copper	0.09505
2" Type M Copper	0.1647

Procedures

Sample Calculations for Estimating Volume of Water in the System

A Consider the piping diagram in the figure.

- The total footage of 1″ type M copper is 290 feet.

- The total volume for the 1″ type M copper is 290 × 0.0454 = 13.17 gallons.

- The total footage of 3/4″ type M copper is 100 feet.

- The total volume for the 3/4″ type M copper is 100 × 0.02685 = 2.685 gallons.

- The total footage of 1/2″ type M copper is 40 feet.

- The total volume for the 1/2″ type M copper is 40 × 0.01319 = 0.53 gallons.

- All radiators are the same and according to the manufacturer's literature, each radiator holds 0.75 gallons of water. We will assume an extra 0.25 gallons of water for the piping that connects each radiator to the main lines. This gives us a total of 1 gallon of water for each radiator.

- Eight radiators, each with 1.0 gallon of water, gives us 8.0 gallons.

- According to the boiler manufacturer, the volume of the cast-iron boiler is 14 gallons.

- Adding the values from steps 3, 5, 7, 9, and 10 gives us an estimate of the total volume of water in the system, Vs, in gallons of water:

13.17 + 2.685 + 0.53 + 8.0 + 14.0 = 38.4 gallons

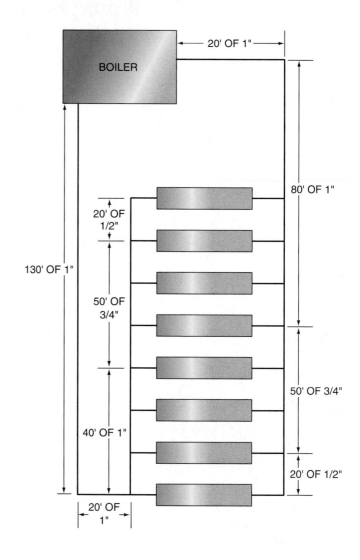

Procedures: Calculating the Minimum Volume, Vt, for the Expansion Tank

- Determine the initial temperature of the cold water entering the boiler, Tc.

- Determine the density of the cold water, Dc, using the following formula:

  $Dc = 62.56 + 0.0003413(Tc) - 0.00006255(Tc^2)$

- Determine the maximum operating temperature of the water in the boiler, Th.

- Determine the density of the hot water, Dh, using the following formula:

  $Dh = 62.56 + 0.0003413(Th) - 0.00006255(Th^2)$

- Estimate the volume of water in the system, Vs, using the previous procedure.

- From the boiler's data sheet, determine the system's pressure relief valve setting, Prv (psig).

- Determine the air-side pressure setting in the expansion tank (refer back to the text for this calculation).

- Substitute the values for Dc, Dh, Vs, Prv, and Pa in the following formula to obtain the required minimum volume for the expansion tank:

  $Vt = Vs[(Dc/Dh) - 1][(Prv + 9.7)/(Prv - Pa - 5)]$

Procedures: Filling and Purging the System

- Measure the difference between the height of the circulator and the height of the highest pipe in the system.

- Divide that result by 2.3.

- Add 3 psig to the new value.

- This figure is the desired system pressure.

- For systems with diaphragm-type expansion tanks, check the air pressure inside the compression tank.

- The air pressure should equal the calculated system pressure.

- Adjust the pressure in the expansion tank as needed. A bicycle tire pump works well here.

A When first filling the system, close the main shutoff valve and open the boiler drain on the first supply-pipe tee to purge the air

- Keep all zones closed except one.

- Use the fill valve to blow the air from that one zone. The water flows first through the system and then into the bottom of the boiler. The air will be pushed through the system piping, through the boiler, and out from the boiler drain at the top of the boiler.

B Repeat the previous two steps for the next terminal unit.

- Repeat the previous steps for each of the remaining zones in the system.

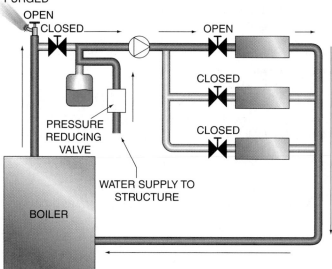

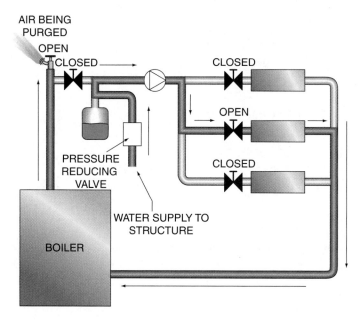

Review Questions

1. **Which of the following is true regarding cast-iron boilers?**
   a. The capacity of the boiler will decrease as the number of sections increases
   b. They typically hold between 10 and 15 gallons of water
   c. They are desirable because they are not affected by long periods of exposure to dissolved air
   d. All of the above are correct

2. **The system component that is responsible for maintaining the water in the boiler at the desired temperature is the**
   a. Aquastat
   b. Low-water cutoff
   c. Bypass valve
   d. Expansion tank

3. **If the hydronic system is equipped with a reset control and the outside temperature increases from 20 degrees F to 50 degrees F, which of the following is possible?**
   a. The temperature of the water in the boiler will change from 170 degrees F to 190 degrees F
   b. The temperature of the water in the boiler will change from 170 degrees F to 150 degrees F
   c. The temperature of the water in the boiler will remain unchanged
   d. Both a and b are correct

4. **When referring to the aquastat settings, which of the following correctly represents the relationship between the cut-in temperature, the cut-out temperature, and the differential?**
   a. Cut in − Cut out = Differential
   b. Cut out − Cut in = Differential
   c. Cut out = Cut in + Differential
   d. Cut in = Cut out − Differential

5. **Which of the following is most likely to occur if a hot water hydronic system is installed with an undersized expansion tank?**
   a. The water temperature in the boiler will be higher than desired
   b. The water temperature in the boiler will be lower than desired
   c. The relief valve will open prematurely
   d. The pressure-reducing valve will fail to feed water to the boiler

6. **As the water temperature in the boiler increases,**
   a. The air pressure in the expansion tank increases
   b. The expansion tank will contain a larger volume of air than water
   c. Both a and b are correct
   d. Neither a nor b is correct

7. **The only point in a hot water hydronic system where the circulator *cannot* affect the pressure in the system is**
   a. The point where the return pipe enters the boiler
   b. The point where the expansion tank is connected to the piping circuit
   c. The point where the water is discharged from the circulator
   d. The point where water from the structure is introduced to the boiler

8. Which of the following can result in a vacuum being pulled in the piping circuit if the expansion tank has an air pressure of 12 psig and the pump creates a 15 psig differential between the pump's inlet and outlet?

a. Both the expansion tank and circulator are in the return line where the circulator is located between the boiler and the expansion tank pumping toward the boiler
b. Both the expansion tank and circulator are located in the supply line where the circulator is located after the expansion tank pumping away from the boiler
c. Both the expansion tank and circulator are in the supply line where the circulator is located between the expansion tank and the boiler pumping away from the boiler
d. All of the above will result in a vacuum being pulled in the piping circuit

9. A pump that has a pressure differential across it of 20 psig has a head of

a. 8.7 feet
b. 20 feet
c. 46 feet
d. 92 feet

10. As the height of the hot water piping in a system gets higher

a. The required air pressure in the expansion tank will increase
b. The required air pressure in the expansion tank will decrease
c. The density of the water will increase
d. The density of the water will decrease

11. Installing the circulator so that it pumps away from the point of no pressure change will

a. Result in the highest possible pressure at the outlet of the pump for that system
b. Help keep air bubbles in solution at the higher pressure
c. Help prevent air from being pulled into the system as the result of a vacuum condition

12. It is required by law that all hot water boilers be equipped with a

a. Pressure-reducing valve
b. Pressure-relief valve
c. Air separator
d. Automatic air vent

13. If the pressure in a hot water boiler rises above safe limits

a. The pressure-relief valve will close
b. The pressure-relief valve will open
c. The pressure-reducing valve will open
d. Both b and c are correct

14. A structure that has separate areas, each with its own means to control the temperature in that area

a. Has a series loop heating system
b. Has only one thermostat
c. Is a zoned system
d. All of the above are correct

15. Gravity flow through loops can be eliminated by using a

a. Zone valve
b. Flow-control valve
c. Both a and b are correct
d. Neither a nor b is correct

16. Two-pipe, direct-return systems have the advantage that

a. The flow through all of the branches is exactly the same
b. The first terminal unit being fed water is the last one to return to the boiler
c. The distance through each branch circuit is exactly the same
d. None of the above are correct

17. The benefit of a series loop system is that

a. The flow through each terminal unit can be controlled separately
b. This system is economical to install
c. Zone valves can be located at each terminal unit
d. All of the above are correct

18 If a one-pipe system has a section of baseboard located below the main loop,

a. One diverter tee should be located on the return side of the terminal unit
b. One diverter tee should be located on the supply side of the terminal unit
c. Two diverter tees should be used
d. No diverter tees are needed

19 On a one-pipe system, placing the tees very close together will

a. Increase the resistance between the tees and increase the flow through the terminal unit
b. Decrease the resistance between the tees and decrease the flow through the terminal unit
c. Decrease the resistance between the tees and increase the flow through the terminal unit
d. Increase the resistance between the tees and decrease the flow through the terminal unit

20 In a two-pipe reverse return system,

a. Balancing valves are typically not needed
b. The first terminal unit supplied is the last one to return
c. Both a and b are correct
d. Neither a nor b is correct

21 In a primary-secondary pumping configuration,

a. The resistance between the tees should be as low as possible
b. There is only one circulator
c. All tees that are used should be diverter tees
d. All of the above are correct

22 Why is an expansion tank not required in the secondary loops of a primary-secondary pumping configuration?

a. The water in the secondary loops will not expand when heated
b. The secondary loops will use the primary loop as the expansion tank
c. The circulator in the secondary circuit would pump all of the water into the expansion tank
d. Both a and b are correct

23 In a primary-secondary pumping configuration, if a secondary pump has a higher capacity than the primary pump

a. The water might flow through the common piping in the opposite direction
b. The system cannot work
c. The secondary loop will be pulled into a vacuum
d. The primary pump will turn in the opposite direction

24 Radiant heat systems are sized to

a. Compensate for the heat lost by the occupied space
b. Regulate the heat loss of the occupants of the occupied space
c. Both a and b are correct
d. Neither a nor b is correct

25 Which of the following is true regarding a slab-on-grade radiant piping circuit?

a. The deeper the tubing is in the concrete, the higher the water temperature needs to be
b. The PEX tubing spacing is typically 12 inches on center
c. There should be a vapor barrier under the slab
d. All of the above are correct

Glossary

adapter a fitting designed to adapt one pipe or fitting to another portion of a piping system

aerator removable threaded housing for a screen attached to a faucet spout that creates a uniform flow stream

air admittance valve one-way valve that allows air to enter a DWV system; used in place of a vent that would normally terminate with another vent or through a roof; abbreviated as *AAV*

air cushion the air above the semi-permeable membrane in an expansion tank

air gap unobstructed vertical space from a device outlet to a point where water could backflow into a piping system

air separator device that removes air from the system

air vent device that removes air from a hydronic system; air vents can be manual or automatic devices

Allen screw a size-specific screw to secure drill bits and saw blades to their operating component; requires a compatible Allen wrench to loosen and tighten

anode rod device installed in a water heater to protect the inside of a storage tank from corrosion

anti-siphon a device that prevents siphoning of contaminants into a piping system

aquastat electrical component that opens and closes its contacts to energize and de-energize electric circuits in response to the temperature sensed by the device

aquifers geologic formations containing water

automatic air vent a fitting that automatically removes air from a hydronic heating system

backfill loose soil placed into an excavated area; also called fill

backflow the dangerous reversal of flow in a piping system that can contaminate the system

back siphon an occurrence caused by a vacuum in a piping system

balancing valve a manually controlled valve used to increase resistance and reduce water flow through a given branch circuit

ball valve type of isolation valve used in most piping systems

band iron thin, perforated metal strapping sold in rolls and used to support pipe

baseboard heating see radiator or terminal unit

bell an enlarged end of a pipe or fitting that receives another pipe end or fitting; may also be called a hub or socket

bend an offset made in tubing on a job site; also, a manufactured offset fitting

bit an abbreviated term to describe a drill bit

boiler equipment designed to heat water using electricity, gas, or oil as a heat source for the purpose of providing heat to an occupied space or potable water

boiler drain drain outlet on a water heater used to drain the storage tank; has a garden hose connection

boiler feed valve valve that reduces the pressure entering the structure to the pressure required by the hydronic system; automatically feeds water into the system to maintain the desired water pressure

brackish lowland water close to the ocean; has high salt content

branch piping of a DWV system that connects to main portions of a system

branch interval vertical distance along a stack equal to one-story height, but no less than 8′; also the area where horizontal branches connect to a stack

branch vent vent that connects one or more individual vents with a stack vent or vent stack

brazing welding process used without flux to weld copper tube; also known as silver soldering

building drain lowest horizontal main drain of a DWV system; conveys wastewater to a building sewer

building sewer conveys wastewater from building drain to point of disposal

burner the main flame assembly that externally heats water in a gas water heater

bushing a compact reducing fitting that inserts into a pipe, fitting hub, or socket

BW&O abbreviation for bath waste and overflow drain assembly used on bathtubs

carpenter individual who installs the wood framing or other woodwork

centrifugal pump pump that moves water through a piping circuit by means of centrifugal force

cheater an unsafe method of gaining extra leverage when using a tool

check valve a one-way directional device installed to protect water from reversing flow in a piping system

chuck the rotating portion of a hand drill in which a drill bit is inserted

chuck key a tool designed to loosen and tighten the drill chuck

circuit electrical wiring system from a power source to equipment or a device

circuit breaker a device that disconnects (isolates) an electrical circuit

circuit vent special vent serving at least two, but no more than eight, fixture traps; begins at its connection to a horizontal branch and terminates at its connection with the vent stack

circulator see centrifugal pump

cleanout a DWV fitting installed to clear obstructions from a drain or sewer

closed loop a system that is closed or isolated from the atmosphere

closet bend a specially designed 90-degree fitting used as the last fitting of a drainage system serving a water closet; one side is 4" and the other is 3"

closet bolts non-corrosive bolts inserted into a closet flange to anchor the toilet to the drainage system

closet flange also called a toilet flange; installed to connect a toilet to the drainage system

collector heats water in a solar water heating system; also called a panel

combo abbreviated term meaning a combination of a DWV wye fitting and eighth bend (45°) fitting

common vent vent serving two fixture drains located on the same floor and connecting either at the same height or at different heights to the drain

compacting the process of compressing loose soil placed back in a trench; also known as tamping

compression a type of connection used for water distribution tubing

compression tank see expansion tank

continuous waste a drain assembly used to connect a double-bowl kitchen sink to a single p-trap

coupling a sleeve that connects two equal sized pipe ends to form a continuous pipe

custom defines homes with fixture upgrades

custom home a house built with fixtures upgraded from the basic fixtures installed in a spec home

D-box trade name for distribution box

desanco another name for a trap adapter

developed length way of measuring the distance a pipe is installed along the centerline of all pipe and fittings

diaphragm-type expansion tank see expansion tank

direct-return configuration of a water heating system in which the first terminal unit supplied with hot water is the first one to return to the boiler and vice versa

distribution box fabricated box or structure to distribute effluent to drain field or other designated location

diverter a device that routes the water from a tub spout to a showerhead, or from a showerhead to a handheld shower unit

drafting triangle a drafting tool used to illustrate straight or angled lines

drain pipe that conveys wastewater from point of entry within a DWV system

drain field area of installation for perforated piping to drain wastewater (effluent); also called leach field

drainage fixture unit (dfu) abbreviated as dfu, it is based on rate of flow measured in gallons per minute or liters per second into a drainage system from a plumbing fixture used to size pipe

drainage, waste, and vent (DWV) complete system draining soil, waste, and wastewater to a point of disposal; circulates air within the system

drawings blueprints

DWV abbreviation for drainage waste and vent system

effluent wastewater that has been separated from solids, but may contain dissolved sewage solids

elbow a fitting used to create an offset; also called a bend

electrolysis a corrosion process caused by directly connecting dissimilar metals

element electrical heating device that internally heats water in an electric water heater

escutcheon flange installed around a pipe to conceal pipe penetrations through a wall, floor, or ceiling

expansion tank device installed in a cold water piping system to absorb the expansion caused by heating water

feet of head the pumping capacity of a pump; 1 foot of head is the equivalent of a 0.433 psig difference between the inlet and outlet of the pump (1 psig = 2.31 feet of head)

female a fitting with internal threads that screws onto a male fitting

filter an accessory that removes particulates from water, but does not purify water

finish the color or polish of a faucet, drain assembly, or other fixture trim item

fitting item in a plumbing system that connects to piping or another fitting to achieve a desired offset or specific connection

fixture branch pipe draining two or more fixture drains to a stack or other drains; in some piping configurations, it can also be known as a horizontal branch

fixture drain drain from a trap serving a fixture that connects to another pipe; also called a waste arm in some design applications

fixture group a bathroom consisting of a tub or shower, toilet, and lavatory

flashing a weather-tight sealing component installed around vent pipes penetrating a roof

floor joist a horizontal wood board used to support plywood or other flooring material

flow check valve see flow control valve

flow-control valve prevents backward and gravity circulation through loops not requiring flow

flue the entire pipe system exhausting fumes from a gas water heater

flush clean a piping system with air or water pressure

flux chemical paste to solder copper tube

foam-core type of non-pressure DWV plastic pipe that has a solid outer layer, a cellular foam middle layer, and a solid inner layer

foot valve a check valve device installed at the bottom of a vertical-drop pipe in a well serving an above-ground pump; ensures that water remains in the pipe after a pumping cycle

gas cock type of isolation valve used in a gas distribution system

gate valve type of isolation valve used mostly in water distribution systems

half bathroom a bathroom consisting of only a toilet and a lavatory

heating element an immersed heating device that heats water in a storage tank when energized with electricity

high limit a safety device serving a water heater that disconnects an energy source when water temperatures approach unsafe temperatures

horizontal branch pipe connecting two or more fixture drains or fixture branches to a main portion of a drainage system; in some piping configurations, it can also be known as a fixture branch

hose bibb often called a sillcock, it is available in numerous designs to serve as a water distribution outlet; has a garden hose connection

hose outlet the point where a hose can connect to a faucet or boiler drain

hub an enlarged end of a pipe or fitting that receives another pipe end or fitting; may also be called a bell or socket

hydraulic gradient vertical distance (rise) from the trap weir to the centerline of the connecting vent fitting

hydronics heating systems that circulate hot water or steam through piping arrangements located in the areas being heated

individual vent vent serving one fixture trap that terminates to open air or connects with another vent; can serve more than one trap in some design applications, such as common venting

instantaneous a water heater that does not use a storage tank; also called tankless

interval equal to one story height, but not less than 8′; relates to a branch connecting to a stack

isolate separate a portion of a piping system by turning a valve or device

isolation valve general term describing that a valve isolates a system or portions of a system

isometric view a three-dimensional view of a piping system indicating the scope of work

joint term to describe the connection point of a pipe and fittings to each other or to fixtures

joist horizontal board that is part of a complete framing system; categorized as floor joist and ceiling joist

knock-out a manufactured portion of a sink or garbage disposer designed to be removed by an installer to install a faucet or dishwasher drain hose

lanyard an approved shock protective line attached to a safety harness and fixed to a secure point or lifeline rope

load-bearing the portion of a structure that bears the weight of the structure, such as a load-bearing wall

loop vent special vent similar to a circuit vent except that it terminates connecting to a stack vent instead of a vent stack and is only used on a top floor or at the highest branch interval

lug a designated raised portion of a valve used in place of a handle and operated with a wrench or tool

male a fitting with external threads that screws into a female fitting

manifold station location of the manifold for radiant heating loops

manual air vent fitting that, when opened manually, will remove air from a hydronic heating system

meter electrical testing tool used to test voltage, amperage, and continuity of an electrical system

monoflo tee see diverter tee

nominal diameter size of pipe and fittings used to order materials; does not indicate exact diameter

nut driver size-specific tool to tighten or loosen nuts, similar to a screwdriver

offset angle that changes a piping route expressed in degrees of a circle such as 90° and 45°

offsets describes all change of direction fittings or the routing of a piping system other than being installed straight

ohm (Ω) a unit of measure describing the resistance of flow of an electrical current

one-pipe hot water hydronic piping configuration that uses a main hot water loop and diverter tees to connect the terminal units to the system

open air outside a building or structure; typically known as vent through roof or VTR

open-end wrench size-specific tool to tighten or loosen nuts and bolts in confined spaces

OSHA Occupational Safety and Health Administration; mandates safety and health regulations

outdoor reset control used on hydronic systems that decreases the temperature of the water as the outdoor temperature increases

partition a non–load-bearing wall designed to separate rooms, not to support a structural load

P-trap nonrestrictive fitting installed at each fixture that does not have an integral trap; uses a water seal to prevent sewer gases from entering occupied areas; often called a trap; see also trap

perc test trade name for percolation test; a method to evaluate percolation conditions of soil

percolation natural drainage ability of soil; also known as perc

PEX tubing cross linked polyethylene; see polyethylene tubing

pilot flame of a gas water heater that ignites gas entering a burner assembly

pipe often used to describe all pipe, tube, and tubing, but is defined as any rigid or hard materials that would break if flexed more than 2% of its diameter

plan view a view of a design from the top; also known as a bird's-eye view

point of no pressure change point in a hydronic hot water system where the expansion tank is connected to the piping system; the pressure at this location cannot be affected by circulator operation

point-of-use water heater a small capacity water heater installed close to the fixture utilizing the water heater

polyethylene tubing tubing material used for buried water loops in radiant heating systems as well as geothermal heat pump systems

pop-up a drain-operating assembly for a lavatory sink, abbreviated as PO

port opening an opening in a fixture, such as a drain or overflow hole that receives drain assemblies, to connect the fixture to the drain system

potable water free from impurities that could cause disease; safe for human consumption

pressure-reducing valve valve that reduces the pressure of the water entering the structure to the desired pressure in the hydronic system

pressure relief valve a reactionary device that protects a water heater against excessive pressure

procure the process of receiving material through ordering or gathering

purification a process to cleanse the water to ensure it is considered potable

radiant heating system heating system that attempts to regulate the heat loss of the individual as opposed to the rate of heat loss of the structure

radiator heat emitters or terminal units that transfer the majority of their heat to the occupied space by radiation

radius half the diameter of a circle; relates to the bend of a fitting or tubing

reactionary valves devices that react to certain conditions within a system to provide protection, such as backflow, pressure, and temperature

reduced-pressure zone valve a reactionary device to protect a potable water system against the possible backflow of undesired water from a connected portion of the system

reducers a fitting used to connect two different pipe sizes together. Reducers are different than busings but accomplishes the goal of reducing a pipe size

relief valve relieves a storage tank or piping system of dangerous conditions such as pressure and temperature or both.

relief vent pipe circulating air between a drainage and a vent system; has several specific areas of installation and is sized based on its use

reverse-return configuration of a water heating system in which the first terminal unit supplied with hot water is the last one to return to the boiler and vice versa

riser diagram an isometric or side view of a large portion or detailed area of a piping system and typically utilized to reflect several stories of a building

rough abbreviation of the term *drain rough-in;* describes the distance to the center of a toilet from the wall behind the toilet

rough-in phase of construction before finish or trim phase when all piping is installed in floors, walls, and ceilings

rough-in sheet information pertaining to specific installation requirements or other unique characteristics of a fixture

sanitary cross a four-way fitting that has limited use in a DWV system. It has the same compact design as a sanitary tee fitting

sanitary tee a three-way fitting that is compact having a sanitary flow pattern and has limited use in a DWV system

schedule classification of pipe indicating the wall thickness of the pipe

section view a view, usually detailed, of a design from the side; also known as a side view

septic tank fabricated holding tank or structure to contain sewage and solids

sewer portion of a drainage system that is installed on the exterior of a building

shank the portion of a drill bit that is secured in a drill chuck

side view a view of a design from the side; also known as a section view

silver solder metal filler used to braze copper tube; it has high silver content and is in stick form

sketch an illustration focusing on a certain portion of an area or piping system

slab concrete floor that defines a building design that does not have a crawlspace or basement

slip joint a type of drainage connection between the rough-in stub-out and tubular p-trap

slope upward or downward installations used to install drainage or venting piping

socket size-specific tool to tighten or loosen nuts and bolts in confined spaces; uses a ratchet handle for its operation

soil stack vertical pipe conveying wastewater that contains fecal matter; the same pipe as a waste stack, can be installed with horizontal offsets

solder metal filler supplied in roll form to solder copper tube; it cannot contain more than 2% lead

soldering process of welding copper tube using flux and solder

spec home a house built using average construction quality and product selection

stack vertical pipe that is at least one story in height; can be installed with horizontal offsets

stack vent vertical pipe that connects to a soil or waste stack; extends to open air or connects with another approved vent; can be installed with horizontal offsets

standard expansion tank see expansion tank

steam boiler equipment that heats water to the point of vaporization for the purpose of providing heat to an occupied space

stop a type of isolation valve

stop-and-waste valve an isolation valve that also has the capability to manually drain the isolated portion of the system.

street a type of offset fitting in which one end has the same outside diameter as a connecting pipe or fitting

stub out pipe that serves a fixture installed during the rough-in phase of construction

stud vertical board to erect walls; 2" $\times$ 4" and 2" $\times$ 6" are the two most common sizes in residential construction

submersible pump a pump motor and impeller assembly that is submersed in a well below ground to provide water to a piping system

submittal data indicating the type of fixture to be used and its unique characteristics; sometimes includes installation information

swing joint a fitting arrangement that creates an offset using two fittings

T&P valve a combination temperature and pressure relief valve

T&S faucet abbreviation for a combination type faucet serving a tub and shower unit

tankless a water heater that does not store water in a storage tank; also called instantaneous

tees a three-way fitting used to connect a branch pipe with a main portion of a piping system and is ordered based on the size of all three sides

terminal where an electrical wire connects to a device; also referred to as a post

terminal unit a radiator or section of baseboard

test ball a testing accessory that is injected with air after being inserted into a drainage pipe to allow the system to be filled with water

test cap used to seal a pipe for testing a DWV system; several types are available

test plug a testing accessory that inserts into a pipe end for testing purposes; several types are available

test tee installed in a DWV system to complete a test and can also be installed to serve as a cleanout in a vertical DWV pipe

thermocouple heat-sensing device to ensure that a gas water heater pilot flame is ignited

thermostat a regulating device to control the temperature of a water heater

torch tool that is ignited and creates a flame to solder or braze copper tube

trap nonrestrictive fitting or device installed at each fixture using a water seal to prevent sewer gases from entering occupied areas; see also P-trap

trap adapter also known as a desanco and is a specialty fitting installed to connect a DWV stub-out pipe to a p-trap

trap distance distance a trap weir is located from its protective vent

trench excavated pocket of soil to install piping; also known as a ditch

trim refers to items that have chrome or other finishes

trim-out finish phase of construction when fixtures are installed; also known as trim

tube less rigid than pipe, but more rigid than tubing; often referred to when ordering copper, i.e., copper tube

tubing flexible or non-rigid materials that can deflect more than 2% of its diameter without breaking; often referred to as pipe

tubular a pipe that is smaller in diameter and has a thinner wall thickness than pipe used for drainage piping

two-pipe system hydronic piping configuration that uses one pipe as the supply and one pipe as the return; can be configured as a direct return or a reverse return

type used to specifically describe different copper tube specifications and to differentiate between various materials, i.e., Type M copper tube or types of plastic pipe

union a three piece fitting installed in a piping system to provide access without cutting the pipe and also used as a termination of the piping system to a piece of equipment

vacuum breaker reactionary device that breaks a vacuum in a piping system when unsafe backflow conditions are present

vacuum relief valve a type of vacuum breaker that is commonly used on a water heater that is piped with a side inlet connection

vent a pipe dedicated to providing airflow, so a drainage system can breathe

vent stack vertical vent pipe that receives other vents and terminates to open air or with stack vent; can be installed with horizontal offsets

volt (V) a measure of electromotive force (EMF) often described as electrical pressure, but a plumber relates it to the amount measured on a voltage meter, such as 120 volts or 240 volts

volume factor provides the number of gallons of water per linear foot of a piping material

volute portion of the circulator housing that carries water from the pump

wall stud vertical wood board or other material used to build a wall

wallboard material that provides a finished wall surface over the wall structure; gypsum board (drywall) is the most common type used in residential construction

waste arm a horizontal fixture drain that begins with the vent connection and terminates at the fixture connection

waste stack vertical pipe conveying wastewater only; the same pipe as a soil stack; can be installed with horizontal offsets

wastewater water that does not contain sewage; term often used instead of the word *effluent* by plumbers

water closet another name for a toilet

water distribution entire potable water piping system, relates to hot and cold water systems

water distribution system though an entire system is distributing water, this refers to the piping inside a house

water service piping on the exterior of a building connecting the potable cold water source, such as a water meter, to the interior water piping system in a building

watts a measure of electrical power or the true power in a circuit, but a plumber relates it to a heating element of an electric water heater element, such as 4500 watts

wax seal a seal that ensures water does not leak and sewer gases do not escape when connecting the toilet to the drainage system

weir the portion of a p-trap where the water flow crests from the trap and enters the connecting horizontal drain

wire cutter tool to cut wire such as electrical wire

wye a three-way DWV fitting with the branch connection being 45° from the main portion of the fitting

zone valve thermostatically controlled valve that opens and closes to regulate the flow of hot water to the terminal units in the occupied space

zoning process of dividing a structure into separate areas, each of which has its own means to regulate the temperature in the space

Index

Note: Items in **bold** indicate table or figure entry.

A

Abbreviations, 202, **202–206**
Above-ground layout, 239
    edge form layout procedure, 246–247
    fixture locations, 240
    floor joist conflicts, 240–241
    manufacturer rough-in sheet, 243–245
    wall layout, 241–243
ABS fittings, 90
ABS pipe, 63. *See also* Plastic pipe
Adapters, 73, 78
Adjustable wrenches, 8–9
Aerator, 136, 146
Air admittance valve, 326, 334
    sizing, 352, 354, **355**
Air compressors, 52
Air cushion, 489, 493
Air gap, 97, 105
Air separators, 489, 498, **499**
Air vents, 489, 498, **499**
    manual air vent, 489
Allen screw, 37, 46
American Society for Testing and Materials (ASTM), 58
Angled jaw pliers, 8
Anode rod, 163, 183
Anti-siphon, 97, 104
Aquastat, 489, 492, **493**
Aquifers, 251, 252
Architectural blueprints, 206–207
    symbols, 207–212, **213**
Architectural symbols, 207–212, **213**
Auger bits, 44–45
Automatic air vent, 489, 498
Aviation snips, 13

B

Backfill, 251
Backflow, 97, 102
Back siphon, 97
Back siphoning, 103
Bagging and tagging, 230–231, **232**
Balancing valve, 489, 503
Ball peen hammer, 10
Ball valves, 97, 99
Band iron, 4, 13, **292**
Band saws, 50
Baseboard heaters, 489
Basin wrenches, 15–16
    use of, 33

Basket strainers
    kitchen sink, 155, **156, 157**
    laundry sink, 156–157, **158**
Basket strainer tools, 16
Bathrooms. *See also* Lavatory sinks
    bathtubs. *See* Bathtubs
    guest bathroom layout, 376–377, **378**
    half bathroom layout, 379–380, **382**
    master bathroom layout, 378–379, **380–381**
    showers. *See* Showers
    symbols, **201**
    toilets. *See* Toilets
Bathtubs, 123–124, **125, 126**
    drain assemblies, 151–154
    faucets, 139–142, **143, 144, 145**
    layout and sizing, 278–279, **280**, 388–390
Bell, 73, 76
Bend, 73, 74
Bidets, 131–132
    drain assemblies, 157–158, **159**
    faucets, 149, **150**
    installation of, 421–422
Bit, 37, 41. *See also* Drill bits
Blueprints
    abbreviations, 202, **202–206**
    architectural blueprints, 206–207
        symbols, 207–212, **213**
    symbols
        architectural, 207–212, **213**
        generally. *See* Symbols
Boiler drain, 97, 103
Boiler feed water valve, 489, 499
Boilers, 488, 489, 491. *See also* Hydronic heating systems
Brackish, 251, 254
Branch, 251, 260, 326, 332
Branch interval, 326, 333
Branch vent, 326, 332–333
    sizing, 350, **352**
Brass fittings, 82–83
Brass pipe, 68
Brazing, 272, 295, 297–298, 314–315
Building drain, 326, 329, 381, **384**
    sizing, 346, **347**
    and vent stack connection, 365
Building sewer, 326, 329
Burner, 163
Bushings, 73, 77
BW&O, 136, 151

C

Cap symbols, 198–199
Carpenter, 228, 240
Cast-iron fittings, 90–91
Cast-iron pipe, 66–67
Centrifugal pumps, 489, 496–498
Chain wrenches, 9
Chalk box, 20–21
Cheater, 4
Check valves, 97, 107–108
Chisels, 15
Chuck, 37, 41
Chuck key, 37, 41
Circuit, 448
Circuit breaker, 448, 453
Circuit vent, 326, 333
Circular saws, 48–49
Circulator, 489, 496
Cleanouts, 73, 87, 326, 330
    drainage, waste and vent systems, 330
    positioning at base of stacks, 361–362
    symbols, 198, **199**
Closed loop, 489
Closet bends, 73, 87, **88**
Closet bolts, 410
Closet flanges, 88–99, 410, 413
Cold chisels, 15
Collector, 163, 182
Combination pliers, 8
Combo fittings, 73, 85
Common vent, 326, 332
Communications, 229
    oral communications, 229–230
    written communications, 229
Compacting, 251
Compass saw, 13
Compounds and sealants, 295
    brazing, 297–298, 314–315
    flaring copper, 298, **299**, 316–319
    flexible tubing
        connection types, 300–301
        working with, 299
    flux, 296
    manifold systems, 299–300
    Material Safety Data Sheets, 295
    pipe dope, 295–296
    plastic pipe
        cutting, 301
        solvent-welding, 301–302, 320
        working with, 301
    solder, 272, 295, 296
        silver solder, 272
    soldering, 272, 295, 296–297
        copper, 310–313
        silver, 295
Compression, 410, 411
Compression tank, 489, 511
Concrete drop-in anchor, **290**
Continuous waste, 410, 418
Copper fittings, 81–82
Copper flaring, 298, **299**, 316–319
Copper flaring tool, 17–18
Copper pipe cutters, 16–17
Copper soldering, 310–313
Copper stub-out bracket, **293**
Copper stub-out pipe, **293**
Copper tube, 65–66
Copper tubing bender, 18
Corrugated stainless steel tube (CSST), 69
Couplings, 73, 76
CPVC fittings, 81
CPVC pipe, 62–63. *See also* Plastic pipe
Crimping tool, 19–20
Curved claw hammer, 10
Custom, 118, 131
Custom home, 136, 137

D

D-box, 326, 359
Desanco, 410
Developed length, 326, 356
Diaphragm-type expansion tank, 489, 494
Direct-return, 489, 506
Dishwashers, 167–168
    installation of, 420
    symbol, **202**
Distribution box, 326, 359
Diverter, 136
Drafting, 212–213
    isometric drafting, 219–220, **221, 222**
        riser diagrams, 220, 222–223, **224**
        triangles, 215–217
    tools. *See* Drafting tools
Drafting paper, 217–218, **219**
Drafting tools, 213
    scale ruler, 213–214
    symbol templates, 217, **218**
    triangles, 194, 215–217
Drain, 61, 73, 326
Drainage, waste and vent (DWV) systems, 60, 83–84, 325, 326, 328
    ABS fittings, 90
    air admittance valve sizing, 352, 354, **355**
    air vents, 489, 498, **499**
        manual air vent, 489
    aquastat, 489
    bathrooms
        guest bathroom layout, 376–377, **378**
        half bathroom layout, 379–380, **382**
        master bathroom, 378–379, **380–381**
    bathtub layout, 388–390
    branch vent, 326, 332–333
        sizing, 350, **352**
    building drain, 326, 329, 381, **384**
        sizing, 346, **347**
        and vent stack connection, 365
    building sewer, 326, 329
    cast-iron fittings, 90–91
    circuit and loop-vent sizing, 350–352, **353**
    circuit vent, 333, **334**

INDEX

cleanouts, 73, 87, 326, 330
    drainage, waste and vent systems, 330
    positioning at base of stacks, 361-362
    symbols, 198, **199**
closet bends, 73, 87, **88**
closet flanges, 88-99, 410, 413
combination waste and vent, 336-337, **338**
combo fittings, 85
continuous waste, 364
crown venting, 369
fittings, 83-91
    closet bends, **88**
fixture branch, 326, 332
    sizing, 342, **344**
fixture drain, 326, 330-331
    sizing, 342, **343**
fixture rough-in, 384-385
hangers, 393-394, **395, 396**
heel inlet 90°, 88
horizontal branch, 326, 332, **333**
    sizing, 342, **344**
individual vent, 326, 332, **333**, 348
    sizing, 348, 350, **351**
island venting, 337, **338**, 367-368
    sizing, 354-355, **356**
kitchen sink layout, 377-378, **379**, 390-392
laundry room layout, 380, **383**
lavatory layout, 386-387, **388**
layout considerations, 375, **376**
    bathtub, 388-390
    building drain, 381, **384**
    fixture rough-in, 384
    guest bathroom, 376-377, **378**
    half bathroom, 379-380, **381-382, 382-383**
    kitchen sink, 377-378, **379**, 390-392
    laundry room, 380, **383**
    lavatory, 386-387, **388**
    master bathroom, 378-379, **380**
    scope of work, 376
    shower pan, 393, **394**
    toilets, 385-386, **387**
    venting system, 381, 384, **385**
    washing machine, 392, **393**
loop vent, 326, 333-334, **335**
major segments, 328-330
master bathroom layout, 378-379, **380-381**
minor segments, 330-335
pipes
    connections to dissimilar pipe, 396-397, **398**
    pipe protection, 394, 396, **397**
P-traps, 73, 89, 326
    symbols, 198
PVC fittings, 90
relief vent, 327, 334-335, 344
sanitary crosses, 73, 86
sanitary tees, 73, 85-86
scope of work, 376
septic systems, 357-360
shower pan, 393, **394**
shower pan liner installation procedure, 403-405
sizing, 337-342
    air admittance valve sizing, 352, 354, **355**
    branch vent sizing, 350, **352**
    building drain and sewer sizing, 346, **347**
    circuit and loop-vent sizing, 350-352, **353**
    fixture branch and horizontal branch sizing, 342, **344**
    fixture drain sizing, 342, **343**
    individual vent sizing, 348, 350, **351**
    island vent sizing, 354-355, **356**
    stack vent and vent stack sizing, 346-348
    unlisted fixture sizing, 363
    waste stack sizing, 344-346
    wet-vent sizing, 352, **354**
stack vent, 327, 330
supports, 393-394, **395, 396**
    wood for pipe support, 406
testing, 397-398, **400, 401**
test tees, 87
toilets, 385-386, **387**
trap adapters, 73, 89-90, 327, 342, 410, 416
trap distance, 327, 356-357
twin elbows, 86
venting system, 381, 384, **385**
vent stack, 327, 330
    and building drain connection, 365
vent-stack and waste-stack connection, 366
vent through roof layout procedure, 402
washing machines, 392, **393**
waste stack, 327, 329-330
    sizing, 344-346
waste-stack and vent-stack connection, 366
wet venting, 336
wet-vent sizing, 352, **354**
wye fittings, 84-85
Drainage fixture unit (dfu), 326, 337
Drain assemblies, 150
    bathtub drain assemblies, 151-154
    bidet drain assembly, 157-158, **159**
    kitchen sink basket strainers, 155, **156, 157**
    laundry sink basket strainers, 156-157, **158**
    lavatory drain assemblies, 150-151, **152**
    shower drains, 154-155, **156**
    symbols, 202
Drain field, 326, 359
Drawings, 194
Drill bits, 44
    auger bits, 44-45
    hole saw bits, 46
    masonry bits, 46-47
    power bore bits, 45-46
    speed bits, 45
    step bits, 47-48
    twist bits, 44
Drilling and notching, 281-282
    hangers, 287-289, **290-294**
    hole diameter, determining percentages of, 307-308
    hole saw finished surface drilling, 53
    joists, 284-285, **286, 287**
    supports, 287-289, **290-294**
    walls, 282-283, **284**

538 INDEX

Drills, 41
    bits. *See* Drill bits
    hammer drill, 43
    hole hawg, 42–3
    pistol drill, 42
    right angle drill, 41–42
Dust masks, 24
DWV, 64, 73. *See also* Drainage, waste and vent (DWV) systems

E

Ear protection, 40, **41**
Effluent, 326, 357
Elbow, 73, 74
Electrical safety, 38
Electric water heaters, 178–182, **183**. *See also* Water heaters
    draining, 469–470
    electrical source, 453–454
    electrical tests, 454–457
    heating elements, 452–453
    high-limit device, 449–450
    installation of, 423–424
    screw-in element replacement, 471–472
    thermostats
        lower thermostat, 451–452
        upper thermostat, 450
    troubleshooting, 449
        electrical source, 453–454
        electrical tests, 454–457
        heating elements, 452–453
        high-limit device, 449–450
        lower thermostat, 451–452
        upper thermostat, 450
Electrolysis, 73, 79
Element, 163, 178
EPA standards, 255
    water filtration, 256–257
    water quality, 255–256
Escutcheons and stops, 97, 100–101, 136, 141, 410
    installation of, 411–413
    installation onto a copper stub-out pipe, 426–429
Expansion tanks, 163, 183, 489, 495
    hydronic heating systems, 493–496
        minimum volume, calculation of, 523
        primary-secondary pumping, 511
    standard expansion tank, 490, 493
    water heaters, 187–188
        replacement of, 479–480
Eye protection, 21–22

F

Face protection, 22–23
Fall protection, 40
Faucets, 135, 137–138
    bathtub and shower faucets, 139–142, **143, 144, 145**
    bidet faucets, 149, **150**
    kitchen sink faucets, 143–145, **146, 147**
    laundry sink faucets, 146–147, **148, 149**
    lavatory faucets, 138–139, **140, 141**
        installation of cultured-marble faucet, 432–437
Feet of head, 489
Female adapters, 73, 78

Filter, 251, 256
Finishes, 136, 137
First aid kits, 24–25
Fittings, 4, 17, 59, 64, 72
    ABS fittings, 90
    brass fittings, 82–83
    bushings, 73, 77
    cast-iron fittings, 90–91
    cleanouts. *See* Cleanouts
    closet bends, 73, 87, **88**
    closet flanges, 88–99
    combo fittings, 73, 85
    copper fittings, 81–82
    couplings, 73, 76
    CPVC fittings, 81
    degree of, 74
    designs, 74–79
    drainage, waste and vent fittings, 83–91
        closet bends, **88**
    female adapters, 73, 78
    galvanized fittings, 82
    male adapters, 73, 78
    offsets, 59, 64, 73, 75
    PEX fittings, 80–81
    polyethylene fittings, 82
    P-traps, 73, 89, 326
        symbols, 198
    PVC fittings, 83
        drainage, waste and vent fittings, 90
    reducers, 73, 76–77
        symbols, 199
    sanitary crosses, 73, 86
    sanitary tees, 73, 85–86
    tees. *See* Tees
    test tees, 73, 87
    trap adapters, 73, 89–90, 327, 342, 410, 416
    twin elbows, 86
    unions, 73, 79
    water distribution fittings, 79–80
        brass fittings, 82–83
        copper fittings, 81–82
        CPVC fittings, 81
        galvanized fittings, 82
        PEX fittings, 80–81
        polyethylene fittings, 82
        PVC fittings, 83
    wye fittings, 73, 84–85
Fixture branch, 326, 332
    sizing, 342, **344**
Fixture drain, 326, 330–331
    sizing, 342, **343**
Fixture group, 374, 375
Fixtures, 117, 119
    bathtubs. *See* Bathtubs
    bidets. *See* Bidets
    drainage, waste and vent rough-in, 384–385
    kitchen sinks. *See* Kitchen sinks
    laundry sinks. *See* Laundry sinks
    lavatory sinks. *See* Lavatory sinks
    layout and sizing
        above-ground layout, 240

INDEX

fixture branch and horizontal branch sizing, 342
fixture drain sizing, 342
unlisted fixture sizing, 363
locations, 240
showers. *See* Showers
symbols, 200–201
toilets. *See* Toilets
Flaring copper, 298, **299,** 316–319
copper flaring tool, 17–18
Flashing, 37, 38
Flat bar, 13–14
Flexible pipe crimping tool, 19–20
Flexible tubing
connection types, 300–301
working with, 299
Floor joist, 37, 41
Floor joist conflicts, 240–241
Flow check valve, 489, 502
Flow-control valve, 489, 502
Flue, 163
Flush, 97
Flushed, 106
Flux, 272, 295, 296
Foam-core, 59, 61
Folding ruler, 6
layout, 29
Foot protection, 24
Foot valve, 448, 461
45° offsets, 195–196
Framing square layout, 30
Framing squares, 7
layout, 30

G

Galvanized fittings, 82
Garbage disposers, 164–167
symbol, **202**
Gas cocks, 97, 102–103
Gas water heaters, 170–178. *See also* Water heaters
draining, 469–470
gas regulator replacement, 474–475
installation of, 424–425
thermocouple replacement, 473
troubleshooting, 457–459
Gate valves, 97, 99–100, **101**
Gloves, 23
Grain, 59
Grinders, 50–51

H

Hacksaws, 12, **13**
Half bathroom, 374, 376
Hammer drill, 43
Hammers, 10
Hand protection, 23
Hand tools, 5
adjustable wrenches, 8–9
aviation snips, 13
basin wrench, 15–16
use of, 33
basket strainer tools, 16

chalk box, 20–21
chisels, 15
copper flaring tool, 17–18
copper pipe cutters, 16–17
copper tubing bender, 18
flexible pipe crimping tool, 19–20
framing square layout, 30
hammers, 10
inside plastic pipe cutter, 11–12
knives, 14
levels, 5–6
use of, 26–27
list of, **5**
metal cutting saw, 12, **13**
nail pullers, 13–14
pipe wrenches, 9–10
use of, 31
plastic pipe cutters, 11
inside cutters, 11–12
plastic pipe saw, 10–11
pliers, 8
plumb bob, 20
screwdrivers, 7–8
squares, 7
framing square layout, 30
tape measures, 6–7
ruler reading, 28
torch regulator assembly, 18–19
torque wrench, 21
tubing cutters, 16–17
Hangers and supports
drainage, waste and vent systems, 393–394, **395, 396**
drilling and notching, 287–289, **290–294**
wood for pipe support, 406
Hard hats, 25
Heating elements, 448, 449, 452–453
Heel inlet 90°, 88
High limit, 163, 178, 448, 449
Hole hawg, 42–3
Hole saw bits, 46
Hole saw finished surface drilling, 53
Horizontal branch, 326, 332, **333**
sizing, 342, **344**
Hose bibb, 97, 103–104
Hose outlets, 97, 103
boiler drains, 103
hose bibb, 103–104, **105**
Hot water boilers. *See* Hydronic heating systems
House connection, 263, **264, 265**
Hub, 73, 74
Hydraulic gradient, 326, 357
Hydronic heating systems, 488
air separators, 498, **499**
air vents, 498, **499**
aquastat, 492, **493**
balancing valves, 503
centrifugal pumps, 489, 496–498
expansion tanks, 493–496
minimum volume, calculation of, 523
primary-secondary pumping, 511

Hydronic heating systems *(continued)*
    filling the system, 520, 524
    firing the system, 520
    flow-control valve, 502–503
    heat source, 491, **492**
    installing the boiler, 518–519
    installing the piping, 519
    low-water cutoff, 493
    one-pipe system, 503, **504**
        sizing diverter tee systems, 505–506
        water flow resistance, 503–505
    pressure-reducing valve, 498–500
    pressure relief valve, 501
    primary-secondary pumping, 508
        circuit piping, 508–509
        circulator pumps, 509–511
        common piping, 508
        expansion tanks, 511
        mixing valves, 511
    purging the system, 524
    radiant heating. *See* Radiant heating systems
    reset control, 492–493
    series loop system, 503
    theory of, 491
    two-pipe direct return, 506–507
    two-pipe reverse return, 507
    water volume, estimation of, 521, 522
    wiring the system, 519–520
    zone valves, 490, 501–502
Hydronics, 489

I

Icemaker box, 169, **170**
Ideal comfort, 512–513
Individual vent, 326, 332, **333**, 348
    sizing, 348, 350, **351**
Inhalation protection, 24
Insert reducing-tee creation procedure, 93
Inside plastic pipe cutter, 11–12
Instantaneous, 163, 170
Internal wrench, 16
Interval, 326, 344
Island venting, 337, **338**, 367–368
    sizing, 354–355, **356**
Isolate, 97, 102
Isolation valves, 97, 98–99, 106
    ball valves, 99
    gas cocks, 102–103
    gate valves, 99–100
    stop-and waste valves, 102
    stops, 100–101
Isometric drafting, 219–220, **221, 222**
    riser diagrams, 220, 222–223, **224**
    triangles, 215–217
Isometric view, 194, 195

J

Jackhammers, 52
Job site sizing, 274–275
Joint, 4, 11, 73, 272

Joists, 194, 206, 228, 239, 272, 278
    drilling and notching, 284–285, **286, 287**
    floor joist conflicts, 240–241

K

Kitchen sinks, 127–128, **129–131**
    basket strainers, 155, **156**, 157
    faucets, 143–145, **146**, 147
    installation of, 418, **419**
        stainless steel sink into a counter top, 438–441
    layout and sizing, 280–281, 377–378, **379**, 390–392
    symbols, **201, 202**
Knee protection, 23–24
Knives, 14
Knock-out, 136, 149

L

Ladder safety, 38–40
Lanyard, 37, 40
Laundry room layout, 380, **383**
Laundry sinks, 128, 131
    basket strainers, 156–157, **158**
    faucets, 146–147, **148, 149**
    installation of, 420–421
    symbols, **201**
Lavatories
    drain installations, 432–437
    installation of, 416–418
    layout and sizing, 276–277, 386–387, **388**
    sinks. *See* Lavatory sinks
    toilets. *See* Toilets
Lavatory sinks, 122–123, **124**
    drain assemblies, 150–151, **152**
        installation of, 432–437
    faucets, 138–139, **140, 141**
        installation of cultured-marble faucet, 432–437
    symbols, **201**
Layout, 234, **235, 236**
    above-ground layout, 239
        edge form layout procedure, 246–247
        fixture locations, 240
        floor joist conflicts, 240–241
        manufacturer rough-in sheet, 243–245
        wall layout, 241–243
    bathtubs, 278–279, **280**, 388–390
    drainage, waste and vent systems. *See* Drainage, waste and vent (DWV) systems
    kitchen sinks, 280–281, 390–392
    lavatories, 276–277, 386–387, **388**
    showers, 279–280
    toilets, 276, 385–386, **387**
    underground layout, 235–236
        trench layout, 238, **239**
        wall layout, 236–238
    wall layout
        above-ground layout, 241–243
        edge form layout procedure, 246–247
        underground layout, 236–238
        water distribution installations, 275–276

water distribution systems, 273
  bathtubs, 278–279, **280**
  kitchen sinks, 280–281, 390–392
  lavatories, 276–277, 386–387, **388**
  pipe sizing, 273
  showers, 279–280
  toilets, 276, 385–386, **387**
  wall layout, 275–276
Levels, 5–6
  use of, 26–27
Load bearing, 194, 209, 228, 240
Locking pliers, 8
Loop vent, 326, 333–334, **335**
Lug, 97, 103

M

Male adapters, 73, 78
Manifold station, 489, 516
Manifold systems, 299–300
Manual air vent, 489
Manufacturer rough-in sheet, 243–245
Masonry bits, 46–47
Material handling, 232, **233**
  vehicle racks, 233, **234**
Material organization, 227, 230
  bagging and tagging, 230–231, **232**
  palletizing, 230, **231**
Material Safety Data Sheet (MSDS), 229, 295
Metal cutting saw, 12, **13**
Meter, 448
  water meter connections, 260–261, **262**
Monoflo tee, 489, 505
Multi-purpose tool, 14

N

Nail-in hanger, **291**
Nail paw, 13–14
Nail pullers, 13–14
National Sanitation Foundation (NSF), 58
Needle nose pliers, 8
Nominal diameter, 59, 60
Notching. *See* Drilling and notching
Nut drivers, 4, 8

O

Offsets, 59, 64, 73, 75
Ohm, 448
One-pipe hot water, 489, 504
Open air, 326, 330
Open-end wrenches, 4, 8
Oral communications, 229–230
OSHA, 37, 38
Outdoor reset, 489, 492

P

Palletizing, 230, **231**
Partition, 272, 282
Percolation, 327, 357
Perc test, 327, 357

Perforated pipe, 68–69
Personal protection equipment, 21
  ear protection, 40, **41**
  eye protection, 21–22
  face protection, 22–23
  first aid kits, 24–25
  foot protection, 24
  hand protection, 23
  hard hats, 25
  inhalation protection, 24
  knee protection, 23–24
PEX fittings, 80–81
PEX pipe, 64. *See also* Plastic pipe
PEX tubing, 489, 513
Pilot, 163, 174
Pipe dope, 295–296
Pipes, 58, 59, 60
  brass pipe, 68
  cast-iron pipe, 66–67
  copper tube, 65–66
  corrugated stainless steel tube, 69
  diameters, 60
  drainage, waste and vent systems
    connections to dissimilar pipe, 396–397, **398**
    pipe protection, 394, 396, **397**
  fittings. *See* Fittings
  perforated pipe, 68–69
  plastic pipe. *See* Plastic pipe
  sizing, 273
    determination of sizes, 305–306
  steel pipe, 67–68
Pipe wrenches, 9–10
  use of, 31
Pistol drill, 42
Plan view, 194, 195
Plastic pipe, 60–61
  ABS pipe, 63
  CPVC pipe, 62–63
  cutters, 11
    inside cutters, 11–12
  cutting, 32, 301
  flexible tubing
    connection types, 300–301
    working with, 299
  PEX pipe, 64
  polyethylene pipe, 63–64
  PVC pipe, 61–62
  solvent-welding, 301–302, 320
  working with, 301
Plastic pipe saw, 10–11
Pliers, 8
Plug symbols, 199
Plumb bob, 20
Point of no pressure change, 489, 495
Point-of-use water heater, 163, 182
Polyethylene fittings, 82
Polyethylene pipe, 63–64. *See also* Plastic pipe
Polyethylene tubing, 489, 513
Pop-up, 118, 123, 136, 150
Portable band saws, 50

Port opening, 136, 151
Potable water, 59, 60
Power-actuated tool, 51
Power bore bits, 45–46
Power tools, 36
    air compressors, 52
    drills. *See* Drills
    grinders, 50–51
    jackhammers, 52
    power-actuated tool, 51
    safety, 38
        ear protection, 40, **41**
        electrical safety, 38
        fall protection, 40
        ladder safety, 38–40
    saws. *See* Saws
Pressure reducing valves, 97, 106, 490, 498–500
Pressure relief valve, 163, 183, 489, 490, 501
Private water systems, 254, **255**
Procure, 230
Procurement, 228
P-traps, 73, 89, 326
    symbols, 198
Public water system, 253, **254**
Purification, 251, 254
PVC fittings, 83
    drainage, waste and vent systems, 90
PVC pipe, 61–62. *See also* Plastic pipe

R

Radiant heating systems, 490, 511
    bypass lines, 517
    cold 70, 512
    concrete-free installations, 515–516
    human body, 511–512
    ideal comfort, 512–513
    manifold station, 516–517
    piping systems, 514
        concrete-free installations, 515–516
        slab on grade, 514–515
        thin slab, 515
        water temperature and direct piping, 517–518
    primary-secondary pumping with high-and low-temperature circuits, 518, **519**
    slab on grade, 514–515
    theory of, 513, **514**
    thermostatic bypass valves, 517, **518**
    thermostatic mixing valve, 517–518
    thin slab, 515
    tubing, 516
    water temperature and direct piping, 517–518
Radiator, 490, 505
Radius, 59, 64
Ratchet copper flaring tool, 17, **18**
Reactionary valves and devices, 97, 104–106, 105
    check valves, 107–108
    pressure-reducing valves, 106
    reduced-pressure zone devices, 97, 109–110
    relief valves, 107
    vacuum breakers, 108
    vacuum relief valves, 109

Reciprocating saws, 48
    flush cuts with, 55
Reduced-pressure zone devices, 97, 109–110
Reducers, 73, 76–77
    symbols, 199
Relief valves, 97, 107, 490
Relief vent, 327, 334–335, 344
Reset control, 492–493
Reverse-return, 490, 507
Right angle drill, 41–42
Riser clamp, **294**
Riser diagrams, 194, 220, 222–223, **224**
Rough, 118, 121, 374, 385
Rough-in, 118, 119, 325, 327
Ruler reading, 28

S

Saber saws, 49
Safety glasses, 21–22
Safety shoes, 24
Sanitary crosses, 73, 86
Sanitary tees, 73, 85–86
Saws
    blade selection, 54
    circular saws, 48–49
    metal cutting saw, 12, **13**
    plastic pipe cutting, 32
    plastic pipe saw, 10–11
    portable band saws, 50
    power saws, 48
        circular saws, 48–49
        portable band saws, 50
        reciprocating saws, 48, 55
        saber saws, 49
    reciprocating saws, 48
        flush cuts with, 55
    saber saws, 49
    wallboard saws, 13
Scale ruler, 213–214
Schedule, 59, 60
Screwdrivers, 7–8
Sealants. *See* Compounds and sealants
Section view, 194, 195
Septic systems, 357–360
Septic tank, 327, 329
Sewer, 59, 61
Shank, 37, 44
Shop drawing, 195
Shower pan, 393, **394**
    liner installation procedure, 403–405
Showers, 124, 126, **127, 128, 129**
    drains, 154–155, **156**
    faucets, 139–142, **143, 144**
    layout and sizing, 279–280
    symbols, **201**
Side view, 194
Silver solder, 272
Silver soldering, 295
Sinks
    kitchen sinks. *See* Kitchen sinks
    laundry sinks. *See* Laundry sinks

lavatory sinks. *See* Lavatory sinks
symbols, 202
Sizing
    drainage, waste and vent systems. *See* Drainage, waste and vent (DWV) systems
    job site sizing, 274–275
    pipe sizing, 273
        determination of sizes, 305–306
    theory of, 274
    water distribution systems. *See* Water distribution systems
Sketch, 194, 212
Slab, 228, 234
Slide square, 214, **215**
Slip joint, 410, 416
Slope, 4, 5
Sockets, 4, 8, 73, 74
Soil stack, 327
Solder, 272, 295, 296
    silver solder, 272
Soldering, 272, 295, 296–297
    copper, 310–313
    silver, 295
Spec home, 136, 137
Speed bits, 45
Squares, 7
    framing square layout, 30
Stack, 73, 89, 327
Stack vent, 327, 330
Standard expansion tank, 490, 493
Steam boilers, 488, 490. *See also* Hydronic heating systems
Steel pipe, 67–68
Steel-toe work boot, 24
Step bits, 47–48
Stop-and waste valves, 102
Stop-and-waste valves, 97
Stops. *See* Escutcheons and stops
Straight claw hammer, 10
Strainer fork, 16
Strapping, 4, 13, **292**
Strap wrenches, 9
Street, 73, 74
Stub out, 118, 119
Stud, 37, 41, 194, 206, 228, 239, 272, 279
Submersible pump, 448, 459
Submittal, 118, 119
Supports. *See* Hangers and supports
Swing joint, 73, 75
Symbols, 195
    architectural, 207–212, **213**
    caps, 198–199
    cleanouts, 198, **199**
    dishwashers, 202
    drain assemblies, 202
    fixtures, 200–201
    garbage disposers, 202
    kitchen sinks, 202
    45° offsets, 195–196
    90° offsets, 195
    piping systems, 198, **199**
    plugs, 199
    P-traps, 198
    reducers, 199
    sinks, 202
    tees, 196, **197**
        perpendicular tee configuration, 197–198
    valves and devices, 199–200
    water heaters, 202
Symbol templates, 217, **218**

T

Tankless, 163, 170
Tape measures, 6–7
    ruler reading, 28
Tees, 73, 75–76, 194
    insert reducing-tee creation procedure, 93
    size identification procedure, 92
    symbols, 196, **197**
        perpendicular tee configuration, 197–198
Tee size identification procedure, 92
Teflon tape, use of, 309
Terminal, 163, 179
Terminal unit, 490, 503
Test ball, 374, 398
Test cap, 374, 398
Test plug, 374, 398
Test tees, 73, 87
Thermocouple, 163, 174
Thermostats, 163, 170
    electric water heaters, 450, 451–452
Threaded rod attachment, **290**
Toggle bolt and threaded rod assembly, **291**
Toilets, 119–122
    ballcock replacement, 481–482
    flush valve replacement, 483–485
    installation of, 413–414, **415**
        two-piece toilets, 429–431
    layout and sizing, 276, 385–386, **387**
    troubleshooting, 466–468
Tools, 3
    drafting tools. *See* Drafting tools
    hand tools. *See* Hand tools
    power tools. *See* Power tools
Torch, 272
Torch regulator assembly, 18–19
Torch strike, 19
Torpedo level, 5
Torque wrench, 21
T&P valve, 97, 107
Trap, 327, 328
Trap adapters, 73, 89–90, 327, 342, 410, 416
Trap distance, 327, 356–357
Trench, 228, 235, 251, 258
Trench layout, 238, **239**
Triangles, 194, 215–217
Trim, 136, 140
Trim-out, 118, 119, 327
Troubleshooting, 447
    toilets, 466–468

544 INDEX

Troubleshooting *(continued)*
    water heaters
        electric water heaters. *See* Electric water heaters
        gas water heaters. *See* Gas water heaters
    well pumps, 459–465
T&S faucets, 136, 139–142, 140, **143, 144**
Tube, 59, 60
Tubing, 59, 60
Tubing cutters, 16–17
Tubular, 410, 411
Twin elbows, 86
Twist bits, 44
Two-hole plate with expandable anchors, **292**
Two-pipe system, 490, 506

U

Underground layout, 235–236
    trench layout, 238, **239**
    wall layout, 236–238
Unions, 73, 79

V

Vacuum breakers, 97, 108
Vacuum relief valves, 97, 109
Valves and devices, 96
    ball valves, 99
    check valves, 107–108
    gas cocks, 102–103
    gate valves, 99–100, **101**
    hose outlets, 97, 103
        boiler drains, 103
        hose bibb, 103–104, **105**
    isolation valves. *See* Isolation valves
    pressure-reducing valves, 106
    reactionary valves and devices. *See* Reactionary valves and devices
    reduced-pressure zone devices, 97, 109–110
    relief valves, 107
    stop-and waste valves, 102
    stops, 100–101
    symbols, 199–200
    vacuum breakers, 108
    vacuum relief valves, 109
Vehicle racks, 233, **234**
Vent, 59, 73, 88
Vent stack, 327, 330
    and building drain connection, 365
Vent systems. *See* Drainage, waste and vent (DWV) systems
Volt, 448
Volume factor, 490
Volute, 490

W

Wallboard, 4
Wallboard saws, 13
Wall layout
    above-ground layout, 241–243
    edge form layout procedure, 246–247
    underground layout, 236–238
    water distribution installations, 275–276

Walls
    drilling and notching, 282–283, **284**
    layout of. *See* Wall layout
Wall stud, 37, 41, 194, 206, 228, 239, 272, 279
Washing machine box, 168, **169**
    layout considerations, 392, **393**
Waste arm, 374, 386
Waste stack, 327, 329–330
    sizing, 344–346
Waste systems. *See* Drainage, waste and vent (DWV) systems
Wastewater, 327, 329
Water closet, 73, 87
Water distribution systems, 59, 251, 256
    bathtub layout, 278–279, **280**
    compounds and sealants, 295–302
    drilling and notching, 281–287
    fittings, 79–80
        brass fittings, 82–83
        copper fittings, 81–82
        CPVC fittings, 81
        galvanized fittings, 82
        PEX fittings, 80–81
        polyethylene fittings, 82
        PVC fittings, 83
    hangers and supports, 287–289, **290–294**
    installations, 271
        compounds and sealants, 295–302
        drilling and notching, 281–287
        hangers and supports, 287–289, **290–294**
        layout and sizing, 273–281
        procedures, 305–322
        testing, 302–303
        wall layout, 275–276
    job site sizing, 274–275
    kitchen sink layout, 280–281
    lavatory layout, 276–277
    layout considerations, 273
        bathtubs, 278–279, **280**
        kitchen sinks, 280–281
        lavatories, 276–277
        pipe sizing, 273
        showers, 279–280
        toilets, 276
        wall layout, 275–276
    pipe sizing, 273
    shower layout, 279–280
    sizing
        job site sizing, 274–275
        pipe sizing, 273
        sizing theory, 274
    testing, 302–303
    toilet layout, 276
    wall layout, 275–276
Water filtration, 256–257
Water heaters
    anode rod, 186
    electric. *See* Electric water heaters
    expansion tanks, 187–188
        replacement of, 479–480

gas. *See* Gas water heaters
gas regulator replacement, 474–475
installation of, 422–423
   electric water heaters, 423–424
   gas water heaters, 424–425
lined pipe nipples, 183–186
pressure switch replacement, 476–478
residential, 170
   electric water heaters. *See* Electric water heaters
   gas water heaters. *See* Gas water heaters
screw-in element replacement, 471–472
solar water heaters, 182–183, **184, 185**
storage tank replacement, 479–480
symbol, **202**
system protection, 183
   anode rod, 186
   expansion tanks, 187–188
   lined pipe nipples, 183–186
thermocouple replacement, 473
Water meter connections, 260–261, **262**
Water quality, 255–256
Water service, 59, 63, 251
Water service installations, 250, 257–258
   burial-depth requirements, 260
   elevation higher than sewer, 266
   elevation same as sewer, 265
   house connection, 263, **264, 265**
   perpendicular installations, 266
   trench safety, 258–260
   trench same as sewer, 263, 265, 267–268
   water meter connections, 260–261, **262**
   well connections, 261–263, **264**
Water sources, 252–253
   private water systems, 254, **255**
   public water system, 253, **254**
Watts, 448
Wax seal, 410
Weir, 327, 328
Well connections, 261–263, **264**
Well pump troubleshooting, 459–465
Wet venting, 336
Wet-vent sizing, 352
Wire cutters, 4, 8
Wood chisel, 15
Wrenches
   adjustable wrenches, 8–9
   basin wrenches, 15–16
      use of, 33
   pipe wrenches, 9–10
      use of, 31
   torque wrench, 21
Written communications, 229
Wye fittings, 73, 84–85

Z

Zone valves, 490, 501–502
Zoning, 490, 502